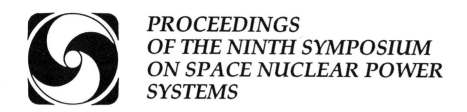
PROCEEDINGS OF THE NINTH SYMPOSIUM ON SPACE NUCLEAR POWER SYSTEMS

EDITORS

Mohamed S. El-Genk
University of New Mexico

Mark D. Hoover
Inhalation Toxicology Research Institute

INSTITUTE FOR SPACE NUCLEAR POWER STUDIES
Chemical and Nuclear Engineering Department
The University of New Mexico
Albuquerque, NM 87131
(505) 277-2813, 277-2814

Co-sponsored by:

NATIONAL AERONAUTICS AND SPACE ADMINISTRATION
    HEADQUARTERS
    LEWIS RESEARCH CENTER
STRATEGIC DEFENSE INITIATIVE ORGANIZATION
UNITED STATES DEPARTMENT OF ENERGY
    ARGONNE NATIONAL LABORATORY
    BROOKHAVEN NATIONAL LABORATORY
    IDAHO NATIONAL ENGINEERING LABORATORY
    LOS ALAMOS NATIONAL LABORATORY
    SANDIA NATIONAL LABORATORIES
UNITED STATES AIR FORCE
    PHILLIPS LABORATORY
    WRIGHT LABORATORY

In cooperation with:

AMERICAN NUCLEAR SOCIETY
    ANS TRINITY SECTION
    ANS ENVRONMENTAL SCIENCES DIVISION
    ANS NUCLEAR REACTOR SAFETY DIVISION
AMERICAN SOCIETY OF MECHANICAL ENGINEERS
    NUCLEAR ENGINEERING DIVISION
    HEAT TRANSFER DIVISION
ASTM, COMMITTEE-10 ON NUCLEAR TECHNOLOGY AND APPLICATIONS
INTERNATIONAL ASTRONAUTICAL FEDERATION
NEW MEXICO ACADEMY OF SCIENCE

Industry Affiliates:

BABCOCK & WILCOX COMPANY
GRUMMAN CORPORATION
ROCKWELL INTERNATIONAL CORPORATION
    ROCKETDYNE DIVISION
WESTINGHOUSE ELECTRIC CORPORATION

Albuquerque Convention Center
Albuquerque, New Mexico
January 12-16, 1992

AMERICAN INSTITUTE OF PHYSICS   NEW YORK

Authorization to photocopy items for internal or personal use, beyond the free copying permitted under the 1978 U.S. Copyright Law (see statement below), is granted by the American Institute of Physics for users registered with the Copyright Clearance Center (CCC) Transactional Reporting Service, provided that the base fee of $2.00 per copy is paid directly to CCC, 27 Congress St., Salem, MA 01970. For those organizations that have been granted a photocopy license by CCC, a separate system of payment has been arranged. The fee code for users of the Transactional Reporting Service is: 0094-243X/87 $2.00.

© 1992 American Institute of Physics.

Individual readers of this volume and nonprofit libraries, acting for them, are permitted to make fair use of the material in it, such as copying an article for use in teaching or research. Permission is granted to quote from this volume in scientific work with the customary acknowledgment of the source. To reprint a figure, table, or other excerpt requires the consent of one of the original authors and notification to AIP. Republication or systematic or multiple reproduction of any material in this volume is permitted only under license from AIP. Address inquiries to Series Editor, AIP Conference Proceedings, AIP, 335 East 45th Street, New York, NY 10017-3483.

L.C. Catalog Card No. 91-58793

Casebound:
ISBN 1-56396-021-4 (Pt. 1)
1-56396-023-0 (Pt. 2)
1-56396-025-7 (Pt. 3)
1-56396-027-3 (Set)

Paperback:
ISBN 1-56396-020-6 (Pt. 1)
1-56396-022-2 (Pt. 2)
1-56396-024-9 (Pt. 3)
1-56396-026-5 (Set)

DOE CONF-920104

Printed in the United States of America.

INTRODUCTION

We are pleased to introduce the Proceedings of the Ninth Symposium on Space Nuclear Power Systems. These volumes contain the reviewed and edited papers that are being presented at the Ninth Symposium in Albuquerque, New Mexico, 12-16 January 1992. The objective of the symposium, and hence these volumes, is to summarize the state of knowledge in the area of space nuclear power and propulsion and to provide a forum at which the most recent findings and important new developments can be presented and discussed.

As in past meetings, the program is proceeded by two-day short courses. This year's topics are Heat Pipe Technology for Space Power Systems and Thermionic Power System Technology. They provide excellent overviews of these fields for symposium attendees, including graduate engineering students. Continuing education has always been one of the most valuable parts of the symposium.

The meeting program formally begins with a full day of plenary presentations which include congressional views and budget prospectives; an update on the Space Exploration Initiative; goals and issues for missions in space; and updates on plans for the SP-100 Program, the U.S. Thermionic Space Nuclear Program, Radioisotope Systems for Exploration of Space, and the use of Nuclear Propulsion for Future Space Missions. The plenary sessions are always a stimulating part of the program, and we are grateful to the congressional and agency leaders who are participating in those sessions. We are grateful to the members of Symposium Advisory Committee for helping to organize the plenary sessions.

At the same time that we and our professional colleagues are learning about the latest views and plans for space nuclear power and propulsion, 200 high school students and 20 teachers from around the state of New Mexico will learn about some of the exciting scientific and technical advances that are being made and about the career opportunities that await them. Irene El-Genk, chairwoman, and Rose Thome, co-chairwoman, are to be commended for their efforts and vision in organizing this important special session. We are grateful to NASA Lewis Research Center for providing the personalized space science folders for the students. We wish the attendees well, and encourage them to carry back their new found insights and enthusiasm to their schools.

This year's proceedings include nearly 200 papers prepared by over 400 authors from more than 50 organizations from many countries. We are grateful to the authors for their cooperation, enthusiasm, and sincere desire to contribute to a clear, useful, and technically sound publication. We gratefully acknowledge the contributions of the members of the Technical Program Committee in helping to organize the technical sessions and review the submissions. Reliable scientific literature cannot be produced without this kind of dedicated effort.

Many other individuals and organizations worked hard and contributed generously to sponsor, plan, organize, and manage this Symposium. They are also listed in the committee sections of these proceedings. They deserve credit for this ninth annual success. We gratefully acknowledge the competent and unflagging help of Mary Bragg and members of the staff of the Institute for Space Nuclear Power Studies in the preparation of these proceedings, and in the professional coordination of many functions and meetings for the preparation of the symposium. In addition, our families continue to provide us with their understanding, patience, and encouragement throughout the publication process. We deeply appreciate that.

In closing, we are sad to note the untimely death on 28 October 1991 at age 55 of Dr. Dudley G. McConnell of the National Aeronautics and Space Administration. He was a dynamic individual who had a distinguished professional career and who lived a caring and productive life as a human being. His leadership of the interagency efforts to launch the Galileo and Ulysses missions have made a lasting contribution to space exploration. Dr. McConnell's picture and obituary are included in the symposium program. He will be missed.

Mohamed S. El-Genk
Technical Chairman

Mark D. Hoover
Publications Chairman

SPACE NUCLEAR POWER PUBLICATIONS

Transactions of 1st Symposium .... out of print

Publications available from UNM's Institute for Space Nuclear Power Studies

Transactions of the 2nd Symposium	$10.00
Transactions of the 3rd Symposium	$10.00
Transactions of the 4th Symposium	$10.00
Transactions of the 5th Symposium	$10.00
Transactions of the 6th Symposium	$20.00
Proceedings of the 7th Symposium	$25.00

Publications available from the American Institute of Physics, c/o AIDC, 64 Depot Road, Colchester, VT 05446, 1-800-445-6638

Proceedings of the 8th Symposium (3-volume hardback set)
ISBN # 0-88318-838-4 AIP Conference Proceedings #217 .... $175.00
Proceedings of the 9th Symposium .... (to be determined)

Publications available from Orbit Book Company, P.O. Box 9542, Melbourne, FL 32902-9542, Phone: (407) 724-9542

Space Nuclear Power Systems 1984	$125.00
Space Nuclear Power Systems 1985	$125.00
Space Nuclear Power Systems 1986	$125.00
Space Nuclear Power Systems 1987	$125.00
Space Nuclear Power Systems 1988	$125.00
Space Nuclear Power Systems 1989	(in preparation)

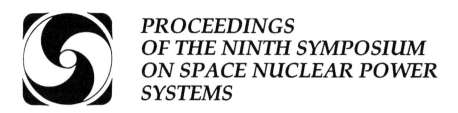

PROCEEDINGS OF THE NINTH SYMPOSIUM ON SPACE NUCLEAR POWER SYSTEMS

ORGANIZING COMMITTEE

HONORARY CHAIRMAN
Honorable Pete V. Domenici
United States Senate (R, NM)

GENERAL CHAIRMAN
Colonel Peter J. Marchiando
Commander, Phillips Laboratory

GENERAL CO-CHAIRMEN

Gary L. Bennett
NASA Headquarters

Wade Carroll
United States Department of Energy

TECHNICAL CHAIRMAN
Mohamed S. El-Genk
Institute for Space Nuclear Power Studies
University of New Mexico

TECHNICAL CO-CHAIRMEN

James H. Lee
Sandia National Laboratories

Ernest D. Herrera
Phillips Laboratory

Donald H. Roy
Babcock & Wilcox Company

PUBLICATIONS CHAIRMAN
Mark D. Hoover
Inhalation Toxicology Research Institute

ADMINISTRATIVE CHAIRWOMAN
Mary Bragg
Institute for Space Nuclear Power Studies
University of New Mexico

ADMINISTRATIVE CO-CHAIRWOMAN

Janice Holloman-Canales
Institute for Space Nuclear Power Studies
University of New Mexico

Maureen Alaburda
Institute for Space Nuclear Power Studies
Univeristy of new Mexico

**SCHOLARSHIP AND HIGH SCHOOL
PUBLIC RELATIONS COORDINATOR**
Irene L. El-Genk
Education and Outreach Advisory Board
Institute for Space Nuclear Power Studies
University of New Mexico

**SECONDARY SCHOOL SPACE SCIENCE
COMPETITION COORDINATOR**
David Kauffman
College of Engineering
University of New Mexico

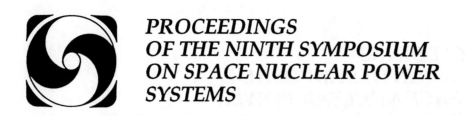

PROCEEDINGS OF THE NINTH SYMPOSIUM ON SPACE NUCLEAR POWER SYSTEMS

STEERING COMMITTEE

CHAIRMAN
Honorable Pete V. Domenici
United States Senate (R, NM)

MEMBERS

J. Sam Armijo
Program General Manager, Space Power
General Electric Company

Thomas J. Hirons
Division Leader, Nuclear Technology
and Engineering
Los Alamos National Laboratory

Richard A. Johnson
Director, Advanced Space Power
Rockwell International/Rocketdyne Division

Stephen J. Lanes
Deputy Assistant Secretary,
Space and Defense Power Programs
United States Department of Energy

Peter J. Marchiando
Commander, Phillips Laboratory

Paul North
Acting Associate Director
Idaho National Engineering Laboratory

Gregory M. Reck
Director, Space Technology
NASA Headquarters

Paul G. Risser
Provost and Vice President for
Academic Affairs
University of New Mexico

Honorable Steven H. Schiff
United States House of Representatives (R,NM)

ADVISORY COMMITTEE

CHAIRMAN
Mohamed S. El-Genk
University of New Mexico

MEMBERS

H. Sterling Bailey
General Electric Company

William J. Barattino
Phillips Laboratory

Gary Bennett
NASA Headquarters

Samit K. Bhattacharyya
Argonne National Laboratory

David L. Black
Westinghouse Electric Corporation

Richard J. Bohl
Los Alamos National Laboratory

David Buden
Idaho National Engineering Laboratory

Wade Carroll
United States Department of Energy

Edmund P. Coomes
Pacific Northwest Laboratory

Roy H. Cooper
Oak Ridge National Laboratory

Richard C. Dahlberg
General Atomics

Tony Gallegos
Senator Domenici's Office

Steve Harrison
National Space Council

Robert Haslett
Grumman Corporation

Ernest D. Herrera
Phillips Laboratory

Mark D. Hoover
Inhalation Toxicology Research Institute

Steven Howe
Los Alamos National Laboratory

Walter Kato
Brookhaven National Laboratory

Ehsan U. Khan
BDM International, Inc.

E. Tom Mahefkey
Wright Laboratory

Patrick J. McDaniel
Sandia National Laboratories

Thomas J. Miller
NASA Lewis Research Center

Joseph C. Mills
Rockwell International Corporation
    Rocketdyne Division

M. Frank Rose
Auburn University

Joseph A. Sholtis, Jr.
United States Air Force

R. Joseph Sovie
NASA Lewis Research Center

Earl J. Wahlquist
United States Department of Energy

Jack Walker
Sandia National Laboratories

Frank L. Williams
University of New Mexico

David M. Woodall
Idaho National Engineering Laboratory

Richard A. Zavadowski
Babcock & Wilcox Company

TECHNICAL PROGRAM COMMITTEE

TECHNICAL CHAIRMAN
Mohamed S. El-Genk
University of New Mexico

TECHNICAL CO-CHAIRMEN

Ernest Herrera	**James H. Lee**	**Donald H. Roy**
Phillips Laboratory	Sandia National Laboratories	Babcock & Wilcox

MEMBERS

Julio C. Acevedo
NASA Lewis Research Center

Douglas Allen
Strategic Defense Initiative Organization

Wayne R. Amos
EG&G Mound Applied Technologies, Inc.

Joseph Angelo
Science Applications International Corp.

H. Sterling Bailey
General Electric Company

Russell M. Ball
Babcock & Wilcox Company

C. Perry Bankston
Jet Propulsion Laboratory

Lester L. Begg
General Atomics

Kenneth J. Bell
Oklahoma State University

Gary L. Bennett
NASA Headquarters

Robert Bercaw
NASA Lewis Research Center

John A. Bernard
Massachusetts Institute of Technology

Frederick R. Best
Texas A&M University

Samit K. Bhattacharyya
Argonne National Laboratory

Walter Bienert
DTX Corporation

David L. Black
Westinghouse Electric Corporation

James B. Blackmon
McDonnell Douglas Astronautics
  Corporation

Richard J. Bohl
Los Alamos National Laboratory

Stanley K. Borowski
NASA Lewis Research Center

Edward J. Britt
Space Power, Inc.

Wayne M. Brittain
Teledyne Energy Systems

John Brophy
Jet Propulsion Laboratory

Neil W. Brown
General Electric Company

R. William Buckman
Refractory Metals Technology

David Buden
Idaho National Engineering Laboratory

David C. Byers
NASA Lewis Research Center

Colin Caldwell
Babcock & Wilcox Company

Wade Carroll
United States Department of Energy

A. Thomas Clark
United States Department of Energy

John S. Clark
NASA Lewis Research Center

viii

TECHNICAL PROGRAM COMMITTEE (CONTINUED)

Edmund P. Coomes
Pacific Northwest Laboratory

Roy H. Cooper
Oak Ridge National Laboratory

Jeffery E. Dagle
Pacific Northwest Laboratory

Richard C. Dahlberg
General Atomics

Vijay K. Dhir
University of California-Los Angeles

Nils J. Diaz
University of Florida

James E. Dudenhoefer
NASA Lewis Research Center

Dale S. Dutt
Westinghouse Hanford Company

Amir Faghri
Wright State University

Gerald H. Farbman
Starpath, Inc.

Mario Fontana
Oak Ridge National Laboratory

Elzie E. Gerrels
General Electric Company

Toni L. Grobstein
NASA Lewis Research Center

Janelle W. Hales
Westinghouse Hanford Company

Bill Harper
Allied Signal Aerospace

Steve Harrison
National Space Council

Richard B. Harty
Rockwell International
    Corporation/Rocketdyne Division

Edwin Harvego
Idaho National Engineering Laboratory

Jack Heller
Sverdrup Technology, Inc.

Eugene Hoffman
United States Department of Energy

Robert S. Holcomb
Oak Ridge National Laboratory

Mark D. Hoover
Inhalation Toxicology Research Institute

Steven Howe
Los Alamos National Laboratory

Maribeth E. Hunt
Rockwell International
    Corporation/Rocketdyne Division

DeWayne L. Husser
Babcock & Wilcox Company

John R. Ireland
Los Alamos National Laboratory

Mohammad Jamshidi
University of New Mexico

Richard A. Johnson
Rockwell International Corporation
    Rocketdyne Division

Albert J. Juhasz
NASA Lewis Research Center

Michael Kania
Oak Ridge National Laboratory

Ehsan U. Khan
BDM International, Inc.

Donald B. King
Sandia National Laboratories

Andrew C. Klein
Oregon State University

Arvind Kumar
University of Missouri-Rolla

James H. Lee
Sandia National Laboratories

E. Tom Mahefkey
Wright Laboratory

Gregory Main
Office of Naval Research

TECHNICAL PROGRAM COMMITTEE *(CONTINUED)*

John Mankins
NASA Headquarters

Albert C. Marshall
Sandia National Laboratories

Charles R. Martin
United States Air Force

John Martinell
EG&G Idaho, Inc.

L. David Massie
Wright Laboratory

Edward F. Mastal
United States Department of Energy

L. David Massie
Wright Laboratory

Edward F. Mastal
United States Department of Energy

Donald N. Matteo
General Electric Company

R. Bruce Matthews
Los Alamos National Laboratory

William H. McCulloch
Sandia National Laboratories

Patrick J. McDaniel
Sandia National Laboratories

Arthur Mehner
United States Department of Energy

Michael A. Merrigan
Los Alamos National Laboratory

Raymond A. Meyer
General Electric Company

Thomas J. Miller
NASA Lewis Research Center

Joseph C. Mills
Rockwell International Corporation
　　Rocketdyne Division

Peyton Moore
Oak Ridge National Laboratory

Ira T. Myers
NASA Lewis Research Center

George F. Niederauer
Los Alamos National Laboratory

William E. Osmeyer
Teledyne Energy Systems

Lewis Peach
NASA Headquarters

Gary Polansky
Sandia National Laboratories

James Polk
Jet Propulsion Laboratory

James R. Powell
Brookhaven National Laboratory

William A. Ranken
Los Alamos National Laboratory

John W. Rice, Jr.
Idaho National Engineering Laboratory

Peter J. Ring
General Electric Company

M. Frank Rose
Auburn University

Richard Rovang
Rockwell International Corporation/
　　Rocketdyne Division

Don Roy
Babcock & Wilcox Company

Paul H. Sager
General Dynamics Space Systems
　　Division

Michael J. Schuller
Phillips Laboratory

Gene E. Schwarze
NASA Lewis Research Center

Stephen Seiffert
Benchmark Environmental Corp.

Joel C. Sercel
Jet Propulsion Laboratory

Harold A. Shelton
General Dynamics Space Systems
　　Division

TECHNICAL PROGRAM COMMITTEE *(CONTINUED)*

Joseph A. Sholtis, Jr.
United States Air Force

James S. Sovey
NASA Lewis Research Center

R. Joseph Sovie
NASA Lewis Research Center

Marland L. Stanley
Idaho National Engineering Laboratory

Walter A. Stark
Los Alamos National Laboratory

Anthony Sutey
Boeing Aerospace Company

Melvin Swerdling
TRW, Inc.

Mark I. Temme
General Electric Aerospace Company

Charles W. Terrell
Phillips Laboratory

Frank V. Thome
Sandia National Laboratories

Robert H. Titran
NASA Lewis Research Center

James A. Turi
United States Department of Energy

Ronald F. Tuttle
United States Air Force Institute
   of Technology

Jan W. Vandersande
Jet Propulsion Laboratory

Charles Vesely
Grumman Corporation

Carl E. Walter
Lawrence Livermore National
   Laboratory

John W. Warren
United States Department of Energy

Joseph A. Weimer
Wright Laboratory

Joseph R. Wetch
Space Power, Inc.

Richard D. Widrig
Pacific Northwest Laboratory

Michael Wiemers
Westinghouse Hanford Company

Ken A. Williams
Science Applications International Corporation

David M. Woodall
Idaho National Engineering Laboratory

Thomas Wuchte
Phillips Laboratory

Francis J. Wyant
SandiaLaboratory Laboratories

SCHREIBER-SPENCE SPACE ACHIEVEMENT AWARD COMMITTEE

CHAIRMAN
Nils J. Diaz
University of Florida/INSPI

MEMBERS

Dale S. Dutt
Westinghouse Hanford Company

Thomas J. Hirons
Los Alamos National Laboratory

Richard A. Johnson
Rockwell International/Rocketdyne
 Division

Thomas J. Miller
NASA Lewis Research Center

Donald H. Roy
Babcock & Wilcox Company

Joseph A. Sholtis, Jr.
United States Air Force

MANUEL LUJAN, JR. STUDENT PAPER AWARD COMMITTEE

CHAIRMAN
Frederick R. Best
Texas A&M University

MEMBERS

Hatice Cullingford
NASA Johnson Space Center

Philip Pluta
General Electric Company

Andrew Klein
Oregon Sate University

Michael Hall
Los Alamos National Laboratory

Ehsan U. Khan
General Electric Company

EDUCATION AND OUTREACH ADVISORY BOARD
(EOAB)

MEMBERS

Mary Bragg
Assistant Director
Institute for Space Nuclear Power Systems
University of New Mexico

Erile Casey
Sandia Preparatory School
Albuquerque, NM

David Eisman
Students for the Exploration
and Development of Space
University of New Mexico

Irene L. El-Genk
High School Special Session Coordinator
University of New Mexico

Dr. Mohamed S. El-Genk
Director
Institute for Space Nuclear Power Studies
University of New Mexico

Steve Kanim
Las Cruces High School
Las Cruces, NM

Dr. David Kauffman
Associate Dean
College of Engineering
University of New Mexico

Michael Lee
Minority Recruitment Coordinator
College of Engineering
University of New Mexico

Harry Llull
Contennial Science & Engineering Library
University of New Mexico

Kathleen Lowry
Students for the Exploration
and Development of Space
University of New Mexico

Joanne Metzler
McKinley County Schools
Gallup, NM

Dr. Stanley Morain
Director
Technicology Application Center
University of New Mexico

Paul Neville
Students for the Exploration
and Development of Space
University of New Mexico

Harry Pomeroy
Yucca Junior High School
Clovis, NM

Laura Reeves
Manzano High School
Albuquerque, NM

Dr. Steve Seiffert
Benchmark Environemntal Corporation
Albuquerque, NM

Rose Thome
High School Special Session Co-Chairwoman
University of New Mexico

Dr. Gail Ward
Director of Student Programs
College of Engineering
University of New Mexico

Lois Weigand
Farmington Junior High School
Framington, NM

HIGH SCHOOL SPECIAL SESSION
Education and Outreach Advisory Board (EOAB)

Co-sponsored by
INSTITUTE FOR SPACE NUCLEAR POWER STUDIES,
University of New Mexico
NASA LEWIS RESEARCH CENTER
NEW MEXICO SPACE GRANT CONSORTIUM
AMERICAN NUCLEAR SOCIETY - TRINITY SECTION

SESSION I

Irene L. El-Genk, Chairwoman
EOAB member
Albuquerque, NM

Rose Thome, Co-Chairwoman
EOAB member
Albuquerque, NM

Welcome
  Irene L. El-Genk, Education Outreach Advisory Board, University of New Mexico, Albuquerque, NM

Introduction of Space Design Contest Winners
  David Kauffman, Associate Dean, College of Engineering, University of New Mexico, Albuquerque, NM

Stellar Evolution
  John A. Bernard, Ph.D., Massachusetts Institute of Technology, Cambridge, MA

Teamwork Makes It Happen
  Sidney Gutierrez, NASA Astronaut, NASA Johnson Space Center, Houston, TX

SESSION II

Joanne Metzler, Chairwoman
EOAB member
Gallup/McKinley County Schools
Gallup, NM

Laura Reeves, Co-Chairwoman
EOAB member
Manzano High School
Albuquerque, NM

UNM College of Engineering Student Programs
  Gail Ward, Ph. D., Director of Student Programs, College of Engineering, University of New Mexico, Albuquerque, NM

Medical and Biological Aspects of Space Travel
  Barbara Lujan, NASA Headquarters, Washington, DC

Space Experiments
  F. Andrew Gaffney, MD, Cardiologist and Payload Specialist on the Space Lab Mission, Southwest Medical School, Dallas, TX

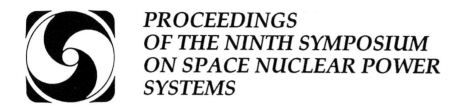
PROCEEDINGS
OF THE NINTH SYMPOSIUM
ON SPACE NUCLEAR POWER
SYSTEMS

TABLE OF CONTENTS - PART ONE

PLENARY SESSION I
CONGRESSIONAL VIEWS AND BUDGET PROSPECTIVES

Chairman,
Honorable Pete V. Domenici
United States Senate (R, NM)
Ranking Republican Member
of the Senate Budget Committee

Co-Chairman,
Honorable Steven H. Schiff,
United States House of Representatives
(R) New Mexico

PLENARY SESSION II
SPACE EXPLORATION INITIATIVE

S. Peter Worden, Chairman
Strategic Defense Initiative Organization
Washington, DC

William L. Smith, Co-Chairman
NASA Headquarters
Washington, DC

Gene Sevin, Director, Offensive and Space Systems, Washington, DC

Fenton Carey, Special Assistant to the Secretary for Space, U. S. Department of Energy Headquarters, Washington, DC

Michael D. Griffin, Associate Administrator for Exploration, NASA Headquarters, Washington, DC

PLENARY SESSION III
GOALS, PERSPECTIVES AND ISSUES

John Russell, Chairman
Phillips Laboratory
Kirtland AFB, NM

Space Nuclear Power
  Richard L. Verga, Strategic Defense Initiative Organization, The Pentagon, Washington, DC

Air Force Missions
  Richard Davis, Los Angeles Air Force Base, Los Angeles, CA

Goals and Technology Issues for Civil Space Power
  Earl VanLandingham, Director, Propulsion, Power and Energy Division, NASA Headquarters, Washington, DC

TABLE OF CONTENTS - PART ONE

PLENARY SESSION IV
SPACE NUCLEAR PROGRAMS

Wade Carroll, Chairman
United States Department of Energy
Germantown, MD

Gary L. Bennett, Co-Chairman
NASA Headquarters
Washington, DC

SP-100 Power System — 1
Vincent C. Truscello, Jet Propulsion Laboratory, Pasadena, CA and Lyle L. Rutger, U.S. Department of Energy, Germantown, MD

An Overview of the Thermionic Space Nuclear Power Program
Colette Brown, U.S. Department of Energy, Washington, DC, and Dan Mulder, Phillips Laboratory, Kirtland, NM

Radioisotope Power Systems for the Exploration of Space
Robert Lange, U.S. Department of Energy, Washington, DC

Nuclear Propulsion for Space Exploration — 24
Thomas J. Miller, NASA Lewis Research Center, Cleveland, OH, and Gary L. Bennett, NASA Headquarters, Washington, DC

POSTER SESSION

Gerald Farbman, Chairman
Starpath, Inc.
Pittsburgh, PA

Mark D. Hoover, Co-Chairman
Inhalation Toxicology Research Institute
Albuquerque, NM

Effect of Spin-Polarized D-3He Fuel on Dense Plasma Focus for Space Propulsion — 30
Mei-Yu Wang and Chan K. Choi, Purdue University, W. Lafayette, IN, and Franklin B. Mead, Jr., Future Technologies, Inc., Edwards AFB, CA

Cesium Feeder for the "Topaz-II" System Study Results and Service Life Characteristics — 35
I. A. Dorf-Gorsky, A. A. Yelchaninov, A. N. Luppov, V. P. Nicitin, and B. G. Ogloblin, Central Design Bureau of Machine Building, St. Petersburg, USSR, and G. V. Sinyutin and V. A. Usov, Kurchatov Institute of Atomic Energy, Moscow, USSR

Special Features and Results of the "Topaz-II" Nuclear Power System Tests with Electric Heating — 41
V. P. Nicitin, B. G. Ogloblin, A. N. Luppov, E. Y. Kirillov, and V. G. Sinkevich, Central Design Bureau of Machine Building, St. Petersburg, USSR, and V. A. Usov and Y. A. Nechaev, Kurchatov Institute of Atomic Energy, Moscow, USSR

Design of Cylindrical Reservoir Concept for Control of a Spacecraft Capillary Pumped Thermal Management System — 47
William J. Krotiuk, General Electric Company, Princeton, NJ

Plasma Instability between Polycrystalline Rhenium Electrodes in a Cesiated Thermionic Diode — 53
Deuk Yong Lee and Dean L. Jacobson, Arizona State University, Tempe, AZ

Cesium Vapor Supply System for "TOPAZ" SNPS — 58a
Georgy M. Gryaznov, Nikolai I. Ezhov, Evgeny E. Zhabotinsky, Victor I. Serbin, Boris V. Slivkin, Yuri L. Trukhanov and Leonid M. Sheftel, SPU "Krasnaya Zvezda", Moscow, USSR

TABLE OF CONTENTS - PART ONE

[1] SPACE EXPLORATION I

Gary L. Bennett, Chairman
NASA Headquarters
Washington, DC

Joe T. Howell, Co-Chairman
NASA Headquarters
Washington, DC

A Split Sprint Mission to Mars — 58
Kyle Shepard, Jack Duffey, Dom D'Annible, Jeff Holdridge, and Walter Thompson, General Dynamics Space Systems Division, San Diego, CA, and Robert C. Armstrong, NASA Marshall Space Flight Center, Huntsville, AL

Requirements for Advanced Space Transportation Systems — 64
William G. Huber, NASA Marshall Space Flight Center, Huntsville, AL

Surface Systems Requirements — 70
John F. Connolly, NASA-Johnson Space Center, Houston, TX

Robotic Planetary Science Missions with NEP — 78
James H. Kelley, Ronald J. Boain, and Chen-wan L. Yen, Jet Propulsion Laboratory, Pasadena, CA

Technology Transfer from the Space Exploration Initiative — 91
David Buden, Idaho National Engineering Laboratory, Idaho Falls, ID

[2] REACTORS AND SHIELDING

Russell Ball, Chairman
Babcock & Wilcox Company
Lynchburg, VA

Carl Walter, Co-Chairman
Lawrence Livermore National Laboratory
Livermore, CA

SP-100 Reactor and Shield Design Update — 97
Nelson A. Deane, Samuel L. Stewart, Thomas F. Marcille, and Douglas W. Newkirk, General Electric Company, San Jose, CA

A Modular Reactor for Lunar and Planetary Base Service — 107
Salim N. Jahshan and Ralph G. Bennett, Idaho National Engineering Laboratory, Idaho Falls, ID

STAR-C Thermionic Space Nuclear Power System — 114
Lester L. Begg, General Atomics, San Diego, CA, Thomas J. Wuchte, Phillips Laboratory, Kirtland AFB, NM, and William Otting, Rockwell International/Rocketdyne Division, Canoga Park, CA

Space-R Thermionic Space Nuclear Power System with Single Cell Incore Thermionic Fuel Elements — 120
Hyop S. Rhee, Joseph R. Wetch, Norman G. Gunther, Robert R. Hobson, Cinian Zheng, Edward J. Britt, and Glen Schmidt, Space Power, Inc., San Jose, CA

Integrated Shielding Systems for Manned Interplanetary Spaceflight — 130
Jeffrey A. George, NASA Lewis Research Center, Cleveland, OH

Heat Pipe Thermionic Reactor Shield Optimization Studies — 141
Vahe Keshishan and Terry E. Dix, Rockwell International/Rocketdyne Division, Canoga Park, CA

TABLE OF CONTENTS - PART ONE

[3] MATERIALS I

J. P. Moore, Chairman
Oak Ridge National Laboratory
Oak Ridge, TN

R. William Buckman, Co-Chairman
Refractory Metal Alloys
Pittsburgh, PA

PWC-11 Fuel Pin Development for SP-100 — 145
Edwin D. Sayre, Robert E. Butler, and Mike Kangilaski, General Electric Company, San Jose, CA

The Production of ASTAR-811C Ingot, Plate and Sheet — 150
R. William Buckman, Jr., Refractory Metals Technology, Pittsburgh, PA, and Peter A. Ring and Mike Kangilaski, General Electric Company, San Jose, CA

Manufacturing SP-100 Rhenium Tubes — 159
Edwin D. Sayre and Thomas J. Ruffo, General Electric Company, San Jose, CA

Weldability of DOP-26 Iridium Alloy: Effects of Welding Gas and Alloy Composition — 164
Evan K. Ohriner, Gene M. Goodwin, and David A. Frederick, Oak Ridge National Laboratory, Oak Ridge, TN

[4] RADIOISOTOPE POWER SYSTEMS

Arthur Mehner, Chairman
United States Department of Energy
Germantown, MD

Richard D. Rovang, Co-Chairman
Rockwell International Corporation/
Rocketdyne Division
Canoga Park, CA

Flight Performance of Galileo and Ulysses RTGs — 171
Richard J. Hemler, Charles E. Kelly, and James F. Braun, General Electric Company, Philadelphia, PA

Modular RTG Status — 177
Robert F. Hartman and Jerry R. Peterson, General Electric Company, Philadelphia, PA; and William Barnett, U.S. Department of Energy, Washington, DC

Design of Small Impact-Resistant RTGs for Mars Environmental Survey (MESUR) Mission — 182
Alfred Schock, Fairchild Space and Defense Corporation, Germantown, MD

Comparison of DIPS and RFCs for Lunar Mobile and Remote Power Systems — 202
Richard B. Harty, Rockwell International/Rocketdyne Division, Canoga Park, CA

TABLE OF CONTENTS - PART ONE

[5] SPACE EXPLORATION II

Thomas J. Miller, Chairman
NASA Lewis Research Center
Cleveland, OH

Richard B. Harty, Co-Chairman
Rockwell International/
Rocketdyne Division
Canoga Park, CA

SP-100 Lunar Surface Power System Conceptual Design — 208
Hubert A. Upton, A. Richard Gilchrist, Robert Protsik, Robert E. Gamble, David W. Lunsford, and Ian R. Temmerson, General Electric Company, San Jose, CA

Lunar Electric Power Systems Utilizing the SP-100 Reactor Coupled to Dynamic Conversion Systems — 216
Richard B. Harty and Gregory A. Johnson, Rockwell International/Rocketdyne Division, Canoga Park, CA,

2.5 kWe Dynamic Isotope Power System for the Space Exploration Initiative Including an Antarctic Demonstration — 222
Maribeth E. Hunt and Richard D. Rovang, Rockwell International/Rocketdyne Division, Canoga Park, CA

Power Options for Lunar Exploration — 228
Judith Ann Bamberger and Krista L. Gaustad, Pacific Northwest Laboratory, Richland, WA

Mars Mission Performance Enhancement with Hybrid Nuclear Propulsion — 234
Jeffery E. Dagle and Kent E. Noffsinger, Pacific Northwest Laboratory, Richland, WA, and Donald R. Segna, U.S. Department of Energy, Richland, WA

[6] RADIATION AND TEMPERATURE EFFECTS ON ELECTRONICS

Gene E. Schwarze, Chairman
NASA Lewis Research Center
Cleveland, OH

Donald B. King, Co-Chairman
Sandia National Laboratories
Albuquerque, NM

A Charged Particle Transport Analysis of the Dose to a Silicon-Germanium Thermoelectric Element due to a Solar Flare Event — 240
Vincent J. Dandini, Sandia National Laboratories, Albuquerque, NM

An Overview of Silicon Carbide Device Technology — 246
Philip G. Neudeck, Ohio Aerospace Institute, Brook Park, OH, and Lawrence G. Matus, NASA Lewis Research Center, Cleveland, OH

A Summary of High Temperature Electronics Research and Developement — 254
Frank V. Thome and Donald B. King, Sandia National Laboratories, Albuquerque, NM

Advanced Thermal Management Techniques for Space Power Electronics — 260
Angel Samuel Reyes, Wright Laboratory, Wright Patterson AFB, OH

SP-100 Position Multiplexer and Analog Input Processor — 266
Akbar Syed, Ray A. Meyer, and Jaik N. Shukla, General Electric Company, San Jose, CA, and Ken Gililand, DSP Consultant, Mountain View, CA

TABLE OF CONTENTS - PART ONE

[7] MATERIALS II

Steven L. Seiffert, Chairman
Benchmark Environmental Corp.
Albuquerque, NM

Robert H. Titran, Co-Chairman
NASA Lewis Research Center
Cleveland, OH

ORR-Sherby-Dorn Creep Strengths of the Refractory-Metal Alloys C-103, ASTAR-811C, W-5Re, and W-25Re 272
  Robert E. English, NASA Lewis Research Center, Cleveland, OH

Evaluation of Properties and Special Features for High-Temperature Applications of Rhenium 278
  Boris D. Bryskin, Sandvik Rhenium Alloys, Inc., Elyria, OH

Effects of Rhenium on the High Temperature Mechanical Properties of a Tungsten-1.0W/O Thoria Alloy 292
  Anhua Luo, Kwang S. Shin, and Dean L. Jacobson, Arizona State University, Tempe, AZ

SP-100 Liquid Metal Test Loop Design 298
  T. Ted Fallas, Gordon B. Kruger, Frank R. Wiltshire, Grant C. Jensen, Harold Clay, Hugh A. Upton, Robert E. Gamble, Christian G. Kjaer-Olsen, and Keith Lee, General Electric Company, San Jose, CA

Development Testing of High Temperature Bearings for SP-100 Control Drive Assemblies 304
  Alfred W. Dalcher, Christian G. Kjaer-Olsen, Carlos D. Martinez, Stanley Y. Ogawa, Dwight R. Springer, and Robert Yaspo, General Electric Company, San Jose, CA

Aluminum Oxide-Matrix Composite Insulators for Thermionic Devices 312
  Emilio Giraldez, General Atomics, San Diego, CA, Jean-Luis Desplat, Rasor Associates, Inc., Sunnyvale, CA., Tony Witt, Thermo Electron Technologies Corp., Waltham, MA, and Leo A. Lawrence, Westinghouse Hanford Company, Richland, WA

TABLE OF CONTENTS - PART ONE

[8] THERMOELECTRIC ENERGY CONVERSION

Jan W. Vandersande, Chairman	Donald N. Matteo, Co-Chairman
Jet Propulsion Laboratory	General Electric Astro-Space
Pasadena, CA	Valley Forge, PA

Resistivity Changes and Grain Growth with Time at Elevated Temperatures of n-type Si(80)Ge(20) with and without GaP and of p-type Si(80)Ge(20) Alloys — 319
  Melvin J. Tschetter and Bernard J. Beaudry, NASA Ames Laboratory, Ames, IA

Status of the Improved n-Type SiGe/Gap Thermoelectric Material — 326
  Jean-Pierre Fleurial, Alex Borschevsky and Jan W. Vandersande, Jet Propulsion Laboratory, Pasadena, CA, and Nancy Scoville and Clara Bajgar, Thermo Electron Technologies Corp., Waltham, MA

Reduced Thermal Conductivity Due to Scattering Centers in p-type SIGE Alloys — 332
  John S. Beaty and Jonathan L. Rolfe, Thermo Electron Technologies Corp., Waltham, MA, and Jan W. Vandersande and Jean-Pierre Fleurial, Jet Propulsion Laboratory, Pasadena, CA

Extrapolated Thermoelectric Figure of Merit of Ruthenium Silicide — 338
  Cronin B. Vining, Jet Propulsion Laboratory, Pasadena, CA

Recent Achievements in Conductively Coupled Thermoelectric Cell Technology for the SP-100 Program — 343
  Donald N. Matteo, Walter R. Kugler, Richard A. Kull, and James A. Bond, General Electric Company, Philadelphia, PA

Development of a High Voltage Insulator, Compatible with Liquid Lithium at High Temperature for Use in the SP-100 Thermoelectric Cell — 353
  James A. Bond and Donald N. Matteo, General Electric Company, Philadelphia, PA

TABLE OF CONTENTS - PART ONE

[9] KEY NUCLEAR TECHNOLOGIES FOR HUMAN EXPLORATION OF THE SOLAR SYSTEM

John Mankins, Chairman
NASA Headquarters
Washington, DC

Joseph Angelo, Co-Chairman
Science Applications International Corporation
Melbourne, FL

SP-100 System Design and Technology Progress — 363
Allan T. Josloff, Herbert S. Bailey, and Donald N. Matteo, General Electric Company, Philadelphia, PA

Space Exploration Mission Analyses for Estimates of Energetic Particle Fluence and Incurred Dose — 372
John E. Nealy, Lisa C. Simonsen, and Scott A. Striepe, NASA Langley Research Center, Hampton, VA

Nuclear Propulsion: A Key Transportation Technology for the Exploration of Mars — 383
Gary L. Bennett, NASA Headquarters, Washington, DC, and Thomas J. Miller, NASA Lewis Research Center, Cleveland, OH

Fast Piloted Missions to Mars Using Nuclear Electric Propulsion — 389
Jeffrey A. George, Kurt J. Hack, and Leonard A. Dudzinski, NASA Lewis Research Center, Cleveland, OH

[10] SPACE POWER ELECTRONICS I

Frank V. Thome, Chairman
Sandia National Laboratories
Albuquerque, NM

Joseph A. Weimer, Co-Chairman
Wright Laboratory
Wright Patterson AFB, OH

MultiMegawatt Inverter/Converter Technology for Space Power Applications — 401
Ira T. Myers and Eric D. Baumann, NASA Lewis Research Center, Cleveland, OH, Robert Kraus, W. J. Schafer Associates, Arlington, VA, and Ahmad N. Hammoud, Sverdrup Technology, Brook Park, OH

Ignition of Cs-Ba Tacitron during Breakdown and Extinguishing Modes — 410
Benard R. Wernsman, Mohamed S. El-Genk, and Christopher S. Murray, University of New Mexico, Albuquerque, NM

An Analysis of Extinguishing Characteristics of a Cs-Ba Tacitron — 417
Christopher S. Murray and Mohamed S. El-Genk, University of New Mexico, Albuquerque, NM, and Vladimir Kaibyshev, Kurchatov Institute of Atomic Energy, Moscow, USSR

Thermionic Power System Power Processing and Control — 427
Kenneth J. Metcalf, Rockwell International/Rocketdyne Division, Canoga Park, CA

Power Management and Distribution Considerations for A Lunar Base — 433
Barbara H. Kenny, NASA Lewis Research Center, Cleveland, OH, and Anthony S. Coleman, Sverdrup Technology, Inc., Brook Park, OH

NPS Options for Lunar Base Power Supply — 440
Nikolay N. Ponomarev-Stepnoi, Vladimir A. Pavshoock, and Veniamin A. Usov, Kurchatov Institute of Atomic Energy, USSR

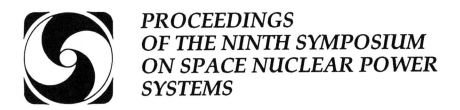

PROCEEDINGS OF THE NINTH SYMPOSIUM ON SPACE NUCLEAR POWER SYSTEMS

TABLE OF CONTENTS - PART TWO

[11] REACTOR AND POWER SYSTEMS CONTROL I

Samit K. Bhattacharyya, Chairman
Argonne National Laboratory
Argonne, IL

Francis J. Wyant, Co-Chairman
Sandia National Laboratories
Albuquerque, NM

Overview of Rocket Engine Control — 446
Carl F. Lorenzo and Jeffrey L. Musgrave, NASA Lewis Research Center, Cleveland, OH

Experimental Study of the TOPAZ Reactor-Converter Dymanics — 456
Anatoly V. Zrodinkov, Victor Ya. Pupko, and Victor L. Semenistiy, Institute of Physics and Power Engineering, Obninsk, USSR

Startup and Control of Out-of-Core Thermionic Space Reactors — 462
Michael G. Houts, Los Alamos National Laboratory, Los Alamos, NM, and David D. Lanning, Massachusetts Institute of Technology, Cambridge, MA

Reactor Dynamics and Stability Analysis for Two Gaseous Core Reactor Space Power Systems — 471
Edward T. Dugan, University of Florida, Gainesville, FL, and Samer D. Kahook, Westinghouse Savannah River Company, Aiken, SC

Concepts and Issues Related to Control Systems for Fusion Propulsion — 479
George H. Miley, University of Illinois, Urbana, IL

[12] THERMIONIC ENERGY CONVERSION I

Richard A. Johnson, Chairman
Rockwell International Corporation
Canoga Park, CA

Richard J. Bohl, Co-Chairman
Los Alamos National Laboratory
Los Alamos, NM

Performance of Fast Reactor Irradiated Fueled Emitters for Thermionic Reactors — 492
Leo A. Lawrence and Bruce J. Makenas, Westinghouse Hanford Company, Richland, WA, and Lester L. Begg, General Atomics, San Diego, CA

TFE Fast Driver Reactor System for Low-Power Applications — 498
Thomas H. Van Hagan, Bryan R. Lewis, Elizabeth A. Bellis, and Mike V. Fisher, General Atomics, San Diego, CA

A Fast Spectrum Heat Pipe Cooled Thermionic Power System — 504
Joseph C. Mills and William R. Determan, Rockwell International/ Rocketdyne Division, Canoga Park, CA, Thomas H. Van Hagan, General Atomics, San Diego, CA and Captain Thomas Wutche, Phillips Laboratory, Kirtland AFB, NM

System Startup Simulation for an In-Core Thermionic Reactor with Heat Pipe Cooling — 510
William R. Determan and William D. Otting, Rockwell International/Rocketdyne Division, Canoga Park, CA

TABLE OF CONTENTS - PART TWO

Special Features and Results of the "Topaz-II" Nuclear Power System Tests   515
with Electric Heating
    V. P. Nicitin, B. G. Ogloblin, A. N. Luppov, E. Y. Kirllov, and V. G. Sinkevich, Central Design Bureau
    of Machine Building, Saint Petersburg, USSR, and V. A. Usov and Y. A. Nechaev, Kurchatov Institute
    of Atomic Energy, Moscow, USSR

[13] SPACE MISSIONS AND POWER NEEDS

Steve Harrison, Chairman   David Buden, Co-Chairman
National Space Council   Idaho National Engineering
Washington, DC   Laboratory

The CRAF/CASSINI Mission: Baseline Plan and Status   516
    Reed E. Wilcox, Douglas S. Abraham, C. Perry Bankston, Sandra M. Dawson, John W. Klein,
    Phillip C. Knocke, and Paul D. Sutton, Jet Propulsion Laboratory, Pasadena, CA

SP-100 First Demonstration Flight Mission Concept Design   532
    Dennis Switick, Charles Cowan, Darryl Hoover, Thomas Marcille, Robert Otwell, and
    Neal Shepard, General Electric Company, San Jose, CA,

Power Beaming Mission Enabling for Lunar Exploration   544
    Judith Ann Bamberger, Pacific Northwest Laboratory, Richland, WA

Space Nuclear Power Requirements for Ozone Layer Modification   550
    Thomas J. Dolan, Idaho National Engineering Laboratory, Idaho Falls, ID

Commercial Power for the Outpost Platform in Orbit   556
    Thomas C. Taylor and William A. Good, Global Outpost, College Park, MD, and
    Charles R. Martin, Gaithersburg, MD

[14] SPACE POWER ELECTRONICS II
(CANCELED)

[15] REACTOR AND POWER SYSTEM CONTROL II

Francis J. Wyant, Chairman   Samit Bhattacharyya, Co-Chairman
Sandia National Laboratory   Argonne National Laboratory
Albuquerque, NM   Argonne, IL

Experimental Demonstration of Automated Reactor Startup with On-Line   562
Reactivity Estimation
    Kwan S. Kwok, John A. Bernard, and David D. Lanning, Massachusetts Institute of Technology,
    Cambridge, MA

SP-100 Controller Development Paradigm   572
    Carl N. Morimoto, Jaik N. Shukla, John A. Briese, and Akbar Syed, General Electric Company,
    San Jose, CA

A Preliminary Feasibility Study of Control Disks for a Compact Thermionic Space Reactor   577
    Scott B. Negron, Kurt O. Westerman, and Lewis C. Hartless, Babcock & Wilcox, Lynchburg, VA

Period-Generated Control: a Space-Spinoff Technology   583
    John A. Bernard, Massachusetts Institute of Technology, Cambridge, MA

TABLE OF CONTENTS - PART TWO

Irradiation Testing of Niobium-Molybdenum Developmental Thermocouple   594
R. Craig Knight and David L. Greenslade, Westinghouse Hanford Company, Richland, WA

Control Aspects of Hydride/BE Moderated Thermionic Space Reactors   604
Norman G. Gunther and Monte V. Davis, Space Power, Inc., San Jose, CA and
Samit K. Bhattacharyya and Nelson A. Hanan, Argonne National Laboratory, Argonne, IL

[16] THERMIONIC ENERGY CONVERSION II

Richard Bohl, Co-Chairman   **Richard A. Johnson, Chairman**
Los Alamos National Laboratory   Rockwell International Corporation
Los Alamos, NM   Canoga Park, CA

Design of a Planar Thermionic Converter to Measure the Effect of Diffusion of Uranium Oxide on Performance   612
Gabor Miskolczy and David P. Lieb, Thermo Electron Technologies, Waltham, MA, and
G. Laurie Hatch, Rasor Associates, Inc., Sunnyvale, CA

Advanced Thermionics Technology Programs At Wright Laboratory   617
Thomas R. Lamp and Brian Donovan, Wright Laboratory/Aerospace Power Division, Wright-Patterson AFB, OH

Barium Interaction with Partially Oxygen-Covered Nb(110) Surfaces   623
Gerald G. Magera, Oregon Graduate Institute of Science Technology,
Beaverton, OR, Paul R. Davis, Linfield College, McMinnville, OR, and Thomas R. Lamp,
Wright Laboratory, Wright Patterson AFB, OH

Sorption Reservoirs for Thermionic Converters   629
Kevin D. Horner-Richardson, Thermacore, Inc., Lancaster, PA, and Kwang Y. Kim,
Wright Laboratory, Wright Patterson AFB, OH

Computation of Dimensional Changes in Isotropic Cesium-Graphite Reservoirs   638
Joe N. Smith, Jr. and Timothy Heffernan, General Atomics, San Diego, CA

Diamond Film Sheath Insulator for Advanced Thermionic Fuel Element   643
Steven F. Adams, Wright Laboratory, Wright-Patterson AFB, OH and Leonard H. Caveny,
Strategic Defense Initiative Organization, Washington, DC

[17] KEY ISSUES IN NUCLEAR POWER AND PROPULSION

Nils Diaz, Chairman   **Edmund P. Coomes, Co-Chairman**
University of Florida   Pacific Northwest Laboratory
Gainesville, FL   Richland, WA

Safety Questions Relevant to Nuclear Thermal Propulsion   648
David Buden, Idaho National Engineering Laboratory, Idaho Falls, ID

Reliability Comparison of Various Nuclear Propulsion Configurations for Mars Mission   655
Donald R. Segna, U.S. Department of Energy, Richland, WA, Jeffrey E. Dagle,
Pacific Northwest Laboratory, Richland, WA, and William F. Lyon, III,
Westinghouse Hanford Company, Richland, WA

The 1981 United Nations Report: A Historical International Consensus on the Safe Use of Nuclear Power Sources in Space   662
Gary L. Bennett, NASA Headquarters, Washington, DC

TABLE OF CONTENTS - PART TWO

The Regulatory Quagmire Underlying the Topaz II Exhibition: The Nuclear Regulatory Commission's Jurisdiction over the TOPAZ II Reactor System   668
    John W. Lawrence, Winston and Strawn, Washington, DC

An Integrated Mission Planning Approach for the Space Exploration Iniviative   675
    Edmund P. Coomes, Jeffrey E. Dagle, Judith Ann Bamberger, and Kent E. Noffsinger, Pacific Northwest Laboratory, Richland, WA

[18] NUCLEAR THERMAL PROPULSION I

John S. Clark, Chairman
NASA Lewis Research Center
Cleveland, OH

James R. Powell, Co-Chairman
Brookhaven National Laboratory
Upton, NY

Fuels and Materials Development for Space Nuclear Propulsion   681
    S. K. Bhattacharyya, Argonne National Laboratory, Argonne, IL,
    R. H. Cooper, Oak Ridge National Laboratory, Oak Ridge, TN,
    R. B. Matthews, Los Alamos National Laboratory, Los Alamos, NM,
    C. S. Olsen, Idaho National Engineering Laboratory, Idaho Falls, ID,
    R. H. Titran, NASA Lewis Research Center, Cleveland, OH, and
    C. E. Walter, Lawrence Livermore National Laboratory, Livermore, CA

Nuclear Thermal Propulsion Test Facility Requirements and Development Strategy   692
    George C. Allen, Sandia National Laboratories, Albuquerque, NM,
    John S. Clark, NASA Lewis Research Center, Cleveland, OH,
    John Warren, DOE/Office of Nuclear Engineering, Washington, DC,
    David Perkins, Phillips Laboratory, Edwards AFB, CA, and
    John Martinell, Idaho National Engineering Laboratory, Idaho Falls, ID

The NASA/DOE Space Nuclear Propulsion Project Plan - FY 1991 Status   703
    John S. Clark, NASA Lewis Research Center, Cleveland, OH

A Unique Thermal Rocket Engine Using A Particle Bed Reactor   714
    Donald W. Culver, Wayne B. Dahl, and Melvin C. McIlwain, GenCorp/Aerojet Propulsion Division, Sacramento, CA

Assessment of the Use of H_2, CH_4, NH_3, and CO_2 as NTR Propellants   721
    Elizabeth C. Selcow, Richard E. Davis, Kenneth R. Perkins, Hans Ludewig, and Ralph J. Cerbone, Brookhaven National Laboratory, Upton, NY

Enabler II: A High Performance Prismatic Fuel Nuclear Rocket Engine   728
    Lyman J. Petrosky, Westinghouse Advanced Energy Systems, Madison, PA

[19] MANUFACTURING AND PROCESSING

Wayne R. Amos, Chairman
EG & G Mound Technologies, Inc.
Miamisburg, OH

Wayne M. Brittain, Co-Chairman
Teledyne Energy Systems
Timonium, MD

A Prototype On-Line Work Procedure System for RTG Production   738
    Gary R. Kiebel, Westinghouse Hanford Company, Richland, WA

Application of PbTe/TAGS CPA Thermoelectric Module Technology in RTGs for Mars Surface Mission   743
    Wayne M. Brittain, Teledyne Energy Systems, Timonium, MD

TABLE OF CONTENTS - PART TWO

Development of a Radioisotope Heat Source for the Two-Watt Radioisotope Thermoelectric Generator   749
    Edwin I. Howell, Dennis C. McNeil, and Wayne R. Amos, EG & G Mound Applied Technologies, Miamisburg, OH

Re-establishment of RTG Unicouple Production   758
    Kermit D. Kuhl and James Braun, General Electric Company, Philadelphia, PA

Transport and Handling of Radioisotope Thermoelectric Generators at Westinghouse Hanford Company   764
    Carol J. Alderman, Westinghouse Hanford Company, Richland, WA

Quality Assurance Systems Employed in the Assembly and Testing of Radioisotope Thermoelectric Generators   770
    William A. Bohne, EG & G Mound Technologies, Inc., Miamisburg, OH

[20] THERMAL MANAGEMENT I

Albert J. Juhasz, Chairman
NASA Lewis Research Center
Cleveland, OH

Maribeth E. Hunt, Co-Chairwoman
Rockwell International/
Rocketdyne Division
Canoga Park, CA

Liner Protected Carbon-Carbon Heat Pipe Concept   781
    Richard D. Rovang and Maribeth E. Hunt, Rockwell International/ Rocketdyne Division, Canoga Park, CA

An Investigation of Natural Circulation Decay Heat Removal from an SP-100 Reactor System for a Lunar Outpost   787
    Mohamed S. El-Genk and Huimin Xue, University of New Mexico, Albuquerque, NM

Rotating Flat Plate Condensation and Heat Transfer   796
    Homam Al-Baroudi and Andrew C. Klein, Oregon State University, Corvallis, OR

High Temperature Thermal Power Loops   800
    Robert Richter, Jet Propuslion Laboratory, Pasadena, CA, and Joseph M. Gottschlich, Wright Laboratory, Wright Patterson AFB, OH

[21] SPACE NUCLEAR SAFETY I: POLICY AND REQUIREMENTS

George F. Niederauer, Chairman
Los Alamos National Laboratory
Los Alamos, NM

Ronald F. Tuttle, Co-Chairman
Air Force Institute of Technology
Wright-Patterson AFB, OH

Nuclear Safety Policy Working Group Recommendations for SEI Nuclear Propulsion Safety   806
    Albert C. Marshall, Sandia National Laboratories, Albuquerque, NM, and J. Charles Sawyer, Jr., NASA Headquarters, Washington, DC

Implications of the 1990 ICRP Recommendations on The Risk Analysis of Space Nuclear Systems   812
    Bart W. Bartram, Halliburton NUS Environmental Corporation, Gaithersburg, MD

TABLE OF CONTENTS - PART TWO

The Peculiarities of Providing Nuclear and Radiation Safety of Space Nuclear High-Power Systems 819
    Albert S. Kaminsky, Victor S. Kuznetsov, Konstantin A. Pavlov, Vladimir A. Pavshoock, and Lev Ja. Tikhonov, Kurchatov Institute of Atomic Energy, Moscow, USSR and Valery T. Khrushch, Institute of Biological Physics, Moscow, USSR

SP-100 Approach to Assure Design Margins for Safety and Lifetime 824
    A. Richard Gilchrist, Richard Prusa, Ernest P. Cupo, Michael A. Smith, and Neil W. Brown, General Electric Company, San Jose, CA

[22] NUCLEAR THERMAL PROPULSION II

Don Roy, Chairman
Babcock and Wilcox Company
Lynchburg, VA

Gary Polansky, Co-Chairman
Sandia National Laboratories
Albuquerque, NM

A Unique Thermal Rocket Engine Using A Particle Bed Reactor 829
    Donald W. Culver, Wayne B. Dahl, and Melvin C. McIlwain, GenCorp/Aerojet Propulsion Division, Sacramento, CA

Assessment of the Use of H_2, CH_4, NH_3, and CO_2 as NTR Propellants 829
    Elizabeth C. Selcow, Richard E. Davis, Kenneth R. Perkins, Hans Ludewig, and Ralph J. Cerbone, Brookhaven National Laboratory, Upton, NY

Enabler II: A High Performance Prismatic Fuel Nuclear Rocket Engine 829
    Lyman J. Petrosky, Westinghouse Advanced Energy Systems, Madison, PA

[23] NUCLEAR TESTING AND PRODUCTION FACILITIES

Michael Wiemers, Chairman
Westinghouse Hanford Company
Richland, WA

Mario Fontana, Co-Chairman
Oak Ridge National Laboratory
Oak Ridge, TN

Nuclear Electric Propulsion Development and Qualification Facilities 830
    Dale S. Dutt, Westinghouse Hanford Company, Richland, WA, Keith Thomassen, Lawrence Livermore National Laboratory, Livermore, CA, James Sovey, NASA Lewis Research Center, Cleveland, OH, and Mario Fontana, Oak Ridge National Laboratory, Oak Ridge, TN

Thermionic System Evaluation Test (TSET) Facility Description 836
    Jerry F. Fairchild and James P. Koonmen, Phillips Laboratory, Kirtland AFB, NM, and Frank V. Thome, Sandia National Laboratories, Albuquerque, NM

Resolving the Problem of Compliance with the Ever Increasing and Changing Regulations 843
    Harley Leigh, Westinghouse Hanford Company, Richland, WA

Modification of Hot Cells for General Purpose Heat Source Assembly at the Radioisotope Power Systems Facility 848
    Betty A. Carteret, Westinghouse Hanford Company, Richland, WA

Tornado Wind-Loading Requirements Based on Risk Assessment Techniques (For Specific Reactor Safety Class 1 Coolant System Features) 854
    Theodore L. Deobald, Garill A. Coles, and Gary L. Smith, Westinghouse Hanford Company, Richland, WA

Design and Equipment Installation for Radioisotope Power Systems Facility Metallic Cleaning Room 162 Subsystem 6130 860
    Harold E. Adkins and John. A. Williams, Westinghouse Hanford Company, Richland, WA

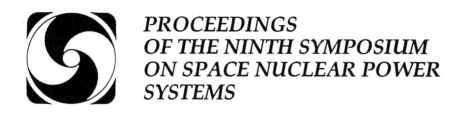

PROCEEDINGS OF THE NINTH SYMPOSIUM ON SPACE NUCLEAR POWER SYSTEMS

TABLE OF CONTENTS - PART THREE

[24] DYNAMIC ENERGY CONVERSION

Robert S. Holcomb, Chairman
Oak Ridge National Laboratory
Oak Ridge, TN

William Harper, Co-Chairman
Allied Signal Aerospace Corporation
Tempe, AZ

SP-100 Reactor with Brayton Conversion for Lunar Surface Applications — 866
Lee S. Mason, Carlos D. Rodriguez, and Barbara I. McKissock, NASA Lewis Research Center, Cleveland, OH, James C. Hanlon, Sverdrup Technology, Inc., Brook Park, OH, and Brain C. Mansfield, University of Dayton, Dayton, OH

A Low-Alpha Nuclear Electric Propulsion System for Lunar and Mars Missions — 878
Edmund P. Coomes and Jeffery E. Dagle, Pacific Northwest Laboratory, Richland, WA

Closed Brayton Power Conversion System Design and Operational Flexibility — 884
Thomas L. Ashe and William B. Harper, Jr, Allied-Signal Aerospace Corp., Tempe, AZ

Free-Piston Stirling Component Test Power Converter Test Results of the Initial Test Phase — 894
George P. Dochat, Mechanical Technology Inc., Latham, NY, and James E. Dudenhoefer, NASA Lewis Research Center, Cleveland, OH

Combined-Brayton Cycle, Space Nuclear Power Systems — 901
Zephyr P. Tilliette, Commissariat a l'Energie Atomique, Gif-sur-Yvette, Cedex, France

[25] SPACE NUCLEAR SAFETY II: METHODS AND ANALYSIS

Albert C. Marshall, Chairman
Sandia National Laboratories
Albuquerque, NM

Neil W. Brown, Co-Chairman
General Electric Company
San Jose, CA

A Methodology for the Risk Analysis of Fission-Reactor Space Nuclear Systems — 907
Bart W. Bartram, Seshagiri R. Tammara, and Abraham Weitzberg, Halliburton NUS Environmental Corp., Gaithersburg, MD

The Studies in the Substantitation of the Nuclear Safety Conception of NPS for Manned Missions — 916
V. Ya. Pupko, F. P. Raskach, V. A. Lititsky, A. G. Shestyorkin, M. K. Ovcharenko, and V. V. Volnistov, Institute of Physics and Power Engineering, Obninsk, USSR, P. I. Bystrov, Yu. A. Sobolev, and N. M. Lipovy, Research and Production Association "Energiya", Kaliningrad, USSR, and F.M. Arinkin, G. A. Batyrbekov, S. U. Talanov, and Sh. Kh. Gizatulin, Institute of Nuclear Physics, Kazahstan Academy of Science, USSR

Nuclear Thermal Rocket Entry Heating and Thermal Response Preliminary Analysis — 923
Leonard W. Connell, Donald L. Potter, C. Channy Wong, and Marc W. Kniskern, Sandia National Laboratories, Albuquerque, NM

TABLE OF CONTENTS - PART THREE

Aging Effects of U. S. Space Nuclear Systems, Currently in Orbit 929
   Bart W. Bartram and Seshagiri R. Tammara, Halliburton NUS Environmental Corp., Gaithersburg, MD

[26] NUCLEAR THERMAL PROPULSION III

Patrick J. McDaniel, Chairman
Sandia National Laboratories
Albuquerque, NM

David Woodall, Co-Chairman
Idaho National Engineering Laboratory
Idaho Falls, ID

Nuclear Thermal Propulsion Engine System Design Analysis Code Development 937
   Dennis G. Pelaccio and Christine M. Scheil, Science Applications International Corp., Torrance, CA, and Lyman J. Petrosky and Joseph F. Ivanenok III, Westinghouse/Advanced Energy Systems, Madison, PA

Computational Study of Nonequilibrium Hydrogen Flow in a Low Pressure Nuclear Thermal Rocket Nozzle 943
   Dana A. Knoll, Idaho National Engineering Laboratory, Idaho Falls, ID

Restartable Nuclear Thermal Propulsion Considerations 950
   Charlton Dunn, Rockwell International/Rocketdyne Division, Canoga Park, CA

An Assessment of Passive Decay Heat Removal in the PeBR Concept for Nuclear Thermal Propulsion 955
   Nicholas J. Morley and Mohamed S. El-Genk, University of New Mexico, Albuquerque, NM

Human Round Trip to Mars: Six Months and Radiation-Safe 967
   Otto W. Lazareth, Eldon Schmidt, Hans Ludewig, and James R. Powell, Brookhaven National Laboratory, Upton, NY

Practical Nuclear Thermal Rocket Stages Supporting Human Exploration of the Moon and Mars 979
   Robert M. Zubrin, Martin Marietta Astronautics, Denver, CO

[27] SIMULATION AND MODELING I

Edwin A. Harvego, Chairman
Idaho National Engineering Laboratory
Idaho Falls, ID

Andrew C. Klein, Co-Chairman
Oregon State University
Corvallis, OR

Modeling the Film Condensate Fluid Dynamics and Heat Transfer within the Bubble Membrane Radiator 990
   Keith A. Pauley, Pacific Northwest Laboratory, Richland, WA, and John Thornborrow, NASA Johnson Space Center, Houston, TX

Numerical Prediction of an Axisymmetric Turbulent Mixing Layer Using Two Turbulence Models 998
   Richard W. Johnson, Idaho National Engineering Laboratory, Idaho Falls, ID

Modeling the Behavior of a Two-Phase Flow Apparatus in Microgravity 1007
   Eric W. Baker and Ronald F. Tuttle, U. S. Air Force Institute of Technology, Wright-Patterson AFB, OH

TABLE OF CONTENTS - PART THREE

"TITAM" Thermionic Integrated Transient Analysis Model: Load-Following of a Single-Cell TFE   1013
   Mohamed S. El-Genk, Huimin Xue, Christopher S. Murray, and Shobhik Chaudhuri, University of New Mexico, Albuquerque, NM

[28] HEAT PIPE TECHNOLOGY I

E. Tom Mahefkey, Chairman
Wright Laboratory
Wright-Patterson AFB, OH

Amir Faghri, Co-Chairman
Wright State University
Dayton, OH

"HPTAM" Heat-Pipe Transient Analysis Model: An Analysis of Water Heat Pipes   1023
   Jean-Michel Tournier and Mohamed S. El-Genk, University of New Mexico, Albuquerque, NM

Moderated Heat Pipe Thermionic Reactor (MOHTR) Module Development and Test   1038
   Michael A. Merrigan and Vincent L. Trujillo, Los Alamos National Laboratory, Los Alamos, NM

Thermionic In-Core Heat Pipe Design and Performance   1046
   William R. Determan, and Greg Hagelston, Rockwell International/Rocketdyne Division, Canoga Park, CA

[29] FLIGHT QUALIFICATION AND TESTING

Thomas J. Wuchte, Chairman
Phillips Laboratory
Kirtland AFB, NM

David Woodall, Co-Chairman
Idaho National Engineering
Laboratory
Idaho Falls, ID

Testability of A Heat Pipe Cooled Thermionic Reactor   1052
   Richard E. Durand, Rockwell International/Rocketdyne Division, Canoga Park, CA, and M. Harlan Horner, General Atomics, San Diego, CA

Test Development for the Thermionic System Evaluation Test (TSET) Project   1060a
   D. Brent Morris and Vaughn H. Standley, Phillips Laboratory, Albuquerque, NM, Michael J. Schuller, Phillips Laboratory, Kirtland AFB, NM

Reactor Test Facilities for Irradiation NPPP Fuel Compositions, Materials and Components   1060
   E. O. Adamov, V. P. Smetannikov, V. I. Perekhozhev, and V. I. Tokarev, Research and Development Institute of Power Engineering, Moscow, USSR, and Sh. T. Tukhvatulin and O. S. Pivovarov, Joint Expedition of SIA "Luch" Semipalatinsk, USSR

Nuclear Facility Licensing, Documentation and Reviews, and the SP-100 Test Site Experience   1067
   Bruce C. Cornwell, Theodore L. Deobald, and Ernest J. Bitten, Westinghouse Hanford Company, Richland, WA

SP-100 Thermoelectric Electromagnetic Pump Development--Electromagnetic Integration Test Plan   1074
   Regina S. Narkiewicz, Jerry C. Atwell, John M. Collett, and Upendra N. Sinha, General Electric Company, San Jose, CA

TABLE OF CONTENTS - PART THREE

[30] NUCLEAR THERMAL PROPULSION IV

Steven Howe, Chairman
Los Alamos National Laboratory
Los Alamos, NM

John Martinell, Co-Chairman
EG & G Idaho, Inc.
Idaho Falls, ID

A Preliminary Comparison of Gas Core Fission and Inertial Fusion for the Space Exploration Initiative 1078
  Terry Kammash and David L. Galbraith, University of Michigan, Ann Arbor, MI

Heat Transfer Model for an Open-Cycle Gas Core Nuclear Rocket 1083
  David I. Poston and Terry Kammash, University of Michigan, Ann Arbor, MI

Nuclear Thermal Rocket Clustering I: A Summary of Previous Work and Relevant Issues 1089
  John J. Buksa, Michael G. Houts, and Richard J. Bohl, Los Alamos National Laboratory, Los Alamos, NM

Nuclear Thermal Rocket Clustering II: Monte Carlo Analyses of Neutronic, Thermal, and Shielding Effects 1097
  Michael G. Houts, John J. Buksa, and Richard J. Bohl, Los Alamos National Laboratory, Los Alamos, NM

A Fission Fragment Reactor Concept for Nuclear Thermal Propulsion 1103
  Ahti J. Suo-Anttila, Edward J. Parma, Paul S. Pickard, Steven A. Wright, and Milton E. Vernon, Sandia National Laboratories, Albuquerque, NM

[31] SIMULATION AND MODELING II

Andrew C. Klein, Chairman
Oregon State University
Corvallis, OR

Edwin A. Harvego, Co-Chairman
Idaho National Engineering Laboratory
Idaho Falls, ID

Thermionic Diode Subsystem Model 1114
  Ralph R. Peters, University of Wisconsin, Madison, WI, and Todd B. Jekel, Sandia National Laboratories, Albuquerque, NM

Modeling the Energy Transport through a Thermionic Fuel Element 1123
  Ronald A. Pawlowski and Andrew C. Klein, Oregon State University, Corvallis, OR

CENTAR Modelling of the Topaz-II: Loss of Vacuum Chamber Cooling during Full Power Ground Test 1129
  Vaughn H. Standley and D. Brent Morris, Phillips Laboratory, Albuquerque, NM, and Michael J. Schuller, Phillips Laboratory, Kirtland AFB, NM

Modeling Space Nuclear Power Systems with CENTAR 1135
  Mark J. Dibben, Taewon Kim, and Ronald F. Tuttle, Air Force Institute of Technology, Wright-Patterson AFB, OH

System Modelling with SALT GPS 1141
  Daniel J. Robbins and Ronald F. Tuttle, Air Force Institute of Technology, Wright-Patterson AFB, OH

TABLE OF CONTENTS - PART THREE

[32] HEAT PIPES TECHNOLOGY II

Walter B. Bienert, Chairman
DTX Corporation
Cockeysville, MD

Michael A. Merrigan, Co-Chairman
Los Alamos National Laboratory
Los Alamos, NM

An Experimental Comparison of Wicking Abilities of Fabric Materials for Heat Pipe Applications — 1147
Timothy S. Marks and Andrew C. Klein, Oregon State University, Corvallis, OR

Sodium Heat Pipe with Sintered Wick and Artery: Effects of Noncondensible Gas on Performance — 1153
Robert M. Shaubach and Nelson J. Gernert, Thermacore, Inc., Lancaster, PA

High Temperature Capillary Pumped Loops for Thermionic Power Systems — 1162
William G. Anderson, Thermacore, Inc., Lancaster, PA, and Jerry E. Beam, Wright Laboratory, Wright-Patterson AFB, OH

Sulfur Heat Pipes for 600 K Space Heat Rejection Systems — 1170
John H. Rosenfeld, G. Yale Eastman, and James E. Lindemuth, Thermacore Inc., Lancaster, PA

[33] APPLIED TECHNOLOGY I: CORE MATERIALS

Janelle Hales, Chairman
Westinghouse Hanford Company
Richland, WA

Walter Stark, Co-Chairman
Los Alamos National Laboratory
Los Alamos, NM

Materials Characterization for the SP-100 Program
Carlos D. Martinez, Mike Kangilaski, and Peter A. Ring, General Electric Company, San Jose, CA

The SP-100 Materials Test Loop
Anthony J. Bryhan, Peter A. Ring, and John R. Zerwekh, General Electric Company, San Jose, CA

Progress in SP-100 Tribological Coatings
Peter A. Ring, Herb Busboom, and Gary Schuster, General Electric Company, San Jose, CA

Secondary Ion Mass Spectrometry (SIMS) Analysis of Nb-1% Zr Alloys
Ashok Choudhury, J. R. DiStefano, and J. W. Hendricks, Oak Ridge National Laboratory, Oak Ridge, TN

[34] NUCLEAR ELECTRIC PROPULSION I

James Sovey, Chairman
NASA Lewis Research Center
McLean, VA

James Polk, Co-Chairman
Jet Propulsion Laboratory
Pasadena, CA

Nuclear Electric Propulsion: An Integral Part of NASA's Nuclear Propulsion Project — 1177
James R. Stone, NASA Lewis Research Center, Cleveland, OH

Nuclear Electric Propulsion Technology Panel Findings and Recommendations — 1183
Michael P. Doherty, NASA Lewis Research Center, Cleveland, OH

TABLE OF CONTENTS - PART THREE

NEP Mission Sensitivities to System Performance — 1192
    James H. Gilland, Sverdrup Technology, Inc., Brook Park, OH

Potassium-Rankine Nuclear Electric Propulsion for Mars Cargo Missions — 1199
    Richard D. Rovang, G. A. Johnson, and Joseph C. Mills, Rockwell International/
    Rocketdyne Division, Canoga Park, CA

Mission Analysis for the Potassium-Rankine NEP Option — 1205
    Elden H. Cross, Frederick W. Widman, Jr., and D. Michael North, Rockwell International/
    Rocketdyne Division, Canoga Park, CA

[35] MICROGRAVITY TWO PHASE FLOW

Zenen I. Antoniak, Chairman　　　　　　　　　　**Frederick R. Best, Co-Chairman**
Pacific Northwest Laboratory　　　　　　　　　　　　Texas A & M University
Richland, WA　　　　　　　　　　　　　　　　　　College Station, TX

Flow Boiling in Low Gravity Environment — 1210
    Basil N. Antar and Frank G. Collins, University of Tennessee Space Institute, Tullahoma, TN,
    and Masahiro Kawaji, University of Toronto, Toronto, Onatario, Canada

Definition of Condensation Two Phase Flow Behaviors for Spacecraft Design — 1216
    Thomas R. Reinarts and Frederick R. Best, Texas A & M University, College Station, TX, and
    Wayne S. Hill, Foster-Miller, Inc., Waltham, MA

Two-Phase Flow Characterization for Fluid Components and Variable Gravity Conditions — 1226
    John M. Dzenitis and Kathryn M. Miller, NASA Johnson Space Center, Houston, TX

Analysis of a Cylindrical Reservoir Concept for Control of a Spacecraft Capillary Pumped Thermal Management System — 1231
    William J. Krotiuk, General Electric Astro Space, Princeton, NJ

[36] SPACE POWER AND PROPULSION TECHNOLOGY I

Dan Mulder, Chairman　　　　　　　　　　　　**Yuri N. Niikolayev, Co-Chairman**
Phillips Laboratory　　　　　　　　　　　　　　　Scientific and Industrial Association
Kirtland AFB, NM　　　　　　　　　　　　　　　(LUCH) Podlosk, USSR

Heat Pipe Cooled Thermionic Reactor Core Fabrication — 1237
    M. Harlan Horner and Thomas H. Van Hagan, General Atomics, San Diego, CA, and
    William R. Determan, Rockwell International/Rocketdyne Division, Canoga Park, CA

Performance Optimization Considerations for Thermionic Fuel Elements in a Heat Pipe Cooled Reactor — 1245
    Elizabeth A. Bellis, General Atomics, San Diego, CA

Radiation Shield Requirements for Manned Nuclear Propulsion Space Vehicles — 1251
    Paul H. Sager, General Dynamics Space Systems Division, San Diego, CA

The Integrated Power and Propulsion Stage: A Mission Driven Solution Utilizing Thermionic Technology — 1259
    Robert M. Zubrin and Tal K. Sulmeisters, Martin Marrietta Astronautics, Denver, CO, and
    Michael G. Jacox and Ken Watts, Idaho National Engineering Laboratory, Idaho Falls, ID

TABLE OF CONTENTS - PART THREE

Cesium Vapor Supply Systems for Lifelong Thermionic Space Nuclear Power Systems   1268
Georgy M. Gryaznov, Evgeny E. Zhabotinsky, Victor I. Serbin, Boris V. Slivkin,
Yuri L. Trukhanov, and Leonid M. Sheftel, SPU "Krasnaya Zvezda", Moscow, USSR

Control Algorithms for "TOPAZ" Type Thermionic Space Nuclear Power Systems of the Second Generation   1274
Irma V. Afanasyeva, Georgy M. Gryaznov, Evgeny E. Zhabotinsky, Gennady A. Zaritzky, and
Victor I. Serbin, SPU "Krasnaya Zvezda", Moscow, USSR

[37] APPLIED TECHNOLOGY II: FUEL MATERIALS

Dale S. Dutt, Chairman   **Dewayne L. Husser, Co-Chairman**
Westinghouse Hanford Company   Babcock & Wilcox Company
Richland, WA   Lynchburg, VA

Preparation of Mixed (U,Zr), (U, Nb), and (U, Nb, Zr) Carbides
A. H. Bremser, R. P. Santandrea, R. R. Ramey, H. H. Meoller, Dewayne L. Husser, and
C. S. Caldwell, Babcock & Wilcox Company, Lynchburg, VA

Determination of High Temperature Phase Relationships in Binary (U, Zr) and (U, Nb) Carbides
R. P. Santandrea, A. H. Bremser, and H. H. Meoller, Babcock & Wilcox Company, Lynchburg, VA

Determination of High Temperature Phase Relationships and Ternary (UZr, Nb) Carbides
R. P. Santandrea, A. H. Bremser, and H. H. Meoller, Babcock & Wilcox Company, Lynchburg, VA

Mixed Carbide Fuel Solidus Melting Temperature
William J Carmack, Richard R. Hobbins, Paul A. Lessing, Dennis E. Clark, and Jon D. Grandy,
Idaho National Engineering Laboratory, Idaho Falls, ID

Physical Properties and Compatability Characteristics of Substoichiometric LiH
Peggy Horton-Campbell and Steve Pawel, Oakridge National Laboratory, Oakridge, TN

SP-100 Fuel Pin Performance based on Data Generated from the Fuel Pin Irradiation Testing Program
Swaminathan Vaidyanathan, Donald C. Wadekamper, and Donald E. Plumlee,
General Electric Company, San Jose, CA

[38] NUCLEAR ELECTRIC PROPULSION II

John Brophy, Chairman   **Jeffrey E. Dagle, Co-Chairman**
Jet Propulsion Laboratory   Pacific Northwest Laboratory
Pasadena, CA   Richland, WA

Multimegawatt MPD Thruster Design Considerations   1279
Roger M. Myers and James E. Parkes, Sverdrup Technology, Inc., Brook Park, OH, and
Maris A. Mantenieks, NASA Lewis Research Center, Cleveland, OH

A Multi-Megawatt Electric Thruster Test Facility   1287
Keith I. Thomassen and E. Bickford Hooper, Lawrence Livermore National Laboratory, Livermore, CA

Resistive Plasma Detachment in Nozzle Based Coaxial Thrusters   1293
Ronald W. Moses, Jr., Richard A. Gerwin, and Kurt F. Schoenberg, Los Alamos National Laboratory,
Los Alamos, NM

TABLE OF CONTENTS - PART THREE

The Concept of a Nuclear Power System with Electric Power of about 2 MW for a Power Transfer Spacecraft — 1304
Nikolay N. Ponomarev-Stepnoi, Vladimir A. Pavshoock, and Michael A. Diachenko, Kurchatov Institute of Atomic Energy, Moscow, USSR

[39] STATIC ENERGY CONVERSION

Elzie E. Gerrels, Chairman
General Electric Company
San Jose, CA

C. Perry Bankston, Co-Chairman
Jet Propulsion Laboratory
Pasadena, CA

HYTEC A High Effeciency Thermally Regenerative Fuel Cell for Space Applications — 1310
Douglas N. Rodgers, Samir A. Salamah, and Prodyot Roy, General Electric Company, San Jose, CA

AMTEC/SHE For Space Nuclear Power Applications — 1316
Thomas K. Hunt, Robert K. Sievers, and Joseph T. Kummer, Advanced Modular Power Systems, Inc. Ann Arbor, MI, and Jan E. Pantolin and David A. Butkiewicz, Environmental Research Institute of Michigan, Ann Arbor, MI

Thermal Modeling of AMTEC Recirculating Cell — 1325
Jerry W. Suitor, Roger M. Williams, Mark L. Underwood, Margaret A. Ryan, Barbara Jeffries-Nakamura, and Dennis O'Connor, Jet Propulsion Laboratory, Pasadena, CA

An AMTEC Vapor-Vapor, Series Connected Cell — 1331
Mark L. Underwood, Roger M. Williams, Margaret A. Ryan, Barbara Jeffries-Nakamura, and Dennis O'Connor, Jet Propulsion Laboratory, Pasadena, CA

Thermoacoustic Power Conversion for Space Power Applications — 1338
William C. Ward and Michael A. Merrigan, Los Alamos National Laboratory, Los Alamos, NM

[40] SPACE POWER AND PROPULSION TECHNOLOGY II

Thomas R. Lamp, Chairman
Wright Laboratory
Wright-Patterson AFB, OH

Hal Shelton, Co-Chairman
General Dynamics Space Systems
Albuquerque, NM

Prediction of the Start-up Characteristics of Thermionic Converter in a Star-C Reactor — 1344
David P. Lieb, Carl A. Witt, Gabor Miskolczy, and Celia C.M. Lee, Thermo Electron Technologies Corporation, Waltham, MA, and John B. McVey, Rasor Associates, Inc., Sunnyvale, CA

Project Thermion: Demonstration of a Thermionic Heat Pipe in Micrograviy — 1351
Frank J. Redd and George E. Powell, Utah State University, Logan, UT

Cascaded Thermionic Converters — 1359
Gary O. Fitzpatrick and Daniel T. Allen, Advanced Energy Technology, Inc., La Jolla, CA, and John B. McVey, Rasor Associates, Inc., Sunnyvale, CA

A Power Propulsion System Based on a Second-Generation Thermionic NPS of the "TOPAZ" Type — 1368
Georgi M. Gryaznov, Eugene E. Zhabotinski, Pavel V. Andreev and Gennadie A. Zartiski, Scientific and Production Association "Krasnaya Zvezda", Moscow, USSR, Anatoly S. Koroteev, Viktor M. Martishin, and Vladimir N. Akimov, Research Institute of Thermal Processes, Moscow, USSR, Nikolai N. Ponomarev-Stepnoi and Veniamin A. Usov, I.V. Kurchatov Institute of Atomic Energy, Moscow, USSR, and Edward J. Britt, Space Power Inc., San Jose, CA

OVERVIEW OF ROCKET ENGINE CONTROL

Carl F. Lorenzo and Jeffrey L. Musgrave
National Aeronautics and Space Administration
Lewis Research Center MS 77-1
Cleveland, OH 44135
(216) 433-3733

Abstract

This paper broadly covers the issues of Chemical Rocket Engine Control. The basic feedback information and control variables used in expendable and reusable rocket engines, such as the Space Shuttle Main Engine are discussed. The deficiencies of current approaches are considered and a brief introduction to Intelligent Control Systems for rocket engines (and vehicles) is presented.

INTRODUCTION

The purpose of this paper is to give a broad overview of Chemical Rocket Engine (CRE) control as background for Nuclear Thermal Rocket Engine control. The paper will discuss the fundamental (underlying) physical issues in CRE control. A brief discussion of modern CREs and their control will follow. This will include a discussion of the Space Shuttle Main Engine (SSME). Recent advanced control approaches for the SSME will be presented along with the benefits which ensue. Current research into Intelligent Control Systems for the SSME which allows high levels of adaptability to engine degradations will be discussed. Finally the connections of current chemical rocket engine controls research to nuclear rocket controls will be explored.

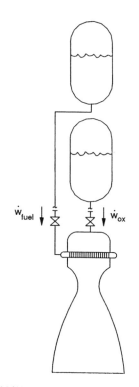

FIGURE 1. Pressure Fed Bi-Propellant Rocket Engine Schematic.

FUNDAMENTALS OF CHEMICAL ROCKET ENGINE CONTROL

The fundamentals of CRE control are best explained by starting with a simplified configuration (Figure 1). This pressure fed rocket engine supplies propellants through appropriate feedlines, control valves and injector elements to a main combustion chamber by pressurizing the supply tanks. The chamber requires the propellants to be delivered in a predetermined ratio (mixture ratio), defined as

$$MR = \frac{\dot{w}_{ox}}{\dot{w}_{fuel}} \quad (1)$$

and at a flow rate level related to the desired thrust. The fundamental function of a rocket engine control is to control thrust (inferred by chamber pressure) and the mixture ratio. The mixture ratio is important since for any propellant combination and pressure level, it sets the combustion temperature and hence the performance and the maximum material temperature. It is also important in terms of propellant utilization.

The basic dynamic equations are found in (Lee et al. 1953). The chamber pressure (P_c) to total weight flow ($\dot{w}_T$) transfer function is given by

where ($c^*/A_T g$) is a proportionality constant, σ is the combustion delay and τ is the chamber fill time ($\tau \sim c^* A_t g V_c / RT_c$) where c^* is the characteristic exhaust velocity, A_T, T_c, and V_c represent the throat area, combustion temperature, and

$$\frac{P_c(s)}{\dot{w}_T(s)} = \left(\frac{c^*}{A_T g}\right)\left(\frac{e^{-\sigma s}}{\tau s + 1}\right) \qquad (2)$$

main chamber volume respectively. In linear form the injectors are flow resistors so that

$$\dot{w}_T(s) = \dot{w}_{fuel}(s) + \dot{w}_{ox}(s) = k_1(P_{i_{fuel}}(s) - P_c(s)) + k_2(P_{i_{ox}}(s) - P_c(s)) \qquad (3)$$

where P_i is the injector pressure. The feedline can be represented in lumped parameter form (continuity and momentum equations) or distributed hyperbolic form (wave equation). In this configuration there are two inputs, namely the valve areas (positions) which control the individual propellant flows and hence the chamber pressure and mixture ratio.

A classical control for this simplified configuration is shown in Figure 2. The following observations are made. Chamber pressure responds to total weight flow. Therefore the chamber pressure flow loop would usually (but not necessarily) go with the propellant having the higher flow rate say O_2 in an H_2 - O_2 engine. Also the two loops tend to be interactive and to minimize excursions of the error signals, one loop is tuned to be the "fast" loop and the other slower. Experience shows that the mixture ratio should be the fast loop. This minimizes excursions in MR away from the set point which in turn keeps the gas and metal temperatures at the design conditions. The chamber pressure is the "slower" loop and its bandwidth is set by thrust response requirements. The type of control shown here would normally require three measurements

P_c, $\dot{w}_{fuel}$, and $\dot{w}_{ox}$ with two control inputs (valve areas) A_{fuel} and A_{ox}. These basic ideas dominate CRE control design for much more complex cycles.

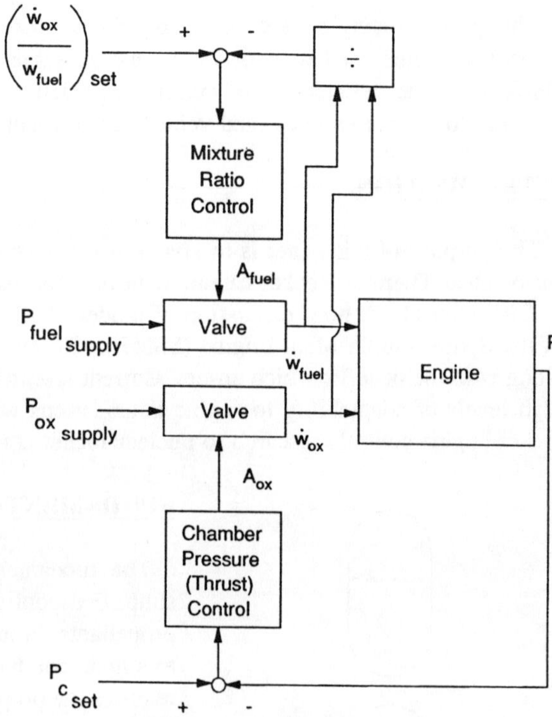

MODERN CHEMICAL ROCKET ENGINES

Turbine driven pumps supply the propellants in most modern rocket vehicles. Numerous cycles are proposed,

FIGURE 2. Classical Control for Bi-Propellant Rocket Engine.

studied, and used in regard to the method that the turbines are powered. The various engine cycles each have their benefits and problems and a discussion of these is well beyond the scope of this paper.

Two representative cycles will be considered. The first of these is the Gas Generator Cycle (Figure 3). The fundamental feature of this cycle is that small amounts of propellant are taken from the main propellant feedlines to be burned in a small auxiliary combustor (gas generator). The generated gases power the turbopumps, and may be used to cool the nozzle and are expelled. The most likely mode of control for this cycle would be to regulate chamber pressure by controlling the gas generator pressure. This would control the speed of both turbopumps and hence the total propellant delivered to the main combustion chamber. Either the oxidizer valve or fuel valve (or both depending on cycle design) could be used to control mixture ratio as the fast loop. From a high level perspective, this control philosophy is very much the same as that detailed in the previous section. In the small, the control designer must assure that the local mixture ratio of the gas generator is controlled to assure gas generator and turbine integrity. Also the flow/power balance between the two turbopumps, that is the propellant delivered by each turbopumps at a given gas generator conditions together with the main chamber cooling requirements will likely determine which main propellant valve will be used to control main chamber mixture ratio.

The second cycle considered is the Expander Cycle (Figure 4). Here, one of the propellants (usually fuel) is circulated as a coolant through the combustion chamber and nozzle. This heated fuel is used to power the turbopumps and then returned to the main chamber where it is injected and burned to create thrust. Note in this cycle most of the fuel is circulated through the turbine. From a controls point of view, this cycle is very similar to the previous cycle. However, overall propellant delivery is set by pump speed which in turn is set by the turbine bypass valve. The oxidizer valve provides control over mixture ratio.

Generally speaking, classical control can be used to design for adequate P_c, and MR control provided loop speeds (bandwidths) are properly accounted. Sensing requirements for P_c and MR performance are minimal, usually being chamber pressure and propellant mass flows.

This discussion does not include treatment of startup or shutdown. Startup is usually a scheduled process based on empirical knowledge of initial ignition, propellant arrival times, and related parameters. Shutdown is also a critical part of the process which must be accurately executed to realize the required mission delta velocity requirement.

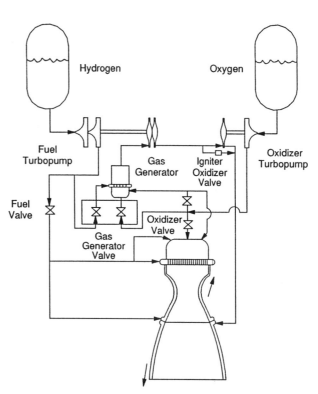

FIGURE 3. Gas Generator Cycle Rocket Engine Schematic.

Many variants of these cycles are possible, and the selection of a cycle for any particular vehicle involves a broad set of considerations such as mission, reliability, manned or unmanned, and maintenance times. However, the high level control philosophy is similar in all cases.

SPACE SHUTTLE MAIN ENGINE CONTROL

The SSME is the first large scale reusable rocket engine developed from a long line of expendable liquid rocket propulsion technology. A two stage combustion process provides the necessary fuel and lox supply pressures to reach the 20684 kPa (3000 psia) chamber pressure resulting in 2091 kN (470,000 lbs) of rated (vacuum) thrust. A propellant flow schematic of the SSME is shown in Figure 5. Hydrogen used to cool the Main Combustion Chamber drives the Low Pressure Fuel Pump (fuel supply) while bleed flow from the High Pressure Lox Pump drives the Low Pressure Lox Pump (lox supply). The fuel and lox preburners acting as the first stage of the combustion process drive the High Pressure Turbopumps which supply coolant flow and lox for the main combustor respectively. The fuel rich gas from the preburners is burned again as primary fuel in the main chamber.

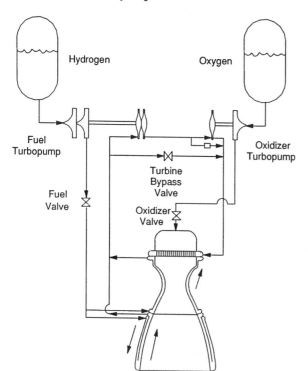

FIGURE 4. Full Expander Cycle Rocket Engine Schematic.

448

Engine control is accomplished through five valves shown in Figure 5, that is Main Oxidizer Valve (MOV), Main Fuel Valve (MFV), Coolant Control Valve (CCV), Oxidizer Preburner Oxidizer Valve (OPOV), and Fuel Preburner Oxidizer Valve (FPOV). In the actual SSME controller (Baseline control), only FPOV and OPOV are used as closed loop control valves. To analytically explore the benefits of enhanced controllability (Musgrave 1991) added the Oxidizer Preburner Fuel Valve (OPFV) and considered the remaining

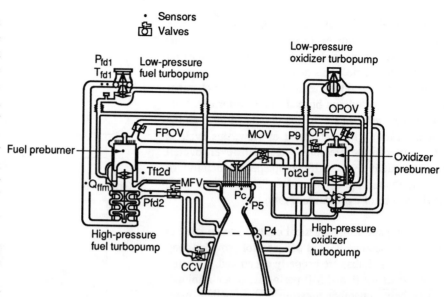

Figure 5. Space Shuttle main engine flow schematic modified by addition of OPFV.

valves to also be closed loop control valves. This actuator configuration is used in the multivariable control (MVC) comparison with the SSME Baseline control which follows the discussion of the baseline controller.

A number of measurement locations are shown in Figure 5 which represent a subset of the SSME ground test sensor suite. Note that the measurements shown are not necessarily Baseline engine control sensors. The discharge pressure and temperature of the Low Pressure Fuel Turbopump (P_{fd1} and T_{fd1} respectively) as well as volumetric fuel flow (Q_{ffm}), and P_c are used for estimating MR in the existing SSME Baseline controller. The discharge pressure of the High Pressure Fuel Turbopump (P_{fd2}), the discharge temperatures of the High Pressure Fuel and Lox Turbines (Tft2d and Tot2d respectively), the pressure of the Fixed Nozzle Heat Exchanger (P_4), the pressure of the Main Chamber Heat Exchanger (P_5), and the fuel supply pressure of the preburners (P_9) are used in conjunction with P_c form the sensor suite for the MVC control to be discussed below.

Engine startup and shutdown are accomplished through open loop scheduling based on extensive computer simulation and test experience. The startup process for a chemical rocket engine is extremely complex and definitive dynamic models have not been created to describe this behavior. Thus, closed loop control has not been attempted in this domain and scheduled valve openings and ignition timing are employed. Closed loop control of the SSME is done via Proportional-Integral (PI) control. A multivariable control approach has been demonstrated in digital computer simulation.

Baseline Controller

The actual SSME controller (Baseline) design philosophy with PI control of P_c and of MR is similar to that discussed for the cycles of Figures 3 or 4. Setpoint control of P_c provides throttling while setpoint control of MR maintains performance and temperature in the main combustion chamber. Regulation of lox flow into the lox preburner and fuel preburner via OPOV and FPOV respectively, adjusts the High Pressure Pump discharge pressures which determine P_c and MR in the main chamber. Lox flow into the lox preburner has an impact on both P_c and MR while lox flow into the fuel preburner has a larger affect on MR only which allows for independent control of both parameters. The CCV is open loop scheduled based on commanded P_c, MOV and MFV are full open, and OPFV is not available on flight hardware making direct control of preburner O/F impossible. Proportional-Integral control has wide acceptance due to the simplicity of design (two parameters), ease of implementation, and speed of computation.

A typical throttle down transient maneuver is shown in Figure 6. The dashed line in Figure 6a represents commanded P_c while the dotted line is the closed loop response for the Baseline control. The Baseline controller

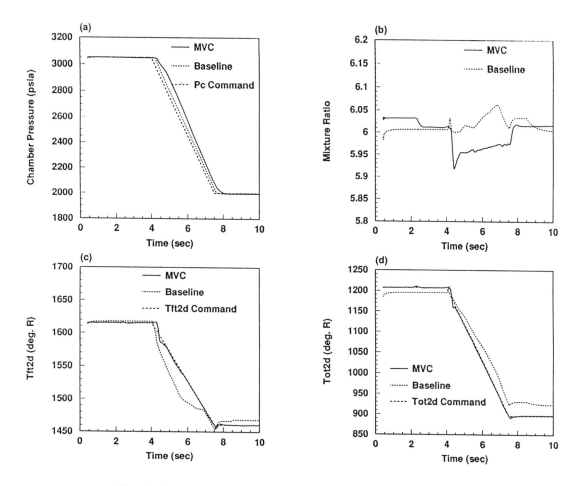

Figure 6. Comparison of Baseline and MVC Responses for the Throttle-Down Transient.

achieves excellent tracking of commanded P_c while minimizing MR excursions from the design point (MR = 6.011) as shown by the dotted line in Figure 6b. The "uncontrolled" turbine discharge temperatures are included in Figures 6c and 6d for the purpose of comparison with the multivariable control later.

Control of P_c and MR only indirectly manage the operation of the four turbopumps which are important life limiting components in the SSME (Cikanek 1987). That is, preburner mixture ratios (temperature) are not directly regulated. This fact may be important relative to the engines not achieving their design life-times. Turbine discharge temperature redlines are used by the Baseline controller to shutdown the engine. Figure 7 shows the closed loop engine response resulting from a change in the High Pressure Fuel Turbine efficiency during mainstage operation. Here, a step decrease of 10% in High Pressure Fuel Turbine efficiency occurs at T = 4 sec while setpoints on P_c and MR are kept constant at design chamber pressure and mixture ratio. The dotted line (Baseline controller) in Figure 7a shows a slight spike in P_c while MR experiences a large increase before returning to setpoint resulting in a temperature spike in the main combustion chamber which is not shown here. The dotted line in Figure 7c shows the discharge temperature of the High Pressure Fuel Turbine rapidly approaching the redline cutoff while Figure 7d shows a rapid drop in High Pressure Oxidizer Turbine discharge temperature. In the next section, the benefits of multivariable control for rocket engines will be discussed in the context of these two examples.

Multivariable Controller

Multivariable control (MVC) methods generally rely on linear state space models of the process to be controlled. A perturbation model of a simplified (39 state) nonlinear dynamic engine model at rated power was used for control design (Musgrave 1991). The linear models of the SSME change very little from the 65% to the 109% power (thrust) level, therefore gain-scheduling was not required.

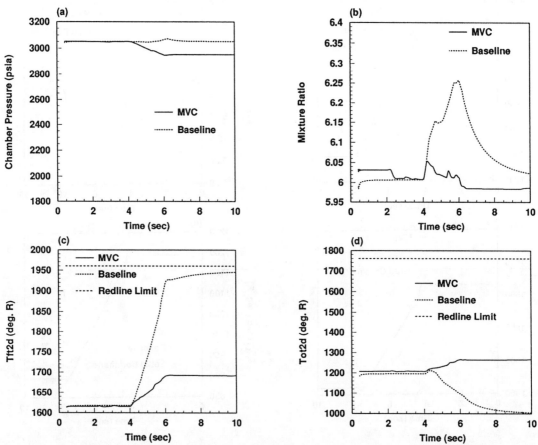

FIGURE 7. Comparison of Baseline and MVC Responses for a 10% Drop in Fuel Turbine Efficiency.

MVC allows the integration of multiple objectives of P_c, Mr, Tft2d, and Tot2d command following while decoupling each of the loops from the others using all six valves in Figure 5 as control valves. Figure 6 shows a multivariable design running at the same sampling rate as the Baseline control (50 hz). The solid line of Figure 6a represents the closed loop MVC response of P_c to reference commands (dashed line). The Baseline controller (dotted line) achieves slightly tighter P_c control than does MVC, however both are satisfactory. Control of MR (solid line) in Figure 6b compares favorably to the Baseline controller with excursions below the setpoint (cool side). The solid lines of Figures 6c and 6d demonstrate the command following capability of the MVC for Tft2d and Tot2d (solid lines) to reference commands (dashed lines).

The benefit of MVC is demonstrated for 10% decrease in High Pressure Fuel Turbine efficiency. In all cases for the MVC of Figure 7 as with Baseline, reference commands for P_c, MR, Tft2d and Tot2d are kept constant at their respective 100% power values. In Figure 7a, we see the controller automatically allowing a slight decrease (3%) in delivered chamber pressure while maintaining mixture ratio (solid line in Figure 7b) thereby avoiding temperature excursions in the main chamber. The dramatic increase (21.9%) of Tft2d from Baseline in Figure 7c is reduced by the MVC to only a 6.25% increase in temperature. This action will preserve the turbine blade life and avoid an unnecessary redline shutdown. Finally, Figure 7d shows only a slight change in Tot2d for the MVC (solid line) compared to the dramatic decrease of the Baseline control (dotted lines). Consequently, the MVC is capable of avoiding a potential redline cutoff which could compromise the mission and/or result in further damage to engine components.

NEW DIRECTIONS

Rocket Engine Intelligent Control

The SSME (Figure 5) is the first rocket engine designed with a philosophy of reusability. The harsh environment encountered in this engine has not allowed realization of the 55 mission design life. Numerous durability problems have been documented for the SSME (Cikanek 1987). These facts together with a desire to space-base some newer rocket engines motivates a technology thrust (Merrill and Lorenzo 1988, and Lorenzo and Merrill 1990) toward Reusable Rocket Engine Intelligent Control. The basic concept of rocket engine Intelligent Control is that using advanced sensors (condition monitoring

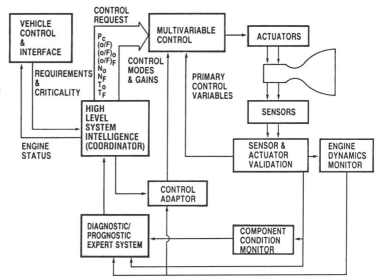

FIGURE 8. REUSEABLE SPACE PROPULSION INTELLIGENT CONTROL SYSTEM FRAMEWORK.

instruments) and on-board diagnostic/prognostic and coordination intelligence an engine with an Intelligent Control System (ICS) can detect and accommodate various sensor, actuator and engine hardware failures. The key functionalities of an ICS are: life extending control, adaptive control, real-time engine diagnostics and prognostics, component condition monitoring, real-time identification, and sensor/actuator fault tolerance. Artificial intelligence techniques are considered for implementing coordination, diagnostics, prognostics, and control reconfiguration functionalities.

A framework for an ICS is shown in Figure 8. The framework provides a rational, top-down basis for the incorporation of system intelligence through the hierarchical integration of the control functional elements. This hierarchy integrates functionalities at the execution level such as the high-speed, closed-loop multivariable controller, engine diagnostics and adaptive reconfiguration with a top level coordination function. The top level coordination function serves to interface the current engine capability with the other engines in the propulsion system, the vehicle/mission requirements, and the crew. It modifies controller input commands and selects various control reconfiguration modes to resolve any conflicts between objectives. A practical baseline framework expanding these ideas for an SSME based Intelligent Control has been proposed (Nemeth 1990). An advanced framework for SSME Intelligent Control is given (Nemeth et al. 1991). The promise of intelligent control is an engine system with greater durability and operability in the face of impending or actual component failure.

Life Extending Control

The concept of Life Extending Control (LEC) compliments that of Intelligent Control discussed above. In LEC the object is to minimize damage accumulation at critical points of the (engine) structure by the way in which the control moves the system through transients (or by the choice of operating domain). Such a concept must also maintain required dynamic performance. In contrast to Intelligent Control, LEC represents what can be done to enhance system durability through the direct control level. LEC is an interdisciplinary thrust between controls and materials/structural science (in particular, fatigue fracture mechanics).

Two broad classes of LEC have been conceptualized by (Lorenzo and Merrill 1991). These are Implicit LEC which uses current technology cyclic based fracture/fatigue damage laws and the Continuous Life Prediction approach which assumes development of continuous differential forms of the damage laws. Only the Implicit LEC concept will be discussed here in order to expose the basic ideas.

The implicit approach to LEC recognizes that current fracture/fatigue science can not predict the differential damage on less than a full cycle of strain. The implicit approach (see Figure 9) selects a sequence of typical command transients (and disturbances) that are representative of those the system would experience in service. Two performance measures are defined: J_p, an objective function that maximizes dynamic performance (possibly by minimizing quadratic state and control excursions) and J_D, a damage measure which uses the best (current) fatigue/fracture theory available to calculate the damage accumulated over the sequence of command transients. An overall performance measure can be defined as $J = J_p + a\, J_D$ where **a** represents the relative importance between performance and life extension. The implicit approach then selects a "best" control algorithm which is applied for the full sequence of command transients. The dynamic performance and damage accumulation over the sequence are optimized (relative to the selected measures) against the

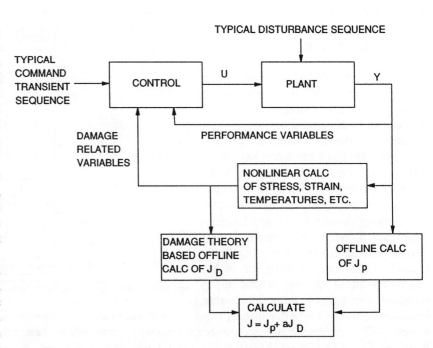

FIGURE 9. Implicit Life Extending Control Approach.

control algorithm parameters. The expectation is to find an algorithm such that the loss in dynamic performance is small ($J_{p,z,min} - J_{p,o,min}$ in Figure 10), for a significant reduction in accumulated damage over the sequence of transients ($J_{D,o,min} - J_{D,z,min}$ is large and life is extended). Here the subscript "o" refers to optimizing for dynamic performance only. An actual operating gain set (point q in Figure 10) is then chosen which satisfies the desired weighting between performance and damage (J).

The mechanics of the implicit approach are detailed as follows. During the design process, two types of feedback variables are considered: (1) the performance variables normally used to manage dynamic performance and (2) nonlinear functions of the performance variables representative of the damage variables (stresses, strains, temperature, and various rates). Various control algorithms are then examined within this feedback structure. That is, the sequence of selected performance and disturbance transients are applied to a simulated system with a trial control and performance J (or J_p and J_D separately) is calculated. A family of algorithms can be developed which are parameterized by the relative tradeoff parameter **a**. The final control can be selected from this set of algorithms with confidence that an effective control and a

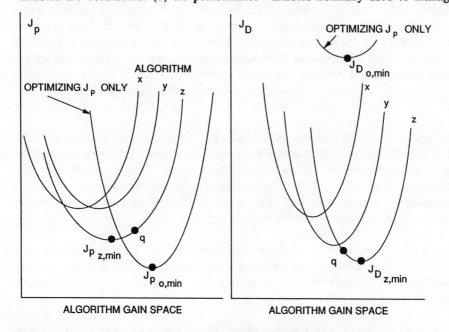

FIGURE 10. Effect of Various Life Extending Control Algorithms on Performance (J_p) and Damage (J_D).

desirable performance/life tradeoff have been established. It is expected that as LEC technology is developed it will find broad application in high performance aerospace systems and elsewhere.

EXTENSIONS TO NUCLEAR PROPULSION

While the Nuclear Thermal Rocket engine is conceptually similar to a chemical rocket engine, it is significantly different in several important ways. One potential Nuclear Rocket Engine cycle is shown in Figure 11. Both systems create thrust by heating a working fluid and expanding it through a convergent-divergent nozzle to supersonic velocities. Additionally, turbomachinery provides the necessary supply pressures for the working fluid. The fundamental difference is the heat source in the nuclear rocket results from the reactor core instead of a chemical combustion process. Specific Impulse (defined as I_{sp} = Thrust / $\dot{w}_T$) for a rocket engine can be expressed as I_{sp} = $K\sqrt{T}$. In a chemical rocket the temperature is set by the propellant combination. In a nuclear thermal rocket the temperature is set by the reactor conditions. Thus I_{sp} is fixed for the chemical rocket but variable for a nuclear thermal rocket limited by core material temperatures.

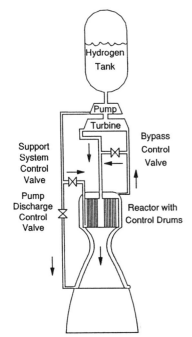

FIGURE 11. Nuclear Thermal Rocket Engine Flow Schematic.

Various studies of nuclear rocket control have been performed (Sanders et al. 1962, Arpasi and Hart 1967, and Hart and Arpasi 1967). The basic control objective of a nuclear rocket is to control thrust level via flow through the turbine and temperature (I_{sp}) via core reaction rates (control drums). Temperature control is similar to mixture ratio control in chemical rockets. However, the fundamental dynamics of the heat addition are quite different. An increase of working fluid (Hydrogen for example) into the reactor core thermally reduces core temperature through heat transfer while simultaneously increasing heat generation by increasing the neutron flux (Crouch 1965). Control of this phenomenon will require anticipatory (lead) action by the reactor control system for good transient performance. Many other issues such as startup, shutdown, and idle mode need also be considered.

Many of the technologies being developed and demonstrated for chemical rocket engine control such as Multivariable Control, Intelligent Control and Life Extending Control will be applicable to nuclear rocket engines in the development of durable, reliable and fault tolerant propulsion systems.

SUMMARY

This paper provided an overview of chemical rocket propulsion control and new technology developments in this area. It is expected that many of these new technologies will find application in the Nuclear Rocket Engine.

Acknowledgments

The authors wish to gratefully acknowledge the technical input from Carl Aukerman, Dale Arpasi, and Albert Powers.

References

Arpasi, D. J. and C. E. Hart (1967) <u>Controls Analysis of Nuclear Rocket Engine at Power Range Operating Conditions</u>, NASA TN D-3978, LeRC, Cleveland, OH., May 1967.

Cikanek, H. A. (1987) "Characteristics of Space Shuttle Main Engine Failures," in <u>23th AIAA/SAE/ASME Joint Propulsion Conference</u>, Paper No. 87-1939, San Diego, CA., 29 June - 2 July 1987.

Crouch, H. F. (1965) *Nuclear Space Propulsion*, Astronuclear Press, Granada Hills, CA., pp. 199-233.

Hart, C. E., and D. J. Arpasi (1967) <u>Frequency Response and Transfer Functions of a Nuclear Rocket Engine System obtained from Analog Computer Simulation</u>, NASA TN D-3979, LeRC, Cleveland, OH., May 1967.

Lee, Y. C., M. R. Gore and C. C. Ross (1953) "Stability and Control of Liquid Propellant Rocket Systems," *American Rocket Society Journal*, March-April 1953, pp. 75-81.

Lorenzo, C. F. and W. C. Merrill (1990) "An Intelligent Control System for Rocket Engines: Needs, Vision and Issues," in <u>Proc. of American Control Conference</u>, San Diego, CA., 23-25 May 1990.

Lorenzo, C. F. and W. C. Merrill (1991) "Life Extending Control: A Concept Paper," in <u>Proc. of American Control Conference</u>, Boston, MA., 26-28 June 1991.

Merrill, W. C. and C. F. Lorenzo (1988) "A Reusable Rocket Engine Intelligent Control," in <u>24th AIAA/SAE/ASME Joint Propulsion Conference</u>, Paper No. 88-3114, Boston, Mass., 11-13 July 1988.

Musgrave, J. L. (1991) "Linear Quadratic Servo Control of a Reusable Rocket Engine," in <u>27th AIAA/SAE/ASME Joint Propulsion Conference</u>, Paper No. 91-1999, Sacramento, CA., 24-27 June 1991.

Nemeth, E. (1990) <u>Reusable Rocket Engine Intelligent Control System Framework Design (Phase I)</u>, NASA Contractor Report CR187043 for NASA LeRC by Rocketdyne Div. Rockwell International Corp., Canoga Park, CA., 6 April 1990.

Nemeth, E., R. Anderson, J. Ols, and M. Olasasky (1991) <u>Reusable Rocket Engine Intelligent Control System Framework Design (Phase II)</u>, RI/RD91-158 for NASA LeRC by Rocketdyne Div. Rockwell International Corp., Canoga Park, CA., 21 June 1991.

Sanders, J. C., H. J. Heppler, and C. E. Hart (1962) <u>Problems in Dynamics and Control of Nuclear Rockets</u>, NASA SP 20, LeRC, Cleveland, OH., December 1962.

EXPERIMENTAL STUDY OF THE TOPAZ REACTOR-CONVERTER DYNAMICS

ANATOLY V.ZRODNIKOV, VICTOR YA.PUPKO, VICTOR L.SEMENISTIY

Institute of Physics & Power Engineering,
249020 OBNINSK, USSR

INTRODUCTION

TOPAZ NPS with in-core thermionic converter is the first complete power system of this type which progressed through the design, development and flight test stage [1].

This NPS has been equipped a three-channel automatic control system with power, current and temperature regulators which ensure its normal operating. Current channel is used for maintaining of the converter current at the normal level by means of influence upon the adjuster of power channel. During longtime reactor operation its power level is periodically corrected by increasing to compensate the degradation of thermionic converter performance. The reactor power correction causes the increase of the operating temperatures. As a sequence, on the reaching of the limiting level of the output coolant temperature, the temperature channel is switching on.

In the course of such type control systems design and verification it is very important to know the reactor dynamical characteristics obtained during the NPS-unit tests. In this report four types of dynamical characteristics are considered: "power/current", "electrical load/current", "power/output temperature" and "temperature/reactivity" ones. They are obtained by special experiments during NPS ground tests. The small perturbation technique was used for the ensuring of minimal influence the normal reactor operation. The perturbations were introduced via the reactor thermal power, electrical load and coolant input temperature.

The obtained experimental data were described by deterministic mathematical models and transfer functions. Both, models and functions were used in the linear forms of the dynamics description suitable for the small perturbation technique data analysis.

The transfer function is a canonical and the most compact form of the dynamical system description which is the ratio:

$$W(s) = Y(s)/X(s),$$

where $Y(s)$, $X(s)$ are the Laplace transformations to the output and input variables of the system, accordingly; the initial conditions are equal to zero, $s = a + jb$ is complex [2].

The applied mathematical model describes the thermal-physics processes in the reactor-converter elements via the set of differential and algebraical equations. The model parameters are defined by means of identification methods [3,4] using the experimental data.

1."POWER/CURRENT" TYPE CHARACTERISTICS.

The corresponding experiment is shown at the Fig.1. The transient process for the electrical current of reactor-converter is satisfactorily described by mathematical model which consists of the single first-order differential equation:

$$\tau_e \frac{d\delta I}{dt}(t) + \delta I(t) = \frac{\partial I}{\partial N} \cdot \delta N(t) \qquad (1)$$

were t -is the time; δ -is the symbol of the deviation; $\delta N(t)$ -is the deviation of the reactor neutronic/thermal power, [kW.]; $\delta I(t)$ -is the deviation of the converter current.

Initial conditions for mentioned deviations are zero; but the stationary magnitudes of the power and current are not zero for the steady-state reactor-converter operation. The meanings of equation (1) parameters are as follows:
$\partial I/\partial N$ -is the partial derivative of the current with respect to the thermal power [A/kW]; $\tau_e = C_e/h_e$ -is the time constant having units of [s], C_e -is the effective heat capacity of the emitter assembly, h_e -is the heat transfer coefficient accounting the heat radiation and conductivity.

The equation (1) is derived on the base of the following (see [4]):
a) The differential equation for the nonstationary temperature distribution in the volume of the emitter assembly;
b) The function $I=I(V)$, where V-is the state vector of the most substantial variables (emitter and collector temperatures, cesium vapour pressure, electrical load resistance).

Numerical estimations for the parameters $\partial I/\partial N$ and τ were obtained with the aid of algorithm which minimizes the discrepancy between the experimental data and model variables using the perturbation theory method. For example, the following estimations were obtained in one of the experiments within TOPAZ ground test program:

$$\partial I/\partial N = (1.324 \pm 0.017) \text{ A/kW.},$$
$$\tau_e = (17.65 \pm 0.16) \text{ s}.$$

Using some simple transformations it may be resulted from equation (1) the following transfer function:

$$W_{NI}(s) = \frac{\partial I}{\partial N} \cdot \frac{1}{\tau_e s + 1} \qquad (2)$$

2. "LOAD/CURRENT" TYPE CHARACTERISTICS.

They were recorded during the stepped change of the reactor-converter electrical load resistance; corresponding transition process for the converter current is shown at Fig.2. The proper mathematical model is obtained as a component of the equation (1). It has the form (see [4]):

$$\tau_e \frac{d\delta I}{dt}(t) + \delta I(t) = -k_I \cdot \frac{\partial I}{\partial N} \cdot \frac{\partial I}{\partial R} \cdot \delta R(t) \qquad (3)$$

The notations correspond to those of (1). Besides it is designated: $\delta R(t)$ [Ohm]-is the change of the resistance; $\partial I/\partial R$ -is the partial derivative of the current with respect to load resistance [A/Ohm]; K_I-is the coefficient of emitter electron cooling [kW./A].

Equation (3) describes the transition process which is called the "emitter electron cooling". In this process the current change causes the change of the energy quantity which is transferred by the current from the emitter to the collector, and naturally it results in the emitter temperature changing.

The magnitudes of the τ_e and $\partial I/\partial N$ equation (3) parameters are mentioned in Sec.1. According to these magnitudes we have $k_I = (0.138 \pm 0.002)$ kW./A, and the complex $\frac{\partial I}{\partial R} \cdot \delta R(t) = \Delta I$ is directly measured in the experiment (see Fig.2). The transfer function which corresponds to equation (3) has the form:

$$W_{RI}(s) = -k_I \cdot \frac{\partial I}{\partial N} \cdot \frac{\partial I}{\partial R} \cdot \frac{1}{\tau_e s + 1} \qquad (3)$$

It may be seen that the converter current transient process caused by the electrical load changing has the same time constant τ_e as in the case of power perturbation.

3. "POWER/OUTPUT TEMPERATURE" TYPE CHARACTERISTICS

The influence of the reactor thermal power was formed by means of the pseudo-random double signal (PRDS) [5]. The responses to such perturbations were recorded in the wide range of the steady-state reactor power and coolant output temperature variations. Then "PRDS/power" and "PRDS/temperature" mutual correlation functions (MCF) were calculated. The MCF were transformed to the mutual Fourier-spectra and further to the frequency characteristics. The latter were approximated by the transfer function which has the form:

$$W_{NT}(s) = \frac{\partial T}{\partial N} \cdot \frac{\exp(-\tau_t s)}{(\tau_e s + 1)(\tau_{tc} s + 1)(\tau_c s + 1)} \qquad (4)$$

The notations made in (4) are the following: $\partial\tau/\partial N$ —is the partial derivative "temperature/power", it is the static coefficient of the transfer function; τ_t—is the transport delay time for the perturbation at the output relating the input perturbation; τ_e—is the time constant for the emitter assembly; τ_{tc}—is the time constant for the thermocouple; τ_c—is the time constant for the coolant. An example of τ_e magnitude is mentioned in the Sec.1.
The other parameter magnitudes from equation (4) are:

$$\tau_t = 3.4s, \quad \tau_{tc} = 8.0s, \quad \tau_c = 2.7s.$$

Effect of the thermal processes within moderator, reflector and reactor vessel with time constants exceeding τ_e, τ_{tc} and τ_c are excluded in the correlation treatment technique by the proper choice of PRDS parameters. The using of the PRDS and the correlation treatment technique allows to decrease the perturbation quantity to the level of the own noises and to obtain the acceptable parameters accuracy in the measurements.

4. "TEMPERATURE/REACTIVITY" TYPE CHARACTERISTICS.

The characteristics of this type are the most important ones from the viewpoint of the reactor-converter controllability. It characterizes the dynamics of the internal feedback existing into the reactor core.

During the TOPAZ reactor tests there was investigated the dynamics of the collector assemblies and of the moderator which have the most strong influence on the reactivity among the other reactor elements. The proper experiment was fulfilled as follows. The reactor-converter was heated to the nominal temperature (590°C) by the coolant which, in its turn, got the heat from an electrical heater. Then the reactor power was set at the minimal controllable level (15W), the power regulator was switched out. The voltage on the electrical heater terminals was "step wise" changed and this caused the deviation of the coolant temperature at the reactor input.

Note that when the temperatures inside reactor are nominal and its thermal power is near zero, the internal "temperature/reactivity" feedback is switched out. The graphical result of one of such experiments is shown at Fig.3 where the parameters are plotted:

$\delta T_{in}(t)$—is the deviation of the input reactor coolant temperature;
$\delta T_{out}(t)$—is the deviation of the output reactor coolant temperature;
$R(t)$—is the reactor reactivity; t—is the time.

The experimental data were treated with application of the proper mathematical model (for the brevity its formulas are omitted). The model parameters were formally adjusted to reach the optimal description of the transient processes obtained. Some of the model adjustment results are the estimations for the partial temperature coefficients of reactivity for collector assemblies $\alpha_c = -(0.089\pm0.002)$ c/K, and moderator $\alpha_m = (0.285\pm0.003)$ c/K.

These parameters define the magnitude and the sign of the corresponding temperature feedback. Common dynamical characteristic for the feedback may be compactly described as follows:

$$W_{TR}(s) = \left[\frac{\alpha_m}{(\tau_m s + 1)} - \alpha_c\right]\exp(-\tau_t s) \qquad (5)$$

Here: τ_m –is the moderator time constant, $\tau_m = (128\pm5)$s;

τ_t –is the transport delay, $\tau_t = (3.7\pm0.3)$s.

It may be seen from Fig.3 and formula (5) that collector assemblies are practically inertialess and have the negative reactivity feedback coefficient. For the moderator the reactivity feedback is positive, it is three times more "strong" than for the collector assemblies and has the significant thermal inertness.

CONCLUSION

The transient processes appearing after the perturbing of the reactor thermal power, electrical load and coolant input temperature are formed by the several reactor elements, namely, by the emitter assemblies, coolant, collector assemblies, moderator and thermocouple measuring the output temperature.

The moderator has the significant in its magnitude positive reactivity temperature coefficient. Nevertheless this negative feature of the reactor is significantly smoothed over by the thermal inertness of the moderator, as well as by practically inertialess negative feedback of collector assemblies. Thus TOPAZ reactor-converter is a quite controllable dynamical system.

REFERENCES

[1] V.A.Kuznetsov et al "Development and Creation of Thermionic TOPAZ NPS" Atomnaya Energiya, v.36, #6, 1974

[2] W.McC.Siebert "Circuits, Signals & Systems" M:.MIR, 1988

[3] P.Eykhoff "System Identification Parameter and State Estimation" John Wiley and Sons, Ltd., 1974

[4] V.Ya.Pupko, A.V.Zrodnikov, Yu.I.Likhachev "The Methods of Adjoint Function for Engineering and Physics Studies" Moscow, Energoatomizdat, 1984

[5] R.E.Urig "Random Noise Techniques in Nuclear Reactor System" The Ronald Press Company, New York, 1970

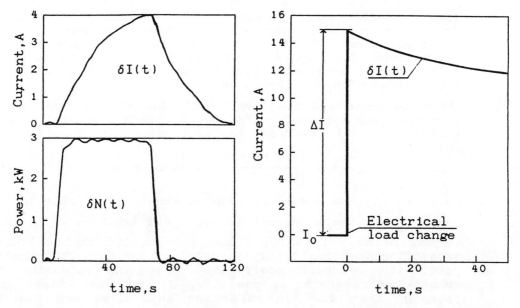

FIG.1. Transient processes caused by power perturbation.

FIG.2. Transient process caused by electrical load changing.

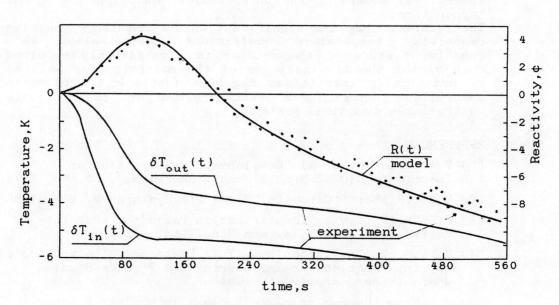

FIG.3. Reactivity deviation caused by cooland temperature perturbation.

STARTUP AND CONTROL OF OUT-OF-CORE THERMIONIC SPACE REACTORS

Michael G. Houts
Los Alamos National Laboratory
Reactor Design and Analysis Group
Los Alamos, NM 87545
(505) 665-4336

David D. Lanning
Massachusetts Institute of Technology
Department of Nuclear Engineering
Cambridge, MA 02139
(617) 253-3843

Abstract

An analysis of out-of-core thermionic space reactor (OTR) startup and control has been performed. The reference OTR chosen for this study is a 75 kWt version of the STAR-C (GA Technologies 1987). The applicability of point kinetics was first verified for the reference OTR system. Point kinetics applicability was verified for core length-to-diameter (L/D) ratios of two and four, and for both subcritical-to-critical and critical-to-supercritical transients. A coupled thermal/point kinetics code was then written, and OTR startup was analyzed. The analyses lead to several observations. First, point kinetics is applicable to the reference OTR for all transients considered. Second, to achieve a 900-s startup the reference OTR must operate at powers well above steady-state rated power during startup. Finally, the large thermal inertia of the radial reflector could be used to reduce radiator temperature during the first several hundred seconds of operation. Further research should be performed on transient heat pipe operation and on off-normal thermionic converter operation.

INTRODUCTION

A schematic of the reference out-of-core thermionic space reactor (OTR) used in this study is shown in Figure 1. The reference OTR chosen for this study is a 75 kWt version of the STAR-C (GA Technologies 1987). The OTR produces 10 kWe at a net efficiency of 13.3%. Heat generated in the core of the OTR is

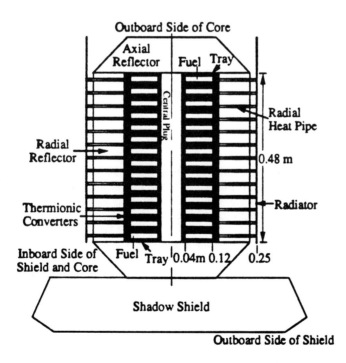

FIGURE 1. Reference OTR showing Inner Core Radius (Ri), Outer Core Radius (Ro), Core Length (L), and the Location of the Central Plug, Fuel Rings, Fuel Trays, Thermionic Converters, Radial Heat Pipes, Radial Reflector, Radiator, Axial Reflector, and Shadow Shield.

conducted to the core's surface, where it is radiated to tungsten heat collectors. The heat is focused (via conduction) to the thermionic emitters located on the back of the heat collectors. Electrons are emitted by the thermionic emitters, travel across a 0.1-mm inter-electrode gap, and are collected by the thermionic collectors. Radial heat pipes transfer heat from the collectors to a radiator that surrounds the radial reflector. Boron carbide control rods located in the radial reflector provide reactor control.

The OTR core is 0.48-m long and has a radius of 0.12 m. The core is composed of 97% enriched UC_2 fuel rings sandwiched between POCO graphite trays. The POCO graphite trays are 9.15-mm thick between a radius of 40 mm and 120 mm, and interlock with adjacent trays to provide core structure and to encase the fuel rings. The interlocks are located between radius 30 mm and 40 mm and between radius 120 mm and 130 mm. Trays are coated with niobium carbide to suppress graphite vaporization, and tray thickness can be varied to provide axial power flattening. The core is surrounded radially by thermionic converters, radial heat pipes, a beryllium radial reflector, and a radiator. Axial neutron reflection is provided by beryllium axial neutron reflectors.

POINT KINETICS APPLICABILITY

QUANDRY Model Development

The applicability of point kinetics to the reference OTR was assessed using the two-group diffusion theory code "QUANDRY" (Smith 1979). QUANDRY solves the two-group, multidimensional, static and transient neutron diffusion equations, and can model transients in the presence of a neutron source.

Two QUANDRY models were developed: one for the reference OTR and one for an analogous OTR with a core L/D ratio of four. The reference OTR was modeled as a rectangular parallelepiped to be compatible with QUANDRY's rectangular coordinate system. The code "MCNP" (Briesmeister 1986) was used to generate two-group absorption and fission cross sections, and the code "COMBINE" (Grimesey 1989) was used to generate transport, total, and Group 1 to 2 scatter cross sections for use in QUANDRY. Finally, discontinuity factors (for deriving the correct face averaged currents from the face and volume averaged fluxes) were added to the QUANDRY model at the interfaces between regions. Further details of QUANDRY OTR model generation can be found in Houts and Lanning (1991).

The accuracy of the QUANDRY model was verified by comparing the k_{eff} calculated by the two-group, rectangular parallelepiped QUANDRY model to that calculated by a continuous energy, geometrically correct MCNP model. For the reference OTR, the difference in the two calculated values for k_{eff} was 0.6%. This level of accuracy was considered adequate for assessing the applicability of point kinetics.

Method of Analysis

The applicability of point kinetics was assessed for both subcritical-to-critical and critical-to-supercritical transients. The applicability of point kinetics was assessed as follows. First, a steady-state QUANDRY calculation was performed to determine the ratio of the flux at a given detector location to the total power being generated by the OTR. This ratio can be thought of as the ratio of detected power to actual power (DP/AP), which by definition is equal to unity at steady-state, zero power critical operating conditions. Thermal feedback is not taken into account in this study, thus DP/AP will equal unity at any power level if the OTR is critical and at steady state. Second, a set of transients were run in which the total power and the flux in the detector cell were tabulated at each time increment. DP/AP was then determined as a function of time for each transient. If DP/AP remains close to 1.0 throughout a transient, point kinetics can be considered applicable for that transient. If DP/AP deviates significantly from 1.0 then point kinetics is not applicable, and a single detector location cannot be used to accurately detect power or reactivity during the transient.

Errors in DP/AP can have two effects. First, if the detected power is less than the actual power, overshoot can result at the end of a transient. Second, if DP/AP is changing, and reactivity is being calculated based on detected power (using inverse point kinetics), reactivity will be calculated erroneously. If DP/AP is decreasing, a non-conservative error in calculated reactivity could occur—actual reactivity could be higher than calculated reactivity.

The effects of source position, detector position, and core length-to-diameter ratio were evaluated for subcritical-to-critical transients. The effects of detector position, core length-to-diameter ratio, and magnitude of reactivity insertion are considered in the critical-to-supercritical transients. The effects of control-rod position are taken into account in all cases.

Results of Analyses

The analyses indicate that for the reference OTR, point kinetics is applicable for severe critical-to-supercritical transients. For the transients evaluated (with L/D=2), the maximum deviation of DP/AP from unity was 2%. For the subcritical-to-critical transients evaluated, DP/AP approaches 1.0 as criticality is approached, although the ratio can differ significantly from unity if the OTR is highly subcritical. For certain source/detector configurations, DP/AP decreases during the subcritical-to-critical transient, possibly leading to a non-conservative error in calculated reactivity. The subcritical-to-critical transients evaluated were severe, with the reactor being brought from a shutdown margin of greater than -3β to critical in one second. Fluctuations in DP/AP decrease as k_{eff} approaches 1.00, reducing possible errors in calculated reactivity. It should be possible to accommodate fluctuations in DP/AP for subcritical-to-critical OTR transients.

Figure 2 plots DP/AP and reactivity as a function of time for a subcritical-to-critical transient in which the reactor is brought from a shutdown margin of greater than -3β to critical in one second. The detector is located at the edge of the inboard axial reflector, and the neutron source is distributed in the radial and axial reflectors. Figure 2 shows that DP/AP approaches unity as reactivity approaches zero. Also, DP/AP is increasing during most of the transient, thus errors in calculating reactivity from DP/AP will primarily be conservative.

For the reference OTR (L/D=2), the most severe critical-to-supercritical transient investigated was a 0.7933β reactivity insertion in one second. The neutron detector was located at the edge of the inboard axial reflector. Figure 3 shows DP/AP during the one second reactivity insertion. DP/AP depends primarily on reactivity, however, the rate of change of DP/AP is of interest in determining possible errors in reactivity measurements. Figure 3 shows that as reactivity approaches 0.7933β, DP/AP drops to slightly below 0.98. During this transient, detected power would thus be only 98% of actual power, and a slight power overshoot could occur at the end of the transient.

OTR STARTUP

Methodology for Analyzing OTR Startup

A transient, one-dimensional, coupled nodal heat transfer / point kinetics code was written to analyze OTR startup. The code, named "TA-OTR" (Transient Analysis of Out-of-Core Thermionic Reactors) accounts for radiation and conduction heat transfer within the OTR, heat generation, and heat storage. Transient heat pipe operation is not modeled, and the time required for heat pipe startup is not taken into account. Heat pipes in the reference OTR are relatively small and may thaw quickly. The radial heat pipes (which heat most of the inboard surface of the radiator) are 0.10-m long.

Table 1 presents a summary of the thermal conductivities and heat capacities (at full power operating temperatures) of the materials used in TA-OTR. Effective core conductivity is obtained by taking a volume weighted average of the values of thermal conductivity for UC_2 (28 W/m-K) and for Poco Graphite (40 W/m-K) given by GA Technologies (1987). The use of high conductivity graphite could substantially increase effective core conductivity. Remaining thermal conductivities and heat capacities presented in Table 1 are taken from Touloukian and Ho (1979).

Table 2 summarizes the effective conductivity and emissivity values that are used at various interfaces within the OTR model. The effective conductivity and emissivity between the TEC and the radial reflector is derived from the heat balance and temperature distribution data presented in Chapter 6 of GA Technologies (1987). Energy transfer between the thermionic emitter and collector is modeled as being by radiation only. The effective emissivity is calculated from the temperature drop between the thermionic emitter and collector that is specified by GA Technologies. Heat transfer within the thermionic converter (between the emitter and collector) also occurs via conduction through the cesium vapor located between the

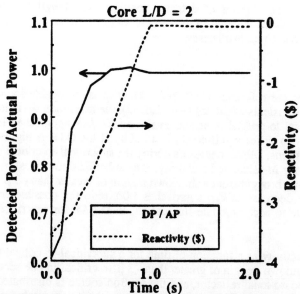

FIGURE 2. Reactivity ($) and Detected Power/Actual Power versus Time (s) Rods withdrawn from Full Shutdown to Critical Position in one Second. Neutron Source Distributed in Axial and Radial Reflector. Detector at Edge of Inboard Axial Reflector.

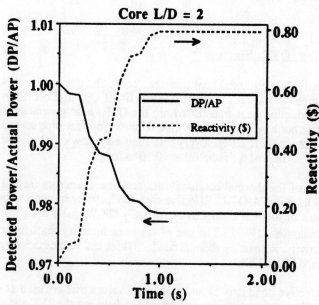

FIGURE 3. Reactivity ($) and DP/AP versus Time (s). Rod Withdrawal resulting in $0.793 Reactivity Insertion in one Second. Detector Located at Edge of Inboard Axial Reflector.

electrodes, and electron cooling. A more detailed transient thermal model of the thermionic converter can be created once transient data for the OTR thermionic converter becomes available. The rate of electron cooling and the effective cesium conductivity as a function of emitter and collector temperature will need to be quantified, and will be design dependent. The effective emissivities of the core/tungsten heat collector-emitter, the reflector/radiator, and the radiator/sink interfaces are taken from the emissivity values of niobium carbide, grain oriented tungsten, beryllium metal, and Nb-1%Zr. The values chosen are supported by GA Technologies (1987) and Gallup (1990).

TABLE 1. Thermal Conductivities and Heat Capacities of Materials used in TA-OTR.

Material	Thermal Cond. (W/m-K)	Specific Heat (J/g-K)
Graphite (C)	40	2.1
Fuel (55%C/45%UC_2 by Volume)	35	0.59
Tungsten (W)	100	0.17
Niobium (Nb)	7	0.25
Beryllium (Be)	80	2.9

TABLE 2. Effective Conductivity and Emissivity Values used at Various OTR Interfaces.

Description of Interface	k (W/m-K)	Emissivity
Core / Tungsten Heat Collector-Emitter	0.0	0.8
Tungsten Emitter - Heat Collector / Thermionic Collector (Effective)	0.0	0.247
Tungsten Emitter-Heat Collector / Reflector	0.132	0.028
Reflector / Radiator	0.0	0.4
Radiator / Sink (Space)	0.0	0.9

GA Technologies (1987) states (pg 6-12) that 269 W of thermal energy is absorbed from the core surface by each thermionic heat collector/emitter. Of that 269 W, 4 W are lost to the heat collector through multi-foil insulation, and 38.9 W are conducted into the thin emitter sleeve (the electrical lead for the emitter). An additional 4.1 W of thermal power is generated by electrical resistance in the sleeve. Of the heat entering the emitter electrical-lead, GA Technologies states that 18.7 W are lost by conduction and 21.1 W are lost by radiation. The thermal transients analyzed in this chapter assume that all heat entering the emitter sleeve/electrical-lead (or generated within the electrical-lead) is transferred to the inner node of the radial reflector, conducted through the radial reflector to its outer surface, and radiated to the inner surface of the radiator. Steady-state OTR temperatures are sensitive to the values of conductivity and emissivity discussed above. The startup transients evaluated during this research are not as sensitive to heat transfer properties.

All startup scenarios analyzed assume that the reactor is initially producing a steady-state power of 1 mW (zero-power criticality has been attained). During an actual startup, additional time will be required to achieve zero-power criticality. Also, the OTR is assumed to initially be at a uniform temperature of 300 K. TA-OTR uses the point kinetics equations to calculate transient power as a function of transient reactivity. The transient reactivity profile is specified by the user. The transient nodal power calculated by the TA-

OTR subroutine is fed back into the thermal analysis portion of the code. Transient temperatures calculated by the one-dimensional thermal analysis are then used for calculating thermal reactivity feedback. The temperature reactivity feedback coefficients used by TA-OTR were calculated using the codes "TRANSX" (MacFarlane 1984) and "TWODANT" (Alcouffe et al. 1984). TRANSX was used to process temperature dependent neutron cross sections from the cross section library "matxs6" (MacFarlane 1984). TWODANT was used (with the TRANSX processed cross sections) to calculate differences in k_{eff} from both changes in cross sections and system thermal expansion. MCNP (Briesmeister 1986) was used to verify the magnitude of the temperature reactivity feedback coefficients. Coefficients were calculated for the core, the tungsten heat collector/thermionic emitters surrounding the core, and the radial reflector.

The startup transient is governed by a combination of limits on maximum average core temperature, maximum reactivity (positive or negative), and maximum power. The control subroutine of TA-OTR ensures that no limits are exceeded. Further details of the assumptions, procedures, and methodology used by TA-OTR can be found in Houts and Lanning (1991).

OTR Startup Results and Observations

Based on the analysis of numerous potential OTR startup scenarios, the following general observations concerning OTR startup can be made. First, because of the poor thermal coupling between the core and the radial reflector and the relatively large thermal inertia of the radial reflector, the radial reflector may take thousands of seconds to approach its steady state operating temperature. Differential thermal expansion will have to be accommodated if a fast startup is required and the reflector is used to provide structural support for the core. Second, a 900-s startup is not possible unless the core is allowed to operate at several times its rated steady-state power during startup. Finally, thermal coupling of the radiator and radial reflector may be desirable to maximize negative feedback during transients, minimize differential thermal expansion concerns, reduce steady-state reflector operating temperature, and reduce radiator operating temperature during the first several hundred seconds of operation. The effects of thermal coupling on heat pipe thaw and thermionic converter operation should be investigated.

In a recent Program Research and Development Announcement (PRDA), the Air Force Space Technology Center solicited designs for out-of-core thermionic space reactors (AFSTC 1990). One of the requirements was to evaluate the impact of a 15-min (900-s) startup on the proposed OTR design. The potential startup scenario presented in this paper assumes that startup must be complete after 900 s. Maximum reactivity is limited to 0.8 β at powers below three times rated power, and 0.5 β at powers above three times rated power. Maximum core thermal power is limited to 750 kWt, which is ten times the rated steady-state power level. Maximum average core temperature is limited to 2194 K, which is the average core temperature at steady state.

Figure 4 plots maximum fuel and average plug, fuel, emitter, collector, and reflector temperature as a function of time for the described startup scenario. Figure 4 shows that average fuel temperature reaches the required value of 2194 K after 280 s, and that collector temperature reaches 900 K after 340 s. The maximum fuel temperature during startup does not exceed the maximum fuel temperature at steady-state operating conditions. Figure 4 also shows that 900 s after startup initiation, radial reflector temperature is approaching 400 K.

Figure 5 plots power and net reactivity as as a function of time for the described startup scenario. After 25 s, power has risen from the 1.0-mW starting value to 750 kW, and is held there because of the limit on maximum thermal power. At 270 s after startup initiation average core temperature is 2180 K. Startup termination begins at this time (as shown by the drop in power and reactivity) to avoid exceeding the desired average core temperature of 2194 K.

Figure 6 plots net reactivity and reactivity from thermal feedback as a function of time for the described startup scenario. Figure 6 shows that cumulative negative reactivity feedback (primarily from core thermal expansion) increases rapidly while average core temperature is rising, and increases more slowly after startup termination begins. Negative reactivity feedback from radial reflector expansion is the main source of feedback after the average core temperature has reached its steady-state value of 2194 K.

CONCLUSIONS

The research that was performed on OTR startup and control leads to the following six conclusions.

1. For the reference system, point kinetics can be considered valid for critical-to-supercritical OTR transients with positive reactivities of up to 0.8 β. Point kinetics may be valid for even more severe transients, although further research needs to be performed.

2. Fluctuations in DP/AP could lead to erroneous reactivity estimates (if inverse point kinetics is used) during severe subcritical-to-critical OTR transients. Fluctuations in DP/AP will not be a concern during more reasonable transients.

3. A 900-s startup is not possible if the OTR cannot operate at powers significantly above steady-state rated power during the startup.

4. Differential thermal expansion will have to be accommodated if the reflector is used to provide structural support for the core. The OTR reflector is thermally de-coupled from the core, and has a large thermal inertia. Radial reflector temperature will thus lag behind core temperature during startup.

5. Thermal coupling of the radiator and reflector may be desirable to maximize negative feedback during transients, to reduce differential thermal expansion concerns, to reduce steady-state reflector operating temperature, and to reduce radiator temperature during the first several hundred seconds of OTR operation.

6. The reference OTR has a large negative temperature coefficient. Increasing system temperature from 300 K to operating temperature results in a 1.6% decrease in k_{eff}.

FUTURE RESEARCH

The primary limitations of the OTR point kinetics validation model are due to the two-group, rectangular geometry modeling of a cylindrical fast reactor. Greater accuracy and detail could be obtained once multi-group versions of QUANDRY become available. The use of cylindrical coordinates would also improve accuracy.

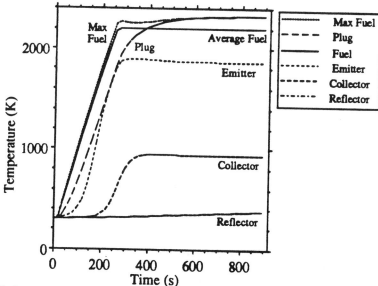

FIGURE 4. Maximum Fuel and Average Plug, Fuel, Emitter, Collector, and Reflector Temperature (K) as a function of Time (s). Below a Power of 225 kWt, Reactivity limited to $0.8; above 225 kWt Reactivity limited to $0.5. Maximum Power 750 kWt.

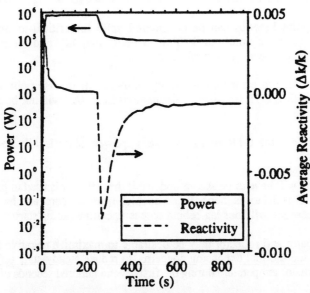

FIGURE 5. Power (W) and Average Reactivity versus Time (s). Below a Power of 225 kWt, Reactivity limited to $0.8; above 225 kWt Reactivity limited to $0.5. Maximum Power 750 kWt.

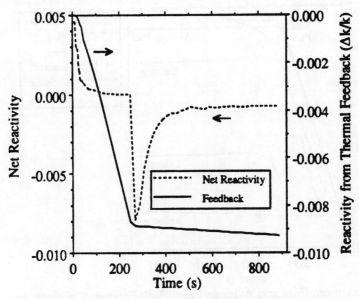

FIGURE 6. Net Reactivity and Reactivity from Thermal Feedback versus Time (s). Below a Power of 225 kWt, Reactivity limited to $0.8; above 225 kWt Reactivity limited to $0.5. Maximum Power 750 kWt.

To further evaluate the potential for rapid OTR startup, the off-normal operating characteristics of OTR thermionic converters need to be quantified. OTR thermionic converter temperatures will approach steady-state values fairly soon after average core temperature has reached its steady-state value; however, the thermionic converters do not completely reach steady-state until several thousand seconds later. If the radiator is thermally coupled to the reflector, thermionic collector temperature will remain significantly lower than its steady state temperature for several hundred seconds. The effects of thermal coupling on heat pipe thaw and thermionic converter operation should be investigated.

Acknowledgments

This research was sponsored by the United States Air Force's Phillips Laboratory, the United States Department of Energy (Division of Technology Support Programs, DE-FG07-90ER12930), and Universal Energy Systems. Research was performed at the Massachusetts Institute of Technology and at the Air Force's Phillips Laboratory. The authors would like to express their gratitude to the numerous individuals at MIT and at the Air Force's Phillips Laboratory who contributed to the success of this research.

References

Air Force Space Technology Center (AFSTC) Program Research and Development Announcement (PRDA) (1990) Thermionic Space Reactor Power System Design, Lt Don Verril, PL/STPP, Kirtland AFB, NM.

Alcouffe, R.E., F. W. Brinkley, D. R. Marr, and R.D. O'Dell (1984) User's Guide for TWODANT: A Code Package for Two-Dimensional, Diffusion-Accelerated, Neutral Particle Transport, LA-10049-M, Rev. 1, Los Alamos National Laboratory, Los Alamos, NM.

Bernard, J.A. (1989) Formulation and Experimental Evaluation of Closed-Form Control Laws for the Rapid Maneuvering of Reactor Neutronic Power, MITNRL-030, Massachusetts Institute of Technology, Cambridge, MA.

Briesmeister, J.F. (1986) MCNP - A General Monte Carlo Code for Neutron and Photon Transport, LA-7396-M, Rev. 2, Los Alamos National Laboratory, Los Alamos, NM.

GA Technologies (1987) Concept Definition Phase of the STAR-C Thermionic Power System for the Boost Surveillance and Tracking System (BSTS), GA-C18676, General Atomics, San Diego, CA.

Gallup, D.R. (1990) The Scalability of OTR Space Nuclear Power Systems, SAND90-0163, Sandia National Laboratories, Albuquerque, NM.

Houts, M.G. and D. Lanning (1991) Out-of-Core Thermionic Space Nuclear Reactors: Design and Control Considerations, MIT-ANP-TR-002, Massachusetts Institute of Technology, Cambridge, MA.

MacFarlane, R.E. (1984) TRANSX-CTR: A Code for Interfacing MATXS Cross-Section Libraries to Nuclear Transport Codes for Fusion Systems Analysis, LA-9863-MS, Los Alamos National Laboratory, Los Alamos, NM.

Smith, K.S. (1979) An Analytic Nodal Method for Solving the Two-Group, Multidimensional, Static and Transient Neutron Diffusion Equations, Masters Thesis, Massachusetts Institute of Technology, Cambridge, MA.

Touloukian, Y.S. and C.Y. Ho (1979) Thermophysical Properties of Matter, IFI/Plenum, New York, NY.

REACTOR DYNAMICS AND STABILITY ANALYSIS FOR TWO GASEOUS CORE REACTOR SPACE POWER SYSTEMS

Edward T. Dugan
Innovative Nuclear Space Power Institute
University of Florida
Gainesville, FL 32611
(904) 392-9840

Samer D. Kahook
Westinghouse Savannah River Company
P.O. Box 616
Aiken, SC 29802
(803) 725-3143

Abstract

Reactor dynamics and system stability studies are performed for two conceptual gaseous core reactor space nuclear power systems. The analysis is conducted using non-linear models which include circulating-fuel, point reactor kinetics equations and appropriate thermodynamic, heat transfer and one-dimensional isentropic flow equations. The studies reveal the existence of some unique and very effective inherent reactivity feedback effects such as the vapor fuel density power coefficient that are capable of stabilizing these systems safely and quickly, within a few seconds, even when large positive reactivity insertions are imposed. However, due to the strength of these feedbacks, it is found that external reactivity insertions alone are inadequate for bringing about significant power level changes during normal operations. Additional methods of reactivity control such as changes in the gaseous fuel mass flow rate, or gaseous fuel core inlet pressure are needed to achieve the desired power level control.

DESCRIPTION OF THE ULTRAHIGH TEMPERATURE VAPOR CORE REACTOR

The Ultrahigh Temperature Vapor Core Reactor (UTVR) is a highly enriched (>85%), BeO externally moderated, circulating fuel reactor with UF_4 as the fissioning fuel. The working fluid is in the form of a metal fluoride such as NaF, KF, RbF, or ^{7}LiF. Side and top view schematics of the UTVR are shown in Figures 1 and 2, respectively.

The UTVR includes two types of fissioning core regions: (1) the central Ultrahigh Temperature Vapor Core regions (UTVC) which contain a vapor mixture of highly-enriched UF_4 and a metal fluoride working fluid at an average temperature of ≈3000 K and a pressure of ≈50 atm, and (2) the boiler columns which contain highly enriched UF_4 fuel. This reactor has symmetry about the midplane with identical top and bottom vapor cores and boiler columns separated by the midplane BeO slab region and the MHD ducts where power is extracted.

The UTVC is surrounded in the radial direction by the wall and wall cooling region. The wall cooling region contains a liquid metal fluoride. By tangentially injecting the metal fluoride into the UTVC, the UTVC walls are maintained at the desired low temperatures (≈2000 K). As the metal fluoride is injected into the UTVC, an annular buffer zone is obtained which aids in maintaining the UF_4 away from the UTVC walls. This reduces the possibility of condensation of uranium or uranium compounds on the UTVC walls. Beyond this buffer zone, the metal fluoride vaporizes and mixes with the UF_4 in the UTVC.

The boiler region, which includes a number of boiler columns, is connected to the UTVC via the UTVC inlet plenums, as shown in Figure 1. The UF_4 liquid is supplied to the boiler columns by means of feedlines. Each boiler column consists of three distinct regions: The subcooled liquid region, the saturated liquid-vapor region, and the superheated vapor region. The UF_4 fluid is vaporized in the boiler columns prior to its entrance to the UTVC.

By configuring the disk MHD generator as an integral part of the reactor (as shown in Figure 1), a significant amount of fissioning occurs throughout the disk MHD generator region; this helps to maintain the required electrical conductivity, despite the relatively low fluid temperatures.

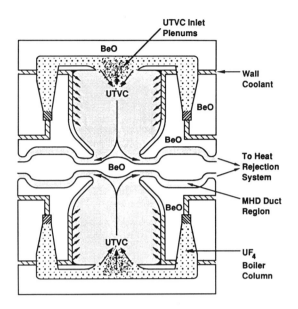
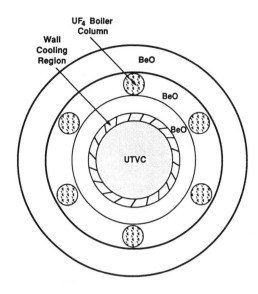

FIGURE 1. Side View Schematic of the Ultrahigh Temperature Vapor Core Reactor.

FIGURE 2. Top View Schematic of the Ultrahigh Temperature Vapor Core Reactor.

The combination of the following three features differentiates the UTVR from other nuclear reactor concepts.

1. The multi-core configuration resulting in a coupled core system by means of direct neutron transport through the media;

2. The circulating fuel and the associated neutronic and mass flow coupling between the UTVC and boiler cores; and

3. The employment of a two-phase (liquid-vapor) fissioning fuel.

UTVR NEUTRONIC ANALYSIS

One- and two-dimensional static neutronic calculations have been performed on the UTVR with S_n transport theory codes. Results from these calculations were used to obtain basic neutronic characteristics and reference configurations for the three-dimensional static neutronic analysis which was performed with Monte Carlo transport theory calculations. UTVR parameters needed for the dynamic neutronic studies such as reactivity, neutron generation time, and core-to-core neutronic coupling coefficients were obtained from the 3-D analysis. Details of the static neutronic analysis results have been previously reported (Kahook and Dugan 1991) and are not repeated here. Instead, the presented results focus on the dynamic neutronic analysis of the UTVR which has not been previously reported.

The goal of the dynamic neutronic analysis is to characterize the UTVR with respect to stability and dynamic response. Circulating-fuel, coupled-core point reactor kinetics equations (lumped parameter models) are used for analyzing the dynamic behavior of the UTVR. The dynamic models treat each fissioning core region (the UTVC and boiler columns) as a point reactor. In addition to including unique reactivity feedback phenomena associated with the individual fissioning cores (such as fuel density feedback of the UTVC and liquid-fuel volume feedback of the boiler columns), the effects of core-to-core neutronic and mass flow (or fluid flow) coupling between the UTVC and the surrounding boiler columns are also included in the dynamic models. The core-to-core neutronic coupling among the fissioning core regions arises both indirectly as a result of the fuel circulating between the two types of fissioning core regions (delayed neutron emission from the decay of the delayed neutron precursors which are carried in the fuel that circulates between the UTVC and boiler columns) and directly by the transport of neutrons through the BeO region separating the UTVC and the surrounding UF_4 boiler columns.

The lumped parameter models are incorporated into COUPL, a special code developed for the dynamic analysis of the UTVR (Kahook 1991). COUPL is constructed in a format suitable for dynamic simulation by the engineering analysis program, EASY5 (Harrison et al. 1988). EASY5 is an interactive program that has the capability to model, analyze, and design large complex dynamic systems defined by algebraic, differential, and/or difference equations. By integrating the differential-difference equations for a period of time and resolving the algebraic equations, EASY5 effectively simulates the behavior of the non-linear system.

The behavior of the UTVC power level (P^U) and the boiler column region power level (P^B) following a $1 positive step reactivity addition to the boiler column region are shown in Figure 3. The reactivity insertion in the boiler region leads to an initial power increase that yields a decrease in the liquid level in the boiler due to increased vaporization. The decrease in the mass of fuel in the boiler then leads to a reactivity decrease and a power decrease in the boiler at t=0.075 seconds. In the UTVC, the reactivity insertion also leads to an initial power increase; the higher pressure and temperature in the UTVC yield an increased mass flow rate for fuel exiting the UTVC and a decreased mass flow rate for fuel entering the UTVC. This fuel decrease then yields a reactivity decrease and a power level decrease in the UTVC at t=0.1 second. The decreased power levels in the boiler region and UTVC eventually lead to increased fuel loadings in the UTVC and boiler region that lead to power increases and a series of damped oscillations ensues. The UTVR system, which was initially at steady state, is seen to rapidly self-stabilize in about 3 seconds without any external reactivity control.

The UTVC and boiler column region power level behavior following a 20¢ positive reactivity step insertion in the UTVC are shown in Figure 4. Once again, the system is seen to rapidly self-stabilize, within about 3 seconds, without any external reactivity control. Figure 5 shows the UTVC and boiler region power level behavior for the same 20¢ positive step reactivity addition to the UTVC when the core-to-core neutronic coupling coefficients are artificially reduced by one order of magnitude. The consequence of this variation is to almost double the time required for the system to self-stabilize. It is apparent that stability of the UTVR is enhanced by a strong core-to-core neutronic coupling. The UTVC and boiler column region power

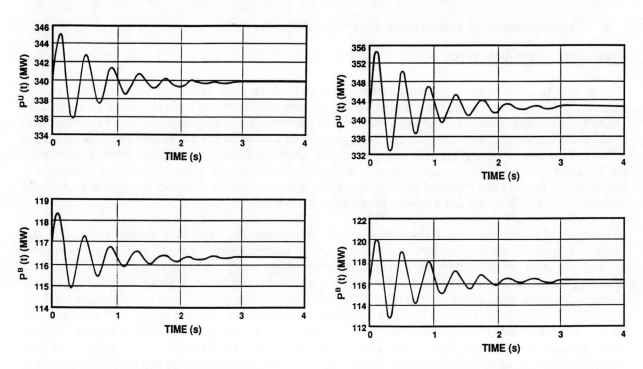

FIGURE 3. UTVC and Boiler Column Regions Power Levels as a function of Time following a $ 1.00 Positive Reactivity Step Insertion Imposed on the Boiler Columns.

FIGURE 4. UTVC and Boiler Column Regions Power Levels as a function of Time following a $ 0.20 Positive Reactivity Step Insertion Imposed on the UTVC.

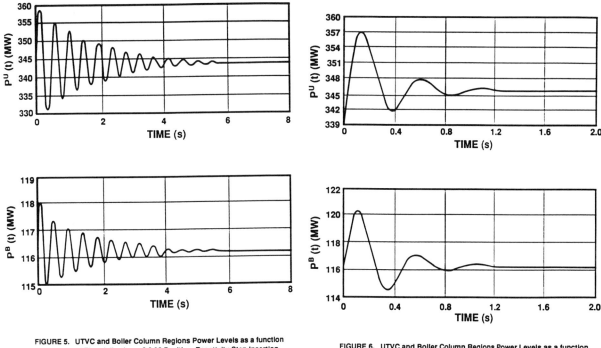

FIGURE 5. UTVC and Boiler Column Regions Power Levels as a function of Time following a $ 0.20 Positive Reactivity Step Insertion Imposed on the UTVC with the Coupling Coefficients Reduced by One Order in Magnitude.

FIGURE 6. UTVC and Boiler Column Regions Power Levels as a function of Time following a $ 0.20 Positive Reactivity Step Insertion Imposed on the UTVC with the UTVC Fuel Mass Reactivity Feedback Coefficient Reduced by a Factor of Five.

behavior for a 20¢ positive step-reactivity addition to the UTVC when the vapor fuel density coefficient of reactivity in the UTVC is artificially reduced by a factor of five are shown in Figure 6. Because the fuel reactivity worth in the UTVC is reduced, more fuel is required to be discharged from the core following the reactivity insertion and this causes the period, and, thus, the amplitudes of the oscillations to increase relative to those shown in Figures 4 and 5. However, because the boiler column fuel mass coefficient is now the dominant feedback mechanism, and because the boiler columns have a larger damping effect due to their liquid fuel, the damping of the oscillations also increases and the time required for the system to self-stabilize is reduced from about 3 seconds to around 1.2 seconds.

DESCRIPTION OF THE SINGLE CORE VAPOR CORE REACTOR

Dynamic neutronic analysis was also conducted on a second Vapor Core Reactor (VCR) concept. The second system consists of a single fissioning gaseous core region surrounded by an external BeO moderating-reflector region. Highly enriched UF_6 or UF_4 in gaseous form is the fissioning fuel and the fuel gas mixture of UF_6 (or UF_4) and helium also serves as the coolant/working fluid. Like the UTVR, the VCR is a circulating-fuel reactor. The fuel gas mixture exits the core through a converging/diverging nozzle which accelerates the fuel/working fluid mixture to a supersonic velocity before it enters the MHD duct. After leaving the MHD duct, the fuel gas mixture passes through a diffuser and a radiator heat exchanger before being pumped back to the core. For the VCR system analyzed in this study, a fuel reservoir-compressor combination act together to maintain the gaseous fuel inlet pressure to the core fixed at the initial steady state value during the examined transients. Fuel flow into the core is through a converging nozzle. A schematic of this VCR power system is presented in Figure 7. As with the UTVR, the MHD duct is configured to be an integral part of the reactor (see Figure 1) and fissioning in this duct helps to maintain the required electrical conductivity. The VCR power system illustrated schematically in Figure 7 operates on a Brayton Cycle whereas the UTVR power system operates on a Rankine cycle.

VCR DYNAMIC NEUTRONIC ANALYSIS

For the dynamic neutronic analysis of the VCR a special computer program, DYNAM, was written that

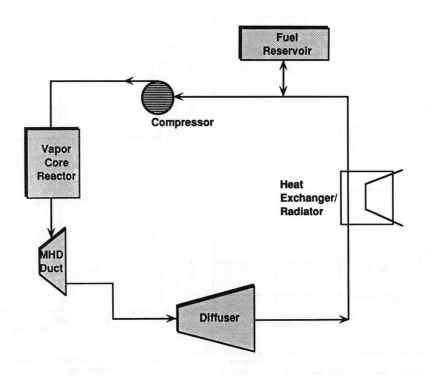

FIGURE 7. Schematic of Vapor Core Reactor Power System Including Balance-of-Plant.

solves the circulating-fuel, point reactor kinetics equations; DYNAM also solves the relevant thermodynamic, heat transfer, and 1-D isentropic flow equations (Kutikkad 1991). Because the VCR does not have multiple, coupled fissioning cores and because there are no two-phase fissioning core regions, the dynamic behavior of the VCR is considerably less complicated than that of the UTVR.

The power behavior of the VCR following a $1 positive step reactivity insertion ($\rho_{EXT}=0.0029$ $\Delta k/k$) is shown in Figure 8. The core transit time for the circulating fuel, τ_ℓ, is 0.08 seconds and the circulation time for the fuel in the remainder of the loop, τ_ℓ, is 0.1 second. The vapor fuel density coefficient of reactivity, α_{fm}, has a value of 0.035 $\Delta k/k$ per kg of fuel. Following an initial power increase, the rise in core pressure and temperature lead to an increased mass flow rate exiting the core and a decreased mass flow rate entering the core. The reduced fuel mass in the core leads to a reactivity decrease and a power decrease at about 0.05 seconds. A series of rapidly damped oscillations ensues and the large negative vapor fuel density power coefficient of reactivity quickly self-stabilizes the system in about 3 seconds without any external reactivity control.

The VCR power behavior for the same positive step reactivity insertion of $\rho_{EXT}=0.0029$ $\Delta k/k$ for the case when τ_ℓ is increased from 0.1 second to 0.5 seconds is shown in Figure 9. The larger τ_ℓ means a relative increase in the number of delayed neutrons decaying in the loop as compared to the vapor core. This yields a reduced effective delayed neutron fraction that in turn leads to an increase in the number, amplitude and duration of the oscillations following the reactivity insertion.

The VCR power behavior for the same $1 positive step reactivity insertion for the case when α_{fm} is artificially reduced to 0.0005 $\Delta k/k$ per kg of fuel is shown in Figure 10. It is apparent that while this reduction in α_{fm} has eliminated the oscillatory behavior, it has also led to a system that is inherently unstable. (Although not shown, when α_{fm} is reduced to only 0.001 $\Delta k/k$ per kg of fuel, the oscillations disappear but the system remains inherently stable.) At the other extreme, too large a value for α_{fm} can also make the system unstable. Figure 11 shows the VCR power behavior following a $1 positive step reactivity insertion for the case when α_{fm} is artificially increased to 0.1 $\Delta k/k$ per kg of fuel. Comparison of this

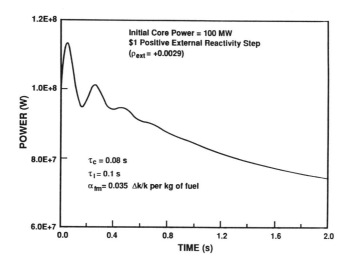

FIGURE 8. Thermal Power versus Time after a One Dollar External Positive Reactivity Insertion.

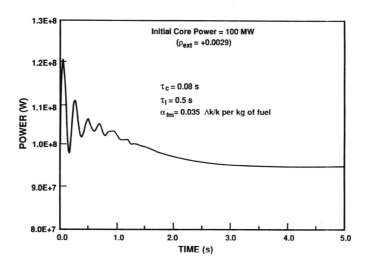

FIGURE 9. BPGCR Thermal Power versus Time after an External Reactivity Insertion - Effect of Varying External Loop Circulation Time.

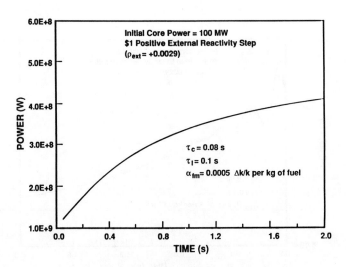

FIGURE 10. Thermal Power versus Time after an External Reactivity Insertion - Effect of Varying the Fuel Mass Feedback Coefficient - Case I.

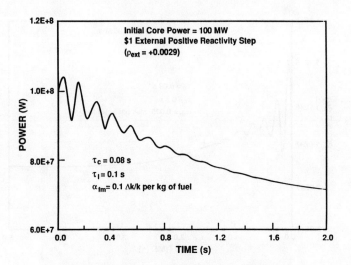

FIGURE 11. Thermal Power versus Time after an External Positive Reactivity Insertion - Effect of Varying the Fuel Mass Feedback Coefficient - Case II.

behavior with that presented in Figure 8 shows more persistent oscillations that take longer to die out. When α_{fm} is increased beyond 0.2 $\Delta k/k$ per kg of fuel, the VCR undergoes undamped oscillations.

CONCLUSIONS

Dynamic analysis of the UTVR and VCR indicates that strong inherent reactivity feedback is capable of stabilizing these systems safely and quickly even when large positive reactivity insertions are imposed. However, due to the strength of the negative reactivity feedback in the UTVR and VCR, external reactivity insertions alone are generally inadequate for bringing about significant power level changes during normal operations (see Figures 3, 4, and 8). Additional methods of reactivity control, such as variations in the mass flow rate of the gaseous fuel or variations in the gaseous fuel core inlet pressure, are needed to achieve the desired power level control. This is especially true for the UTVR which possesses not only the negative reactivity feedback associated with the vapor fuel density but also the large negative reactivity feedback due to liquid fuel volume changes in the boiler cores. Because of this latter effect, the UTVR exhibits good dynamic stability even when the vapor fuel density feedback is suppressed. This is not the case for the VCR. In fact, oscillatory behavior and stability in the VCR can be completely controlled by the value of α_{fm} established by the selected fuel gas density in the vapor core.

Acknowledgments

This work was performed under contract NAS3-26314, managed by NASA Lewis Research Center for the Innovative Science and Technology Office, Strategic Defense Initiative Organization. Computer support was provided by the National Science Foundation/San Diego Supercomputer Center and by the University of Florida and the IBM Corporation through the Research Computing Initiative at the University of Florida Northeast Regional Data Center.

References

Harrison, J., P. Kamber, and R. Hammond (1988) <u>EASY5 Engineering Analysis System</u>, User's Guide, 20491-0516-R2a, The Boeing Company, Seattle, WA.

Kahook, S.D. (1991) <u>Static and Dynamic Neutronic Analysis of the Uranium Tetrafluoride, Ultrahigh Temperature, Vapor Core Reactor System</u>, Ph.D. Dissertation, University of Florida, Gainesville, FL.

Kahook, S.D. and E.T. Dugan (1991) "Nuclear Design of the Burst Power Ultrahigh Temperature UF_4 Vapor Core Reactor System", in <u>Proc. Eighth Symposium on Space Nuclear Power Systems</u>, CONF-910116, M.S. El-Genk and M.D. Hoover, eds., American Institute of Physics, New York.

Kutikkad, K. (1991) <u>Startup and Stability of A Gaseous Core Nuclear Reactor System</u>, Ph.D. Dissertation, University of Florida, Gainesville, FL.

CONCEPTS AND ISSUES RELATED TO CONTROL SYSTEMS FOR FUSION PROPULSION

George H. Miley
Fusion Studies Laboratory
University of Illinois
Urbana, IL 61801
(217) 333-3772

ABSTRACT

Fusion engines will provide flexible, very high specific impulse propulsion systems. Variable thrust is provided by mixing propellant with high velocity plasma exhaust. Such a system would place strong and unique demands on the fusion reactor control system. Major control system parameters include: a fast start-up capability that minimizes fuel and propellant requirements, a demand following capability for maneuvers, variable fuel/propellant mixing over a wide range, stabilization against thermal instabilities, rapid shutdown and standby capability. These parameters and conceptual control concepts are discussed within the framework of candidate fusion reactors based on both magnetic and inertial confinement.

INTRODUCTION

Fusion offers the highest energy release per unit mass of fuel of any potential propulsion power source short of anti-matter (for example, see Miley 1988a; Forward and Davis 1988). Thus fusion propulsion is a prominent candidate for missions beyond the moon, especially for deep space travel requiring high specific impulses (Schulze 1991). Consequently, as briefly outlined here, a variety of conceptual design studies for fusion rockets have been reported. However, the issue of control, the subject of this paper, has received negligible attention. Yet, for space propulsion, the need to provide fast start-ups with minimum fuel expenditure, to offer emergency shutdown and restart capability, and to provide responsive thrust control for maneuvers, along with conventional shutdown and idling capabilities, represent critical aspects of the design issue.

The splendid progress made in both magnetic and inertial fusion research (see, for example, Crandall 1989) puts energy break-even experiments near. Hence, this is an appropriate time to look forward towards the practical application of fusion for space flight propulsion systems in order to influence the route for R & D into fusion concepts in a timely mannner.

APPROACHES TO FUSION PROPULSION

Broadly viewed, two main approaches to fusion have been considered: Magnetic Fusion Energy (MFE) devices and Inertial Confinement Fusion (ICF).

Preliminary conceptual designs using MFE devices include: Bumpy Tori (Roth et al. 1972; Roth 1988); Spherical Tokamak (Borowski 1987); Tandem Mirror (Santarius 1986, 1988); Field Reversed Configuration (FRC) (Miley 1988b; Chapman et al. 1988); Colliding Torus (Haloulakos and Bourque 1988). ICF concepts have generally envisioned a pulsed laser driver (Hyde 1983; Orth 1988) while earlier studies (Project Orion) considered small thermonuclear explosives. In addition, various hybrid concepts have been considered such as: The Migma (Ho 1977), Inertial Electrostatic Confinement (Miley et al. 1991a,c; Bussard 1990) Dense Plasma Focus (Miley et al. 1991b), magnetically confined ICF (Kammash 1987), magnetic dipole (Hasegawa 1991).

While the details of these various concepts vary widely, some general observations are possible. First, all involve a directed fusion-heated plasma exhaust where a propellant such as hydrogen is mixed in to vary the specific impulse. The means to uniformly mix the hot plasma with the cold hydrogen requires much research. Mixing has been recognized as a key issue ever since the original studies by Englert (1967). Schematics of two types of approaches are shown in Figs. 1,2.

In the FRC example, mixing occurs as plasma leakage from the closed field region flows into the open field region. In ICF, the fusion target is surrounded by a frozen shell of hydrogen propellant.

In addition, the choice of the fusion fuel cycles poses new issues. Aneutronic fuels offer important reduction in neutron hazards and activation (Miley 1986; Ho 1977), but result in the need for improved confinement (or larger driver energies). In the attractive case of D-^{3}He, a new source of ^{3}He via breeding (Miley 1988) or lunar mining (Wittenberg et al. 1986) is required.

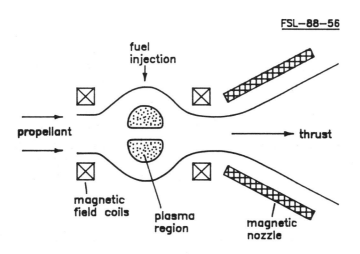

Figure 1. FRC Propulsion Concept. Plasma escaping the closed magnetic field region is mixed with propellant introduced along the open field region.

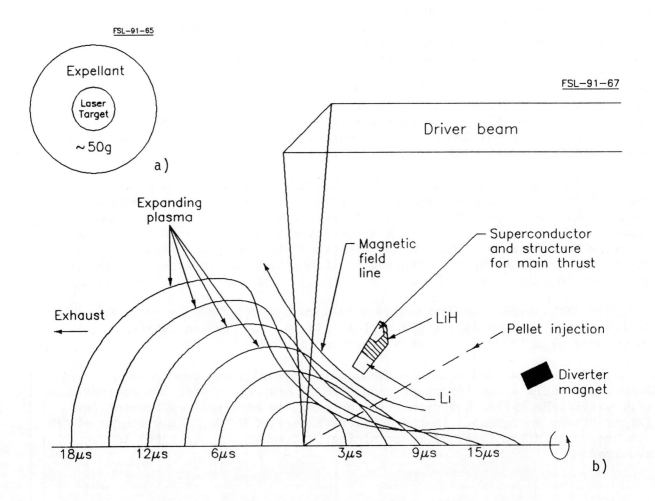

Figure 2. ICF Propulsion Target Concept. As shown in 2a the fusion target is surrounded by a frozen propellant layer. 2b shows use of a magnetic field to obtain a directed exhaust. (from Orth 1988).

Systems engineering aspects of space flight propulsion systems are of sufficiently great importance that it is essential that we consider control aspects concurrently with confinement and specific impulse/thrust developments. Control requirements will vary depending upon the choice of reactor, i.e. ICF or MCF and there will be differences within those broad confinement approaches.

While there are an abundance of potentially attractive concepts, a serious concern about fusion propulsion, however, is the developmental time scale. Present fusion programs (for example, ITER -- see Crandall 1989) anticipate long lead times to develop commercial fusion power on earth. Thus if space propulsion is viewed as being done as a serial development, the time scale is simply too long. On the other hand, development should be done in parallel to expedite the process. Many requirements are different and this, plus simplifications offered by the space environment, lead to the view that space power can be developed faster than commercial terrestrial power (Teller 1991).

MAJOR COMPONENTS

A conceptual comparison of the major components for the control of a fission propulsion system vs. a fusion rocket is shown in Fig. 3. In comparison with terrestrial fusion power stations, a flight propulsion system adds several important new components involving the fusion fuel, autogenous start-up energy, and driver energy. We will briefly comment on each of these.

First, the fusion fuel must be stored and fed (or injected, depending on the fuel form, gaseous vs. pellets) into the fusion chamber at the desired rate. The fueling rate will be determined by the power level desired and the leakage rate from the confinement system as well as the desired fuel/propellant ratio required to achieve the specified impulse/thrust ratio. Thus, for example, if pellet injection into a magnetically confined plasma is involved, the control may use parameters like the pellet rate, pellet size and velocity. For an ICF system, similar control of target parameters could be required. In addition, if the propellant is introduced as a frozen outer layer, cf Fig. 2, the target fuel/propellant ratio becomes a key variable. This then implies the ability to select and inject targets from stockpiles with different parameters.

Another unique feature of a fusion system is that a significant input energy is required to start up the reactor; that is, to heat the plasma up from cold to reacting conditions. In a steady-state system, this occurs infrequently; in a pulsed system such as ICF, however, this energy must be supplied by the driver for each pulse. In the latter case, once the reactor is running, part of the output energy can be recycled to the driver. Still the initial start-up requirement would dominate energy storage requirement as in the case of steady-state devices.

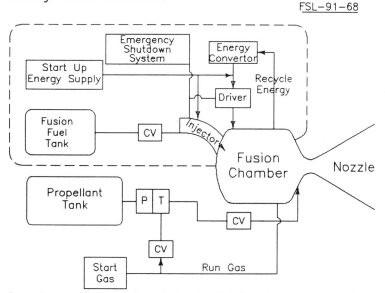

Figure 3. Major Components of a "Typical" Fusion Rocket System. Components within dashed lines are <u>added</u> the typical fission rocket of Fig.7-1, Bussard and DeLauer (1958). The control rod has been removed. A controller (not shown) must now coordinate/control the driver energy and fuel injector along with the propellant flow.

Determination of the preferred start-up energy source requires trade studies, and the value of the energy level would be heavily dependent on the confinement concept. Energy store candidates vary from capacitive energy stores to a small SNAP type reactor. A possible alternative approach is a modular fusion reactor where one module would always remain running as a back-up supply for re-starts.

The recycle energy and its conditioning for re-injection also requires study. Magnetic systems typically involve "injectors" such as radio-frequency supplies, neutral beam accelerators, inductive current drives, and so on, while ICF involves pulsed lasers or particle beam accelerators. In either case, if the fusion energy is extracted for recycle via a thermal cycle (via a heat exchanger on the nozzle coolant leading to an electrical generator) electrical energy "conditioning" components are required. All of these subcomponents must then, in turn, be coordinated through the master control system.

Another feature that distinguishes the fusion rocket from the fission system is that the direct equivalent of a control rod does not exist. Its replacement is strongly dependent on the confinement system. Table I lists some possible "equivalents." For example, one possible scenario for magnetic systems would be to control power levels via fuel/energy injection rates and provide rapid shutdown by impurity injection. A pulsed ICF rocket would be more straight forward in this respect since the target injection rate (pulse rates) provide a direct control technique. Implementation may not be so straight forward, however, since the target trajectory control system becomes involved when target parameters as well as repetition rates are changed.

TABLE I. SOME CONTROL "ROD" EQUIVALENTS

```
Operational Control

    Fuel Injection Rate
    Driver Power Level
    Impurity Injection
    Combinations of Above and
    Fuel/Propellant Ratio

"Safety-rod" Equivalents

    Driver/Injector Disconnect
    Impurity Injector
```

The propellant part of the system would have the same general characteristics as for a fission rocket. Thus, as discussed by Bussard and DeLauer (1958), this includes start-up gas for the turbopump followed by "boot-trapping" from the reactor exhaust during normal operation. All of these constraints, plus compensation for tolerances, insurance of transient stability, and so forth must now be applied to the combine propellant, fuel and energy driver system.

CONTROL SYSTEMS

An overview of the typical control system architecture is shown in Fig. 4. Appropriate sensors provide real time input data to the rocket controller which in turn "oversees" control of vehicle functions and the fusion rocket system control "effectors." The sensors envisioned for fission rockets (Bussard and DeLauer 1958; M. G. Millio 1991) must now be supplemented by a diagnostic array that monitors fusion plasma conditions such as densities, temperatures, magnetic fields, and so forth.

For reliability, it is important to develop a simple, rugged sensor array. This implies single sensors measuring the power level via neutron flux and perhaps a line average density by laser scattering. On the other hand, some systems may require more detailed measurements, including plasma density, temperature and current profiles, resulting in a large array of sensors. In any case, these considerations only add to the complexity of sensor issues. Already, for a fission rocket, Thorne and King (1991) have concluded it will require a new generation of sensors that are 2 to 3 orders of magnitude more radiation tolerant and which can operate dependably for years at high temperatures and yet tolerate extreme cold. Reduction of radiation-induced noise in cabling poses another goal.

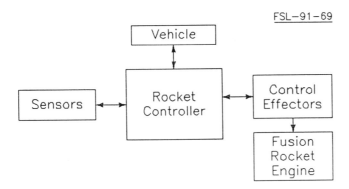

Figure 4. Fusion Rocket Control System Architecture.

In some respects the fusion plasma may simplify sensor problems. For example, Thorne and King (1991) cite a "critical need" for sensors to measure fission rocket exhaust temperatures. In the case of the fully ionized fusion plasma exhaust, conventional diagnostics involving laser probes could be adopted in a fairly straight forward way.

The issue control "rod" equivalents has already been noted (Table I). One effect of this is to require added control channels and added synchronization if multiple effector elements are used.

CONTROL METHODS: TOKAMAK EXAMPLE

Fusion propulsion control systems will have to be tailored for the specific fusion system employed. Even for terrestrial fusion reactors, little work along these lines have been reported to date. However, we can identify several generic characteristics. First, a multi-input -- multi-output (MIMO) system appears essential to coordinate the action of the variety of subsystems and components. Second, the exact behavior of the fusion plasma appears to be very difficult to model so that adjustment of the controls based on on-line operational data becomes highly desirable, for example, via modern adaptive control methods (Astrom and Wittenmark 1989). Preliminary work along these lines has involved control studies for Tokamak reactors. Thus, it is instructive to consider this work to illustrate concepts while recognizing that the Tokamak per se, due to the massive magnets required, is not a viable candidate for fusion propulsion.

The Tokamak nuclear island control components involved are illustrated in Fig. 5. Vertical field control is necessary to establish an equilibrium plasma position in the magnetic field. Once a fusion burn is initiated, added controls become necessary. Fueling and heating controls are fundamental and, along with current driver controls, can potentially provide the necessary control over plasma profiles (to optimize power, avoid instabilities), thermal instabilities, and fusion product ("ash") build-up. Other components such as impurity injection provide added control over thermal instabilities and emergency shutdown. These components would have to be integrated with space craft components and subsystems (Fig. 4). For illustrative proposes, however, we will only focus on the nuclear "island" part of the system.

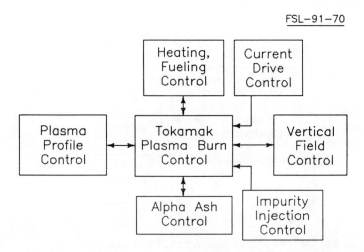

Figure 5. Control Components for Tokamak Nuclear Island.

PRIOR TOKAMAK CONTROL STUDIES

Approaches to fission "thermokinetic" control have been reviewed by Sager (1988). Those studies along with more recent ones for ITER (Haney et al. 1990; Mandrekas and Stacy 1991) have used a simplified profile-averaged 0-D model. Relatively straight forward static position schemes are found to be able to control the burn instability in that case. Tokamak thermokinetics can also be controlled dynamically by compression and expansion of the plasma (Ohnishi et al. 1984). However, implications of this approach for the separatrix position and the diverter power profile need to be assessed. Impurity seeding can be effective in controlling the positive power excursions and this approach has also been addressed in various studies (Vold et al. 1987; Sager 1988).

All of these studies have ignored the problem of plasma profile control, avoidance of instabilities, and the need to integrate burn and plasma position control. Thus, in the next section we briefly consider a more comprehensive case.

APPROACH VIA MIMO ADAPTIVE SELF-TUNING

Plasma control is well suited to treatment by formalism in order to incorporate the variety of control components. In future reactors, the density, temperature, and q profiles need to be closely monitored and controlled to derive optimum plasma performance and to prevent undesirable plasma disruptions. Profile control can be employed to manipulate the current density, plasma density and temperature profiles in order to achieve pressure and safety factor profiles favorable to MHD stability.

EXAMPLE CONTROL ALGORITHM

In this paper we narrowly focus on the profile control algorithms. Then we consider an illustrative case that involves demand following control.

Many engineering systems, including the tokamak reactor exhibit dynamical behavior that must be modelled by partial differential equations. These are called distributed parameter systems (DPS), and in general their state of space description involves an infinite dimensional description of the profiles. Considerable control theory has been developed to deal with DPS; see the surveys by Brogan (1968) and Russell (1978). The control algorithm can be developed via finite dimensional descriptions using techniques such as collocation, Galerkin, and the finite element methods. However, practical DPS feedback control must be reduced to small number of actuators and sensors using a compact control algorithm which can be implemented by an on-line digital controller.

In the present work (Varadarajan and Miley 1991) the DPS description of plasma density n, temperature T, and deuterium fraction in D-T fuel have been reduced to a finite dimensional form based on nonlinear basis functions and an error minimization condition using Galerkin formalism. The resulting first order ordinary differential equations for the discrete parameters are linearized to derive an MIMO control model than set in a self-tuning adaptive control model. This choice is motivated by the fact that plasma transport can only be approximately described. Hence, the system dynamics is better treated by a parameter adaption scheme where the control law makes use of updated parameters from on-time diagnostics.

Figure 6 shows an overall flow diagram for the technique. The governing equations are advanced by Runge-Kutta-Verner algorithm, and the control is adaptively derived at every time step of 0.05 second.

EXAMPLE CASE WITH LOAD DEMAND FOLLOWING

Varadarajan and Miley (1991) have examined a variety of cases to evaluate the adaptive self-training control system described above. For present purposes, we only consider one example to illustrate power load following of the type that would be important for maneuvering a space vehicle. For this calculation,

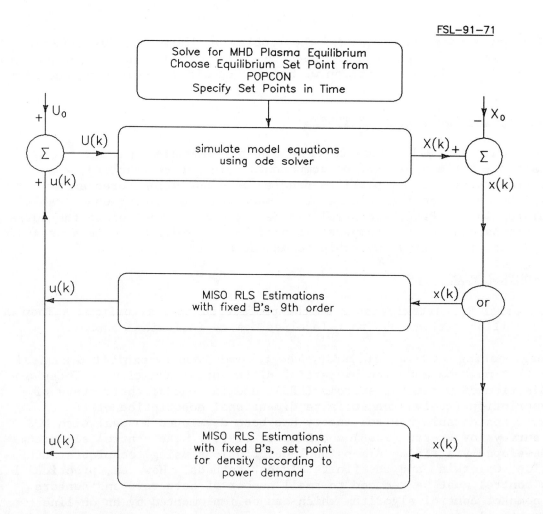

Figure 6. Flow Diagram for the Adaptive MIMO Self-Tuning Control System.

we consider an ITER-type Tokamak with 5T magnetic field on axis, 6-m major radius, 2-m minor radius, 22-MA plasma current, an average density of 5.5 x 10^{19} m^{-3} and a density weighted average temperature of 20-keV. The 1989 recommended ITER energy confinement scaling is employed (ITER Report, IAEA, 1989) along with other ITER physics design criteria.

In this illustration, as shown in Fig. 7, the net fusion power is first ramped down 50 MW and then quickly ramped back up over a time scale of ~7 seconds. The fast ramp-down maneuver is particularly challenging and has not been possible with most prior control concepts. The maneuver is successfully accomplished by judicious selection of the set points for density profile variables. The main control effector involves the particle injection source (S_n) which is distributed across the minor radius at six locations described by $S_n = \Sigma\ S_k\ f_{nk}$ for k = 1...6 where f_{nk} determines the profile shape. In this case we have not attempted to control the temperature profile, but rather allowed it to adjust to the other control action. The response of the injection sources to this maneuver is shown in Fig. 8. For this particular design set point the power density is quite sensitive to the density profile, making fuel injection a very effective "effector." This is not always the case, however, so multiple effectors may be required in other situations.

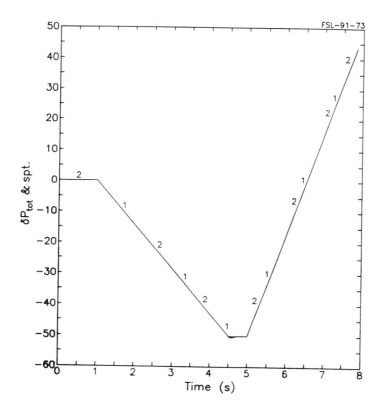

Figure 7. Fusion Power Ramp-down and Up for Demand-Following Example.

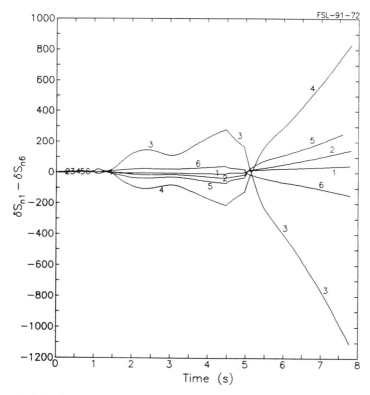

Figure 8. Individual Particle Input Amplitudes Corresponding to the Maneuver of Fig. 7. Numbers correspond to the radial location of the source injection (for example, 10 f_{nk}, k =1.66).

In summary, while this example leaves many questions unanswered, it does illustrate a strong potential for use of MIMO adaptive control techniques to achieve the type of performance needed for fusion rocket operation. The example illustrates a power following capability of the type needed to provide maneuvering capability. Other innovative design features are necessary to meet requirements such as fast start-up with minimum fuel/propellant expenditure, emergency shutdown, and so on.

CONCLUSION

A variety of control issues have been outlined for fusion rocket system. These issues pose strong constraints in future designs. Innovative designs are needed to meet some key points such as re-start, emergency shutdown. Also, a control system capable of coordinating a variety of complex inter-related subsystem components is essential. Precise modelling of some components such as the fusion plasma transport appears difficult. In view of these considerations, an adaptive MIMO self-tuning system is proposed as a potential approach. Its application to a Tokamak example was presented as an illustration, and the results are encouraging. Thus, while much more effort on control is essential, some possible routes have been identified.

ACKNOWLEDGEMENTS

This work was partly sponsored by Contract Number DOE/F290691-C-0006 and by Rockford Technology Associates, Inc. V. Varadarajan contributed much of the work on Tokamak control. Valuable discussions and suggestions by Dr. N. Schulze, NASA, are greatly acknowledged

REFERENCES

Aström, K. J., and B. Wittenmark (1989) *Adaptive Control*, Addison-Wesley Publ. Co., New York.

Borowski, S. K. (1986) "A Physics/Engineering Assessment of a Tokamak-Based Magnetic Fusion Rocket," AIAA 86-1759, June, 1986.

Borowski, S. K. (1987) "Nuclear Propulsion - A Vital Technology for the Exploration of Mars and the Planets Beyond," Case for Mars III Conference, 1987.

Brogan, L. (1968) "Optimal Control Theory of Systems Described by Partial Differential Equations," in *Advances in Control Systems*, 6, 221, C. T. Leondes, Ed., Academic Press, New York.

Bussard, R. W. and R. D. DeLauer (1958) *Nuclear Rocket Propulsion*, McGraw-Hill, New York, 244-276.

Bussard, R. W. (1991) "Fusion as Electric Propulsion," J. Propulsion and Power, *Am. Nucl. Society*, 6: 567-574, Sept.-Oct., 1990.

Chapman, R., G. H. Miley, M. Heindler, and W. Kernbichler (1989) "Fusion Space Propulsion with a Field-Reversed Configuration," *Fusion Technology*, 15: 1154.

Crandall, D. (1989) "The Scientific Status of Fusion," *Nucl. Instr. and Methods in Phys. Res.*, B42: 409-418.

Englert, G. W. (1967) "High-Energy Ion Beams Used to Accelerate Hydrogen Propellant Along Magnetic Tubes of Flux," NASA TND-3656, NASA Lewis Laboratory, Cleveland, OH.

Forward, R. L. and J. Davis (1988) Mirror Matter, John Wiley & Sons, Inc., New York.

Haloulakos, B. E., and R. F. Bourque (1988) "Mission and Propulsion Requirements," in Miley, 1988a.

Haney, S., and L. J. Perkins (1990) "Active Control of Burn Conditions for the International Thermonuclear Experimental Reactor," Fusion Tech. 18: 606-617

Hasegawa, A., L. Chen, M. Mauel (1991) "Fusion Reactor Based on a Dipole Magnetic Field," Nuclear Fusion 31: 125.

Ho, R. (1977) "Advanced Fuel Fusion Application to Manned Space Propulsion,"J. Nucl. Instr. & Methods, 144: 69-72

Hogan, W. J., W. R. Meier, N. Hoffman, K. Murray and R. Olson (1988) "Inertial Fusion Power for Space Applications," in Miley, 1988a.

Hyde, R. A. (1983a) "A Laser Fusion Rocket for Interplanetary Propulsion," UCRL-88857, 27 Sept. 1983.

ITER Conceptual Design Interim Report, IAEA, October, 1989.

Kammash, T., and D. Galbraith (1987) "A Fusion Reactor for Space Applications," Fusion Tech. 12: 11-21.

Mandrekas, J., and W. M. Stacey (1991) "Evaluation of Different Control Methods for the Thermal Stability of the International Thermonuclear Experimental Reactor," Fusion Tech. 19: 57-77.

Miley, G. H. (1976) Fusion Energy Conversion, Am. Nucl. Society, LaGrange, IL.

Miley, G. H., ed., (1988a) Notes from Minicourse on Fusion Appl. in Space, "8th ANS Topical Meeting on The Tech. of Fusion Energy, Salt Lake City, UT 9 Oct. 1988.

Miley, G. H., ed., (1988b) "Potential Use of Field Reversed Configurations for Space Applications," in Miley, 1988a.

Miley, G. H. (1988c) "^{3}He Sources for D-^{3}He Fusion Power," Nucl. Instr. and Methods in Phys. Res., A271: 197-202.

Miley, G. H., J. H. Nadler, T. Hocherg, O. Barnouin, and Y. B. Gu (1991a) "An Approach to Space Power," Vision-21 Symposium, NASA Conf. Publ. 10059, 141, Cleveland, OH, 1991.

Miley, G. H., R. Nachtrieb, J. Nadler, and C. Choi (1991b) "Use of a Plasma Focus Device for Space Propulsion," AIAA -91-3617, AIAA/NASA/OAI Conf. on Advanced SEI Tech., Cleveland, OH, 4-6 Sept. 1991.

Miley, G. H., J. H. Nadler, T. Hochberg, Y. Gu, and O. Barnouin (1991c) "Inertial-Electrostatic Confinement: An Approach to Burning Advanced Fuels," Fusion Tech. 19: 840-845.

Millio, M. G. (1991) "Tech. Readiness Assessment of Advanced Space Engine Integrated Controls and Health Monitoring," AIAA 91-3601, AIAA/NASA/OAI Conf. on Advanced SEI Tech., Cleveland, OH, 4-6 Sept. 1991.

Ohnishi, M., A. Saki, and M. Okomoto (1984) "Space-dependent Analysis of Feedback Control to Suppress Thermal Runaway by Compression-Decompression," Nuclear Technol./Fusion, 5, 326-333.

Orth, C. D.,(1988) "VISTA-- A Vehicle for Interplanetary Space Transport Applications," in Miley, 1988a.

Park, G. T., and G. H. Miley (1986) "Application of Adaptive Control to a Nuclear Power Plant," Nucl. Sci. Eng., 94: 145-156.

Roth, R., W. Rayle, and J. Reinmann (1972) "Fusion Power for Space Propulsion," New Scientist, 125.

Roth, R. (1988) "Space Applications of Fusion Energy," in Miley, 1988a.

Russell, D. (1978) "Controllability and Stability Theory for Linear Partial Differential Equations: Recent Progress and Open Questions," SIAM Rev., 20: 371.

Sager, G. (1988) "Tokamak Burn Control," (DOE/ER/52127-36) Fusion Studies Laboratory Report.

Santarius, J. F. (1986) "D-^{3}He Tandem Mirror Reactors on Earth and in Space," APS Bulletin 31: 1499.

Santarius, J. F. (1988) "Lunar 3He, Fusion Propulsion, and Space Development," UWFDM-764, Fusion Tech. Inst., U. of Wisconsin.

Schulze, N. R. (1991) "Fusion Energy for Space Missions in the Twenty First Century," TM-4298, NASA Headquarters, Washington, DC.

Teller, E. (1991), invited talk, ICENES '91 Conference, Monterey, CA, June, 1991.

Thorne, F. V., and D. B. King, (1991) "Instrumentation and Control for Space Nuclear Propulsion," AIAA 91-3633, AIAA/NASA/OAJ Conf. Adv. SEI Tech., Cleveland, OH, 4-6 Sept. 1991.

Varadarajan, R., and G. H. Miley (1991) "Adaptive MIMO Self-Tuning Control of Tokamak Thermokinetics, to be published Fusion Tech..

Vold, E. L., T. K. Mau, and R. W. Conn (1987) "Tokamak Power Reactor Ignition and Time-Dependent Fractional Power Operation," Fusion Tech., 12: 197-229.

Wittenberg, L. J., J. F. Santarius, and G. L. Kulcinski, (1986) "Lunar Source of ^{3}He for Commercial Fusion Power," Fusion Tech., 10: 167-178.

PERFORMANCE OF FAST REACTOR IRRADIATED FUELED EMITTERS FOR THERMIONIC REACTORS

Leo A. Lawrence and Bruce J. Makenas
Westinghouse Hanford Company
P. O. Box 1970, Mail Stop L5-02
Richland, WA 99352
(509) 376-5543/5447

Les L. Begg
General Atomics
P. O. Box 85608
San Diego, CA 92138
(619) 455-2482

Abstract

UO_2-fueled W emitters are being irradiated in a fast neutron spectrum under both real-time and accelerated burnup conditions with emitter surface temperatures of approximately 1800 K to establish performance for use in thermionic reactors with power ranges from tens of kilowatts to multimegawatts. Examinations at interim and terminal burnup levels are establishing performance parameters such as emitter deformation. Emitters at 4 at.% burnup in accelerated tests and 2 at.% burnup in real-time tests have shown excellent performance.

INTRODUCTION

Fueled emitters are being irradiated in a fast neutron spectrum to establish performance for use in thermionic reactors with power ranges from tens of kilowatts to multimegawatts with lifetimes up to 7 years. UO_2-fueled W emitters are being irradiated in the Experimental Breeder Reactor No. II (EBR-II) under both real-time and accelerated burnup conditions at prototypic operating temperatures. Tests are removed periodically for interim examinations to measure emitter growth and to determine fuel structure evolution with increasing burnup. Select emitters were also removed for destructive examinations. Emitter deformation as a result of fuel burnup has been identified as a feasibility issue limiting thermionic converter lifetime (Bohl and Ranken 1989).

The Thermionic Fuel Element (TFE) Verification Program is structured to include both individual component tests and integral TFE tests (Bohl et al. 1991). The Uninstrumented Fueled Accelerated Component (UFAC) tests are providing first-time data on the performance of fueled emitters irradiated under prototypic conditions to goal burnups.

IRRADIATION

Fueled emitters were encapsulated in stainless steel capsules for irradiation in the EBR-II (Lawrence and Veca 1990). The emitter was surrounded by a Nb sleeve to simulate operation in a TFE and to control emitter operating temperatures with a He/Ar gas mixture in the W to Nb gap (Figure 1). A large fission gas plenum was included to minimize fission gas pressure loading of the emitter. Tests in the EBR-II included three different test assemblies (Figure 2). Accelerated burnup testing was accomplished by reducing the diameter of the emitter, thereby accelerating the accumulation of burnup. A factor of two reduction in emitter diameter resulted is a factor of four acceleration in the accumulation of burnup, all other factors held constant. Test variables, in addition to irradiation time and burnup, included fuel form (solid pellet, annular pellet, insulated, and wafered), emitter thickness, and irradiation temperature. Insulated fuel consisted of replacing the outer region of the enriched UO_2 fuel pellet with an annulus of depleted UO_2. Wafered fuel contained alternating layers of UO_2 and W to reduce overall fuel temperatures. Both concepts are being evaluated for reducing emitter deformation.

EXAMINATIONS

Capsules were neutron radiographed at interim burnup levels and emitter diameters were measured from the radiographs. Terminal examinations of selected emitters included physical diametral measurements, which agree well with corresponding measurements from the radiographs. Diameter measurements are shown in Figure 3 for two accelerated burnup emitters that were irradiated under similar conditions at interim levels up to approximately 12,500 hours. One emitter had significantly less deformation than the other. The emitter with less deformation was 25% thicker than the companion.

Neutron radiographs also showed fuel structure evolution (Figures 4 and 5). Fuel redistribution in the annular pellet fuel resulted in an oval shaped central void consistent with expected operating conditions. The insulated fuel had a smaller central void compared to the annular pellet fuel. The insulated fuel has a higher fuel smeared density, i.e., total UO_2 per unit volume in the emitter, compared to the annular fuel. The higher smear density is the reason for the smaller central void. The wafered fuel shows a series of lens shaped voids in the fuel separated by the W disks. The shape of these center voids are consistent with the calculated axial and radial heat transfer for the wafer concept.

Metallographic examinations of accelerated burnup emitters at half goal burnup indicated excellent performance (Figure 6). The fuel showed extensive restructuring and columnar grains, as expected, and there was no measurable interaction of the UO_2 with the W.

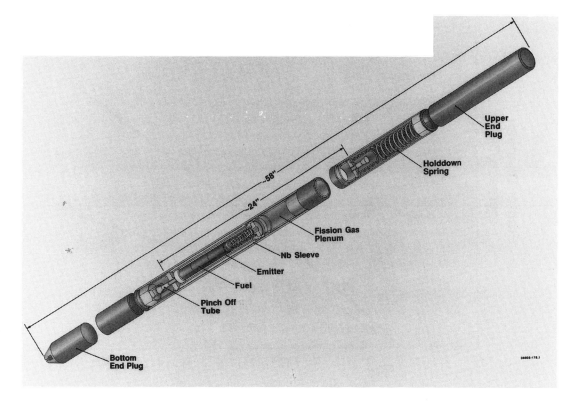

FIGURE 1. Schematic of UFAC Capsule Assembly for Fueled Emitter Testing in EBR-II

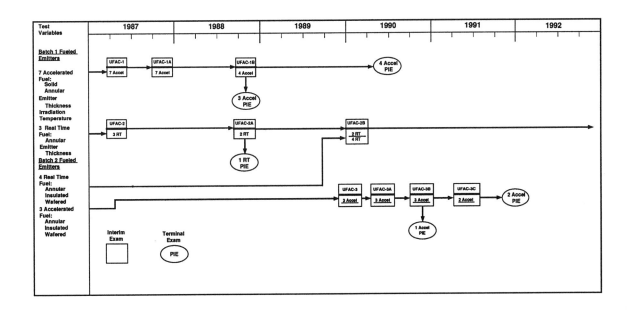

FIGURE 2. UFAC Irradiation Schedule for Fueled Emitter Testing in EBR-II

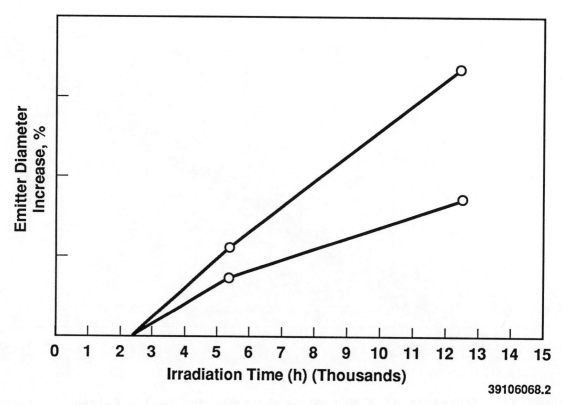

FIGURE 3. Emitter Deformation in Two Accelerated Burnup Emitters after 12,500 Hrs of Irradiation. [The Lower Curve Corresponds to a Thicker Walled (i.e., 25%) Emitter]

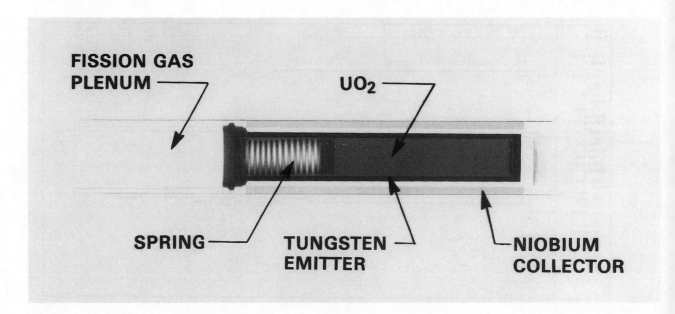

FIGURE 4. Preirradiation Neutron Radiograph of an Encapsulated Fueled Emitter

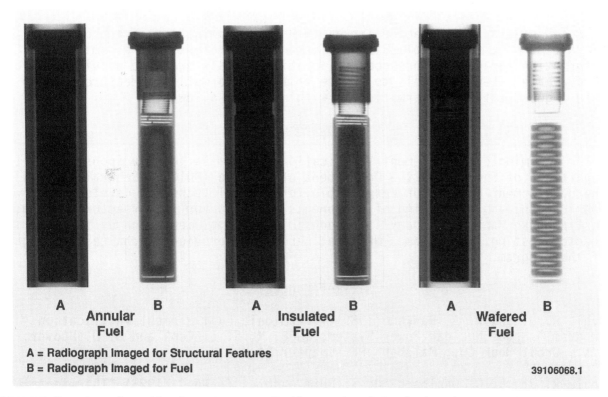

FIGURE 5. Postirradiation Neutron Radiograph of Fueled Emitters with Annular, Insulated, and Wafered Fuel

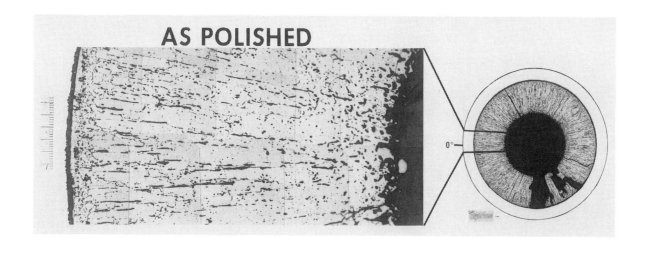

FIGURE 6. Postirradiation Fuel and Emitter Structures at Half Goal Burnup in an Accelerated Burnup Emitter

CONCLUSIONS

Fueled emitter fast reactor irradiations and examinations are providing first time data for establishing performance parameters such as emitter deformation for performance code calibration. Tests are also providing data to evaluate alternate fuels to reduce emitter deformation. Measurements to date suggest emitter deformation is within design allowance.

Acknowledgements

The Thermionic Fuel Element Verification Program is jointly sponsored by the Department of Energy and the Department of Defense (SDIO). General Atomics is the development contractor responsible for overall technical development, fabrication, and processing of components and TFEs for fast reactor testing. Westinghouse Hanford Company is responsible for implementation of the fast reactor testing. Los Alamos National Laboratory provides technical oversight of the program.

References

Bohl, R. J. and W. A. Ranken (1989) "Thermionic Fuel Element Verification Program," Space Nuclear Power Systems 1988, M. S. El-Genk and M. D. Hoover, Eds., Orbit Book Co., Malabar, FL, Chapter 34.

Bohl, R. J., R. C. Dahlberg, D. S. Dutt, and J. T. Wood (1991) "Thermionic Fuel Element Verification Program-Overview," in Proc. Eighth Symposium on Space Nuclear Power System, CONF-910116, M. S. El-Genk and M. D. Hoover, Eds., American Institute of Physics, New York, pp. 636-640.

Lawrence, L. A. and A. R. Veca (1990) "Fast Reactor Testing Thermionic Fuel Element Components for Space Power," Trans. American Nuclear Society, 62: 245-247.

TFE FAST DRIVER REACTOR SYSTEM FOR LOW-POWER APPLICATIONS

Thomas. H. Van Hagan, Bryan R. Lewis,
Elizabeth A. Bellis and Mike V. Fisher
General Atomics
P.O. Box 85608
San Diego, CA 92186-9784
(619) 455-3603

Abstract

This paper addresses reactor design considerations for an in-core thermionic system concept proposed for emerging space power applications with electric power requirements in the 10 to 50-kWe range. At this power level an in-core thermionic reactor core requires a combination of thermionic fuel elements (TFEs) and driver fuel elements to achieve nuclear criticality. A pumped liquid-metal loop cools the reactor, transporting the reject heat to a heat pipe radiator. The system concept is a straightforward derivative of the thermionic 100-kWe system designed during the SP-100 Phase 1 program. Combining existing thermionic technology with LMFBR fuel technology and pumped-loop waste heat removal defines a concept that has the advantages of reliability, scalability, technology maturity, and design flexibility.

INTRODUCTION

The SP-100 Phase 1 program studies (General Atomics 1985) showed that in-core thermionics is a very attractive space nuclear power system candidate for power levels of 100 kWe and above. At 100 kWe the reactor is just above the nuclear criticality limit with an all-TFE core. Therefore, extrapolating this approach to power levels significantly below 100 kWe must be accomplished either by turning down the output of a critical, all-TFE core or by removing TFEs and replacing them with driver fuel or moderator pins to maintain criticality. The latter option--a driver reactor-- is clearly superior from the standpoints of mass, size, and thermionic performance.

Whether to design the driver reactor for a fast or a thermal neutron spectrum is a choice that affects the system efficiency and power scalability. Table 1 compares the general characteristics of fast vs. moderated TFE driver reactor systems. In general, adding moderator to the core improves efficiency and reduces fuel loading but introduces fuel burnup issues, hydrogen management considerations, lower radiator temperature (tending to offset the efficiency improvement), and larger core size. These issues have little impact at very low power levels (e.g., 10 kWe) but become rapidly dominant as output power increases beyond 20 to 30 kWe. The fast driver system is preferred because it has a competitive mass and size at 10 kWe, when compared to other low-power reactor systems, and is gracefully scalable over a wide power range, extending well beyond 100 kWe.

TABLE 1. TFE Driver Reactor Characteristics.

Characteristic	Fast Driver	Thermal Driver
Emitter OD	1.1 in.	< 1.1 in.
TFEs Needed	Less	More
Criticality assurance	UO_2 or UN fuel rods	Moderator blocks
Drivers Needed	More	Less (or none)
Fuel Loading	Higher	Lower
Fuel Burnup	Lower	Higher
Fast Fluence	Higher	Lower
System h	Lower	Higher
Radiator Temp.	Higher	Lower

CONCEPT DESCRIPTION

Figure 1 illustrates the key features of the fast driver concept. The example shown is a 40-kWe system that thermally couples an in-core thermionic reactor to a fixed (non-deployable) radiator by means of a single, pumped liquid-metal coolant loop. Table 2 summarizes the reactor design features for a typical application. An annular induction EM pump circulates eutectic NaK-78 at a maximum temperature of 1050 K through the core. A thermoelectromagnetic pump provides passive, diverse decay heat removal. The reactor vessel and piping material can be either refractory alloy Nb-1Zr, or stainless steel, depending upon the desired heat rejection temperature level, and a redundant, carbon-carbon, sodium heat pipe radiator rejects the waste heat to space. A redundant power processing and control approach based on Space Station Freedom technology completes the major subsystems in the concept.

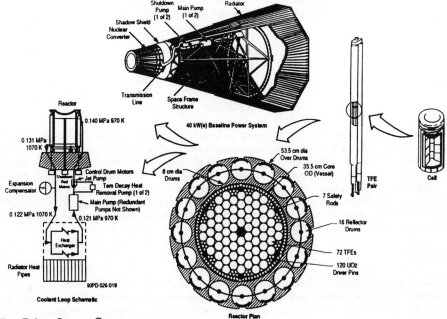

FIGURE 1. Fast Driver System Concept.

TABLE 2. Reactor Design Summary

Dimensions/Physical Feature		Materials	
Core/reactor OD, cm	36/54	Vessel, wetted parts	Nb-1Zr
Core/reactor H, cm	51.3/91.3*	Radial reflector	Be
Radial/axial reflector thickness, cm	9/9	Axial Reflector	BeO
No. of TFEs/driver pins	72/120	Safety rod absorber	B_4C
No. of control drums/safety rods	16/7	Safety rod follower	BeO
TFE OD x pitch, cm	3.339/3.344	Nuclear fuel	UO_2
Driver OD x pitch, cm	1.04/1.06		
Element spacing method	wire wrap		
Vessel thickness	0.2 cm		
Nuclear		Thermal Power	
Fuel load, U-235	125	TFE total, kWt	479
Uranium enrichment, %U-235	93	TFE reject, kWt	431
Burnup, %	2.2	Driver reject, kWt	285
		Reactor power, kWt	764

* ~40 cm safety rod extension not included

The TFE design, which is shown in Figures 2 and 3, is based on the UO_2-fueled, "F-series" thermionic converter developed in the earlier thermionic programs (General Atomics et al. 1985 and Gulf General Atomic Co. 1973) and selected as the reference for the SP-100 Phase 1 thermionic system study. The driver pins can use either UO_2 or UN, depending upon program motivations.

The TFE fast driver system concept is technologically well-founded and can draw upon the following data bases:

- TFE: F-series converter data from AEC programs,
- Nb-1Zr system: SP-100 program; earlier SNAP programs,
- UO_2 driver fuel technology: LMFBR program,
- UN driver fuel: SP-100 program,
- EM pump: SNAP-8 program,
- Decay heat removal pump: SNAP-10A, and
- Carbon-carbon heat pipe radiator: reference approach for SP-100.

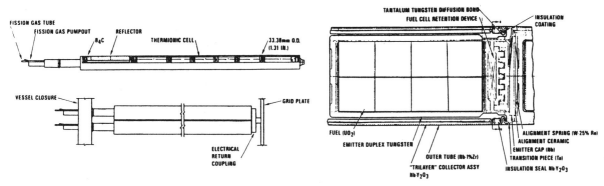

FIGURE 2. Thermionic Fuel Element Assembly.   FIGURE 3. Thermionic Cell Assembly.

REACTOR CORE ARRANGEMENT

The reactor configuration shown in Figure 1 has the core matrix of TFEs and driver pins housed within a cylindrical pressure vessel. Inside the vessel the axial reflector is incorporated within the end sections of the individual elements. Reactivity is controlled by rotating poison-backed drums within the radial reflector outside the vessel. The reflector and control arrangement is typical of in-core thermionic space power reactors.

The optimum arrangement of the TFEs and the drivers depends on several considerations, the most important of which is the TFE thermal power. Increasing this reduces the parasitic system mass and size influence of the drivers. Other key criteria include core size, fuel loading, and TFE power profile. It is desirable to have the TFE power profile as uniform as possible because power flattening involves local removal of TFE fuel, which can result in a core size increase to maintain reactivity. There are three general TFE/driver arrangement schemes that can be considered:

- Annular core - TFEs in the center of the core with the driver pins outboard,
- Mixed core - TFEs and drivers intermingled throughout the core, and
- Internal core- Drivers in the center of the core; TFEs outboard.

An analytical examination of these options confirms that locating the TFEs in the center of the core, where the thermal power is highest, has the biggest payoff. This comparison considered a TFE fast driver reactor in the 20 to 30-kWe range. Table 3 presents the results, which were generated with the MCNP code (Los Alamos National Laboratory 1986).

TABLE 3. Core Arrangement Comparisons

	Annular Core	Mixed Core	Internal Core
TFE Type	"F" Series	"F" Series	"F" Series
TFEs in Core (6 cells each)	54	54	54
No. of Safety Rods (in-core)	7	7	8
Core Height (cm)	49	49	49
Driver Fuel Type (0.5 cm radius)	UN	UN	UN
TFE Fuel Volume (cm^3)	8618	8618	8618
Radial Reflector Thickness (cm)	10	10	10
Axial Reflector Height (cm)	10	10	10
Core Radius (cm)	16.5	17.25	17.0
Driver Fuel Volume (cm^3)	3626	5335	3242
Driver Enrichment (%)	97	80	97
TFE Enrichment (%)	97	97	70-97 (zoned)
% Power in TFEs	56.6	55.7	57.7 (zoned)
Standard Deviation in TFE Peaking Factors	0.02	0.05	0.07 (unzoned) 0.04 (zoned)
K-effective	1.05	1.05	1.03 (zoned)
Shutdown Margin (K_{eff} all rods in)	0.84	0.84	0.82

Figure 4 shows the MCNP-generated graphical representations of the three core patterns and their associated power peaking profiles. The large circles with thick boundaries are the TFEs, the large circles with thin boundaries are core safety (shutdown) rods, and the small circles represent the driver rods. The TFEs are identified by number to relate the core patterns to the bar charts. The cross-hatched areas in the core patterns represents solid beryllium zones, which have been used to moderate and reflect neutrons and to maintain approximately equivalent coolant flow areas in the core. The k-effective values for the three cases are close enough to permit a valid comparison. The core parameters for all three are roughly equivalent with the exception of the driver fuel volume which is much larger in the mixed core arrangement. The driver fuel in this case, however, is enriched to a lesser extent. The mixed core was not considered primarily because it has the largest diameter, poorest radial profile and may present possible complexities in coolant flow management. The internal core has the largest percentage of power in the TFEs but requires fuel zoning to maintain a relatively flat radial power profile. This percentage would decrease if additional fuel zoning were required. The annular core arrangement has slightly less power in the TFEs but has more uniform power generation and a better radial profile in the abscence of any fuel zoning. Placement of the driver elements in an annular arrangement tends to buffer the adverse effects of the radial reflector and control drums on the core radial power profile. The annular core also has a smaller diameter than the internal core arrangement. The difference in diameters would become more pronounced if the internal core arrangement was adjusted to bring k-effective up to the value of 1.05 exhibited in the annular core. An added advantage of the annular arrangment is the ability to adjust core reactivity as the design evolves. It is easy to add driver volume to the annular driver core by simply adding driver elements to the periphery. Adjustments in reactivity for the internal core, however, would require that a new core pattern be devised.

REACTOR THERMAL CONSIDERATIONS

The reactor concept shown in Figure 1 assumes that the NaK-78 coolant makes one axial pass through the core. However, it should be noted that having the TFEs and driver pins in separate regions permits consideration of a two-pass arrangement that could significantly reduce the size of the radiator. In this variant the NaK would pass first over the TFEs and then turn to pass over the driver pins. With this routing the NaK would be heated by the drivers to a higher temperature than that achievable with the one-pass configuration, thus reducing the parasitic effect of the driver heat load on the radiator. This option is more appropriate to Nb-1Zr systems and will involve some additional mechanical and thermal (interpass heat leak prevention) considerations.

The TFE fast driver reactor is characterized thermohydraulically by a low heat flux into the coolant (on the order of 20,000 W/m^2) and a high heat transfer coefficient (on the order of 35,000 W/m-K), which combine with the high thermal conductivity of the TFE outer layers to allow close spacings of the TFEs without producing large circumferential temperature gradients. The driver pins, which are considerably smaller in diameter, have similar characteristics. The elements are wire-wrapped to achieve a close-packed array, and the wire pitch can be adjusted to promote uniform flow distribution.

Fuel temperatures are within acceptable limits. The UO$_2$ fuel in the TFE has a substantial radial thermal gradient during normal operation. At representative (≈40 Wt/cc) power densities this gradient can be on the order of 700 K, resulting in peak fuel temperatures of about 2400 K. Gradients in the driver fuel are generally considerably less, owing to the smaller pin diameter. Going to UN, which has a much higher thermal conductivity than UO$_2$, reduces the peak driver fuel temperature to within a few degrees of the NaK coolant temperature.

The TFE fast driver reactor also has the generic attribute of fast-spectrum, in-core thermionic reactors for withstanding a loss-of-coolant-accident (LOCA) without melting the fuel or breaching the fuel containment. Finite-element transient thermal analysis of the annular core reactor geometry shown in Figure 4 indicates that the reactor elements will experience a mild overshoot (less than 200 K) during the first 10 minutes of the transient and will diminish rapidly thereafter. These results are shown below in Figure 5. The analysis assumes that the reactor experiences a total, instantaneous loss of coolant and shuts down simultaneously. Heat transfer within the reactor core is assumed to take place by radiation only. At the core boundary the heat is conducted to the outer surface of the radial reflector, where it is radiated to space.

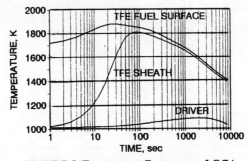

FIGURE 5. Temperature Response to LOCA.

Peaking Factors - Annular Core
56.6 % Power in TFEs

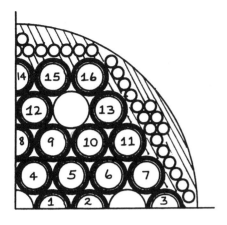
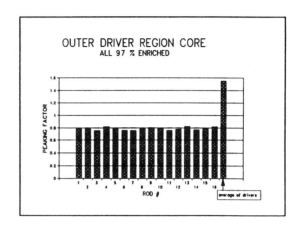

Peaking Factors - Mixed Core
55.7 % Power in TFEs

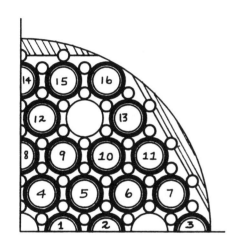
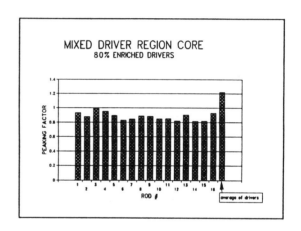

Peaking Factors - Internal Core
57.7 % Power in TFEs (zoned core)

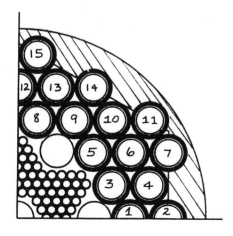
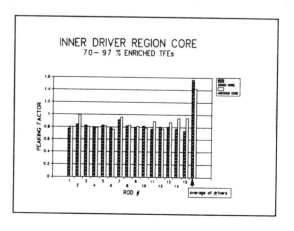

FIGURE 4. TFE/Driver Arrangements and Power Peaking Factors for Several Core Designs.

THERMIONIC PERFORMANCE

Independent variables that characterize thermionic performance include emitter temperature, collector temperature, interelectrode gap spacing, cesium pressure and electrode surface work function. For the TFE fast driver concept the collector temperature will likely be set above the thermionic optimum by radiator size considerations. Optimizing the electrical output of the TFE fast driver reactor involves a strong analytical interaction between core neutronics, thermionic performance, and TFE lifetime. Fuel thermal power density is a key parameter that links these considerations: dependent upon the reactivity insertion, it governs the combination of emitter temperature and current density achievable with the TFE. These parameters, in turn, control the rate of emitter radial deformation that governs the TFE lifetime. For a seven-year lifetime and an interelectrode gap of 0.038 cm., previous work (General Atomics 1985) indicates that the system mass will tend to minimize when the thermionic parameters approach an emitter temperature of 1750-1775 K, a collector temperature of 950-1000 K, and a power density of 40-45 Wt/cc.

The TFE converter design can be tuned to compensate for nonuniformities in the axial power distribution within the TFE. If left uncorrected, this power distribution follows a cosine shape, which results from the variation of the neutron population within the core. This produces a corresponding temperature profile in the TFE emitters, reducing the electrical output of the outboard, lower-temperature converters. This temperature profile can be flattened by graduating the lengths of the individual converters within the TFE. The graduated emitter lengths take advantage of the electron cooling afforded by the thermionic electron emission process to control emitter temperature. The shorter converters are placed in the center of the TFE, placing the smallest emitters (highest current densities) where the thermal power is highest, and vice versa.

CONCLUSION

The TFE fast driver reactor system concept combines mature technology with excellent scalability to provide mission planners with a flexible power system that can be tailored to their specific needs. The fast driver core easily scales from a few kilowatts to megawatts, and the system concept readily accommodates alternate technologies. The TFE fast driver appears competitive at low power even though the system efficiency is low, and it becomes quite attractive as the power level increases. The concept has the inherent safety features of in-core thermionic space power reactors, including the ability to maintain its integrity during a loss-of-coolant-accident. The ability to use the same design concept over an extremely wide power range offers an avenue for reducing program development and qualification costs.

Acknowledgements

This paper presents work from an internal study sponsored jointly by General Atomics and Rocketdyne Division of Rockwell International Corp.

References

General Atomics (1985) *GES Baseline System Definition and Characterization Study Final Report for the Period December 1984 through July 1985*, GA-C18062, General Atomics, San Diego, CA, August 1985.

General Atomics et al. (1985) *SP-100 Thermionic Technology Program Annual Integrated Technical Progress Report for the Period Ending September 30, 1985*, GA-A18182, General Atomics, San Diego, CA, November 1985.

Gulf General Atomic Co. (1973) *Development of a Thermionic Space Power System, Final Summary Report*, Gulf-GA-A12608, Gulf General Atomic Co., San Diego, CA, June 1973.

Los Alamos National Laboratory (1986) MCNP--A General Monte Carlo Code for Neutron and Photon Transport, LA-7396-M, Rev. 2, Los Alamos National Laboratory, Los Alamos, NM.

A FAST SPECTRUM HEAT PIPE COOLED THERMIONIC POWER SYSTEM

Joseph C. Mills and William R. Determan
Rockwell International, Rocketdyne Division
6633 Canoga Ave
Canoga Park, CA 91303
(818) 718-3357

Thomas H. Van Hagan
General Atomics
P. O. Box 85608
San Diego, CA 92186-9784
(619) 455-3603

Captain Thomas Wuchte
Air Force Phillips Laboratory
PL/STPP
Kirkland AFB, NM
87117-6008
(505) 846-2677

Abstract

This paper summarizes the design and performance characteristics of a heat pipe cooled thermionic (HPTI) power system being developed by a team headed by Rockwell International and General Atomics (GA). The design utilizes multicell, in-core thermionic fuel elements (TFEs) in a fast spectrum reactor core that is passively cooled by in-core heat pipes. The fast spectrum promotes competitive mass scalability over the power range of interest for future military application of 10 to 100 kWe without changing basic components or technologies. The number of TFEs and companion uranium nitride fuel elements are merely varied to achieve the critical mass requirements for each power level. The redundant in-core heat pipes in conjunction with an internally redundant heat pipe radiator help assure meeting key design goals for no single point failures and high survivability to both natural and hostile threats. These attractive attributes are achieved using already developed or under development technology.

INTRODUCTION

The Air Force Phillips Laboratory, in August of 1990, launched a 1-year thermionics system design effort to produce detailed conceptual designs of several candidate thermionic reactor power systems. The objective was to provide a database to guide technology development; guide further systems design and analysis efforts; and better define the military utility of thermionic space power. The Rocketdyne/GA team proposed several concepts (Mills and Dahlberg, 1991) and was awarded two of the three system design contracts funded by the Phillips Laboratory. One of these two study contracts addressed the fast spectrum, heat pipe cooled thermionic power system, which is the subject of this paper.

Figure 1 shows the schedule of activities performed during the 1-year design study of the HPTI concept. The first 3 months of the program were devoted to design option evaluations and system trade-offs leading toward the eventual selection and confirmation of a baseline 40-kWe design at the mid-point of the program. The results of these early design evaluations were presented by Mills (1991). The remainder of the program was devoted to refining this baseline design, examining excursion points (for example, 10- and 100-kWe power levels), and formulating a detailed development, qualification, and flight acceptance program including cost and schedule estimates. The purpose of this paper is to describe the baseline design that has evolved as a result of this one year effort, the rationale behind some of the key design features, the performance characteristics of the design, and the key drivers that most strongly influence design and/or performance. Many of the specific results from this 1-year HPTI program are being or have been published separately (Bellis 1991, Determan et al. 1992, Determan 1992, Dix 1992, Keshishian et al. 1992, Horner et al. 1992, Lieb et al. 1992, Metcalf 1992, and Durand et al. 1992).

REQUIREMENTS

The key requirements for the HPTI program are listed in Table 1. The values for the baseline design are listed first followed by excursion points for some parameters whose impact was to be evaluated, albeit with less rigor. Quantitative requirements for system mass, area, or efficiency were not specified, and thus the design effort was focused toward achieving the minimum system mass that best satisfied all the other requirements. An important objective of the HPTI program was to determine which of these requirements were significant drivers in the design evolution process.

BASELINE DESIGN

The 40-kWe HPTI system is pictorially depicted in Figure 2 and shown in the launch (stowed in the Titan-IV vehicle) and deployed configurations in Figures 3 and 4, respectively. A fast spectrum design was chosen over a moderated concept after design and analyses indicated an approximate 600-kg system mass advantage for the fast spectrum concept at 40 kWe. Uranium nitride fuel drivers were selected over uranium oxide fuel after Monte Carlo (MCNP) analyses indicated an approximate 0.05 greater k_{eff} for comparable geometries with uranium nitride. The larger diameter (2.794-cm emitters) F-Series TFEs were selected for the baseline as they permit

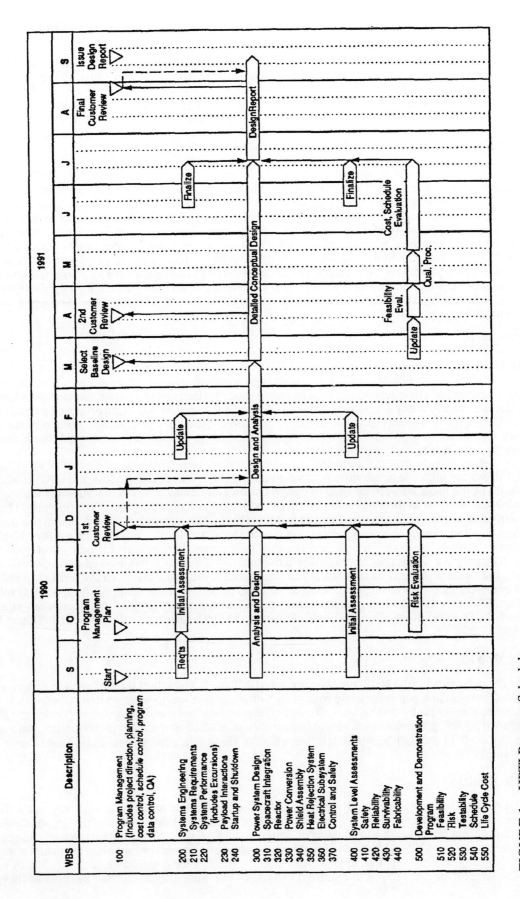

FIGURE 1. HPTI Program Schedule.

TABLE 1. System Design Requirements.

Criteria	Requirement
Performance	
Nominal power	40 kWe EOL; evaluate scalability between 10 and 100 kW
Lifetime	10 y
Operating altitude	1,000 to 36,000 km; any inclination
Packaging	Titan IV/Centaur payload fairing with a 7 m long x 4.5-m diameter payload. Evaluate 13-m-long payload
Separation distance	5 m. Evaluate 15-m distance
Mass	Minimum consistent with requirements
Payload Interactions	
Dose plane	4.5-m circular diameter
Neutron dose	10^{15} nvt (1 MeV equivalent)
Gamma dose	10^7 Rad (Si)
Thermal flux	0.14 W/cm^2
Start-Up and Shutdown	
Start-up time	24 h; evaluate impact of 15-min start-up time
No. start-up/shutdowns	At least 10
Reliability	95% for 10 y
Environment	Natural space-debris, meteorites, thermal cycles, Van Allen belts, cosmic, reactor induced radiation
Single point failures	Identify and determine system impact
Safety	Subcritical under all credible launch accident scenarios; pass INSRP review
Survivability	SUPER requirements and goals
Testability	
Qualification/Acceptance Launch environment	Identify test program to meet MIL-STD-1540B Titan IV/Centaur launch vehicle

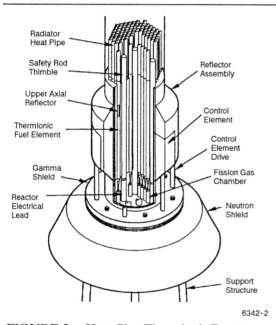

FIGURE 2. Heat Pipe Thermionic Reactor.

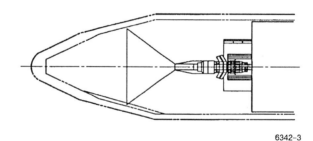

FIGURE 3. HPTI Power System in Launch Configuration.

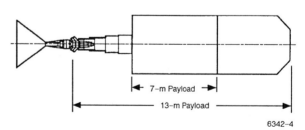

FIGURE 4. HPTI System Deployed.

scaling over the entire 10 to 100-kWe power range without changing the basic TFE component. The key operating parameters for the thermionic converters were determined via comprehensive optimization analyses using state-of-art computational tools LIFE4 and TECMDL (Bellis 1991). The optimized design employs graduated emitters (5.19, 5.65, and 8.05-cm lengths) with a nominal emitter temperature of 1750 K, a 15-mil interelectrode spacing, a 1106 K cesium reservoir graphite temperature, a maximum electrode current density of 5.3 A/cm^2, and an overall TFE efficiency of 10.2%. Redundant sliding reflectors were selected over rotating drums due to their greater nuclear effectiveness. Redundant in-core safety rods provide an additional shutdown system for safety purposes. The reactor shield, after a comprehensive set of MCNP design option analyses, is a dished, layered concept of LiH/W/LiH that was configured for minimum mass (Keshishian et al. 1992). The in-core heat pipes use a bowtie-shaped configuration that was formulated based on significant multidimensional thermal analyses which indicate that multiple failures of both the in-core and the attached ex-core radiator heat pipes can be tolerated with only modest increases in core temperatures and negligible loss in power output (Determan 1992). A forward facing, conical carbon-carbon heat pipe radiator was selected over a rearward facing design after a system studies indicated a 1,500-kg shield mass and 50% radiator area advantage for the forward facing option. The forward facing option requires a system deployment mechanism that is achieved using a corrugated sandwich structure boom with movable concentric cylinders for low mass and adequate reliability. Table 2 summarizes how this design concept satisfies all the key contract requirements.

PERFORMANCE SUMMARY

Table 3 depicts the key performance parameters for the baseline 40-kWe design. The net system efficiency is only 4.8% but is consistent with and overshadowed by the more important system characteristics of a low radiator area (20 m^2) for packaging purposes and a competitive system mass as summarized in Table 4. Two system mass estimates are presented: one for the baseline separation distance (from power system to payload) of 5 m and the second for the recommended separation distance of 10 m that was derived from MCNP shielding design option analyses (Keshishian et al. 1992). The HPTI system specific power of 12 to 13 W/kg at 40 kWe is extremely

TABLE 2. HPTI Satisfies All Key Requirements.

Parameter	Requirement	Comment
Performance	4-kWe baseline; 10- and 100-kWe excursion 10 y life Titan IV launch vehicle	Fully scalable over 10-100 kWe with same components; Titan IV integration with deployable system
Payload interactions	5 m; 15-m separation distances 4.5-m diameter dose plane 10^{15} nvt; 10^7 rad	LiH/W/LiH shield meets 5-m requirement; no W shield at 15 m; other payload impacts insignificant
Start-ups and shutdowns	10 start-ups; 24 h to 15 min	Meet all start-ups from critical conditions. Cold start-up for >1 h
Reliability	95%	No obvious mission ending single point failures. 0.96 reliability estimate
Safety	Pass INSRP review Subcriticality assurance	K_{eff} for water flooded worst case = 0.92
Survivability	SUPER requirements and goals	Hostile projectile requirement dictates radiator configuration
Testability	MIL-STD-1540B	Full compliance with 1540B

TABLE 3. HPTI 40-kWe Performance Summary.

Parameter	Value
Net electrical power (kWe)	40
Thermal power (kWt)	840
Net system efficiency (%)	4.8
Emitter temperature (K)	1750
Collector temperature (K)	1085
Radiator temperature (K)	975
Overall length (m)	6.5
Main radiator area (m^2)	31.5
One-sided area (m^2)	20

TABLE 4. HPTI System Mass Estimate.

	April 1991 Mass (kg)	
Subsystem	10 m*	5 m*
Reactor converter	975	959
Radiation shield	700	1,445
Heat rejection	346	388
Electrical	218	168
Control and safety	145	139
Spacecraft Integration	554	210
System total	2,938	3,309

*Separation distance

competitive with hardened solar photovoltaic and SP-100 thermoelectric power systems when compared on an equivalent basis (for example, same radiation dose and boom structural design requirements).

Note that the above HPTI system performance can be further improved via the incorporation of certain design features that use more advanced technology. Nuclear studies indicate that modest changes in the TFE design could improve system efficiency by reducing the number of fuel driver pins required for criticality purposes. Increasing the emitter diameter by 2 to 3 cm or reducing the intercell gap from 2 cm to 1.5 cm would each increase k_{eff} by more than 0.02. A program technical risk assessment also identified higher strength emitters that have higher operating temperatures and permit smaller interelectrode gaps as a promising advanced alternative with significant performance benefit potential.

Table 5 summarizes how the baseline 40-kWe design would scale down to 10 kWe and up to 100 kWe with the same basic components and technology. Scalability is achieved by varying the number of TFEs and fuel driver elements. At 10 kWe the relative number of fuel driver elements is increased for criticality purposes leading to a lower system efficiency. The specific power is accordingly reduced and, although not overly attractive, is still competitive with other power system options (solar, thermoelectrics, other thermionic concepts). At 100 kWe the need for fuel drivers for criticality purposes is obviated, and thus the system efficiency is increased and the specific power then becomes very attractive when compared to alternatives.

TABLE 5. HPTI Scalability With Same Components.

	10 kWe	40 kWe	100 kWe
Number of TFEs	36 (12 x 3)	72 (12 x 6)	156 (12 x 13)
UN fuel drivers	26	36	0
Safety rods	11	13	13
Converter η (%)	7.2	10.2	10.2
Net system η (%)	2.7	4.8	8.0
Radiator area (m²)	14.5	31.5	36.7
System mass (kg)	2,122	3,309	4,738
	*1,850	*2,938	*4,373
Specific power (W/kg)	4.7	12.1	21.2
	*5.4	*13.6	*22.9

*Optimum separation distance

Other in-depth system level studies were performed to both guide the design process and to demonstrate the capability of the HPTI concept to satisfy the full spectrum of issues for space nuclear systems. Both qualitative (failure modes and effects criticality analyses) and quantitative reliability analyses were performed to assure the presence of no credible mission-ending single point failures and to indicate the potential for satisfying the 0.95 reliability requirement. MCNP neutronic analyses of potential accident configurations were conducted to indicate subcriticality in all credible instances, and reentry analyses were performed to demonstrate intact reentry, both as part of demonstrating safety sufficient to gain launch approval. A testability assessment was conducted, leading to the identification of a viable approach to conduct system flight acceptance testing on a multicell TFE so as to comply with the requirements of MIL-STD-1540B. A fabricability assessment indicated the ability, using proven hot isostatic pressing techniques, to fabricate and assemble the HPTI reactor while allowing for incremental testing at several interim steps to verify quality.

DESIGN/PERFORMANCE DRIVERS

Table 6 summarizes the three key requirements which most strongly influence the HPTI design and/or performance. Greater separation distances reduce the shield mass but are offset by increases in both the boom and power conditioning and control system cable masses. The baseline conical forward radiator, which is the recommended configuration for the HPTI concept, is only partially survivable to the low speed hostile projectile threat. An alternative design, a dodecagon forward facing concept, would be fully survivable but would result in the indicated 500-kg mass penalty and would require a radiator deployment in addition to the system deployment currently in the baseline design. The contract design requirements treated the payload as a smooth boundary envelope. Titan IV system integration studies indicate that permitting the power system to invade isolated portions of this envelope can significantly reduce shield mass and help satisfy the stringent Titan IV center of gravity (CG) restrictions.

CONCLUSIONS

An extensive, 1-year, conceptual design effort has indicated that the fast spectrum HPTI design is a versatile concept for future military applications. The HPTI design is competitive on a system mass and area basis for the entire range of anticipated power levels (for example, 10 to 100 kWe) and survivability requirements. The concept is scalable with the same components and technology, and the technology utilized is amenable to supporting a 2000 flight date. System level studies have indicated that the design is fully capable of successfully gaining

TABLE 6. HPTI Key Design/Performance Drivers.

Design Driver	Impact	Comment
Separation distance	5-m separation distance results in very heavy gamma shield	500-kg mass saving possible at optimum payload separation distance at 10 m
Hostile projectile threat (low speed)	Generically requires significant armoring and many, shorter heat pipes in exposed radiator; large armor mass and cost impact	Developed two closely related designs for (1) meteoroid, debris, and moderately dense pattern and (2) high-density pattern threat
Payload envelope geometry	Relaxing dose limits at outer periphery can reduce shield mass; permitting power system stowage within portion of payload envelope can simplify deployment and satisfy Titan IV cg limitations	Can improve system performance by fully integrating TI power system with payload

approval to launch by the Interagency Nuclear Safety Review Panel (INSRP), can satisfy the testability requirements of MIL-STD-1540B, and can be fabricated and assembled using proven straightforward approaches. A key lesson from the design study is that detailed rather than a cursory conceptual design activity is necessary to fully flesh out promising concepts and design features if they are to comply with the full spectrum of often conflicting requirements.

Acknowledgments

The work described herein was performed under Air Force Phillips Laboratory contract F29601-90-C-0059 under the direction of Project Officer Captain Thomas Wuchte. Other organizations participating on the Rocketdyne/GA team were Thermo Electron Technologies Corporation, Rasor Associates, and S-CUBED who provided valuable assistance in the areas of testability, cesium reservoir design, thermionic transient performance, and survivability assessment.

References

Bellis, E. A. (1991) "Performance Optimization Considerations for Thermionic Fuel Elements in a Heat Pipe Cooled Thermionic Reactor," 26th IECEC Conference, Boston, MA, August 1991.

Determan, W. R. et al. (1992) "System Startup for an In-Core Thermionic Reactor," in Proc. 9th Symposium on Space Nuclear Power Systems, CONF-920104, to be held in Albuquerque, NM, 12-16 January 1992.

Determan, W. R. (1992) "Thermionic In-Core Heat Pipe Design and Performance," in Proc. 9th Symposium on Space Nuclear Power Systems, CONF-920104, to be held in Albuquerque, NM, 12-16 January 1992.

Dix, T. (1992) "Reentry Analysis for Thermionic Systems," in Proc. 9th Symposium on Space Nuclear Power Systems, CONF-920104, to be held in Albuquerque, NM, 12-16 January 1992.

Durand, R. and H. Horner (1992) "Testability of a Heat Pipe Cooled Thermionic Reactor," in Proc. 9th Symposium on Space Nuclear Power Systems, CONF-920104, to be held in Albuquerque, NM, 12-16 January 1992.

Horner, H. et al. (1992) "Heat Pipe Cooled Thermionic Reactor Fabrication and Assembly," in Proc. 9th Symposium on Space Nuclear Power Systems, CONF-920104, to be held in Albuquerque, NM, 12-16 January 1992.

Lieb, D. et al. (1992) "Prediction of Start-up Characteristics of a Heat Pipe Cooled Thermionic Fuel Element (TFE)," in Proc. 9th Symposium on Space Nuclear Power Systems, CONF-920104, to be held in Albuquerque, NM, 12-16 January 1992.

Metcalf, K. (1992) "Power Management and Distribution (PMAD) for Thermionic Power Systems," in Proc. 9th Symposium on Space Nuclear Power Systems, CONF-920104, to be held in Albuquerque, NM, 12-16 January 1992.

Mills, J. C. and R. C. Dahlberg (1991) "Thermionic Systems for DOD Missions," Proceedings of 8th Symposium on Space Nuclear Power Systems, CONF-910116, M. S. El-Genk and M. D. Hoover, eds., University of New Mexico, Albuquerque, NM, 6-10 January 1991.

Mills, J. C. (1991) "Survivability and Other Aspects of the Heat Pipe Cooled Thermionic Power System," Classified Session, 8th Symposium on Space Nuclear Power Systems, Albuquerque, NM, 6-10 January 1991.

SYSTEM STARTUP SIMULATION FOR AN IN-CORE THERMIONIC REACTOR WITH HEAT PIPE COOLING

William R. Determan and William D. Otting
Rockwell International Corporation
Rocketdyne Division

6633 Canoga Avenue
Canoga Park, CA 91303
(818) 718-3375

Abstract

The heat pipe cooled thermionic (HPTI) reactor relies on in-core sodium heat pipes to provide a redundant means of cooling the 72 thermionic fuel elements (TFEs) which comprise the 40-kWe reactor core assembly. In-core heat pipe cooling was selected for the reactor design due to a requirement for multiple system on-orbit restarts over its lifetime. Powering up the reactor requires the in-core and radiator heat pipes to undergo a thaw cycle with a rapid ascension in power to their operating temperatures. The present study considers how fast the thaw-out and power ascension cycle can be safely accomplished within a reactor core. As part of the study, a transient startup simulator model of the heat pipe cooled reactor system was developed. Results of the startup transient simulation are provided.

INTRODUCTION

The U.S. Air Force Phillips Laboratory is sponsoring research studies of small, compact space nuclear power systems based upon thermionic power conversion technology for military mission applications. The requirements for these missions are numerous and quite stringent. Two of the most important requirements are the need for a power system with multiple restart capability with potential startup times as short as one hour, and a system design which possesses no-mission-ending single point failures. In response to the request for proposal from the Phillips Laboratory, the General Atomics/Rocketdyne team proposed to develop a conceptual design of a heat pipe cooled thermionic reactor concept (HPTI). The HPTI concept is illustrated in Figure 1. The basic thermionic cell is assembled in a six-cell series connection into a TFE. Two TFEs are then connected in series to form a TFE pair. The TFE pairs are then inserted into a core honeycomb structure, which is composed of receiving cavities for the TFEs and sodium in-core heat pipes with a cross-sectional shape in the form of a bow tie. Six in-core heat pipes surround each TFE and transfer their waste heat axially to the radiator heat pipes located above the reactor's lower axial reflector. Only three in-core heat pipes are required to transfer the TFE's waste heat. The top cavity formed by the six in-core heat pipes and each TFE lead provide a 2.54-cm diameter 17.5-cm high thermal well that receives the evaporator section of a roughly 4-m long radiator heat pipe. Figure 2 illustrates the HPTI reactor/shield assembly with its radiator interface. Sodium bonding is employed to thermally bond the radiator heat pipes and the TFEs into their respective core cavities during the final core

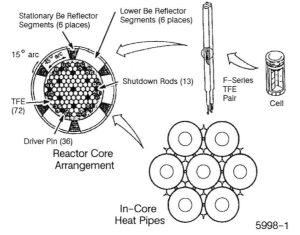

FIGURE 1. Heat Pipe Cooled TFE Concept.

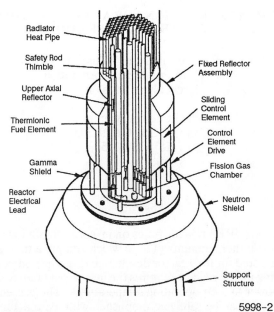

FIGURE 2. Heat Pipe Thermionic Reactor.

assembly process. When the TFE leads are connected into a 12x6 series/parallel electrical arrangement, the HPTI design exhibits no single point failures.

TRANSIENT SIMULATOR DEVELOPMENT AND BENCHMARKING

In order to evaluate the start-up characteristics of the HPTI system, a system transient simulator model was developed. A key part of the model is the performance simulation of the core heat pipes and radiator heat pipes. Liquid-metal heat pipe transient behavior has been under investigation for more than 15 years, but at very limited funding levels. In the mid-1980s, the SP-100 program sponsored some transient heat pipe research at the Los Alamos National Laboratory (LANL). This work involved transient tests of a 4-m lithium/molybdenum heat pipe, as well as development of a computer model to accurately predict heat pipe transient behavior. Funding limitations never permitted completion of the computer model, but the transient test results were documented in an SP-100 program briefing (Merrigan 1985). These transient tests included rapid start-up from a frozen state to full power operation at 1300 K, shutdown of the pipe to a frozen state, and restart-up of the frozen lithium heat pipe. The results from these tests were used as the basis for developing and benchmarking a simple heat pipe transient simulator model.

The LANL start-up test results are shown in Figure 3. The heat pipe was instrumented with thermocouples at eight positions along its condenser length. The condenser section was coated with a high emissivity material ($\epsilon \sim 0.7$) to reject heat to a cold wall. The heat pipe was thawed and raised to 13-kW throughput within two hours and 20 minutes. Analysis of the lithium heat pipe's performance limits over the 1000 to 1300 K evaporator exit temperature range showed that the pipe was sonic limited throughout its condenser length during the entire start-up transient. Therefore, the heat pipe simulator model was developed assuming that the power transfer limit on each heat pipe node down the length of the pipe is based upon the local sonic limit at the temperature of the node. The sonic limits for the heat pipe as a function of temperature were defined using the Los Alamos steady-state heat pipe analysis code, HTPIPE. The HTPIPE results and the Los Alamos data both indicated that the lithium pipe had a "turn-on" temperature of about 1000 K. This is the evaporator temperature at which the lithium vapor transport becomes the dominating heat transfer mechanism.

The initial runs of the simulator assumed that the vapor temperature, which sets the transfer power down the pipe, was the same as the condenser wall temperature at each condenser node. This gave performance predictions in agreement with the test data at low thermal power, but as the thermal power of the pipe was increased, the simulator predicted high temperatures in the evaporator and first condenser sections. In order to adjust for this, a temperature differential between the local condenser wall temperature and the condenser vapor temperature was defined for each node. The temperature differential was estimated by referencing the vapor-wall temperature difference to the heat pipe's "turn-on" temperature. As the temperature difference between the node temperature and the "turn-on" temperature increased (higher vapor transport), the vapor-wall temperature difference for that node also increased.

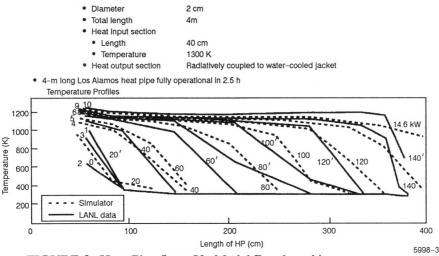

FIGURE 3. Heat Pipe Start-Up Model Benchmarking.

The new power transfer limits on each condenser node were based upon the local sonic limit using the higher adjusted vapor temperature, thus giving increased transport power capability as compared to the initial model. Using this relatively simple methodology, the heat pipe transient simulator model predicted performance which agreed well with the Los Alamos test data. The condenser wall temperatures as a function of time, as predicted by the simulator, are shown in Figure 3. This heat pipe simulation methodology was then applied to the HPTI system start-up transient simulator model.

HPTI TRANSIENT MODEL

The transient simulator model for the heat pipe cooled thermionic reactor core consisted of: (1) a single thermionic fuel element model with its integral cesium reservoir and upper and lower axial reflector segments, (2) its three associated in-core heat pipes which were lumped together into a three-node array, and (3) a seven-node radiator heat pipe model with four condenser nodes with variable fin heights for each node radiating to the space environment. This 17-node model is illustrated in Figure 4. Heat transfer admittances and thermal capacitances (including heats of fusion to account for the liquid metal melting), were developed from the specific component geometries for each node in the model. Results from an MCNP nuclear model (MCNP 1990) of the HPTI core were used to develop gamma and neutron heating rates in each of the nodes as a function of core neutronic power level. Electron cooling of the emitter node and joule heating of nodes 1, 2, and 6 through 10, as shown in Figure 4, were included within the TFE submodel. Performance limits for the three in-core heat pipes and the single radiator heat pipe (both types use sodium working fluid) were generated over the 750 to 1100 K evaporator exit temperature range. The in-core heat pipes were shown to have a "turn-on" temperature of 800 K and were sonic limited throughout the temperature range of interest. The radiator heat pipe was sonic limited in the lower temperature range, but capillary limited at 1100 K. These limits are shown in Figure 5 for comparison, and they indicated that the core heat pipes would be isothermal above 800 K since the radiator heat pipe's limits were lower than the in-core values at comparable temperatures.

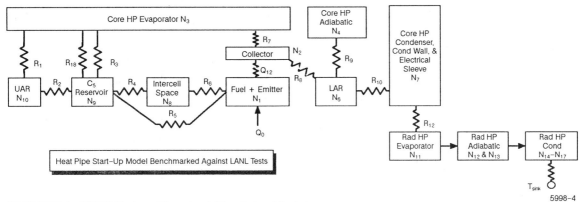

FIGURE 4. HPTI System Transient Simulator Model.

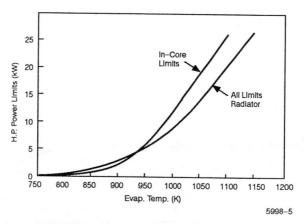

FIGURE 5. Radiator Heat Pipe Limits.

ONE HOUR START-UP RESULTS

A preprogrammed start-up scenario was developed for the TFE with 50 min of nuclear heating of the core prior to thermionic cell ignition. The adequacy of the response of the cesium reservoir's temperature profile for safe thermionic cell ignition during a rapid start-up transient was addressed in a parallel study (Miskolczy 1991). Following ignition of all the thermionic cells, the current density was linearly ramped to its full power value within 10 min. The predicted nodal temperature profiles as a function of time are shown in Figure 6. TFE thermal input power, radiator pipe reject power, and TFE electrical power output are shown in Figure 6a for the one-hour start-up transient. The temperature profiles of the integral cesium reservoir, core heat pipes, and the seven radiator heat pipe nodes are shown in Figure 6b. The last condenser node for the radiator heat pipe does not reach its operating temperature until 20 min after the TFE has reached its full thermal power output. A subsequent transient was also performed simulating a reactor scram from full power and eventual radiator freeze. The results of the shutdown simulation for the radiator heat pipe are shown in Figure 7. The 2.54-cm OD radiator heat pipe consisted of a 17.5-cm long evaporator section, two 72.25-cm long adiabatic sections, and four 72.25-cm long condenser sections with a 0.07-cm tapered carbon-carbon fin along these condenser sections. The temperature profiles of each section are shown as a function of time along with the evaporator thermal input power. These results indicate that the pipe's liquid metal inventory will freeze (371 K line in the figure) progressively from the cold end of the condenser (C-4) to the adiabatic sections. This would leave the evaporator and adiabatic section's liquid inventories to freeze last (i.e., assuring restart capability within the radiator heat pipe).

CONCLUSIONS

The results from the HPTI simulator evaluations indicate that system thaw and power ascension to full electrical output is not limited by the heat rejection subsystem for a one-hour start-up scenario. Refreezing of

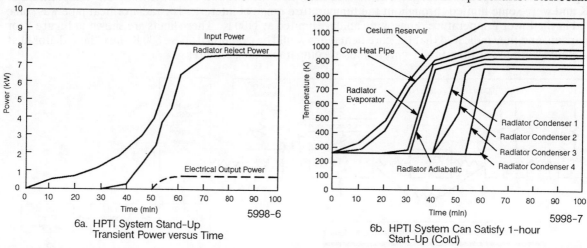

FIGURE 6. Predicted Nodal Temperature Profiles as a Function of Time.

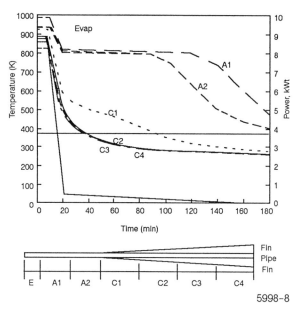

FIGURE 7. HPTI Radiator Shutdown Transient Temperature versus Time.

the radiator heat pipe following a reactor scram from full power resulted in a restartable system with the radiator evaporator node being the last node to freeze due to reactor decay heat generation. These results are preliminary in nature and cannot adequately address all of the technology concerns on liquid metal heat pipe use in a space environment. Long term effects of running a radiator heat pipe with a frozen condenser section during periods of zero power production remain to be addressed. The results do indicate the potential benefits provided by the HPTI design.

Acknowledgments

The authors wish to thank Mike Merrigan of the Los Alamos National Laboratory for his useful comments and insights on the liquid metal heat pipe tests, Gabor Miskolczy of ThermoElectron Technologies Corporation for his support in the thermionic model development, and Brian Lewis of General Atomics for the nuclear modeling input to this effort. This work was sponsored by the U.S. Air Force Phillips Laboratory under the AFSTC contract F29601-90-C-0059.

References

Merrigan, M. (1985) "Heat Transport System Feasibility Issues Report to SP-100 System Selection Committee," Los Alamos National Laboratory, Los Alamos, NM, 21 June 1985.

Miskolczy, G. et al. (1991) "Prediction of the Startup Characteristics of a Heat Pipe Cooled Thermionic Fuel Element (TFE)," 26th IECEC, Vol. 3, pp. 105–109.

"MCNP—A General Monte Carlo Code for Neutron and Photon Transport, Version 4," LA-7396-M, Los Alamos National Laboratory, Los Alamos, NM, 26 June 1990.

SPECIAL FEATURES AND RESULTS OF THE "TOPAZ II" NUCLEAR POWER SYSTEM TESTS WITH ELECTRIC HEATING

V. P. Nicitin, B. G. Ogloblin, A. N. Luppov
E. Y. Kirillov, V. G. Sinkevich
Central Design Bureau of Machine Building
Saint Petersburg, 195272, USSR

V. A. Usov, Y. A. Nechaev
Kurchatov Institute of Atomic Energy
Moscow, 123182, USSR

THE TEXT OF THIS PAPER IS PRESENTED IN PART ONE OF THESE PROCEEDINGS ON PAGES 41-46 IN THE POSTER SESSION SECTION.

THE CRAF/CASSINI MISSION: BASELINE PLAN AND STATUS

Reed E. Wilcox, Douglas S. Abraham, C. Perry Bankston,
Sandra M. Dawson, John W. Klein, Philip C. Knocke, and Paul D. Sutton

Jet Propulsion Laboratory
California Institute of Technology
4800 Oak Grove Drive
Pasadena, CA 91109
(818) 354-1173

Abstract

Mission, spacecraft, and power subsystem design considerations for the Comet Rendezvous Asteroid Flyby (CRAF)/Cassini (Saturn Orbiter) Project encompass scientific/performance and environmental/safety considerations. Based on the previous experience with the Galileo and Voyager spacecraft, a CRAF/Cassini baseline design has been proposed that seeks to optimize mission performance and reliability while achieving a level of environmental safety commensurate with previous RTG-powered missions. Concurrently, in support of NASA's satisfaction of National Environmental Policy Act (NEPA) requirements, alternative designs capable of mitigating any significant environmental impacts associated with the baseline design are being identified. A comparison of the baseline and alternative designs will be reported in a publicly available Environmental Impact Statement (EIS) scheduled for release in draft form this Spring. The baseline design supporting these impact determinations and alternative comparisons currently rely, for CRAF, on a Venus-Venus-Earth gravity assist trajectory to the comet Tempel 2 and, for Cassini, on a Venus-Earth-Jupiter gravity assist trajectory to Saturn. The spacecraft involved in this baseline make use of a common cylindrical core design comprised of virtually identical subsystems and of a host of certain mission-specific instruments. Due to the low solar intensity at the rendezvous distances associated with both missions and the past success associated with RTG use on deep space missions, three Radioisotope Thermoelectric Generators (RTGs) have been selected as the baseline power source.

INTRODUCTION

In an attempt to better understand the birth and evolution of our solar system, NASA has been pursuing a three-stage investigative strategy -- reconnaissance, exploration, and detailed study -- for each of our planetary system's three components: the inner solar system (Mercury, Venus, Earth, and Mars), the primitive bodies (comets and asteroids), and the outer solar system (Jupiter, Saturn, Uranus, Neptune, and Pluto). The first stage of the investigative strategy, reconnaissance, has made use of flyby missions to provide a cursory examination of every planet except for Pluto. The second stage of this strategy, exploration, makes use of orbiter missions to enable prolonged observation of specific solar system components. Orbiter missions have already been completed for all of the inner solar system except Mercury. However, the outer solar system and primitive bodies have yet to be explored with orbiters. Galileo, currently enroute to Jupiter, constitutes the first step in this stage of exploration. The next step will be the CRAF and Cassini missions -- scheduled for launch later this decade. CRAF, the Comet Rendezvous Asteroid Flyby mission, will be the first primitive-body orbiter mission. Cassini will be a Saturn orbiter mission. In the future the third investigative stage, detailed study, will entail surface investigation and sample return missions.

The following pages describe NASA's design for the CRAF/Cassini missions, spacecraft, and power subsystems. The paper discusses the evolution of CRAF/Cassini science objectives and their associated requirements, the importance of National Environmental Policy Act (NEPA) considerations, and the formulation of a baseline mission, spacecraft, and power subsystem plan in the context of these objectives, requirements, and considerations.

CRAF/CASSINI SCIENCE OBJECTIVES

The Science Objective Selection Process

The scientific impetus for a primitive body or planetary mission begins several years before Congress actually grants a project an official start. The process is initiated when sufficient interest in the science community prompts science advisory committees to NASA's Office of Space Science and Applications, such as the Space Science and Applications Advisory Committee (SSAAC), to recommend some general science objectives. NASA then forms a science working team to define a mission and associated strawman payload capable of meeting these objectives. Over a period of several years, NASA then refines the strawman payload and associated science objectives down to specific instruments and associated goals. At this point, NASA submits an Announcement of Opportunity, AO, (i.e., a request for proposals) to the science community and subjects the proposals it receives to an intensive peer review. During this period, NASA further develops conceptual mission, spacecraft, and power subsystem designs to provide a basis for requesting an official start from Congress. Such a request is usually made after the peer review and selection of principle investigators.

The general science objectives driving instrument selection and mission, spacecraft, and power subsystem design for the CRAF/Cassini baseline are enumerated below.

Overview of the CRAF Science Objectives and Their Associated Requirements

The CRAF mission has two principal components: (1) the comet rendezvous and (2) the asteroid flyby. The principal objectives of the comet rendezvous portion of the mission are (1) to determine the composition and character of a comet nucleus and (2) to characterize the changes that occur in a comet over a significant portion of the comet's orbit. The target comet must be sufficiently active (referring to the amount of gas and dust expelled from the nucleus as it approaches the sun), and it must have an adequate history of ground-based observations. It must also be in an orbit which allows the spacecraft to rendezvous with the comet.

The asteroid flyby portion of the mission will endeavor to characterize the physical and geological structure of a main-belt asteroid. Associated with this objective are certain requirements regarding the target asteroid. First, the asteroid should be primitive, a state usually indicated by a low albedo. Second, the asteroid's size and the circumstances of the flyby must permit an accurate determination of the asteroid's mass from ground-based observations of the spacecraft's trajectory.

Overview of the Cassini Science Objectives and Their Associated Requirements

The Cassini mission has five primary objectives: (1) investigation of Saturn's atmospheric properties; (2) exploration of Titan, a moon having an atmosphere and harboring organic compounds; (3) investigation of Saturn's icy satellites; (4) examination of Saturn's rings; and (5) measurement of Saturn's magnetosphere. As part of the Titan investigation, the Cassini spacecraft will deploy a probe, the Huygens Probe, into Titan's atmosphere and will map its surface with radar. The spacecraft must also fly by Titan, on multiple occasions, to achieve the gravity assists needed for navigating Saturn's rings and icy satellites.

Secondary mission objectives include a flyby of an asteroid and investigation of Jupiter's magnetotail. As with CRAF, it is desired that the asteroid be large and primitive.

NEPA AND THE CHOICE OF BASELINE ALTERNATIVES

Like the science objectives, NEPA requirements constitute another important context within which the mission, spacecraft, and power subsystem baseline must evolve and be evaluated against competing alternatives.

NEPA Requirements

NEPA requires that an environmental assessment or environmental impact statement (EIS) be prepared for any federal projects with potentially significant environmental impacts. NASA's guidelines for implementing NEPA regulations require that an EIS be prepared for the "development and operation of nuclear systems, including reactors and thermal devices used for propulsion and/or power generation" (Code of Federal Regulations, 14 CFR part 1216.3, 1990). Based on NASA's preliminary plans to use RTGs as the on-board power for these missions, this process is currently underway for those missions that comprise the NASA Primitive Bodies and Outer Solar System Exploration Program (PB & OSSEP), including the CRAF and Cassini missions.

A major goal of the EIS is to identify and assess the reasonable alternatives to the proposed actions that will "avoid or minimize adverse effects of these actions upon the quality of the human environment" (Council on Environmental Quality, 40 CFR part 1502.1, 1978). The EIS discusses the purpose and need for the proposed action as well as reasonable alternatives for accomplishing this purpose and need. The environmental impacts of the various alternatives, including a no-action alternative, are carefully evaluated in the draft and final EISs.

NEPA and the Evolution of a Mission, Spacecraft, and Power Subsystem Baseline

NEPA requires that an EIS be prepared "at the earliest possible time to ensure that planning and decisions reflect environmental values" (Council on Environmental Quality, 40 CFR part 1501.2, 1978). To adequately discuss potential environmental impacts, however, a baseline design has to be developed to a degree that allows both a meaningful characterization of potential environmental hazards and a comparison with reasonable alternatives.

The Role of the EIS in Considering Alternatives to the Baseline

NEPA requires that the Agency preparing an EIS consider reasonable alternatives to this baseline. NASA defines alternatives to the baseline according to the best information available on a program's potential environmental impacts, and on public responses to the agency's notice of intent to prepare an EIS. In the case of CRAF/Cassini, the principal source of potential environmental impact is the risk of a release of radioactive material from the RTGs in the event of certain types of launch accidents or an inadvertent reentry during an Earth flyby. Hence, the alternatives of interest, pending further public comment, are (1) non-RTG power sources such as solar arrays (to eliminate any hazard from the RTGs during all mission phases), (2) mission designs that do not involve Earth gravity assists (to eliminate any reentry hazard), and (3) the "no Action" alternative mentioned above (which constitutes a decision not to proceed with the Program). This final alternative is required by NEPA guidelines (Council on Environmental Quality, 40 CFR part 1502.14, 1978).

The EIS will examine the alternatives to the RTGs and to the mission. To facilitate NASA's comparison of alternatives and their impacts, JPL is developing a set of supporting studies. Information from this three-volume set will be incorporated into the EIS. Volume 1 of the supporting studies will contain a description of the baseline mission and spacecraft. Volume 2 will identify and compare mission and power alternatives. And, Volume 3 will address Earth impact probabilities and the Earth avoidance strategy.

Status of Investigation Into Mission and Power Alternatives to the Baseline

The set of EIS supporting studies currently being prepared by JPL to examine the baseline and alternative missions, power sources, and their potential impacts will be released as separate JPL technical documents at the time of the draft EIS release in the Spring of 1992. A 45-day comment period will follow this release during which the public can comment on any aspect of the EIS. At the end of the comment period, NASA will prepare the final EIS, which will include responses to questions raised during the public comment period. The final EIS is scheduled for release in late fall, 1992.

The following sections are present to provide more specific information on the mission, spacecraft, and power subsystem baseline against which the alternatives will be compared.

THE BASELINE CRAF/CASSINI MISSION DESIGN

Mission Selection

The CRAF and Cassini science objectives, discussed previously, result in a number of mission requirements and restrictions which impact trajectory availability and mission selection. For CRAF, the orbital characteristics of the cometary target, the asteroid flyby requirements, and the desired comet exploration strategy all have performance implications which must be considered in evaluating a potential mission. Cassini trajectory selection is affected by the Saturn exploration requirements, the satellite tour design, and the delivery requirements of the Huygens probe. The limitations of the launch vehicle and on-board spacecraft propulsion systems are vital discriminators in selecting viable trajectories for either mission. Additionally, programmatic issues such as risk management, schedule, and cost must be addressed.

The Titan IV (SRMU)/Centaur has been chosen by the project as the CRAF/Cassini launch vehicle, due to its high planetary injection performance. With the current spacecraft configuration, mass, and propellant requirements, the performance of the Titan IV(SRMU)/Centaur does not permit direct trajectories to Saturn, or to any of the eight short period comets being considered for CRAF. Instead, it is necessary to use multiple planetary gravity assists to give the spacecraft sufficient heliocentric energy to send it to its target. Several gravity assist trajectories have been found for CRAF and Cassini. Williams, et al. (1991) includes a detailed description of trajectory options and mission analysis for CRAF.

CRAF Primary Mission

The current primary CRAF mission was selected from among several options presented at the CRAF/Cassini Baseline Confirmation Review in January, 1991. This trajectory, illustrated in Figure 1, starts with a launch from the Cape Canaveral Air Force Station in early February, 1996. The spacecraft loops through the inner solar system several times, flying past Venus twice and Earth once, before gaining sufficient energy to reach its target, Comet Tempel 2. Between the second Venus gravity assist and the Earth gravity assist, the spacecraft performs a close flyby of 739 Mandeville, a primitive main belt asteroid with a radius of 55 km. During this flyby, examinations of the asteroid by spacecraft instruments, and ground-based observations of the spacecraft's trajectory will permit detailed assessments of the composition and mass of the asteroid.

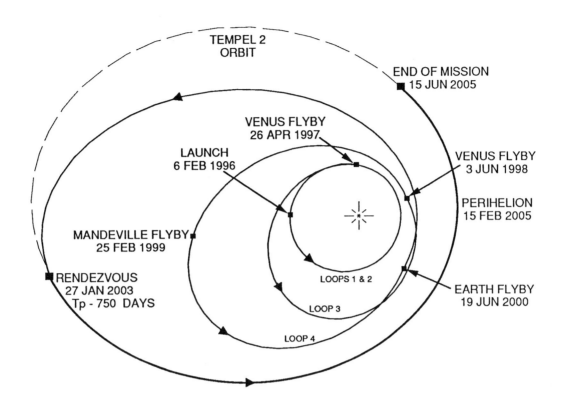

FIGURE 1. Craf Primary Trajectory.

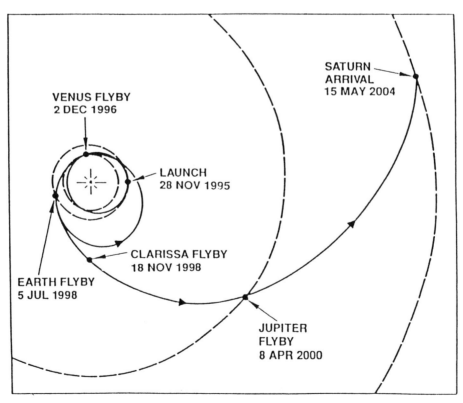

FIGURE 2. Cassini Primary Trajectory.

Rendezvous with Comet Tempel 2 occurs approximately 7 years after launch, in January of 2003. After an initial series of cautious flybys, the spacecraft enters orbit about the quiescent nucleus, measuring the nucleus mass and determining its composition and rotation state. As the comet moves closer to the sun and becomes more active, the spacecraft backs away from the nucleus to explore the surrounding envelope of gas and dust (coma) of the comet, and the developing tail. End of mission occurs four months after comet perihelion (approximately 2.5 years after rendezvous) in June, 2005.

Cassini Primary Mission

The Cassini primary mission, also selected at the Baseline Confirmation Review, launches in November, 1995. The interplanetary trajectory, illustrated in Figure 2 (Peralta, 1991), involves a flyby of Venus, followed by an Earth gravity assist, a close flyby of asteroid 302 Clarissa, and a Jupiter gravity assist. Saturn orbit insertion occurs in May, 2004. (A recent revision of this trajectory delayed Saturn arrival until June, 2004, and changed the asteroid flyby target to 1989 UR-1.)

Approximately two months after arrival, the spacecraft's orbit will be altered to allow a close flyby of Saturn's largest moon, Titan. Shortly before the Titan flyby, the Huygens probe will be deployed into the atmosphere of Titan. In addition to the measurements of Titan's atmosphere by the probe, the surface of Titan will be observed by the orbiter's radar mapper. Using repeated Titan encounters to alter the spacecraft's trajectory, the orbiter will perform an extensive tour of the Saturnian satellite system. Detailed examinations of the Saturnian atmosphere and magnetosphere will also be performed during the four year interval between Saturn arrival and end of mission.

CRAF/Cassini Launch Rescheduling

Recent budgetary restrictions may force a rescheduling of both CRAF and Cassini launches, and attendant changes in the mission profile. One launch scenario involves a CRAF launch in May, 1997. Instead of two Venus gravity assists and one Earth gravity assist, this mission calls for a flyby of Venus, followed by two Earth gravity assists. Two asteroids (406 Erna and 379 Huenna) are examined on the way to the December, 2005 rendezvous with Comet Kopff. End of mission occurs in September, 2009. The Cassini spacecraft would be launched in October, 1997. After two Venus flybys, followed by an Earth gravity assist and a Jupiter gravity assist, Saturn orbit insertion would occur in June, 2004. For both CRAF and Cassini, the general nature of the mission plan after encounter with the primary target is unchanged.

THE BASELINE CRAF/CASSINI SPACECRAFT DESIGN

Key Requirements and Constraints

The baseline CRAF/Cassini spacecraft design is driven by various Project requirements and constraints. The highest level design requirements imposed by the Project are, first, that the design be reliable, safe, and provide the required science return. The design cost must also be accommodated by the fixed price cost ceiling imposed by Congress for the development phase. Other key Project requirements influencing spacecraft design are discussed below:

o   Commonality and modularity: the basic design must be useable for potential follow-on missions with little or no redesign.

o   Long life and high reliability: design the spacecraft for a 13.5-year mission life.

- Wide range of environments and mission requirements: design a common spacecraft that can accommodate the large quantity of propellant needed for orbit insertion or rendezvous, plus small and large trajectory trim maneuvers. The design must also be able to accommodate a wide thermal range, since solar distances may range between 0.6 AU (perihelia for Venus flyby trajectories) and 9.3 AU. The spacecraft must have sufficient radiation tolerance, and be able to withstand the impact of cometary dust and gas, as well as small ring particles.

- High mission return: high precision pointing control and stability must be available. The spacecraft must have large data storage capability, and be able to maintain high data rates.

- Robust design: margins must be able to accommodate design evolution and in-flight performance degradation.

- Large, sophisticated science payload: the spacecraft engineering subsystems must be able to support instrumentation packages which enable maximum science return for minimum project cost.

Design Process

The present CRAF/Cassini baseline spacecraft design has evolved considerably since the mid-1980s when first CRAF and, later, Cassini mission/spacecraft studies were initiated. The merger of CRAF and Cassini into a common spacecraft design occurred in 1988. The initial activities concentrated on spacecraft architectural studies wherein, for example, configuration and payload were evaluated for the lowest cost, maximum science payload capability. Initial power source (solar, RTG, solar/RTG) trade-offs were performed during this period. These studies are normally conducted to define these and other spacecraft architectures; for CRAF/Cassini, these studies are faciliting the Environmental Impact Statement (EIS) process in the definition of alternatives, as described previously.

The Spacecraft Design Team is an interdisciplinary team that supports the evolution of the spacecraft design. Design and implementation problems/refinements are evaluated and reviewed by the Team. The Team evaluates the competing forces for reducing cost and increasing performance while maintaining the norms of reliability, safety, and long-life within a fixed Project cost. The Team, in turn, provides recommendations to the Project.

The Spacecraft Design Team has developed a baseline CRAF/Cassini design which responds to the previously described key project requirements. The details are provided in Appendix A. The design makes use of a common cylindrical core design comprised of virtually identical subsystems and a host of certain mission-specific instruments. This includes the baseline Power and Pyrotechnic Subsystem, described in Appendix B.

The baseline power source, sized to meet current mission power requirements (Appendix B, Table 3), comprises three Radioisotope Thermoelectric Generators (RTGs). The selection of the RTG as the baseline design is based upon the planned trajectories for the CRAF and Cassini missions which would take both spacecraft to at least 4 A.U. distance from the sun. In the past, these distances have normally precluded the use of solar power due to low solar intensity. In addition, RTGs have proven to be extraordinarily reliable as demonstrated in the Pioneer, Voyager, Galileo, and Ulysses missions.

The spacecraft and power system design are now being evaluated against alternative approaches and optimized to achieve the greatest reliability and functionality for the spacecraft. Of particular importance to the EIS process is the selection of the power source. Studies are underway to evaluate alternate power sources for comparison with the baselined RTG, including: radioisotope sources based on other conversion technologies (Brayton, Stirling, AMTEC, etc.), reactor-based systems, solar photovoltaic conversion (conventional planar and concentrator), solar thermal conversion, and primary stored energy systems (batteries and fuel cells). The power source and distribution design tradeoffs are being evaluated against the following criteria: technology readiness, launch vehicle constraints, environments, and science impacts.

Technology readiness relates to the ability of a subsystem component to provide the power level required, meet the launch date, and meet the lifetime and reliability requirements needed for the mission. For example, can advanced, ultra-light weight solar arrays be flight qualified and their reliability demonstrated consistent with the time frame for satisfying the missions' objectives? Can more efficient thermal-to-electric conversion technologies be qualified to optimize power or reduce radioisotope fuel requirements?

Launch vehicle constraints are associated with the mass and volume capabilities of the launch vehicle, while the environments encountered involve taking into account the range of the spacecraft from the sun, solar occultations, particles and fields, and launch conditions, to name a few. The distance from the sun dictates the thermal environment as well as the solar flux available if solar arrays are used. This, in turn, affects the size and mass of solar arrays or collectors. Solar occultations determine battery requirements in a solar powered system, which may impact both the mass and lifetime of the spacecraft. And, the relative performance of silicon and gallium arsenide cells in radiation environments, at low temperatures, and low intensity light impact array size and spacecraft mass.

Finally, tradeoffs impacting science investigations include such items as maintaining instrument fields of view, instrument pointing accuracy, minimizing spacecraft turn times, etc.

CONCLUDING REMARKS

As discussed previously, JPL is currently preparing a set of EIS supporting studies to examine the baseline and alternative missions, power sources, and their potential impacts. These studies will be released as separate JPL technical documents at the time of the draft EIS release in the Spring of 1992.

Acknowledgments

This work was conducted at the Jet Propulsion Laboratory, California Institute of Technology, under contract with the National Aeronautics and Space Administration in support of the CRAF/Cassini Project. The authors wish to thank Sylvia Miller, Roger Diehl, Bill Nesmith, and Ray Goldstein for their participation in reviewing this report. Much of the work described in this paper represents the combined efforts of many individuals on the various mission and spacecraft teams. Special thanks are therefore extended to the members of the CRAF Mission Design Team, the Cassini Mission Design Team, the CRAF/Cassini Spacecraft Design Team, and the Power and Pyrotechnics Subsystem Design Team.

References

Code of Federal Regulations (1990). Title 14, Aeronautics and Space. Subpart 1216.3 -- Procedures for Implementing the National Environmental Policy Act. Washington, D.C.: U.S. Government Printing Office.

Peralta, F. (1991), personal communication, JPL.

U.S. Council on Environmental Quality (1978), Executive Office of the President, <u>Regulations for Implementing the Procedural Provisions of the National Environmental Policy Act.</u> 40 CFR Parts 1500-1508. November 29, 1978.

Williams, S., Knocke, P., Bright, L. (1991), *Interplanetary Mission Design for the Comet Rendezvous Asteroid Flyby Mission,* AAS 91-396, AAS/AIAA Astrodynamics Specialist Conference, Durango, Colorado, August 19-22, 1991.

Appendix A

Baseline Spacecraft Design Architecture

Configuration: Cylindrical monocoque core structure for propulsion simplicity; 12 bay electronics bus configuration with JPL standard dual shear plate electronics construction; standard boom designs adaptable to either of three science platforms; upper and lower equipment modules with standard field joint interfaces to facilitate integration; pyrotechnic separation from launch vehicle adapter; structural design margins suitable for Titan IV/Centaur launch vehicle.

Communication: 4.0 m diameter high gain antenna ; two low gain antennas; X-band uplink & downlink; antenna feeds adaptable to S, X, Ku, and Ka-bands.

Avionics: MIL-STD-1553B command and data bus/protocol; Consultive Committee for Space Data Standards (CCSDS) packetization command and telemetry uplink/downlink protocol; common MIL-STD-1750A flight computers (ADA Language) for Attitude Control and Command and Data Subsystems; common avionics interfaces via a standard bus interface unit (BIU); 1.8 Gbit redundant Solid-State Recorder; common cabling.

Attitude/Articulation: Star updates using the wide angle imaging camera; reaction wheels for on-orbit precision momentum compensation/pointing control; thruster control for coarse attitude control (cruise, reaction wheel unloading); modular high precision actuator (one module for scanning, another for turntable), sized for largest payload application; low precision actuator for coarse pointing applications.

Thermal Control: Standard passive techniques (radiators, multi-layer blankets, paint, radioisotope heater units - RHUs); shades for inside 1AU; and electrical heaters.

Power: Three Galileo-class radioisotope thermoelectric generators (RTGs); 30 Vdc power distribution via solid-state power switches.

Propulsion: 4300 kg maximum bi-propellant capability for main engine maneuvers; 140 kg maximum monopropellant capability for small thruster maneuvers; redundant 490 N (110 lbf) main engine; redundant 0.5 N (0.11 lbf) thrusters, 8 per string.

Fault Protection: Block redundant, Class A (highest quality standards) with cross-strapping internal to subsystems when cost effective and dictated by increased reliability; Project-approved single point failures permitted when analysis indicates probability of failure is small.

Common Science: Imaging Science (1 narrow angle camera, and redundant wide angle cameras - shared with Attitude Control); common magnetometer and boom; standard instrument cover actuator.

Mission Specific Spacecraft Design Characteristics

The previous discussion of the baseline spacecraft-design architecture described a common, modular engineering bus design. There are also mission-specific spacecraft-design architectural characteristics. These characteristics include:

- a High Precision Scan Platform (HPSP) structure sized for the mission-unique science (CRAF or Cassini),
- a Low Precision Pointing Platform (LPPP) for dust collection instruments (CRAF only),
- a turntable for fields and particles instruments (Cassini only), and
- Ka-band communications, Huygens Probe, and Probe support equipment including the Probe Relay Antenna (Cassini only).

Scientific Payloads

The scientific payloads for CRAF/Cassini are summarized in Tables 1 and 2.

TABLE 1. CRAF Instrument Payload

Instrument Name	Investigator
Cometary Ice & Dust Experiment	G. Carle, NASA/Ames
Cometary Dust Environment Monitor	M. Alexander, Baylor Univ.
Cometary Matter Analyzer	J. Kissel, Max Planck Inst.
Coordinated Radio & Electron Wave Experiment	J. Scudder, NASA/Goddard
Cometary Retarding Ion Mass Spectrometer	T. Moore, NASA/Marshall
Imaging Subsystem	J. Veverka, Cornell Univ.
Magnetic Field Investigation	B. Tsurutani, JPL
Neutral Gas & Ion Mass Spectrometer	H. Niemann, NASA/Goddard
Suprathermal Investigation of Planetary Environments	J. Burch, Southwest Research Inst.
Thermal Infrared Radiometer Experiment	F. Valero, NASA/Ames
Visible Infrared Mapping Spectrometer	T. McCord, SETS
Radio Science	D. Yoemans, JPL

TABLE 2. Cassini Instrument Payload

Instrument Name	Investigator
Cassini Plasma Spectrometer	D. Young, Southwest Research Institute
Cosmic Dust Analyzer	E. Grun, Heidelberg, FRG
Composite Infrared Spectrometer	V. Kunde, NASA/Goddard
Ion & Neutral Mass Spectrometer	TBD
Imaging Subsystem	C. Porco, JPL
Magnetometer Field Experiment	D. Southwood, UK
Magnetoatmospheric Imaging Instrument	S. Krimigis, John Hopkins Applied Physics Lab
Cassini Radar	C. Elachi, JPL
Radio & Plasma Wave Science	D. Gurnett, Univ. of Iowa,
Radio Science	A. Kliore, JPL
Ultraviolet Imaging Spectrograph	L. Esposito, Univ. of Colorado
Visible & Infrared Mapping Spectrometer	R. Brown, JPL

Appendix B:

POWER SUBSYSTEM DESIGN

The baseline CRAF/Cassini power and pyrotechnic subsystem is designed to satisfy the requirements of both the CRAF and Cassini mission as well as other outer planet missions of the future. In order to keep costs low, only one baseline design is utilized for all missions. In addition it must be flexible enough to add energy storage and additional primary power sources at a later date if the mission requirements change or spacecraft load management is required.

Baseline Design

Key requirements assuming three RTGs as the baseline power source are:

(1) The Power and Pyrotechnics Subsystem (PPS) shall maintain bus voltage within specification at all RTG power levels.
(2) The PPS shall have the capability to shunt from zero to full RTG power output. The spacecraft temperature control design shall allow PPS to shunt any variation from 600 Watts to full RTG power for up to 2 hours; or shunt up to 600 Watts indefinitely.
(3) The PPS shall distribute power nominally as regulated 30 Vdc +0.25 Vdc to -1.50 Vdc.
(4) All loads shall be controlled through Solid State Power Switches.
(5) The PPS shall develop circuitry suitable to fire standard pyrotechnic devices.

The block diagram of Figure 4 meets these requirements. A detailed discussion of the baseline design follows. The design is being carried out by JPL and flight hardware fabrication will be performed by LORAL/EOS, Pasadena, California.

Radioisotope Thermoelectric Generators (RTGs)

The baseline power source used for the CRAF/Cassini missions is three RTGs for each mission. Plans call for two of the six generators to come from previous missions with four new generators being fabricated for these missions. Generator F5 is the spare from the Galileo and Ulysses missions. Generator F2 is the qualification generator from the Galileo Mission. Current plans have the remaining four generators F6, F7, F8, and F9 being fabricated by GE under contract to the Department of Energy.

The fuel for the new generators will be taken from existing stockpiles. Two-thirds of this stockpile will be reblended with enriched Plutonium to develop the maximum thermal load and thus the maximum beginning of mission (BOM) power. Table 3 shows an estimate of each generator and its power for the various missions.

TABLE 3. Projected RTG power production during mission.

	Generator	BOM	EOM
CRAF	F2, F5, F6	794	612
Cassini	F7, F8, F9	813	639

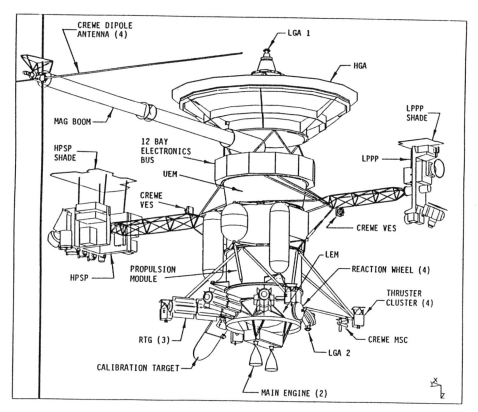

CRAF Deployed Rear Trimetric View

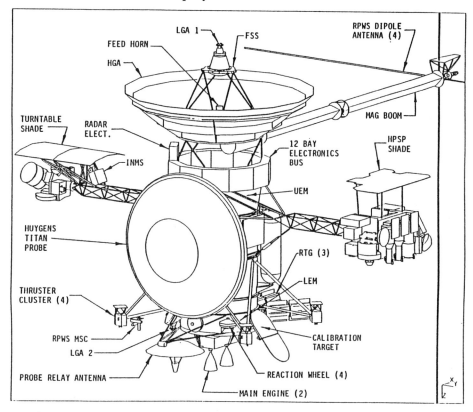

CASSINI Deployed Front Trimetric View

FIGURE 3. Deployed CRAF and Cassini Spacecraft Showing Key Configuration Elements.

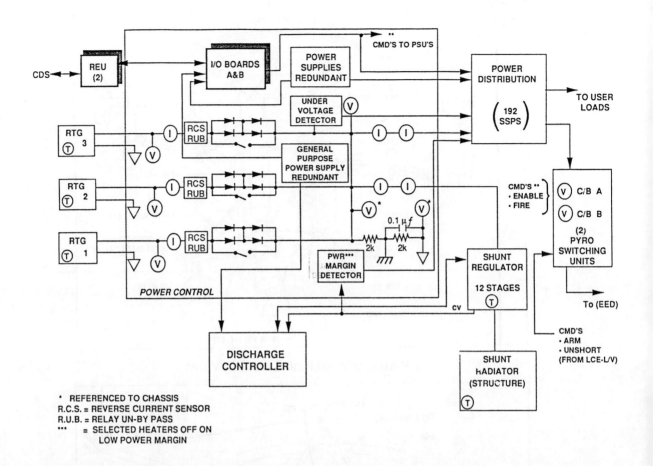

FIGURE 4. CRAF/Cassini Power and Pyrotechnic Subsystem Block Diagram.

Bus Regulation

In order to assure maximum life of the RTGs, excess power not required by the load is shunted or radiated to space. The Shunt Regulator Assembly (SRA) performs this task in conjunction with the Shunt Radiators located on the outside of the spacecraft. The SRA consists of 12 stages of linear regulation that overlap in order to assure smooth regulation under all operating conditions. Twelve stages were needed in order to minimize the thermal load within the spacecraft bay. The spacecraft bus voltage is used as a measure of the excess power. By controlling the bus voltage to 30Vdc, the power from the RTGs is assured to be maintained at a maximum. The control loop is tight enough to ensure that in the steady state the bus voltage is maintained between 30.25 Vdc and 28.5 Vdc. During transients when excess load is added almost instantly, small amounts of additional energy is required in order to hold up the bus. The discharge controller performs this function. It consists of 12,100 µF of capacitance that is charged to the bus voltage. When the bus falls below a certain value, the Discharge Controller Assembly (DCA) injects the needed excess energy onto the bus until the fault protection can commence.

Power Distribution

Once the power has been generated and the bus regulated, the power must be distributed to the loads. CRAF/Cassini will implement a new device in order to perform this function. The Solid State Power Switch (SSPS) is a solid state circuit breaker that includes many specialized functions necessary to the converters by controlling the amount of voltage applied to the controlling switch. In addition, each switch has a preprogrammed trip point that determines when the circuit breaker trips. This value can be set between 375mA and 3A. The SSPS is commanded on/off from the Command and Data Subsystem (CDS). Each switch has a unique address or code so that there is no chance that a switch is turned on or off by accident. Because of the large number of loads (which includes instrument power converters and actuators, other subsystem power converters and actuators, and heaters) the SSPS can be implemented in groups of 16. If needed, a full set of 256 switches is available. Due to the complex nature of the SSPS, a hybrid implementation is proposed for CRAF/Cassini. CTS of West Lafayette, IN has been selected as the hybrid manufacturer. To reduce the size even further, the entire digital portion of the SSPS will be fabricated as a gate array and then integrated with the analog die to produce the final hybrid.

Fault Protection

The SSPS is a critical element of the fault protection philosophy for the PPS. The ability to switch off a load ensures that the bus will remain within regulation. If the preprogrammed trip levels fails to switch off a faulted load, failsafe backup circuitry will ensure that the load is switched from the bus within 200 mS when the load current reaches 6 A for 1 mS. This is critical for the DCA. If the faulted load is not removed within this period of time, the DCA will not have enough energy to hold the bus at 30 Vdc. In the case where a fault occurs and the SSPS does not detect the fault, an undervoltage detection circuit will remove all nonessential loads from the bus if the bus voltage falls to 26.50 Vdc. Nonessential loads are those loads not considered necessary for the health and safety of the spacecraft. The essential loads on CRAF/Cassini are the Attitude and Articulation Control Flight Computers, the Command and Data Subsystem Electronics, the PPS Remote Engineering Unit (used to communicate with CDS), and the PPS General Purpose Power Supply needed to power the SSPS and PPS electronics. There is also circuitry to protect against RTG shorts. Because the mission is being designed such that minimal science can be returned if only 2 RTGs are available, diodes are placed in series with the RTG. To reduce the power losses associated with the diodes, relays are placed in parallel. The control for the relay is the current from the RTG. If this current reverses, then an internal fault within the RTG is present. The relay will then open isolating the faulted RTG.

Pyrotechnic Switching Unit (PSU)

There are a number of pyrotechnic devices that must be fired for successful deployment of instruments or structure. Further, various propulsion operations require pyrotechnic actuation. The PSU supplies the necessary energy and safety for firing these devices. The PSU is made up of a capacitor bank that stores the energy required to fire the pyrotechnic devices. In addition, various arm and enable switches are included in order to ensure no misfires occur during the mission lifetime.

Past RTG Power System Performance

This type of power source design has been in existence since the mid-60s. While the individual components have changed, the overall design concept has not changed since Voyager. In fact, the Voyager and Pioneers are still performing their task after 14 and 18 years respectively. Based upon this success, the Galileo and Ulysses power subsystems were developed. In essence, their functional block diagrams were exactly the same. Again, the RTGs and baseline power subsystem have performed well; 2 years of continual operation for Galileo and 1 year for Ulysses. While the initial power output of the RTGs was somewhat lower than originally anticipated, the RTGs have performed as predicted since that time. It is postulated that the longer than normal storage time for Galileo and Ulysses due to the Challenger accident is the primary cause for the initial power discrepancy. As the missions proceed, it is anticipated that Galileo and Ulysses will perform in the same manner as Voyager and will last a similar period of time.

SP-100 FIRST DEMONSTRATION FLIGHT MISSION CONCEPT DESIGN

Dennis Switick, Charles Cowan, Darryl Hoover
Thomas Marcille, Robert Otwell, Neal Shepard

General Electric Company
P.O. Box 530954
San Jose, CA 95153-5354
(408) 365-6456

Abstract

A conceptual design study was performed to define a SP-100 Space Reactor Power System (SRPS) compatible with a set of proposed first demonstration flight objectives and constraints. The principal mission objective is the confirmation and demonstration of the SP-100 technology and its design application maturity for space missions. The objective is to be achieved by SRPS assembly, ground acceptance test, launch vehicle integration, and on-orbit demonstration of the full range of SRPS performance capabilities. A proposed second objective is to confirm the functional integration and active interfaces of a state-of-the-art electric propulsion thruster subsystem with an SRPS power source, and to demonstrate its on-orbit performance. Key mission requirements and constraints are the following:

- Economical Atlas AS launch into an 1100 km orbit;
- SRPS gross output of 30 Kwe for 3 year mission lifetime;
- Spacecraft mass less than 7000 Kg; SRPS less than 2500 Kg;
- Accommodate users by order of magnitude reduction in SRPS fluences; and
- SP-100 program performance & safety requirements apply.

An SRPS and spacecraft conceptual design was devised which meets all mission requirements. User ability to meet the adopted radiation limits for the GFS user plane interface, and the use of a larger launch vehicle would allow a significant SRPS mass reduction (400 Kg) for early 30 kWe missions.

INTRODUCTION

A conceptual design study was performed to define a SP-100 Space Reactor Power System (SRPS) compatible with a set of proposed first demonstration flight objectives and constraints.

For the very first SRPS in orbit, the principal mission objective is the confirmation and demonstration of the SP-100 technology and its design application readiness for space missions. It is proposed that a ancillary mission objective be to confirm the functional integration and interfaces of a state-of-the-art electric propulsion (EP) subsystem with an SRPS power source and spacecraft bus, and to demonstrate the EP on-orbit performance.

A draft functional requirements document was prepared for the SRPS wherein the unique nature of this first mission is reflected. Additional constraints were self-imposed to maximize the spacecraft compatibility with cost and user capabilities for near term execution:

- Integrate with an economical and proven launch vehicle and spacecraft bus;
- Provide an electrical output of 30 kWe for a 3 year lifetime;
- Reduce the SRPS fluences at the user plane to accommodate the radiation tolerance of near-term user electronics; and
- Provide a solar auxiliary power unit for mission flexibility and investment protection.

MISSION PLAN

The spacecraft will be placed in a nuclear-safe 1100 km circular orbit, with housekeeping power supplied by a solar auxiliary power system whenever the SRPS is not operating. The SRPS will first be put through its planned test program of operational maneuvers, autonomous control, and response to anticipated faults. After confirmation of the SRPS behavior and any additional tests, EP system testing will be initiated. The reactor power will be held constant during EP testing, with excess electrical power dissipated by the SRPS shunt. Mission time is reserved for additional testing following analysis of the data. A final full power system run is planned to complete the test program, with the EP system operated until available fuel is exhausted.

To demonstrate SP-100 technology readiness the mission plan includes the assembly, ground acceptance testing and integration of the SPRS with the payload and launch vehicle, as well as on-orbit testing of the full range of SRPS capabilities. Diagnostic flight instrumentation is added to confirm critical SRPS design and performance parameters:

- Temperatures   reactor shield internal
                           reflector drive motor
                           multiplexer environment at specified components
                           infared camera

- Pressures   reactor shield internal
                           accumulators

- Boom shape   optical camera & tabs

- Radiation   neutron & gamma flux at specific locations
    fields

At the end of the test program the SRPS will be permanently shut down and its status monitored. The spacecraft will continue to be powered by the solar auxiliary power system until ground commanded to terminate the mission.

SPACECRAFT CONFIGURATION

The overall spacecraft configuration results from meeting the mission objectives and allocated system level functional requirements, integration with the prescribed Atlas II AS launch vehicle (General Dynamics 1990), and on-orbit operations interfaces. The major spacecraft elements are the Space Reactor Power System (SRPS), the nuclear electric propulsion (NEP) module, and the Multimission Modular Spacecraft (MMS) bus (NASA 1986). The Atlas launch vehicle and MMS were chosen for their cost and proven flight capabilities.

Based on the Atlas heavy lift payload envelope and a previous 30 kWe SRPS study (Armijo et al. 1991), the SRPS heat rejection radiators were configured to allow for axial compression of the SRPS and maximum space for the MMS and NEP systems. This resulted in a trade-off between the SRPS radiator/shield configuration and axial space for the MMS and NEP modules. Key issues for packaging of the MMS and NEP modules were the stowage of the solar panel and communications dish, area for the NEP power processor waste heat radiator, compatibility with the MMS thermal management and attitude control functions, and the desire to locate the EP thruster near the spacecraft longitudinal axis.

The resulting spacecraft launch configuration is shown in Figure 1, including the relationship to the Atlas payload envelope. The reactor shield cone half-angle was increased to 17 degrees to allow sufficient room for SRPS components and primary structure. Without the launch packaging constraints the shield cone half-angle could be reduced to 13 degrees, saving mass while satisfying the deployed shielding requirements. The 2.6 meter diameter NEP module represents the maximum allowable within the packaging constraints.

Once placed in orbit by the Centaur upper stage the only deployments required are the SRPS boom, followed by the solar panel and high gain antenna (HGA). Figure 2 shows the spacecraft deployed configuration. The solar panel is not required when the SRPS is providing power to the housekeeping bus, and is designed to transition to a Standby Mode orientation, with a minimum cross section to the reactor radiation field.

Table 1 provides the spacecraft mass summary including three alternate electric propulsion thruster subsystems. The primary and secondary structures allowance of 1000 kg was based on current GE experience for a range of spacecraft. Based on the component sizes and masses, it is possible to package and demonstrate any of the alternate thruster subsystems on the demonstration flight.

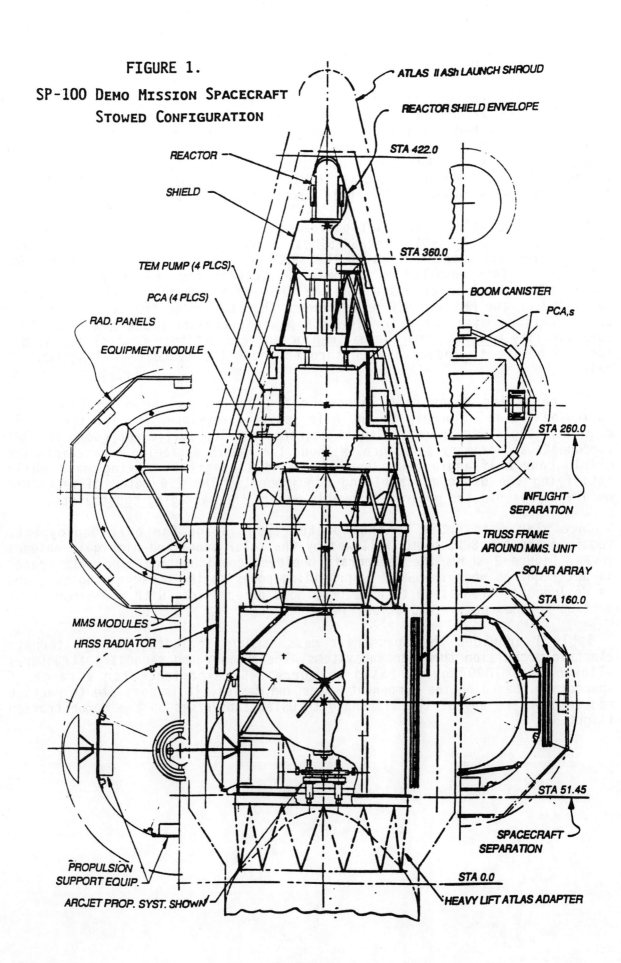

FIGURE 1. SP-100 Demo Mission Spacecraft Stowed Configuration

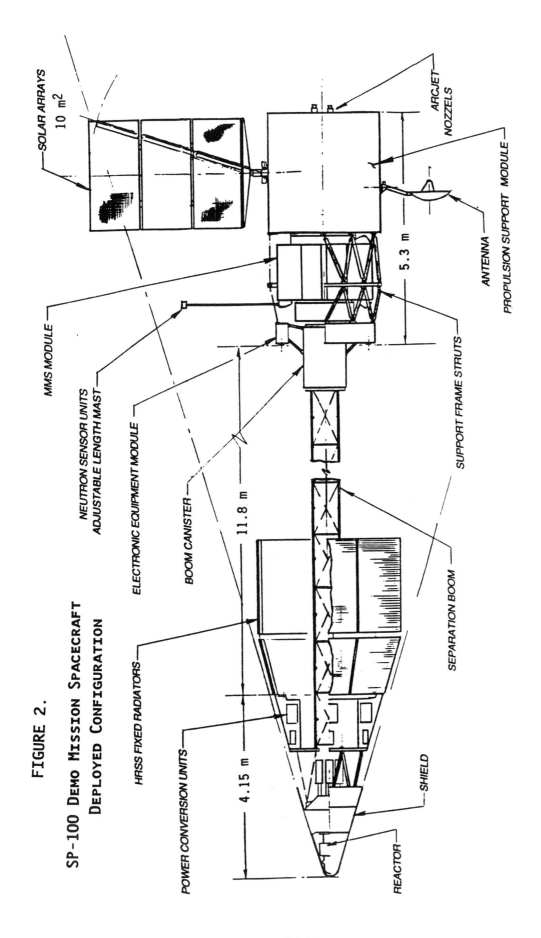

FIGURE 2.
SP-100 Demo Mission Spacecraft Deployed Configuration

TABLE 1. SRPS 30 kWe Demonstration Spacecraft Mass Summary (kg) Integration with Alternate NEP Options

System	Component	H2 Arcjet	NH3 Arcjet	Kr Ion
NEP Module	Power Processor	70	70	145
	Power cabling & voltage control	10	10	35
	PPU radiator	5	5	10
	Propellant	250	1000	285
	Propellant tankage	175	150	170
	Propellant dist. system	10	10	10
	Thruster/gimble	40	40	35
	Controller, sensors	10	10	10
	Heaters/thermal control	10	10	15
	EP diagnostics	50	50	50
		630	1355	765
Structural interface	Launch vehicle integration	250		
	primary structure	750		
	secondary structure	250		
		1250	1250	1250
Utility bus	Nominal MMS + PM-1	950		
	Solar array & drive	75		
	X-Link antenna/gimble	75		
	SRPS diagnostics	175		
		1275	1275	1275
SRPS @ 30kWe	Reactor	325		
	Shield	895		
	Primary heat transport	215		
	Reactor I&C	245		
		1680	1680	1680
	Power conversion	130		
	Heat rejection	260		
	Power cond., control & dist.	145		
	Mechanical/structural	250		
		785	785	785
	SRPS mass ==	2465		
	Spacecraft mass =	5620	6345	5755

SRPS DESIGN

The current study focused on updating the SRPS nuclear subsystems. The results of the previous 30 Kwe study were used directly for the space subsystems. The nuclear subsystems of the SRPS include the reactor, shield, primary heat transport, and reactor instrumentation and control. The reentry thermal shield is also included. These subsystems were scaled from the SP-100 Generic Flight System (GFS) design (Josloff et al. 1991) to meet the specific objectives and requirements for this initial flight system; in particular:

- The reactor thermal power was reduced from 2415 kWt to 790 kWt,
- The fluence/dose limits at the user plane were reduced by an order of magnitude leading to a neutron fluence limit of 1×10^{12} neutron/cm^2, and a gamma dose limit of 5×10^4 rads, and
- The reactor lifetime and integral power was specified as three years on-orbit and three full-power-years, as compared to 10 years on-orbit and 7.3 full-power-years for the GFS.

The demonstration flight system was designed to achieve a 1375 K reactor coolant outlet temperature with a 65 K core temperature rise. In performing the optimization analyses design rules are used to size the major components based upon an evaluation of the key design-limiting performance characteristics.

The principal objective of the optimization study for this initial SRPS flight system was to minimize the mass of the nuclear subsystems, while satisfying the system performance, control, and safety requirements. The optimization was primarily focused on the mass sensitivities and correlations for the reactor and shield subsystems, and to a lesser degree upon the mass sensitivities for the primary heat transport, and instrumentation and control subsystems.

The nuclear subsystems were designed to incorporate all the basic features of the GFS configuration. However, in several instances it was possible to take advantage of the smaller 30 kWe size to reduce the overall system mass. Thus, for example, advantage was taken of the increased radiative heat loss from the smaller core to accommodate the consequences of the loss of coolant accident without an Auxiliary Cooling Loop. Similarly, the use of internal honeycomb ducts to constrain the movement of the fuel pins is not required and was eliminated. It is also possible to utilize ex-core, rather than in-core safety rods for this design. As a consequence, the control system for the demonstration flight reactor is designed to incorporate a combination of large rotating drums and movable safety rods within fixed reflectors. By eliminating the in-core rods, the fuel mass and core size are significantly reduced.

The key features of the reactor and shield subsystems are shown in Figure 3. Reactor control and shutdown is provided by two independent control systems. The normal startup, shutdown and power operation functions are achieved by rotating the drums to vary the location of the boron carbide segment. Six small-diameter rods are also used in combination with the drums to achieve reactor shutdown. Because of the small size of the core and the lower system lifetime of three-full-power years, only a single enrichment zone was required for the fuel.

The Primary Heat Transport Subsystem (PHTS) piping layout for this design was based on the earlier 30 kWe GFS design study. Four primary loops are utilized for this system with direct coupling to four secondary loops. The primary loop pressure drop was reduced to under 3 psi by increasing the minimum spacing between the core and the reactor vessel wall, and by optimization of the pipe diameters in the primary loop. The instrumentation and control components were scaled from the GFS configuration utilizing simple design relationships.

SPACECRAFT BUS & ELECTRIC PROPULSION

The flight-qualified MMS was chosen to fulfill the function of the spacecraft bus. The MMS is designed for and assigned the following functions:

- Spacecraft attitude control;
- Communications and data handling;
- Housekeeping power;
- Communications link to ground; and
- Mission control.

To complete these functions on the demonstration mission spacecraft, various antennae and a solar array are provided as ancillary components and interfaced with the MMS. The PM-1 propulsion module is included for spacecraft attitude and maneuvering test needs. The full complement of three hydrazine tanks may not be required, but this has not been finalized. The MMS is electronically interfaced with the SRPS controller, data handling systems, and the SRPS housekeeping bus power supply. It also interfaces with the NEP module controller and the two diagnostic packages to perform data processing and transmission. The solar auxiliary power system supplies power to the MMS bus whenever the SRPS is unavailable.

The MMS is supported by the spacecraft primary structure and does not carry any of the SRPS or NEP module loads. Special care is taken to properly integrate the spacecraft structures so as to not interfere with the MMS sensors or attitude control thrusters. The high gain antenna (HGA) is a one meter dish designed to communicate through a geosyncronous communications satellite. For launch packaging constraints the HGA is mounted on the NEP module.

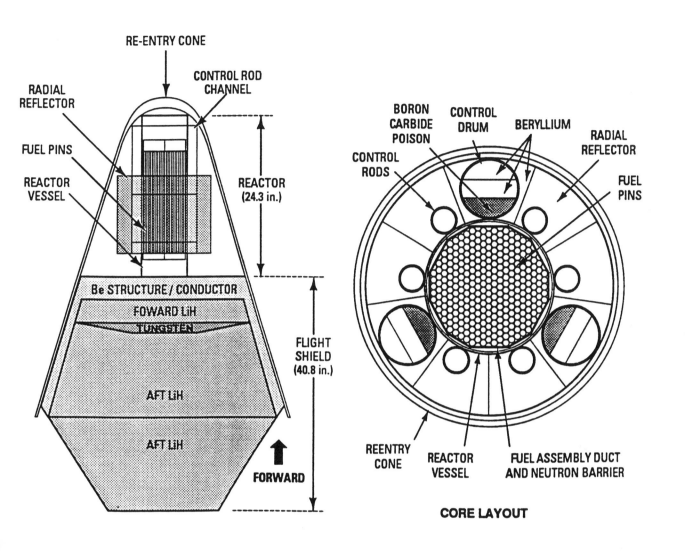

FIGURE 3. Nuclear System for the 30 kWe Flight Demonstration.

The solar auxiliary power system is sized, at 10 square meters, to provide about 750 watts (gross) for spacecraft housekeeping. This auxiliary power source provides mission scheduling flexibility and time for trouble shooting, should it be necessary. To minimize spacecraft mass by reducing the reactor shield cone angle, the deployed solar array is designed to fold back within the shielded region whenever the reactor is providing housekeeping power. This configuration eliminates the SRPS-induced radiation scattering off that part of the solar array which would have been outside the shielded region, and reduces back scatter during SRPS operation. The solar array is mounted on the NEP module for packaging and both thermal and view factor interface constraints with the remainder of the spacecraft.

The electric propulsion module was developed for any of three alternate thruster options, each at about a 27 kWe input power level. The options selected are a hydrogen arcjet, a krypton ion engine, and an ammonia arcjet. The sizing, mass and interfaces for each thruster subsystem were estimated by a propulsion expert and are addressed in the following subsections. A common radiator design, based on a low mass ammonia heat pipe concept (Armijo et al. 1991), is used to reject the power processing unit (PPU) waste heat. Attributes of the 27 kWe thruster subsystems are compared in Table 2.

Krypton ion thruster (Borphy and Garner 1991):

Krypton was selected as the propellant due to concerns over the long term availability of xenon supplies for the space program. The Krypton is stored as a super-critical fluid at about 2500 psi, with a planned supply of 200 days at full power thrust. Fuel supply is not limiting in either volume or mass; a 60% tankage mass fraction was used. A factor of three increase in the krypton supply (20 months at full thrust) could readily be accommodated within a 1 meter diameter tank, and still be within the NEP mass allocation limit.

The ion thruster PPU is more massive and complex than for the arcjet systems. Three PPUs are employed in this design to handle the 27 kWe of SRPS input power. Failure of one of the processors would result in operation at lower thrust, but valuable test data would still be obtained. The thruster size is about 90 cm in diameter and does not require any advancement to the state-of-the-art in grid technology.

Hydrogen arcjet thruster (Curran 1991):

This thruster technology is a present focus for efficient near-earth orbital transfer propulsion. The hydrogen is stored as a subcooled fluid at about 50 to 200 psi, with an available supply of 10 to 25 days at full power thrust. The fuel supply is limited by the volume available on the spacecraft. A flight qualified hydrogen storage tank is not expected to be available to optimally meet the SP-100 demonstration mission requirements (Kroeger 1991). The employed tank is based on the scaling algorithms of Donabedian 1987, and is near the maximum volume which can be integrated with the spacecraft. An estimate of the volumetric heating of the hydrogen by the SRPS-induced radiation field is 3 milliwatts, which is <1% of the calculated total tank heat leakage. Hence, SRPS radiation heating of the stored hydrogen is not a design constraint for the anticipated mission. The PPU is based on a three-phase buck regulator design under development by the Air Force.

Ammonia arcjet thruster :

Rocket Research recently reported the successful test of a 30 kWe ammonia arcjet at a specific impulse (ISP) of >800 sec (AW&ST 1991). For our mission the propulsion system component characteristics are judged essentially the same as the hydrogen arcjet except for the propellant tankage and were further defined from the literature (Karass 1991, Zafran 1989, and Deininger and Vondra 1987). The ammonia is assumed stored at about 100 to 150 psi in a titanium tank. The amount of propellant is constrained by mass rather than volume, as in the case of hydrogen. We have chosen a 1000 hour (40 day) supply, and estimated the tank and distribution mass as 15% of the propellant mass.

TABLE 2. Attributes of NEP Thruster Subsystems.

		Krypton Ion	H2 Arc Jet	Ammonia Arc Jet
ISP	(sec)	4800	1200	>800
Thrust duration	(days)	200[a]	10-25[b]	40[c]
Thrust	(N)	0.72	3->1.2	2.2
Subsystem mass	(Kg)	765	630	1355

a Easily extended to 600 days with 1665 Kg mass
b Volume constrained
c Mass constrained

CONCLUSIONS

An SRPS and spacecraft conceptual design, using a proven spacecraft bus and economical launch vehicle was devised to meet all mission requirements. A test program and diagnostic data requirements were defined that, in conjunction with the ground based qualification program, will confirm the SP-100 technology readiness for mission applications.

The conceptual SRPS spacecraft incorporates an order of magnitude reduction in the adopted DOE/NASA/DOD radiation limits for the GFS user plane interface. A user ability to meet GFS standards, and the use of a larger launch vehicle would allow a significant SRPS mass reduction (400 Kg) for early 30 kWe missions.

An ancillary mission objective to confirm the flight worthiness and integration of an EP module with the SRPS can be effectively accomplished as part of the same SRPS demonstration mission. The mission would accommodate any one of the defined 27 kWe class electric propulsion thruster concepts; hydrogen arc jet, ammonia arc jet, or krypton ion engine.

The SRPS demonstration spacecraft concept, once confirmed, could readily serve as a template for realistic, near term missions involving an SRPS power source and/or electric propulsion.

Acknowledgment

This work was performed under GE AstroSpace Independent Research and Development Project H-31, 1991.

References

Armijo,J.S., et al. (1991) "SP-100 Progress", Proc. of Conference on Advanced SEI Technologies, American Institute of Aeronautics and Astronautics, Cleveland OH, Sept. 4-6, 1991

Aviation Week & Space Technology (1991) page 56, April 8, 1991

Borphy,J. and C.Garner (1991) Private communication, Jet Propulsion Laboratory

Curran,F. (1991) Private communication, NASA, Lewis Research Center

Deininger,W.D. and R.J.Vondra (1987) "Spacecraft and Mission Design for the SP-100 Flight Experiment", AIAA 87-2026, June 1987

Donabedian,M. (1987) "A Computer Program for Analysis of Cryogenic Fluid Storage Systems", AIAA paper 87-1496, June 1987

General Dynamics Commerical Launch Services, Inc, "Mission Planner's Guide for the Atlas Launch Vehicle Family", Rev 2, July 1990

Josloff,A.T., D.Matteo and H.Bailey (1991) "SP-100 System Design & Technology Progress", Proc. 26th Intersociety Energy Conversion Conference, Boston, MA, August 4-9, 1991

Kroeger,E. (1991) Private communication, NASA, Lewis Research Center

NASA S-700-10, Multimission Modular Spacecraft Systems Specification, Rev A, April 1986

Zafran,S. (1989) "Conceptual Arcjet System Design Considerations for the SP-100 Mission", AIAA paper 89-2596, July 1989

POWER BEAMING: MISSION ENABLING FOR LUNAR EXPLORATION

Judith Ann Bamberger
Pacific Northwest Laboratory
P.O. Box 999 M/S K7-15
Richland, Washington 99352
(509) 375-3898

Abstract

This paper explores several beam power concepts proposed for powering either lunar base or rover vehicles. At present, power requirements to support lunar exploration activity are met by integral self-contained power system designs. To provide requisite energy flexibility for human expansion into space, an innovative approach to replace on-board self-contained power systems is needed. Power beaming provides an alternative approach to supplying power that would ensure increased mission flexibility while reducing total mass launched into space. Providing power to the moon presents significant design challenges because of the duration of the lunar night. Power beaming provides an alternative to solar photovoltaic systems coupled with battery storage, radioisotope thermoelectric generation, and surface nuclear power. The Synthesis Group describes power beaming as a technology supporting lunar exploration. In this analysis beam power designs are compared to conventional power generation methods.

INTRODUCTION

Providing continuous power to support exploration and manned activites on the moon for one full lunar-day stay (28 Earth days) presents significant design challenges because of the length of the lunar night. Continuous power to activities on the lunar surface can be supplied using electrochemical energy storage devices, photovoltaic energy, radioisotope decay, or nuclear fission reactors (Synthesis Group 1991). To enhance exploration flexibility on the moon, the ability to distribute power to remote sites is critical. Power beaming is a technology to distribute power to remote sites from the generator source. In the Synthesis Report (1991) power beaming is described as a technology supporting extended lunar exploration. The Synthesis Group (1991) reports "Power beaming for surface-to-surface power distribution may greatly reduce the mass of rovers and other mobile surface systems, assuming line of sight constraints can be met. If nuclear electric propulsion is developed for use in the lunar or Mars cargo vehicle, the orbiting transfer vehicle may be a convenient power source for surface operations. If power beaming can be demonstrated at a reasonable cost, long term development could provide attractive benefits." Several power beaming concepts have been proposed for providing surface base power and remote power for lunar missions. This paper explores several beam power concepts proposed for powering either lunar base or rover vehicles.

EARTH-BASED LASER POWER BEAMING CONCEPT

Energy storage comprises a significant fraction of the mass of a lunar power system such as a solar photovoltaic array. The energy storage requirements associated with the photovoltaic system can be significantly reduced or eliminated by illuminating the arrays by an earth-based laser system. Landis (1991) describes an earth-based system designed to provide power to the lunar photovoltaic system during the lunar night. Rather (1991) also describes an earth-based system for providing continuous lunar power.

The system involves locating large lasers on high-elevation cloud-free sites at multiple ground locations to beam power to the moon. Large lenses or mirrors with adaptive optics would reduce beam spread due to diffraction or atmospheric turbulence. System components are detailed in Table 1. In this application, lasers must operate in or near the visible spectrum in which the atmosphere is nearly transparent.

TABLE 1. Components of a Ground-Based Laser Beamed Power System

Power Requirements on Earth	Power Requirements on Moon
2.5 MW laser power required options • one 2.5 MW free electron laser • 50-50 kW conventional lasers 0.53 μm wavelength	50 kW power required at night

Earth-Based Transmitter	Lunar-Based Receiver
10 m diameter lens or mirror with adaptive optics	750 m^2 GaAs photocells at 20% efficiency

Significant mass savings are derived by reducing the amount of energy storage on the moon. Additional benefits derived from this concept center around the location of the laser power system on earth. This location permits considerable design simplification for the laser power beaming system; terrestrial systems can be easily repaired; therefore, highly redundant systems are not required. Disadvantages include the large transmission distance and the need for adaptive optics to compensate for atmospheric turbulence.

SPACE-BASED LASER POWER BEAMING CONCEPT

Power needs projected for lunar exploration missions on the lunar surface range from a few kilowatts for initial manned outposts and rovers to hundreds of kilowatts for permanent bases and in-situ resource utilization. Operation during the lunar night presents a formidable energy storage challenge for non-nuclear power systems because of the length of those nights. The most massive components of the non-nuclear power systems are those used for energy storage. This excessive mass limits non-nuclear technologies that require energy storage to low power missions. By eliminating or reducing the need for energy storage the competitiveness of non-nuclear power systems with nuclear power systems would be enhanced. Gill et al. (1991) investigates two types of space-based beam powered systems and compares these systems with conventional lunar-based technologies requiring energy storage.

The reference concept for providing lunar surface power is a photovoltaic array (PV) coupled with a regenerative fuel cell (RFC). To supply 100 kW of power during the lunar night, the system mass associated with the regenerative fuel cell is estimated to be 50,000 kg. The photovoltaic array mass is estimated to be 5,000 kg, less than 10% of the total system mass. An alternative to the regenerative fuel cell that involves power beaming includes locating a beam power system at the lunar libration point (L1), a distance of 56,000 km from the moon. Two types of laser systems are analyzed for this application: a solar pumped iodine laser and a free electron laser.

In comparing the lunar surface power system mass for the three systems at surface power levels above 25 kW, both laser systems require less mass on the lunar surface than the photovoltaic/regenerative fuel cell system. At a nighttime surface power level of 100 kW, the PV/RFC system requires 55,000 kg, the PV/FEL system requires 30,000 kg, and the PV/solar-iodine system requires 25,000 kg. When considering the results of this analysis, power beaming offers a lower mass option than fuel cell technologies. This mass advantage may diminish when a beam power system is compared to a nuclear power system sited on the lunar surface; however, other issues such as safety and transmission flexibility may dominate.

Nuclear electric propulsion (NEP) offers synergism with beaming power to the lunar surface. Initially, NEP systems can be used to transport cargo to the lunar surface; after completing this mission, the NEP is stationed at a libration point and power is beamed to the lunar surface. This alternative has been investigated by Pacific Northwest Laboratory (Bamberger and Segna 1991 and Coomes et al. 1991). The laser system design is an AlGaAs laser diode beaming power to a GaAs photovoltaic energy conversion system (Coomes et al. 1990). At L1, a 10 MW nuclear electric power system could provide 5 MW of electrical power for use on the lunar surface.

LUNAR ROVER POWER SYSTEMS

The power system constitutes a significant fraction of the mass of a lunar rover vehicle. Replacing the on-board power system with a receiver and beaming the power to the rover vehicle would significantly reduce the vehicle mass. De Young et al. (1991) presents a conceptual design of a high-power long-duration lunar rover powered by a laser beam power system.

The concept involves three lunar-orbiting, laser diode array power systems, each powered by an SP-100 reactor beaming power to a lunar rover located on the lunar surface. The laser diode array is phase locked to produce a single, coherent beam with a 0.62-m² spot size at the rover vehicle at the maximum transmission distance with a pointing accuracy of 0.2 μ radians. System components are detailed in Table 2.

TABLE 2. Components of a Lunar-Orbit Based Laser Beamed Power System to Power a Lunar Rover Vehicle

Power Requirements on Orbit	Power Requirements on Moon
50 kW laser power required • 3 SP-100 reactors • 1815 km orbit	22.5 kW power required for rover vehicle
Lunar Orbit-Based Transmitter	**Lunar-Based Receiver**
0.8 μm wavelength 8 m² aperture with gas lens 0.2 μ radians pointing accuracy	GaAlAs photovoltaic converter 0.62 m² maximum spot size 23 kg/kW specific mass

Benefits derived from this concept center around the reduced mass and increased mobility of the rover vehicle. The laser power transmission concept provides a lunar power infrastructure that is highly flexible in its ability to provide power to any point on the lunar surface. Disadvantages include the need for the rover to continuously track the beam, even while the rover vehicle is in transit. System enhancements may be obtained by providing some energy storage capacity on the rover vehicle.

Meador (1991) has investigated several power beaming systems that could be used to power lunar surface exploration vehicles (SEV). These vehicles are pressurized, serve a crew of four, and provide a mobile habitat/laboratory with a range of more than 1000 km for lunar exploration. In the analysis, Meador (1991) designs several beam power systems capable of providing 25 kW to the SEV. The lunar orbiting laser beam power stations would be either pumped by the sun or powered by SP-100 nuclear reactors. In comparison to other on-board power systems, the beam power system requires only a photovoltaic energy conversion system which is very light. Competing on-board power technologies and their disadvantages include

- photovoltaic arrays with battery storage - very large for high power
- nuclear systems radioisotope thermoelectric generation (RTG) and dynamic isotope power system (DIPS) - intrinsically low power; plutonium fuel shortage
- regenerative fuel cell coupled with photovoltaic arrays with battery storage - massive energy storage capacity required for operation during the lunar night
- SP-100 on-board (design would be considered a two vehicle SEV, one for exploration, one for the power system) - mobility, maneuverability, mass, and safety problems.

Meador (1991) concludes that laser power beaming is mission enabling for lunar exploration. Two laser technologies are proposed. Photodissociation iodine systems, with t-C_4F_9I as the gaseous iodine lasant are close to being technologically ready for lunar applications. Longer term laser alternatives include electrically pumped laser-diode arrays and laser-diode-pumped neodymium slabs. Optical focusing, pointing, and tracking technologies have been partially developed and tested, and photovoltaic conversion systems are being developed. Research programs to develop these concepts are underway; laboratory databases are being generated, and full-scale system design may be complete by 1997.

CONCLUSIONS

A variety of power beaming concepts have been proposed for providing power to the lunar surface to support extended lunar base operations and surface exploration. The concepts center around laser energy transmission from a remote power system located either on Earth or in lunar orbit. Several laser technologies are under consideration. Each of these concepts provides significant mass savings when compared to power system alternatives sited on the lunar surface because the latter require significant mass for energy storage during the lunar night.

Acknowledgments

This work was performed at the Pacific Northwest Laboratory operated for the U.S. Department of Energy by Battelle Memorial Institute under Contract DE-AC06-76RLO 1830.

References

Bamberger, J. A. and D. R. Segna (1991) "Exploration Mission Enhancements Possible with Power Beaming," in Proceedings of the Eighth Symposium on Space Nuclear Power Systems, CONF-910116, M. S. El-Genk and M. D. Hoover, eds., American Institute of Physics, New York, 1991, Vol. 1, pp 240-244.

Coomes, E. P., J. A. Bamberger, and L. A. McCauley (1990) An Assessment of the Impact of Free Space Electromagnetic Energy Transmission of Strategic Defense Initiative Systems and Architecture, PNL-6932, Pacific Northwest Laboratory, Richland, Washington.

Coomes, E. P., J. E. Dagle, J. A. Bamberger, and K. E. Noffsinger (1991) "An Integrated Mission Approach to the Space Exploration Initiative will Ensure Success," in Proceedings of the Eighth Symposium on Space Nuclear Power Systems, CONF-910116, M. S. El-Genk and M. D. Hoover, eds., American Institute of Physics, New York, 1991, Vol. 2, pp 551-556.

De Young, R. J., M. D. Williams, G. H. Walker, G. L. Schuster, J. H. Lee (1991) "A Lunar Rover Powered by an Orbiting Laser Diode Array" in Proceedings of the Eighth

Symposium on Space Nuclear Power Systems, CONF-910116, M. S. El-Genk and M. D. Hoover, eds., American Institute of Physics, New York, 1991, Vol. 1, pp 253-257.

Gill, S. P., J. Rosemary, and J. Arnold (1991) "Beaming Power Systems for Lunar Surface Applications," in Power Beaming Workshop Supplementary Information for the May 14-16, 1991 Workshop held at Pasco, Washington, PNL-SA-19599, E. P. Coomes, Chairman, Pacific Northwest Laboratory, Richland, Washington, pp C6.1-C6.8.

Landis, G. A. (1991) "Space Power by Ground-Based Laser Illumination," in Proceedings of the 26th Intersociety Energy Conversion Engineering Conference, American Nuclear Society, La Grange Park, Illinois, 1991, Vol. 1, pp 1-6.

Meador, W. E. (1991) "Laser Power Beaming Concepts" in Power Beaming Workshop Supplementary Information for the May 14-16, 1991 Workshop held at Pasco, Washington, PNL-SA-19599, E. P. Coomes, Chairman, Pacific Northwest Laboratory, Richland, Washington, pp C7.1-C7.11.

Rather, J. D. G. (1991) "Project SELENE Original Conception and Project SELENE" in Technology Workshop on Laser Beamed Power, From Earth to the Moon and Other Applications, Proceedings of the February 5, 1991 Workshop, NASA Lewis Research Center, Cleveland, Ohio, pp A-1 to B-20.

Synthesis Group (1991) America at the Threshold -- America's Space Exploration Initiative, The Synthesis Group, U.S. Government Printing Office, Washington D.C.

SPACE NUCLEAR POWER REQUIREMENTS FOR OZONE LAYER MODIFICATION

Thomas J. Dolan
Idaho National Engineering Laboratory
P.O. Box 1625
Idaho Falls, ID 83415-1550
(208) 525-5619

Abstract

This work estimates the power requirements for using photochemical processes driven by space nuclear power to counteract the Earth's ozone layer depletion. The total quantity of ozone (O_3) in the Earth's atmosphere is estimated to be about 4.7×10^{37} molecules. The ozone production and destruction rates in the stratosphere are both on the order of 4.9×10^{31} molecules/s, differing by a small fraction so that the net depletion rate is about 0.16 to 0.26% per year. The delivered optical power requirement for offsetting this depletion is estimated to be on the order of 3 GW. If the power were produced by satellite reactors at 800 km altitude (orbit decay time ~ 300 years), some means of efficient power beaming would be needed to deliver the power to stratospheric levels (10-50 km). Ultraviolet radiation at 140-150 nm could have higher absorption rates in O_2 (leading to production of atomic oxygen, which can combine with O_2 to form O_3) than in ozone (leading to photodissociation of O_3). Potential radiation sources include H_2 lasers and direct nuclear pumping of ultraviolet fluorescers.

INTRODUCTION

Recently the problem of depletion of the Earth's ozone has been receiving increased attention. The atmospheric ozone layer provides a useful service by absorbing most of the "Damaging Ultraviolet" radiation (290-320 nm) before it reaches the earth's surface. Without this protective layer, some animal and plant species would have difficulty surviving, and the incidence of human skin cancer and cataracts would be much greater (National Academy of Sciences 1979). A major effort has been launched to reduce emission of chlorinated fluorocarbons from aerosols and industrial processes. However, many other chemicals contribute to ozone depletion, such as hydrocarbons emitted by ruminants, bogs, and volcanoes. These sources will be difficult to curtail. It is worthwhile investigating the possibility of increasing the ozone production rate in atmosphere. The purpose of the present work is to estimate the power requirements for counteracting ozone depletion in the stratosphere using photochemical processes driven by space nuclear power.

OZONE DYNAMICS

Production

Ozone is produced mainly by the reaction

$$O + O_2 + M \rightarrow O_3 + M \tag{1}$$

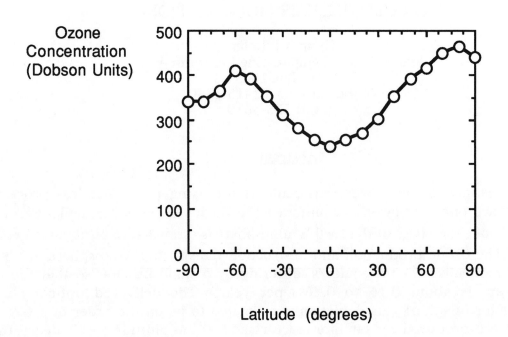

FIGURE 1. Variation of Longitudinally-Averaged Ozone Concentration with Latitude.

where M is a third body required to carry away energy released by the combination reaction. The free oxygen atoms are produced mainly by solar radiation

$$O_2 + h\nu \rightarrow O + O \qquad (2)$$

in the stratosphere (altitude 11-50 km), and by

$$NO_2 + h\nu \rightarrow NO + O \qquad (3)$$

in the troposphere. (The troposphere is that part of the atmosphere below the tropopause, the height of which varies from 8-18 km, with an average value of about 11 km).

Ozone Distribution

According to April 1966 satellite data (Warneck 1988, Figure 3-2) the time-averaged ozone concentration in the northern hemisphere had three peaks at 70° north latitude and 0°, 110°, and 280° longitude. In the southern hemisphere (October 1970 data) there was one broad peak at 60° south latitude from 140° to 200° longitude. The concentrations varied from about 240 Dobson units at the equator to about 500 Dobson units at the peaks. (One Dobson unit = 2.69×10^{20} molecules/m^2.) In order to estimate the total amount of ozone present in the atmosphere, we have computed the longitudinal-average concentration from those data. The result is shown in Figure 1. Taking into account the cosinusoidal variation of land area with latitude, we calculate the global average ozone concentration for this case to be 321 Dobson units, which is about 1.34 times the value at the equator.

The ozone concentration also varies with altitude z above the earth's surface. We have made an approximate fit to the low-latitude experimental data (which have ± 25% variation about the median):

$$n \approx n_{max} \exp[-(z-z_{max})^2/a^2] \qquad z > 10 \text{ km} \qquad (4a)$$
$$n \approx n_o \qquad z < 10 \text{ km} \qquad (4b)$$

where $n_{max} = 3 \times 10^{12}$ molecules/cm^3, $z_{max} = 26$ km, $a = 9$ km, and $n_o = 0.5 \times 10^{18}$ m^{-3} (based on data from Figures 3-3 and 3-5 of Reference 2). Integrating this low-latitude data vertically, we find

$$\int_0^\infty dz\, n(z) \approx 6.84 \times 10^{22} \text{ molecules/m}^2 = 254 \text{ Dobson units}, \qquad (5)$$

which is consistent with Figure 1. Assuming the global average value is 1.34 times as great, the total quantity of ozone in the earths atmosphere is given by

$$N \approx 1.34(4\pi R^2) \int_0^\infty dz\, n(z) = 4.7 \times 10^{37} \text{ molecules (ozone)}, \qquad (6)$$

where $R = 6.378 \times 10^6$ m is the earth's radius (perfect sphere model).

Ozone Destruction

It is desirable to minimize the ozone loss, in order to protect the world ecosystem and human health. Ozone is destroyed by photodissociation

$$O_3 + h\nu \rightarrow O_2 + O, \qquad (7)$$

(the free oxygen can produce more ozone, however), and by collisions with free oxygen

$$O + O_3 \rightarrow O_2 + O_2. \qquad (8)$$

The dominant destruction process in the stratosphere is now known to be

$$\begin{aligned} X + O_3 &\rightarrow XO + O_2 \\ XO + O &\rightarrow X + O_2 \\ \text{net:} \quad O + O_3 &\rightarrow O_2 + O_2, \end{aligned} \qquad (9)$$

where the catalyst X represents H, OH, NO, Cl and Br compounds. Ozone is also lost from the stratosphere by transport downward to the troposphere, where additional destruction processes occur.

Modeling the ozone concentration involves following the atmospheric life cycles of molecules based on hydrogen, carbon, nitrogen, and halogens, including interactions with solar radiation, plants, soil, and water. These molecule groups control the concentrations of ozone-destructive catalysts. About 150 chemical reactions and photochemical processes involving 50 different species must be accounted for as functions of altitude, latitude and time. The approximate destruction rates of ozone by various catalyst processes are shown in Table 1.

Improved knowledge of reaction rate coefficients is gradually improving the ozone depletion estimates. According to a recent NASA Report (Watson et al. 1990), "The analysis of the total-column ozone data from ground-based Dobson instruments show measurable downward trends from 1969 to 1988 of 3 to 5% . . . in the Northern Hemisphere (30° to 64° N latitudes) in the winter months that cannot be attributed to known natural processes." This represents an average ozone depletion rate $\lambda = 0.16$ to 0.26% per year. Here we assume $\lambda \approx 0.21\%/\text{year} = 6.7 \times 10^{-11}/\text{s}$.

Table 1. Globally Integrated Stratospheric Loss Rates of Ozone by Various Reactions (from Warneck 1988 Table 3-4.)

Catalyst	Loss Rate 10^6 mole/s
O	21.1
ClO	15.2
NO$_2$	27.5
OH	17.6
Total	81.4 = 4.9×10^{31} molecules/s

PROSPECTS FOR INCREASING THE OZONE LAYER

Reducing the Depletion Rate

Ozone preservation may be accomplished by affecting either the source or depletion rates, using either chemical or electrical means. The main chemical technique for reducing the depletion rate would be international regulations on production of some ozone-destroying chemicals. If CFC production were cut by 90-100% in the year 2000, the chlorine concentration in the atmosphere would gradually decrease from about 5 ppb (2000) to about 2 ppb (2070), primarily by interaction with tropospheric OH, and the ozone depletion rate would be significantly reduced.

We could also attempt to reduce the ozone destruction rate electrically, by ionizing or chemically altering the catalysts which destroy ozone, such as chlorine. The density of chlorine is about two orders of magnitude lower than the ozone density, so the required energy is potentially lower. However:

1. There are many different catalysts (H, OH, NO$_2$, ClO, BrO, ...). Although it may become technically feasible in the future to tune the energy sources to multiple frequencies to remove or deactivate many of these ozone-destroying catalysts, a way to do so with low energy cost is not yet apparent;

2. Efficient, tunable uv lasers do not exist; and

3. Even if a catalyst were altered (for example, by ionizing chlorine), the energy supplied to alter a molecule does not remove it from the atmosphere. Later, it might change back to its original form or to other hazardous forms by various chains of reactions.

A technique more immediately available would be to produce ozone electrically in the stratosphere, in an effort to increase the source term.

Ozone Production

Ozone can be produced by electrical discharges in air. (Some O$_2$ molecules become dissociated into atomic oxygen, which then combines with O$_2$ to form O$_3$.) In the stratosphere we might use energy from a nuclear-powered or solar-powered satellite to enhance the dissociation of O_2 molecules. In order to have an orbit decay time of 300 years, the satellite would need to have an initial altitude of about 800-900 km. The energy would have to be transmitted down to the stratosphere, possibly by the following means:

1. As a beam of ultraviolet photons from a nuclear pumped lamp or laser beam, which could dissociate O_2 molecules;

2. As a beam of radio waves at about 1.4 MHz, which could heat electrons by cyclotron resonance in the earth's magnetic field; and

3. By power-beaming to secondary stations in the stratosphere where the resulting electrical power could be used to produce ozone by an electrical discharge or by uv photon generation.

Assuming that the energy can be tuned so that it dissociates O_2 but does not appreciably destroy O_3, we will estimate how much energy would be required in order to make a significant contribution to global ozone production. The tuning problem is left for future study.

Energy Required to Replace Destroyed Ozone

The destruction rate of ozone molecules was shown in Table 1. The rate of change of the total quantity of ozone in the atmosphere may be represented by a simple equation

$$dN/dt = S - D \approx -\lambda N, \tag{10}$$

where the source rate S and destruction rate D are both about 4.9×10^{31} molecules/s (Warneck 1988 Tables 3-1 and 3-5). The slight difference between these rates gives rise to a gradual depletion of the ozone. The power required to counteract this depletion is

$$P = W_3 (-dN/dt) = W_3 \lambda N \tag{11}$$

where W_3 = the average energy required to produce one ozone molecule. Since dissociation of one O_2 molecule produces two oxygen atoms, which can produce two ozone molecules, the minimum energy required under ideal conditions would be about half the dissociation energy, or 2.6 eV. However, some of the atomic oxygen is lost via other reactions, so not all of it produces ozone. Some of the delivered power may also contribute to ozone destruction, so the effective energy requirement per net ozone molecule produced will be higher than the minimum. Here we will assume that $W_3 \sim 6$ eV. The delivered optical power required to counteract the assumed ozone depletion rate is

$P \sim$ (6 eV/molecule)(1.6×10^{-19} J/eV)(6.7×10^{-11}/s)(4.7×10^{37} molecules) = 3 GW.

If the efficiency of coupling electrical power to chemical reactions (via an ultraviolet lamp, laser beam, electrical discharge, or radiowaves) were on the order of 30%, then the electrical power required would be $P_e \sim 10$ GW_e. Alternatively, electrical power production requirements could be minimized by using a nuclear pumped laser or fluorescer (Prelas et al., 1988).

Satellite Power Plant Requirements

If nuclear reactors were used for a satellite power station, then the required radiator area would be

$$A = P_e (1-\eta)/\eta \, e \, \sigma \, T^4 \qquad (12)$$

where P_e = electrical power (W), e = surface emissivity, η = efficiency of converting thermal to electrical power, $\sigma = 5.67 \times 10^{-8}$ W/m^2 K^4, and T = radiator temperature (K). To produce $P_e = 10$ GW$_e$ with thermionic cells at $\eta = 0.1$, with T = 1000 K and e = 0.8, the required total radiator area A = 2×10^6 m^2.

If photovoltaic cells with efficiency η were used for satellite power stations, then the required area with incident heat flux q would be

$$A = P_e/\eta q. \qquad (13)$$

For an incident solar flux q = 1400 W/m^2 at efficiency $\eta = 0.1$, the solar panel area required to produce 10 GWe would be A = 7×10^7 m^2, a factor of 35 higher than the radiator area for reactors with thermionic cells. Such a large area of photovoltaic cells would be cumbersome to launch, install, and maintain.

CONCLUSION

If an effective means of coupling electrical energy into ozone production could be achieved, the ozone depletion rate could be slowed significantly by about 3 GW of delivered optical power. If space nuclear power with 10% conversion efficiency were used, the required area for heat rejection radiators would be on the order of 2×10^6 m^2. Use of nuclear-pumped ultraviolet fluorescers should also be investigated.

Acknowledgements

I am grateful to D. M. Woodall and D. Buden for helpful discussions. This work was partially supported by the U.S. Department of Energy, Office of Energy Research, under Contract DE-AC07-76IDO1570.

References

Miller, A. S. and I. M. Mintzner (1986) "The Sky is the Limit: Strategies for Protecting the Ozone Layer", World Resources Institute Research Report #3.

National Academy of Sciences (1979) "Protection against Depletion of Stratospheric Ozone by Chlorofluorocarbons," Washington DC.

Prelas, M. A., F. P. Boody, J. F. Kunze, and G. H. Miley (1988) "Nuclear Driven Flashlamps," *Lasers and Particle Beams*, 6(1): 25-53.

Warneck, P. (1988) *Chemistry of the Natural Atmosphere*, Academic Press, San Diego.

Watson, R. T., M. J. Kurylo, M. J. Prather, and F. M. Ormond (1990) "Present State of Knowledge of the Upper Atmosphere 1990: an Assessment Report," NASA Office of Space Science and Applications, Reference Publication 1242.

**COMMERCIAL POWER FOR THE
OUTPOST PLATFORM IN ORBIT**

Thomas C. Taylor, President
William A. Good, Director of Marketing
Global Outpost, Inc.
335 Paint Branch Drive
College Park, MD 20742-3261
Phone 301-314-9858

Charles R. Martin
20438 Meadow Pond Place
Gaithersburg, MD 20879
Phone 301-590-8737

ABSTRACT

GLOBAL OUTPOST, Inc., in cooperation with NASA, proposes to place a Space Shuttle External Tank (ET) in orbit for commercial purposes. The OUTPOST External Tank derived platform discussed in this paper is designed to service a commercial market and will also be available for governmental use. The platform requires power for its operation and will sell power to customers as one of 68 different commercial services offered. A number of alternatives for power systems are under study. The company expects to purchase a power subsystem from the available alternatives. The proposed facility will require 4 to 10 KW of power and remain in orbit for five or more years. The plan will combine private sector funding with NASA assistance to create a business location in orbit. While the ET normally falls just short of orbital velocity, it can be placed in orbit with little additional energy. The platform proposed is a minimum cost, man visited facility aimed at attracting commercial ventures seeking a low cost way to use the attributes of space.

INTRODUCTION

GLOBAL OUTPOST plans to create a simple commercial space service platform in orbit for industry and government. The concept salvages the External Tank of the Space Shuttle and is anticipated to be in orbit in the early to mid 1990's. Initially, it is a very simple platform placed in orbit by the space shuttle and providing power, communications and other subsystems. Figure 1 depicts the ET with the added trusswork which will comprise the OUTPOST platform. This platform will be compatible with, but separate from the Space Station.

GLOBAL OUTPOST, Inc. and NASA signed an Enabling Agreement in 1990 for the use of five external tanks as low earth orbit space platforms conditioned on meeting safety and engineering requirements and demonstration of a sound business program for its use. NASA (1990) This agreement was an outgrowth of the commercial space initiatives developed during President Reagan's Administration. Preliminary engineering for the platform has been completed by the Martin Marietta Corporation under contract to GLOBAL OUTPOST, Inc. In exploring the commercial markets for such a platform it became apparent that part of the OUTPOST Truss could be used as a space power testbed and accommodate several different types of power. The platform is available to act as a load source for the power testbed. The company is willing to provide power system testbed locations in orbit and credit the power system owner for the power used by the platform. The platform services are available to commercial customers, government agencies and others. The platform could be used for research, long term technology testing storage, EVA assembly tests, flight operations tests, service/maintenance tests, robotic demonstrations, transportation operations and logistics unloading dock.

GLOBAL OUTPOST continues to work within the NASA Enabling Agreement to work out the technical details related to the placement of an External Tank in orbit and the assembly of the first OUTPOST Platform. Senior NASA officials charged with the responsibility of implementing a more aggressive agency program in support of commercial space activity have been helpful in assisting the company break down the barriers for a commercial service platform.

Reducing the cost of transportation from the surface to orbit will be one of the major accelerators of mankind's expansion into the space frontier. Fletcher (1988) Another major accelerator will be low cost commercial operations in orbit. The commercial development of each segment of the space business emerges as the costs and technology required for each specific industry reach a certain enabling point. The sequence of development is predictable. First, communications satellites, second, commercial remote sensing, third, commercial launch vehicles and next is commercial services in orbit. Bishop (1988)

Copyright © 1991 by GLOBAL OUTPOST, Inc. Published by others with permission

POWER REQUIREMENTS

The company has hired the Martin Marietta Corporation to confirm the technical feasibility of the proposed venture. The GLOBAL OUTPOST Power Subsystem has been studied in several forms and a number of different power alternatives appear possible with existing technology. The minimum power requirements for the platform are as follows:

- Minimum Power Required 4 KW
- Existing Flight Qualified Subsystem
- 6 each Nickel-Cadmium Batteries
    - 110 Amp-Hour Capacity (each)
    - Discharge Rate 22 Amps (each)
    - Weight 230 lbs. (each)

	MASS (kg)	AREA (sf)	Dev. Unit $/M ea.	Production Unit $/M
SOLAR (PV)	1000	1440	none	10
SOLAR DYN.	1800	60	270	30
NUCLEAR	1300	14	260	70
RTG	1300	25	130	125
BEAMED	700	7700	NO EST.	NO EST.

Table 1 Power Alternatives

POWER ALTERNATIVES

The power requirements are based on assumptions about potential users of the GLOBAL OUTPOST Platform. Unless the platform were used as a testbed for electric propulsion, the power requirements are not likely to exceed 10kWe. The reasonable alternatives include photovoltaic cells coupled with batteries, solar dynamic, a small nuclear reactor, radioisotope thermoelectric generators (RTG) and power beaming. Each option has advantages and disadvantages.

Solar is currently available, but the arrays are large resulting in significant drag at these low altitudes. Also the low duty factor for solar at low altitudes implies high battery requirements if power must be available continuously. A small reactor has the advantage of capacity for growth, small area (low drag), 100% duty cycle and no pointing requirements. However, nuclear will be expensive to develop, expensive for each operational unit, there will be resistance to using reactors at Shuttle orbital altitudes and they will have a negative impact on scientific missions studying x or gamma radiation. The only RTG in this power class is the Dynamic Isotope Power System (DIPS). This system is still in development, but past RTG's have had high reliability and, for this power range, looked technically superior to solar power and reactor power for many missions. However, there are some non-technical drawbacks for the DIPS. The large fuel loading of Plutonium (40-50 kg) which would be required for this power level raises questions about fuel availability, cost and public safety. Finally, power beaming offers some of the advantages of nuclear without some of its disadvantages. A reactor, or a constellation of reactors at a respectably safe altitude could beam power at 300 GHz to customers at ranges up to 600 km. The reactor operator assumes all of the development and deployment costs and risks and simply offers power to a number of customers on a reimbursable basis. The disadvantage is that unless the system is co-orbital (and in the present case it would not), a single reactor would result in a low and irregular duty factor. Table 1 compares the key characteristics for these systems.

Given the criteria, it would appear that solar is the clear choice for this platform; however, there are some additional considerations. It might be possible to combine an arcjet demonstration experiment with a small reactor flight test experiment. The high development costs of such a reactor system would be largely borne by the government in this case, but could be partially amortized by selling useful power to customers once the platform reaches final operational orbit. The significant issue is reactor start-up at parking orbit altitude. It might be acceptable to simply buy power from a reactor power supply owned and operated separately and fed thru a tether as opposed to being beamed. In this case the reactor and the GLOBAL OUTPOST Platform would operate at the same altitude. The issue here is continuous reactor operation at low orbit. A final hybrid possibility is to use an arcjet engine in concert with a reactor and tether feed. Taking advantage of gravity gradient stabilization, it may be possible to raise the system orbit without dynamic instabilities. This will be the subject of a later paper.

BACKGROUND

The utilization of the External Tank in orbit was suggested as early as the mid 1970's. Dr. Gerard K. O'Neill (1976) suggested the use of the ET as reaction mass for mass drivers in space in a 1976 book called The High Frontier: Human Colonies in Space.

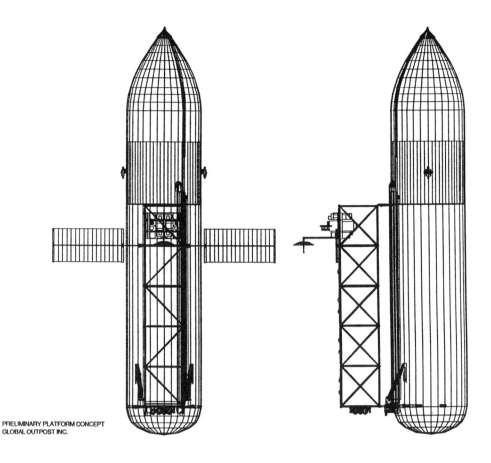

Figure 1 Front and Side View of the OUTPOST Platform Showing the Solar Arrays and Subsystem Package Used to Salvage the External Tank in Orbit.

In most remote surface facility developments on the earth's surface, power is a major consideration. Taylor (1981) It will be the same in orbit and on the planets and moons to be developed in the solar system. On the Alaska North Slope in the mid 1970's, for example, the camp used various forms of power in the development of the North Slope Oil Development Facilities. The oil companies used the partially refined fuel in the form of naptha cracked from crude oil pumped from the field. This power source was used to burn in the vehicles and for heat. It was crude oil that was the reason for the development and the most logical source of immediate power in Alaska.

Mankind will not be as fortunate to find an easy source of power on the Moon or Mars. The nights are 14 days long. Mankind must take a reliable power source from earth to these new remote outposts on the Moon and Mars. Each pound of cargo transported to the north slope cost the oil companies and their customers about a $ 1.00 per pound to fly it into the Arctic Slope. The cost of cargo to the Moon will be 10,000 to 100,000 times as much. This means our power source must be low weight. It means mankind must begin to develop these power systems now and qualify the hardware in space.

In a remote base on the surface of earth, where transportation costs are high, the economic forces stimulate many cost reduction techniques including the use of packing containers as tools sheds. Taylor (1981) The reuse of external tank in orbit has a parallel to Mr. Taylor's North Slope experience. The nuclear power systems that propel mankind to the nearby planets may also be the power plants required for the surface bases.

Taylor (1979) approached Martin Marietta Corporation (MMC) with some ideas for the utilization of the ET in orbit. Some of the concepts were too far in the future and evolved into surface based projects. Spencer (1980) His near term ideas resulted in a contract in 1979 with MMC, Advanced Programs near New Orleans, LA, which outlined a number of potential uses of the ET. Taylor (1979) The platform was suggested as a load source for the power testbed and other uses. In similar contracts in the early 80's, Taylor (1980), Taylor (1981) and Taylor (1982a and 1982b) assisted MMC in several areas of ET Applications in orbit. These studies resulted in a variety of papers. Mr. Taylor's contracts with other aerospace organizations also studied the ET in orbit. Taylor (1984), Hedgepath (1984) In 1980, Taylor (1980) presented "Commercial Operations for the External Tank in Orbit" at the Eighteen Goddard Memorial Symposium in Washington, D.C. This began the evolutionary development of the OUTPOST Concept which emerged almost ten years later as an unmanned commercial service platform.

THE OUTPOST PLATFORM

The External Tank offers an opportunity for a cost effective platform for commercial operations in low earth orbit. GLOBAL OUTPOST, Inc. proposes a simple platform placed in orbit with NASA's cooperation. The use of the ET in orbit started with President Reagan's commercial space policy and evolved to an Expression of Interest submitted to NASA. Reagan (1988), CBD (1988) A number of technical groups continue to assist the company through the signed agreement and the technical discussions on the assembly of the platform are currently proceeding with NASA-Johnson Space Center in Houston and the Marshall Space Flight Center in Huntsville, AL. NASA (1990)

POWER IN SPACE

In the development of facilities in remote locations on the earth surface, it is power that paces the development. In the jungle it was diesel fuel trucked 200 miles into the jungle. In Alaska, it was crude oil cracked slightly to produce a volatile form of liquid energy you could not spill on your hands. In space it must be much more innovative. The OUTPOST Platform will probably start with Solar Arrays, but the amounts of power required by the evolving operations really require much more power than a simple solar array system will produce.

A COMMERCIAL SPACE PLATFORM OPPORTUNITY

A series of simple commercial services are possible using the ET after providing the basic subsystems to the ET in orbit. The services are the low end of the complexity spectrum. Some are precursors to more complex research and experimentation that could be performed on the Space Station.

POTENTIAL ACTIVITIES ON PLATFORM IN ORBIT

The following orbital operations or services could potentially be offered.

1. Docking of Commercial Launch Vehicles
2. Storage & Maintenance of Launch Vehicles
3. Tether Re-Entry Service
4. Payload Handling and Storage
5. Protected Storage of Cargo
6. Consolidation of Cargo from Surface
7. OMV and OTV Service and Docking
8. Cryogenic Propellant Transfer & Storage
9. Electrical Power, Water, Thermal Maintenance & Other Consumables
10. Cargo on Demand Rates to Other Orbits including GEO
11. Satellite Servicing and Repair
12. Tether Deployment of Vehicles/Payloads to Higher Orbit
13. Large Space Structures Assembly, Construction, Testing and Deployment
14. Free Flyer Platform Outfitting & Refurbish
15. LDEF Type Tray Rent, Active and Passive

The introduction of services at the platform is expected to be a phased development program. Services will be market driven.

TECHNICAL DETAILS

The ET is the largest and heaviest (when loaded) element of the space shuttle. NASA (1975), NASA (1988) The External Tank is manufactured at the Michoud Assembly Facility near New Orleans, LA, by the Martin Marietta Corporation under contract with NASA through the Marshal Space Flight Center, Huntsville, Alabama. The External Tank is discarded after every flight. The External Tank attains ~98% of full orbital velocity on a typical shuttle flight and can be taken to orbit with little extra energy.

FUTURE EXTERNAL TANK UTILIZATION

Someday more of earth inhabitants will live off the planet than on it. Sounds impossible, but it is like Columbus saying more Europeans will live in the new world than in Europe some day. Eventually, the Space Exploration Initiative will happen and it will require extensive transportation infrastructure. The External Tank in orbit is the first of a series of resources required for mankinds exploration and later migration off the planet. Much of the drive that has pushed mankind to explore and develop new resource locations has been economic in nature in the past. This economic resource driven exploration is likely to continue and the ET can evolve in the future into a node for transportation services. The ET Scenario is broad enough to stimulate creativity and innovation within the Free Enterprise System. It has generated a variety of other concepts including aerospikes, various platform configurations and science platforms. Taylor (1982), Crain (1984) An ET in orbit is likely to accelerate the innovation and creativity in and around the core platform.

CONCLUSIONS

The OUTPOST platform is a man visited, low earth platform derived from the External Tank. The system requirements include those needed to convert a discarded ET in orbit into an orbital platform. Power is anticipated to be 4 KW minimum. The platform can be a power testbed in orbit. Additional subsystems are to be added incrementally to the platform to expand its capability, satisfy emerging market requirements and to generate revenue. If the costs can be kept low, then more commercial organizations can afford to start space research and technology development work in orbit.

The far term opportunities of the platform include the transportation node potential of the OUTPOST concept. Taylor (1986) The power required could approach 50 KW. It would use a tether technique to enhance the efficiency of space launch vehicles and the ET weight offered by the ET would provide a mass heavy platform able to effectively enhance most space vehicles. Adequate power could open up new forms of propulsion and permit expansion into new types of industries. A space power testbed could prepare mankind for the challenge of remote bases on other bodies in our solar system. These far term opportunities require additional research, much more power and most of the tether research must be performed in orbit to be effective. This research activity could stimulate creation in the future of the equivalent of a harbor or an airport in orbit with all the economic implications that such advanced systems imply. The transportation node aspects of the platform may provide economical services in future projects on the moon and beyond.

The bases placed on other bodies in our solar system will require power systems capable of operating in the 14 day long nights on the Moon and locations on Mars that can not be resupplied even in emergencies in under 9 months. The camp in Alaska hurt when the weather cut it off for 5 days, imagine a base on Mars where a new part for the power system is critical to your survival. It is not too early to start testing the space power systems required. OUTPOST, it is believed, can play a role in a program of innovative government and industry cooperation to help commercial organizations as well as government to participate in the space science, microgravity research and technology development plus position this country in the international space commerce of the future. Power should be the focus of a major technology development program and the OUTPOST Platform would like to play a part in the development and testing of power systems in space. The platform is prepared to become a space power testbed and provide incentives to technology development managers testing power subsystems on the platform. External Tanks, as selectively employed in the OUTPOST concept, can become an additional tool to expand space power and commercial activities in space in the 1990's.

Acknowledgments

This work was performed by GLOBAL OUTPOST, Inc. under internal funding support. The company has funded studies with Martin Marietta, at the Michoud Assembly Facility and Eagle Engineering Inc. in Houston, Texas. The company acknowledges the support from the SDI Organization through a funded study on a power technology orbital test on the platform.

REFERENCES

Bishop, Dr. Peter C. (1988) Space Business, Space Business Research Center, Houston, TX 77058-1090, market data

Crain, W.K., Fleming, J., Ganoe, W.H. and Taylor, T.C. (1984) "Performance Enhancement of the Space Transportation System," Lausanne, Switzerland IAF, IAF-84-08. Proposed modification to the External Tank of the Space Transportation System to provide payload weight enhancement.

Commerce Business Daily (1988) Special Notice, Wed., June 1, 1988, "NASA Support for President's Commercial Space Initiative Involving Private Sector Use of Space Shuttle's External Tanks," Page 32, Issue No. PSA-9602

Fletcher (1988) President's Reagan Commission on Space Report, National Commission on Space, Washington, DC, Isolates the cost of transportation to orbit as barrier and ET as a resource

Hedgepeth, J.M., Mobley, T.B. and Taylor, T.C. (1984) "Construction of Large Precision Reflectors Using the Aft Cargo Carrier," Lausanne, Switzerland IAF, IAF-84-389. Proposed using Aft Cargo Carrier to carry Large Deployable Reflector and backup structure.

NASA (1975) System Definition Handbook, Nov, Space Shuttle External Tank, Martin Marietta

NASA (1988) National Space Transportation System Reference, Volume 1, Systems and Facilities, June, NASA, Page 59

NASA (1990), Enabling Agreement, Rev. 1, Signed 13 Feb. 1990 leads to five ET's's in orbit with Launch Services Agreement Negotiations expected to start early 1992 for a 1995-6 launch by space shuttle

O'Neill, Gerard K. (1976) The High Frontier: Human Colonies in Space

Reagan, R. (1988) "The President's Space Policy and Commercial Space Initiative to Begin the Next Century," Feb. 11, 1988, Page 3, Fact Sheet

Spencer, J. (1980) "Space Shuttle External Tank Habitability Study," Space Systems Development Group, Study Director, LA, Eight ET-LH2 tanks in a Torus

Taylor, T. C. (1979) "ET Applications in Space," Study Contract for Martin Marietta, Denver Aerospace, Michoud Operations Advanced Programs, Dec, Taylor & Associates, Inc. Generation of ET Concepts

Taylor, T.C. (1980) "Commercial Operations for the External Tank in Orbit," Eighteenth Goddard Memorial Symposium, Washington, D.C., AAS 80-89, March 1980. Proposed an ET Derived Commercial Service Platform in orbit; similar to the OUTPOST concept.

Taylor, T. C. (1981) Personal Experience 1975-1979, Civil Engineer - Construction Supervisor, North Slope Oil Development Facilities, Prudhoe Bay, Alaska, various technical papers on remote harsh facility construction environments

Taylor, T. C. (1982a) TPS Inspection Mission Contract, Consultant to Martin Marietta Aerospace, Michoud, LA, ET Mission alternatives obtaining TPS samples in orbit.

Taylor, T.C. (1982b) "Orbital Facility Operations Through an Assured Market Scenario," Paris IAF, IAF-82-33. Proposed prepurchase of orbital services.

Taylor, T.C. (1983) "The External Tank Scenario: Utilization of the Shuttle External Tank for the Earth to Mars Transit," Taylor & Associates, Inc. Mars Conference, Boulder, CO, Proposed an ET Derived Commercial Service Platform in orbit as jump off for Mars Mission

Taylor, T.C. (1984) NASA-Ames Research Center study "Large Deployable Reflector (LDR) Concepts Using the Aft Cargo Carrier (ACC)," Taylor & Associates, Inc., R. Bruce Pittman, NASA Study Director, Explored LDR configurations in ACC to orbit.

Taylor, T.C. (1986) "The Transportation Node Platform in Orbit," AFSC/NSIA Cost Reduction and Cost Credibility Workshop, Denver, CO. Proposed a Tether Platform based on SBIR Air Force Aerospike-Tether research by Taylor & Associates, Inc.

Tewell, J.F. (1981), Anderson, J.W. of MMA and T.C. Taylor "A Commercial Construction Base Using the External Tank," AIAA-0460, 2nd AIAA Conference on Large Space Platform, Feb. 2-4, 1981, San Diego, CA, Taylor & Associates, Inc., Applied remote construction base techniques to orbit

Witek, N. J. (1980) and Taylor, T.C. "The External Tank as a Large Space Structure Construction Base," IAF Tokyo, Japan, IAF-80-A41

EXPERIMENTAL DEMONSTRATION OF AUTOMATED REACTOR STARTUP WITH ON-LINE REACTIVITY ESTIMATION

Kwan S. Kwok, John A. Bernard, and David D. Lanning
Nuclear Reactor Laboratory
Massachusetts Institute of Technology
138 Albany Street
Cambridge, MA 02139
(617) 253-4211/4202/3843
FAX (617) 253-7300

Abstract

A generic method for performing automated startups of nuclear reactors described by space-independent kinetics under conditions of closed-loop digital control was developed, implemented, and tested on the 5-MWt MIT Research Reactor. The technique entails first obtaining a reliable estimate of the reactor's initial degree of subcriticality and then substituting that estimate into a model-based control law so as to permit a power increase from subcritical on a demanded trajectory. The estimation of subcriticality is accomplished by application of the 'Perturbed Reactivity Method' which was developed in the course of this research. The shutdown reactor is perturbed by the insertion of reactivity at a known rate. Observation of the resulting period permits determination of the initial degree of subcriticality. A major advantage to this method is that repeated estimates are obtained of the same quantity. Hence, statistical methods can be applied to improve the quality of the calculation. In addition to describing the perturbed reactivity method, information is given on the selection and architecture of the digital computers and associated instrumentation utilized to conduct the automated startups. Also presented are experimental results in which the efficacy of this technology for the performance of automated reactor startups was demonstrated.

INTRODUCTION

This paper reports the development, implementation, and testing of a generic method for performing automated startups of nuclear reactors under conditions of closed-loop digital control. A unique feature of the technology described here is that the reactor's initial degree of subcriticality is determined on-line at the outset of the startup and then incorporated in a model-based control law so as to allow a rise to power on a demanded trajectory. This capability is of particular importance for spacecraft reactors because they will be launched in a shutdown condition and not made critical until placed in orbit (Bennett and Schnyer 1991). The research reported here on automated reactor startup required that three related but separate tasks be performed. These were the design and installation of the Advanced Control Computer System (ACCS) which consists of the digital hardware and instrumentation needed to demonstrate automated startup technology, the theoretical derivation of the 'perturbed reactivity method' which is the technique developed here for the on-line estimation of subcriticality, and the experimental demonstration of the technology on the 5-MWt MIT Research Reactor. Each of these tasks is described in the ensuing sections of this paper. The paper begins with a brief overview of existing approaches to the startup of spacecraft reactors.

EXISTING APPROACHES TO SPACECRAFT REACTOR STARTUP

The startup of nuclear reactors that are deployed in earth orbit has of course been extensively studied in both the United States (ROVER/NERVA Nuclear-Powered Rocket Program (Strait and Hohmann 1966)) and the Soviet Union (TOPAZ-2 Advanced Soviet Space Nuclear Power System

(Nickitin et al. 1991 and Makarov et al. 1991)). As regards the ROVER/NERVA program, reactor power was adjusted by proper positioning of the control devices. For rapid increases of power, this was done closed-loop in response to the difference between the observed and demanded power signals. For automated startups, this was done open-loop by moving the control devices along a preset trajectory. Such operation was restricted to one particular set of initial conditions because it was assumed that rotation of the drums that served as the control devices by a prescribed amount would cause the neutronic power to rise by about one decade, not create an exceedingly short period, and result in the reactor's being slightly below critical. Control would then be transferred to a closed-loop power comparator circuit which would bring the system to full power. The utility of this approach was limited by the range of initial conditions to which it was applicable. If those conditions were not met, then the predetermined trajectory used for the open-loop portion of the cycle might not produce the anticipated result. Operation of the TOPAZ-2 is also achieved through a combination of open and closed-loop actions. The startup sequence of the reactor is as follows. Once the required orbital altitude has been achieved, electrical power to the control drive system is turned on by radio command from Earth. The startup of the reactor consists of adding reactivity at a rate of not more than 2.5×10^{-4} $\Delta K/K/s$ by ground-command control of the rotation of the control drums. This rate of reactivity insertion is maintained until criticality is reached at which point it is decreased by a factor of ten. This reduced rate of reactivity insertion is then continued until 10 kW of thermal power is attained. At this time, an analog regulating circuit is turned on to control the thermal power automatically. The thermal power regulator is programmed to operate between 10 kWt and 150 kWt, and to provide a desired electrical output of up to 6 kWe.

The technology reported in this paper on automated reactor startup differs from both that of the ROVER/NERVA and the TOPAZ-2 programs as well as from earlier work done on this topic at the Massachusetts Institute of Technology (Bernard et al. 1991) in that the reactor's initial degree of subcriticality is determined on-line as part of the startup process. This feature both enhances safety by providing an independent corroboration of the anticipated effect of control device movement and facilitates the use of model-based control laws by generating accurate reactivity estimates.

ADVANCED CONTROL COMPUTER SYSTEM

Since the late 1970s, MIT has been engaged in a program to develop and demonstrate advanced techniques for the instrumentation and control of nuclear reactors. In 1981, in partnership with the Charles Stark Draper Laboratory, an LSI-11/23 mini-computer was purchased and installed in the reactor control room. This single computer was used to perform all required control functions including data acquisition, signal validation, supervisory control, calculation of the actuator signal, and signal output. It was recognized in 1988 that this system would not suffice for research on automated startups. Problems included the need to monitor reactor power over many decades of operation, the ever-increasing demand for high numerical throughput, and the desirability of separating software related to safety from that associated with control. The former is virtually invariant while the latter is subject to frequent change. Accordingly, the Advanced Control Computer System or ACCS was designed and installed (Kwok et al. 1991). It consists of five interconnected computers, linked in a multiple-computer/single-task architecture. Figure 1 is a block diagram showing the configuration and purpose of each component. These are:

1. <u>Rack-Mount 80386</u>: This data acquisition IBM-AT computer is assigned three major tasks. First, it collects data from a maximum of thirty-two sensors, performs signal validation on the collected data, outputs the validated information to up to four other computers, and displays the validated information on the console CRT monitor. Second, it computes the maximum allowed control signal using the supervisory reactivity constraint algorithm (Bernard et al. 1988) as well as limits of other MITR technical specifications, receives the requested control signal from the other computers, compares that signal with the one calculated by the supervisory algorithm, outputs the more conservative signal to the control rod motors, and displays the control

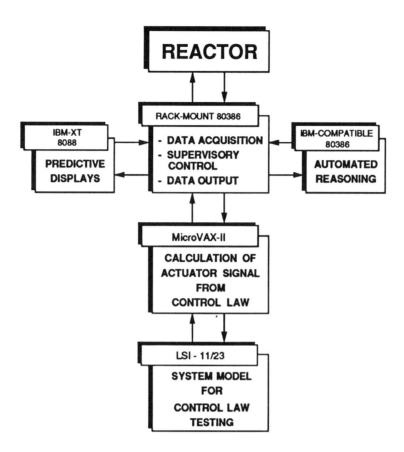

FIGURE 1. Configuration and Functions of MITR-II Advanced Control Computer System.

decision on the screen. This computer's third function is to write the desired data to the permanent disk. Changes to this computer's software are rare.

2. <u>MicroVAX-II</u>: The VAXstation II/GPX is a machine dedicated for intensive floating-point computations. Engineering and control calculations such as are required for the MIT-SNL minimum time control laws (Bernard et al. 1990), are performed on this machine. The MicroVAX-II receives validated information on the reactor from the data acquisition system (Rack-Mount 80386). It then calculates the demanded control signal from whatever control law is being tested, and exports that signal to the data acquisition system for output to the control rod motors. Changes in this system's software are expected to be frequent.

3. <u>IBM-Compatible 80386</u>: This is a high-speed machine on which computer programs are first edited, compiled, and finally linked to form an executable module. This machine is capable of supporting automated reasoning using PROLOG, LISP, or C. It is designed to be compatible in all details with the Rack-Mount 80386 data acquisition system.

4. <u>IBM-XT 8088</u>: This computer's role is to receive validated signals from the data acquisition computer and to display model-based predictive information (Lau et al. 1988) or a safety parameter display on its screen.

5. **LSI-11/23**: This unit is connected to the MicroVAX-II for the purpose of providing an independent machine on which a model of a reactor can be run. This permits new controllers to be programmed on the VAXstation II/GPX and tested against a simulation model running on the LSI-11/23 prior to the performance of actual closed-loop runs on the reactor. This approach has the advantage that new programs are tested under realistic conditions. In particular, signals must be passed between two computers as is done for actual implementations. Previously, the new control law and the model ran on the same unit.

Integration of components within the data acquisition computer was accomplished through a passive back plane which is basically a non-intelligent bus that allows only lines such as data, status, and timing to be passed. Integration of the five separate computers was achieved through use of RS-232 serial communication.

In addition to the above five computers, the ACCS was equipped with a broad-range power sensor that spanned about seven decades. This was essential for the performance of automated startups because the digital system would have to monitor power levels that ranged from a few watts at source level to 5 MW at full power. Accordingly, a boron-lined, gamma-compensated ion chamber which correctly indicated the reactor's neutron flux level from full shutdown to full power was made available. The output of this chamber is connected to a KEITHLEY Model 485 auto-ranging picoammeter which is equipped with a Model 4853 IEEE-488 interface. This unit is basically a 4-1/2 digit (4 significant digits with sign) auto-ranging picoammeter with seven DC current ranges. The design of this system allows the Rack-Mount 80386 data acquisition system to receive an on-scale reading of the output of the neutron-sensitive compensated ion-chamber in a digitized form. This makes the system less prone to electrical interference or noise than if a signal were first obtained in analog form and then converted for digital operation. A further advantage is that the output of the Model 485 is digitized.

In summary, the ACCS's multiple-computer/single-task architecture with auto-ranging picoammeter offers several advantages over single computer systems. These include separation of safety-related software that is in finished form from control software that is under development, transmission of both the scale and reading of a power instrument so as to permit operation over many decades, the acquisition of digitized signals directly from the neutron sensors, and the performance of interactive simulations.

AUTOMATED REACTOR STARTUP

The basis of the technique described here for automated startups is the 'Perturbed Reactivity Method,' which was developed for the estimation of subcriticality (Kwok 1991). The insight that led to this method was the realization that comparison of the observed reactor period in a perturbed shutdown reactor with the period calculated to exist in a similarly perturbed critical reactor would provide a means of estimating the degree by which the real reactor was actually shutdown. The mathematical method used to implement the perturbed reactivity method does not explicitly make this comparison. But it does depend on making a perturbation of known magnitude to the subcritical reactor and observing the resulting instantaneous period. Once the initial degree of subcriticality has been determined, it is substituted into a model-based control law, such as the alternate MIT-SNL Period-Generated Minimum Time Control Law (Bernard et al. 1989), and that law is then used to perform the power ascension on a demanded trajectory. The degree of subcriticality is updated at every time step in order to account for any feedback reactivity or reactivity that is inserted by causes that are unknown to the digital controller. Thus, automated startups are a two-step process. Both steps require the use of a model of the reactor dynamics. The perturbed reactivity method uses the model to generate an expected response from a known change in reactivity. The control law uses the model to determine the actuator signal needed to cause the reactor power to move along the demanded trajectory. The combination of the two permits automated startups in which reactor power is raised on a demanded trajectory.

Perturbed Reactivity Method

The perturbed reactivity method was developed, as part of the research reported here, to provide an accurate means for estimating a reactor's initial degree of subcriticality. This technique, which is applicable to reactors described by space-independent kinetics, entails perturbing a shutdown reactor by the insertion of reactivity at a known rate and then estimating the initial degree of subcriticality from observation of the resulting reactor period. This technique is in certain respects similar to inverse kinetics (Sastre 1960). However, it differs in two important aspects. First, the perturbation must be made at a known rate. Second, the net reactivity, $\rho(t)$, present in the reactor is treated separately using superposition as:

$$\rho(t) = \rho_{ukn}(t) + \rho_{kn}(t) \tag{1}$$

where $\rho_{ukn}(t)$ is the reactivity present in the core excluding the reactivity associated with the known perturbation, and

$\rho_{kn}(t)$ is the reactivity associated with the known perturbation.

The quantity ρ_{kn} would normally be generated by moving a calibrated control device connected to a digital controller. Also, it would normally be computed by means of a balance using data from previously performed calibrations. It is this separation of the net reactivity into two components that makes the perturbed reactivity method useful for automated startups. Specifically, there are no significant reactivity feedback mechanisms in a subcritical core. Hence, in a shutdown reactor, the quantity ρ_{ukn} will be the initial degree of subcriticality and it will remain constant during the startup. Thus, repeated estimates can be made of this quantity and signal smoothing techniques can be applied to improve accuracy.

The perturbed reactivity method requires a model of the reactor dynamics for its implementation. Thus far two have been considered. These are the point kinetics equations and the alternate dynamic period equation (Bernard et al. 1988). Both accurately describe reactors characterized by space independent kinetics. The difference between the two is that the rate of change of reactivity is explicitly represented in the latter. The advantage of the second approach is therefore that it results in a more accurate estimation of the unknown reactivity during transient conditions under which reactivity is inserted at a substantial rate. This occurs because the effect of the perturbation is apparent in the numerical implementation of the model based on the dynamic period equation.

When the perturbed reactivity method is implemented with the point kinetics model, the unknown reactivity is given by:

$$\rho_{ukn}(t) = l^*\omega(t) + \overline{\beta} - \rho_{kn}(t) - \frac{l^*}{T(t)} \sum_{i=1}^{N} \lambda_i C_i(t) - \frac{l^*}{T(t)} Q(t). \tag{2}$$

When implemented with the alternate dynamic period equation as the model, the unknown reactivity is given by:

$$\rho_{ukn}(t) = \left\{ \omega(t)(\overline{\beta} - \rho_{kn}(t)) + l^*\left[\dot{\omega}(t) + (\omega(t))^2 + \lambda'_e(t)\omega(t)\right] - \right.$$

$$\dot{\rho}_{ukn}(t) - \dot{\rho}_{kn}(t) - \lambda'_e(t)\rho_{kn}(t) - \sum \overline{\beta}_i\left(\lambda_i - \lambda'_e(t)\right) -$$

$$\left. l^*\left[\lambda'_e(t)Q(t) + \dot{Q}(t)\right]/T(t) \right\} / \left[\lambda'_e(t) + \omega(t)\right] \tag{3}$$

where the quantity λ_e' is the alternate multi-group decay parameter which is defined as:

$$\lambda_e'(t) \equiv \sum \lambda_i^2 C_i(t) / \sum \lambda_i C_i(t) \qquad (4)$$

and where other symbols are defined as:

- l^* is the prompt neutron lifetime,
- $\omega(t)$ is the inverse of the dynamic reactor period,
- $\bar{\beta}$ is the effective delayed neutron fraction,
- $T(t)$ is the amplitude function and is a weighted integral of all neutrons present in the core,
- N is the number of groups of delayed neutrons, including photoneutrons,
- λ_i is the decay constant of the ith precursor group,
- $C_i(t)$ is the concentration of the ith precursor group normalized to the initial power,
- $Q(t)$ is the effective source strength,
- $\dot{\omega}(t)$ is the rate of change of the inverse of the dynamic reactor period,
- $\bar{\beta}_i$ is the effective fractional yield of the ith group of delayed neutrons,
- $\dot{\rho}_{kn}(t)$ is the rate of change of the known reactivity, and
- $\dot{Q}(t)$ is the rate of change of the effective source strength.

Simulation studies were performed to evaluate the efficacy of the perturbed reactivity method. Figure 2 shows the results of a simulated automated startup in which the reactor was initially subcritical by 2000 mβ. The power was to be increased from 1 kW to 1 GW on a demanded period of 0.30 s. The perturbed reactivity method (Equation (3)) was used to estimate the reactivity and the alternate MIT-SNL Period-Generated Minimum Time Control Law (Bernard et al. 1990) was used to generate the control signal. No limitations were placed on the available rate of insertion or removal of reactivity. Shown are the resulting power profile, the net reactivity present in the core, and the estimate generated by the perturbed reactivity method of the initial reactivity that was present in the core. This latter curve is labeled 'Initial Reactivity.' As is evident from the figure, this simulated transient was completed satisfactorily. (Note: The spike present in the estimate of the initial reactivity at the moment of transient termination was the result of the numerical method associated with calculating prompt effects.) Apparent are the characteristic features of the MIT-SNL laws including rapid adjustments of reactivity both to initiate and to terminate the transient. Of significance is that the perturbed reactivity method correctly estimated the initial reactivity that was present in the reactor as –2000 mβ. Moreover, it provided this estimate throughout the entire transient.

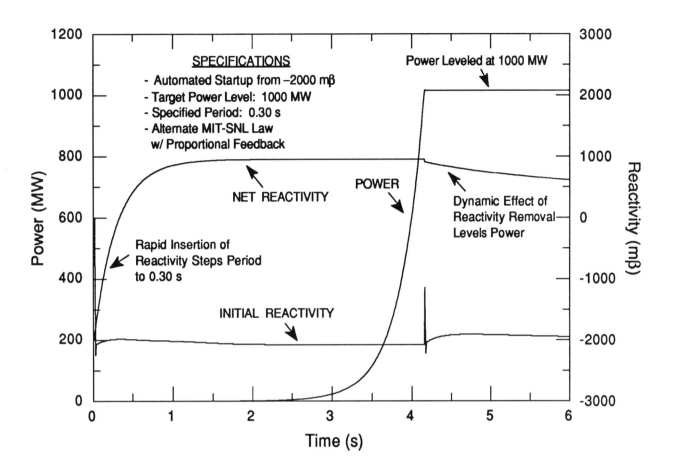

FIGURE 2. Simulation Study illustrating Use of Perturbed Reactivity Method for an Automated Startup from -2000 mβ on a 0.30-s Period.

Experimental Demonstration

Implementation of the perturbed reactivity method requires that the source term be accurately characterized. If, as is valid for the MIT Research Reactor, a distributed photoneutron source can be assumed then its strength can be calculated using:

$$Q_o = -\frac{T_o}{l^*}\rho_o \tag{5}$$

where symbols are as previously defined except that the subscript (o) denotes the shutdown condition. This relation was used to obtain the source strength for a variety of power histories. Once a sufficient number of calculated source values and power histories had been accumulated, an empirical correlation was developed that gave the source strength as a function of the power history. In this way, it was not necessary to know the initial degree of subcriticality (ρ_o) in order to estimate the source which was in turn needed to implement the perturbed reactivity method.

Experimental evaluations of the perturbed reactivity method in conjunction with both automated startups and other applications were conducted on the MIT Research Reactor from 28 December 1990 to 8 April 1991. A total of one hundred and seventy-five separate closed-loop runs were

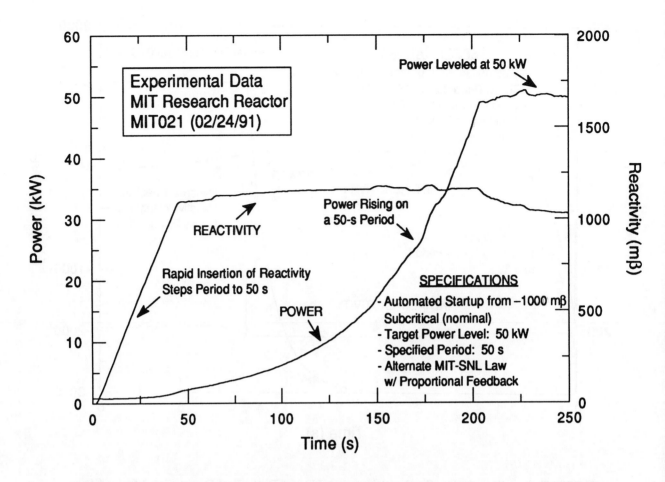

FIGURE 3. Demonstration of Perturbed Reactivity Method for an Automated Startup with Reactor Initially Subcritical by 1000 mβ.

performed in nineteen sessions during this time interval. The primary objective of the experiments was to evaluate the perturbed reactivity method and in particular to show its application to automated startups in which the initial degree of subcriticality was not known. As part of these experiments, sensitivity studies were conducted to compare signal smoothing techniques, to evaluate numerical methods, and to assess the efficacy of the reactor models used to implement the perturbed reactivity method. Automated startup experiments were then performed in which reactor power was increased from source level to 50 kW on a 50-s period. Experiments were initiated with the reactor exactly critical and then repeated with the reactor subcritical by a greater amount for each successive run. Automated startups were also done with the reactor initially in a fully shutdown ($-$ 8350 mβ) state. The technique was shown to be both effective and reliable.

Figures 3 and 4 are from one of the experiments that was performed to evaluate the perturbed reactivity method for automated startups. Shown in Figure 3 are the power and reactivity profiles obtained during an automated startup in which the reactor was initially subcritical by 1000 mβ. The reactivity plotted is that inserted by the control device rather than the net reactivity. The power was leveled at the demanded value of 50 kW and the demanded trajectory, a 50-s period, was attained. Figure 4 shows the same power profile together with the estimate of the initial degree of subcriticality. As is evident from this figure, the perturbed reactivity method correctly determined that quantity throughout the transient.

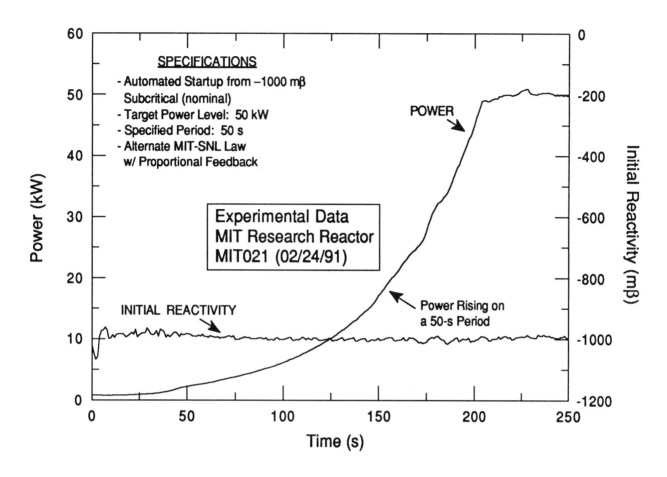

FIGURE 4. Use of Perturbed Reactivity Method to Estimate Initial Reactivity Present for Automated Startup shown in Figure 3.

CONCLUSIONS

A generic method for the automated startup of reactors described by space-independent kinetics has been developed and implemented on the 5-MWt MIT Research Reactor. An important aspect of this technology is the perturbed reactivity method which allows a reactor's initial degree of subcriticality to be determined on-line during the startup. This information is then supplied to a model-based control law which in turn permits the power ascension to be achieved on a demanded trajectory. The perturbed reactivity method could be used, as was reported here, for automated startups. Alternatively, its role could to confirm the results of existing reactivity estimation practices such as balances.

The research reported here should be of interest to those designing both spacecraft reactors and terrestrial multi-modular power plants. Relative to the former, current practice has been to use either predetermined sequences of control signals or ground-based commands. As mission complexity increases and power demands rise, more autonomous methods of operation will be essential. Relative to the latter, licensed human operators will remain in charge. However, reliability can be improved and costs minimized through the incorporation of a high degree of automation in these facilities. The significance of the present work to both spacecraft and terrestrial reactors is that all necessary components for the performance of automated reactor startups have been identified, assembled, and demonstrated on an actual reactor under conditions of closed-loop digital control.

Acknowledgments

Appreciation is expressed to Ms. Georgia Woodsworth and Mr. Shing Hei Lau who assisted with paper preparation, and to Mr. Ara Sanentz who performed the layout and prepared the graphics. This research was supported by the U.S. Department of Energy under Contract No. DE-FG07-90ER123.90.

References

Bennett, G. L. and A. D. Schnyer (1991) "NASA Mission Planning for Space Nuclear Power," in *Proc. 8th Symposium on Space Nuclear Power Systems*, CONF-910116, M. S. El-Genk and M. D. Hoover, eds., American Institute of Physics, New York, NY, pp. 77-83.

Bernard, J. A., A. F. Henry, and D. D. Lanning (1988) "Application of the 'Reactivity Constraint Approach' to Automatic Reactor Control," *Nucl. Sci. Eng.*, 98(2): 87-95.

Bernard, J. A., K. S. Kwok, P. T. Menadier, F. V. Thome, and F. J. Wyant (1990) "Experimental Evaluation of the MIT-SNL Period-Generated Minimum Time Control Laws for the Rapid Adjustment of Reactor Power," in *Space Nuclear Power Systems 1988*, M. S. El-Genk and M. D. Hoover, eds., Orbit Book Co., Malabar, FL, pp. 495-508.

Bernard, J. A., K. S. Kwok, T. Washio, F. J. Wyant, and F. V. Thome (1991) "Experimental Demonstration of the MIT-SNL Period-Generated Minimum Time Control Laws for Rapid Increases of Reactor Power from Subcritical Conditions," in *Space Nuclear Power Systems 1989*, M. S. El-Genk and M. D. Hoover, eds., Orbit Book Co., Malabar, FL.

Kwok, K. S. (1991) *Automated Startup of Nuclear Reactors: Reactivity Estimation, Computer System Development, and Evaluation*, Ph.D Thesis, Department of Nuclear Engineering, Massachusetts Institute of Technology, Cambridge, MA, June 1991.

Kwok, K. S., J. A. Bernard, and D. D. Lanning (1991) "Design, Assembly, and Initial Use of a Digital System for the Closed-Loop Control of a Nuclear Research Reactor," in *Proc. 6th Intersociety Energy Conversion Engineering Conference (IECEC)*, held in Boston, MA, August 1991, Vol. 5, pp. 7-12.

Lau, S. H., J. A. Bernard, K. S. Kwok, and D. D. Lanning (1988) "Experimental Evaluation of Predictive Information as an Operator Aid in the Control of Research Reactor Power," in *Proc. American Control Conference*, held in Atlanta, GA, June 1988, Vol. 1, pp. 214-220.

Makarov, A. N., M.S. Volberg, G. M. Grayznov, E. E. Zhabotinsky, and V. I. Serbin (1991) "The Operating Regimes and Basic Control Principles of SNPS 'TOPAZ,'" in *Proc. 8th Symposium on Space Nuclear Power Systems*, CONF-910116, M. S. El-Genk and M. D. Hoover, eds., American Institute of Physics, New York, NY, pp. 662-665.

Nickitin, V. P., B. G. Ogloblin, A. N. Luppov, N. N. Ponomarev-Stepnoi, V. A. Usov, Y. V. Nicolaev, and J. R. Wetch (1991) "'TOPAZ-2' Thermionic Space Nuclear Power System and the Perspectives of Its Development," in *Proc. 8th Symposium on Space Nuclear Power Systems*, CONF-910116, M. S. El-Genk and M. D. Hoover, eds., American Institute of Physics, New York, NY, pp. 631-635.

Sastre, C. A. (1960) "The Measurement of Reactivity," *Nucl. Sci. Eng.*, 8: 443-447.

Strait, B. G. and G. L. Hohmann (1966) "Automatic Startup for Nuclear Reactor Rocket Engines," in *Proc. 1st IFAC Symposium on Automatic Control in the Peaceful Uses of Space*, Plenum Press, New York, NY, pp. 415-421.

SP–100 CONTROLLER DEVELOPMENT PARADIGM

Carl N. Morimoto, Jaik N. Shukla, John A. Briese, and Akbar Syed
GE Aerospace, SP–100 Programs
P.O. Box 530954
San Jose, CA 95153–5354
(408) 365–6600

Abstract

To facilitate the development of the Space Reactor Power System (SRPS) controller, a rapid prototyping and multi–phased development methodology is being utilized. The rapid prototyping environment used in the development models both the controller and the system being controlled. Since the validation of the SRPS control strategies is a long lead activity to ensure the required safety and control features, the SRPS controller development is carried out in phases, starting with normal modes of operation and followed by transient and off–normal modes. In every phase, the rapid prototyping of the control strategies is used (1) to establish well–defined controller requirements, (2) to perform fast identification of changes and refinement of the strategies, and (3) to conduct in–phase correction and optimization of the strategy and component development. This approach allows the validation of the component controllers individually and evaluation of the effect of the component integration to identify functional and safety issues in a timely manner. This SRPS controller development paradigm can significantly reduce the development risks and lowers total cost by providing a solid foundation for requirements definition and a coordinated optimization of hardware and software.

INTRODUCTION

The objective of the SP–100 Space Reactor Power System (SRPS) project is to develop safe and reliable nuclear power system for various space applications. The instrumentation and control (I&C) subsystem of the SRPS is critical to the operation of the SRPS. The I&C subsystem controls the reactor by adjusting the reactor radial reflectors in response to power demands from the controller. The adjustments are controlled by a combination of preprogrammed reactivity additions (during start–up) and reactor outlet temperature. The controller provides the temperature setpoints for start–up, normal operation, and shutdown sequences, provides the signals to the control drive motors, receives and conditions signals from sensors in the nuclear subsystems, maintains the reactor power at the demand value by moving the reflectors to maintain reactor temperature at the setpoint value, and provides the automatic shutdown and the end–of–mission shutdown function.

The development of the controller involves designing and producing controllers whose target components are still undergoing technology development. However, the controller development cannot be delayed until the completion of the SRPS components because the controller development requires a long lead time to implement a reliable system and to coordinate in–phase development of controller and components in the different stages of their development.

To develop a reliable SRPS controller, a development paradigm is being utilized that uses a rapid prototyping methodology and a phased development approach. Both of these approaches are well established and are in wide use. When applied together, they are particularly suited for the development of the SP–100 SRPS controller, where requirements (such as system and various components, controller hardware, and control strategies requirements) are not static. The Power Autonomous Control Environment (PACE) laboratory, which contains the computer technology (hardware and software) platform for rapid prototyping, has been established to perform this development work. The initial configuration of the prototyping platforms is operational, with capability (functional and performance) upgrades being planned incrementally as component controller hardware becomes available.

This paper describes the technology platform established for the prototyping and the development approach that applies the prototyping to the four development phases of the SRPS controller.

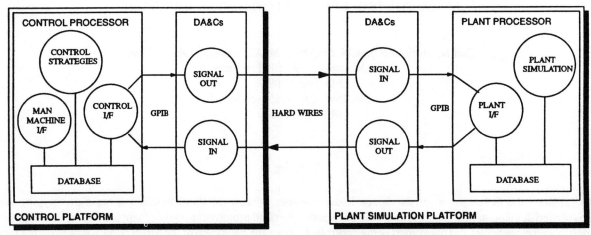

FIGURE 1. Prototyping Platform Functional Block Diagram.

PROTOTYPING PLATFORMS

The simplified architecture of the prototyping platform is show in Figure 1. This configuration consists of two platforms linked by hard wires between the Data Acquisition & Control (DA&C) units. The platforms are:
- **CONTROL PLATFORM** – implements the process logic for the SRPS controller and its component controllers (This platform provides the simulation of controllers and man machine interface (MMI) including control inputs and monitoring displays that are needed to operate the rapid prototype processes.); and
- **PLANT SIMULATION PLATFORM** – simulates the interfacing units (such as the actuators and sensors) and the process system being controlled (such as reactor, heat transport systems, and associated components).

The Control Processor and Plant Processor in Figure 1 are Sun SPARCserver 4/470 workstations with GPIB links to the Hewlett Packard HP3852 DA&C units.

The technology selected to support these platforms' functions includes:
- **Real–Time Expert System**, which provides a knowledge base to accumulate the control strategies being prototyped, object and rule based processing, MMI tools (include menus, buttons, and displays) and simulation of all inputs and output signals including control, feedback, and sensor signals;
- **Relational Database Management System**, which provides the common data structure to be shared among all application software; and
- **ARIES Simulation programs**, which provides detailed simulation of various systems such as a Generic Flight System (GFS) and Nuclear Assembly Test (NAT). This simulation program is also being developed under the SP–100 program.

The prototyping platform is initially a generic, non–specific tool. As a particular component or system is rapid prototyped, the platform acquires its functionality and evolves toward the fully functional simulator of both the controller and plant processor. The prototyping platform will be used in four arrangements as shown in Figure 2.
- Configuration A – allows controller/plant simulation execution with the communication between the controller and plant simulation being handled by the database (This configuration is used to rapid prototype the controller rules and procedures where actual routing of the data signal through hard wire is not needed.);
- Configuration B – allows full controller/plant simulation execution with input/output data routed through DA&C hard wires, which provide data signal transmission similar to the "real" cases;
- Configuration C – allows the verification of the fabricated controller operation using the plant simulator prior to testing on the actual plant system or components; and
- Configuration D – allows the operation of the plant system or components using the Control Platform to control the operation, where the controller strategies can be fine tuned on an actual system.

The chief uses of the prototyping platforms are for the development of the control strategies and the refinement of the process control rules.

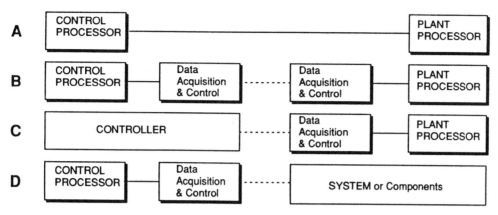

FIGURE 2. Different Configurations for using Prototyping Platforms.

PARADIGM

The SRPS controller interfaces with a complex set of technologies and components, which include sensors, static and moving parts (such as motors, clutches, bearings and resolvers) in a high temperature and radiation environment. The controller and associated I&C electronics must also operate in a radiation environment. Figure 3 shows the I&C interfaces between the reactor controller and the sensors and control drive assemblies. The design and interface specifications involving these components are part of the SP–100 technology development. Since the design of plant system/components and controller components are not static, the method of developing a controller from a priori system definition for these complex system and interfaces does not work. The requirements change during the technology development and become moving targets in the development effort. The method of iterative requirements analysis and development has been demonstrated to be a successful approach to accommodating changing requirements.

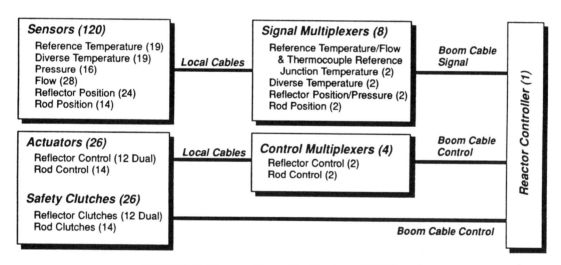

FIGURE 3. Reactor Controller Portion of I&C Interfaces.

The paradigm applies the elements of rapid prototyping and multi–phases to the SRPS controller development. The approach for each of the multi–phases consists of the steps described below:
- Start with the idealized model of the plant system/components and the high level controls specified in the preliminary design specifications;
- Expand the preliminary control strategies in term of manual control steps;
- Develop an executable prototype of the manual controls, based on the preliminary control strategies;
- Review the execution to determine whether the plant model behaves as expected;
- Change either the manual control strategy or the model of the plant system/components being controlled (if the plant model is not behaving the way anticipated);

- Update the manual control strategies and the plant model as required;
- Develop an executable prototype of the automatic controls, based on the previously proven manual control prototype;
- Refine further the simulation model when more details become known or change the prototype to reflect the requirement changes and/or concept changes; and
- review and analyze the execution for re–analysis of requirements.

Through iterative prototyping and analysis, the control requirements and control strategy prototypes evolve toward more detailed and reliable products.

The four–phase development approach coordinates the controller software development and controller hardware development such that the prototyping platforms can be incrementally upgraded to integrate the controller hardware being developed in–phase with the control strategy development. Since the SRPS control strategies cannot be all generated at the start of the controller development, the control strategies will be developed by building a new strategy on top of the previously prototyped strategies in four phases as described below:

- PHASE 1 – Develop and verify software definition of integrated strategies for NORMAL SRPS conditions, where the strategies include autonomous start up, standby, and operating conditions;
- PHASE 2 – Develop and verify software definition of strategies for SAFETY, SCRAM AVOIDANCE, and OFF–NORMAL conditions, where the strategies include coverage for failure of sensors and actuators;
- PHASE 3 – Develop and verify strategies for SIGNAL VALIDATION and COMMAND VALIDATION, where command validation determines if a command is valid for the state of the SRPS and is safe to implement; and
- PHASE 4 – Develop and verify software definition of strategies for PLATFORM FAULT TOLERANT mode, where the strategies perform failure isolation and management.

The prototyping of the control strategies for each of the development phases establishes well defined controller requirements and enables fast identification of changes and refinement of the strategies in the next phase when the introduction of more strategies may change existing conditions or introduce new conditions on the execution of the existing controls.

The prototyping of the control strategies in concurrence with the development of SRPS components also allows in–phase correction and optimization of the strategy and component. The knowledge base will be built up as individual control strategies are validated. Generation of this knowledge base in–phase with the development of other SRPS components will minimize the frequency of discovery of performance and safety issues later in the development cycle and ensure that all significant safety issues are identified in a timely manner.

Specific examples of the advantages of using this development approach are described below for some significant development issues and safety and reliability concerns.

Periodic Evaluation

Top level safety and control strategies need to be evaluated, expanded, and periodically re–evaluated by using prototyping platforms. Requirements need to be revised to ensure safety and performance requirements and changes in design features or performance resulting from component development programs must be examined periodically to ensure integrated system performance. In the component development program, the development work is being conducted at many locations in the country. The development is based on preliminary requirements and design choices are being made by individual developers which may impact integrated system performance. Without periodic evaluation of integrated system performance, based on up–to–date design and performance data of the components at a given time in development cycle, there is a risk that when developed components are put together they can not be integrated to meet the performance, safety, and reliability requirements. Rapid prototyping of the SRPS using the prototyping platform will provide a snapshot of integrated SRPS performance at any point in time during the technology development cycles. Periodic evaluation allows optimization and correction of component development by addition, deletion, modification, or reallocation of development and interface requirements.

Signal Validation

The paradigm will be applied to develop and test control strategies to conduct signal validation and diagnostics in real time to avoid unnecessary scrams using the prototyping platform with analytic sensor models and plant simulation models to detect and compensate for sensor noise, drift, and degradation. On-board SRPS sensor calibration is limited and over the life of the SRPS, the sensors will drift and degrade. The issue is further complicated by EMI effects on the boom cable or other random noise. Control strategies must be developed to validate these data based on signal heuristics, current mode of the SRPS, value of other measured variables, value of analytical sensors, and simulation model predictions.

Scram Avoidance

The paradigm will be applied to develop and test diagnostic and recovery strategies for SRPS component failures and performance degradations and for special operations (for example, exercising control drives) to enhance the life of SRPS. The overriding SRPS requirement is safety. In the absence of a controller knowledge base of scram avoidance strategies, the controller will default to shutdown and the SRPS component duty cycle may increase beyond what the developed component can accommodate. This will result in either reduced SRPS lifetime or additional unplanned technology development. Control strategies may also enhance the SRPS life, provided they are validated in a timely manner and appropriate features are included in the identified components and test data obtained during the component development. One such feature is periodic exercising of the control rod drives to reduce the possibility of self welding. Therefore, in-phase controller development is an integral part of achieving required SRPS lifetime.

In order to satisfy the high mission availability requirement, the SRPS must be designed to operate with failed and degraded components during its lifetime. Therefore, the controller knowledge base must include diagnostic and recovery (from failures and degradation) strategies. Implementation (automatic or manual) of these strategies will place requirements on other SRPS components (new sensors, sensor performance, SRPS system configurations, etc.) which directly impact the component's technology development program. Lack of adequate coverage (the conditional probability that the system will recover given that a fault has occurred) will result in an unreliable SRPS that does not meet requirements. Likewise, recovery time must be fast to avoid adversely impacting the SRPS availability and mass of auxiliary power needed for restart.

Controller Software Quality

In order to have a reliable controller with acceptable software defects requires a long lead activity involving early and on-going software requirements definition and development. Prototyping platforms will be used to define software requirements and develop and test software modules to meet SRPS operating and safety requirements. The SRPS controller prototyping platform provides an efficient and effective tool for systematically testing and correcting software errors to reduce latent defects. Out-of-phase controller development does not allow sufficient time to reduce software latent defects to an acceptable level. Excessive latent defects must be corrected to avoid low mission availability and a potentially unsafe SRPS. The SRPS controller software definition must be an on-going process, in phase with the development of the rest of the SRPS technologies to develop flight quality software with an acceptable number of latent defects.

CONCLUSION

The SRPS controller is critical to nuclear safety and SRPS lifetime and reliability. The development and validation of controller strategies utilizing a methodology of rapid prototyping and feedback evaluation in multiple phases can significantly reduce the development risks and lowers total cost by providing a solid foundation for requirements definition and a coordinated optimization of hardware and software. Furthermore, SRPS controller performance and reliability are enhanced by incremental and periodic evaluation of system performance using the PACE laboratory.

Acknowledgment

This work is being performed by the GE Astro-Space under the Depart of Energy contract identified as the SP-100 Flight System Qualification (FSQ) contract (no. DE-AC03-86SF16006) for a Space Reactor Power System.

A PRELIMINARY FEASIBILITY STUDY OF CONTROL DISKS FOR A COMPACT THERMIONIC SPACE REACTOR

Scott B. Negron, Kurt O. Westerman, and Lewis C. Hartless
B & W Space and Nuclear Systems
Rt. 726, Mt. Athos Road
Lynchburg, VA 24506-1165
(804) 522-5085

Abstract

Results of a preliminary feasibility study of control disks are presented. These results show improved performance in compact space reactor systems through a more even fuel burnup due to improved core power profiles. Reactivity worth and axial power profile calculations have been performed for a fully reflected epithermal core typical of a compact thermionic reactor. Similar calculations were performed for the same reactor employing control rods for performance comparison. The neutronics calculations were performed in three dimensions utilizing the MCNP computer code due to the axial power dependance on rotational geometry changes. The analysis shows that control disks would provide enough worth to compensate for the built in excess reactivity for an estimated 10 year lifetime and still can take the system subcritical at end of life. Axial flux profiles show the advantages over the same system employing control rods in the form of a more even fuel burnup.

INTRODUCTION

During the last decade, a number of feasibility studies have been performed on the use of nuclear power reactors for various space applications. Thermionic reactors have been emphasized in many of these studies due to their low mass and high reliability. These studies have also shown that in order to compete with chemical and solar alternatives in the desired 10's to 100's of kWe range, the primary design criteria of a space reactor system must be achieving high power to weight ratio while maintaining a high overall system reliability. Control disks are intended to improve the performance of such a system by helping to achieve these requirements through minimization of control system mass, enhancing core lifetime through a more even fuel burnup, and providing a highly reliable control mechanism for power regulation and shutdown.

REACTOR CONTROL THEORY

In order to achieve a desired core lifetime, a nuclear reactor must be designed with the appropriate amount of excess reactivity to compensate for fuel burnup and poison production. Consequently, the reactor control mechanism(s) must be designed to compensate for this excess reactivity and provide for safe shutdown. Control mechanisms may be broadly classified into three groups depending on their useful range of worth. Safety controls are employed to provide enough negative reactivity to ensure sub-criticality in the event of an emergency (such as accidental water immersion) and to ensure sub-criticality during transport. Regulating controls are employed to regulate power through it's operating range. Separate shim controls are sometimes independently employed (such as a depletable isotope) to compensate for prolonged fuel burnup and poison production.

Control mechanisms may be further classified according to their method of changing system reactivity. Each of these are listed below with a brief description of the operating principle.

1. Insertion/removal of material - The most common means of changing reactivity is the insertion or removal of either a fissile or absorber material. Control rods are the most common example, usually being made of B_4C or other refractory materials containing absorbing isotopes such as Hf, Gd, or Ta.

2. External reflection - In a unreflected reactor, leaking neutrons are lost to the system and must be balanced by the number of fission neutrons produced in order to maintain criticality. Variable reflection of these neutrons back into the system is a common form of reactor control for space power systems. This is especially true for smaller cores having a higher leakage. Control drums and shutters are two familiar examples of reflective control devices.

3. Internal geometry changes - More commonly referred to as "flux shadowing", internal geometry changes may vary reactivity by moving internal components to areas of higher or lower worth. To illustrate this, consider a piece of absorbing material near the center of a reactor. If the geometry of the piece is changed to produce a high surface area to volume ratio, decreased resonant self shielding causes absorption to increase, increasing the negative worth of the material. This phenomenon can change local flux profiles and "shadow" fuel elements, ultimately regulating reactor power by reducing the fission rate in the shadowed region. The familiar separation or "Godiva" type control scheme may also be classified as a geometry change, however a change in leakage causes reactivity to change .

CONCEPT DESCRIPTION

Control disks belong to the class of control devices which operate by the flux shadowing principle. The theory behind this principle is that reactivity may be controlled by changing the internal arrangement of some of the core components to cause a variable internal self shielding.

A typical control disk assembly is shown in Figure 1. The disk contains radial arc segments of a neutron absorber (such as B_4C), such that rotation of one of the disks can align or dis-align the pattern. In the dis-aligned configuration, the core halves are isolated from one another and are each subcritical. As the disk is rotated to the "open" position, leakage from one core half to the other steadily begins to increase k_{eff} until criticality is reached.

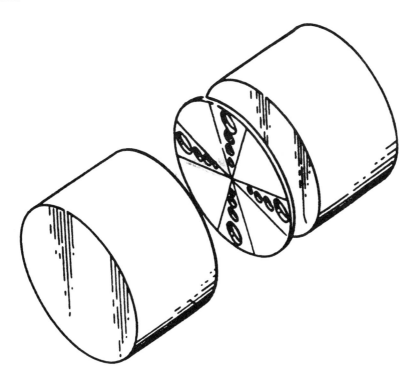

Figure 1. Control Disk Configuration .

Control disks offer many advantages over existing control schemes, especially in compact space reactor designs. These advantages are best illustrated by comparison with the other types of control devices employed in space reactors. Control drums have an advantage over control rods in that they free up internal core volume for other devices such as thermionic diodes or heat pipes. However, the mass of most control drum systems often approaches the mass of the core itself and greatly increases the total system mass. This is especially true for faster spectrum reactors, where the backscattering of the leaking surface flux is highly anisotropic due to the neutron's kinetic energy. In addition, the drive system needed to synchronize multiple drum elements can become extensive and costly to design when volume and system mass are of major consideration.

Control rods also become impractical in very compact systems due to the high level of flux distortion caused by the asymmetric presence of a strong absorber. Figure 2 illustrates the axial flux distortion caused by the insertion of a highly absorbing control rod. This distortion results in uneven fuel burnup and lower overall efficiency due to a higher flux peak to average ratio. As will be seen later, the axial profile produced by the control disk contains two flux peaks on either side of the assembly, thus decreasing the peak to average ratio.

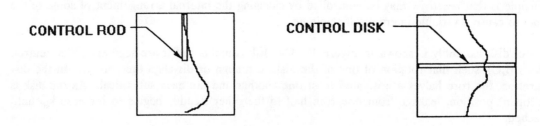

Figure 2. Axial Flux Shape, Control Rod vs. Control Disk.

Control disks eliminate both of these problems by containing the mechanism within the core boundaries (reducing system size) while providing a symmetrical and more uniform flux profile. While control rods require space outside of the core for retraction (such as a guide tube), only a small servo motor is needed to actuate the disks. The need for complex synchronizing elements is also eliminated.

ANALYSIS

A cylindrical homogenized reflected core having a centrally located control disk assembly was modeled using MCNP in order to evaluate the performance of the disk assembly. Two identical fully enriched graphite moderated UC_2 core halves were modeled with axial and radial reflectors containing a typical control disk assembly located at the center. The active fuel region was 18 cm in diameter and 14 cm in height. Figure 3 shows the axial and radial cross sections of the system, generated by the plotting routine of MCNP. The radial segments, illustrated in Figure 3, are made of alternating regions of B_4C and aluminum. Several different radial configurations were analyzed with different materials.

Axial flux profiles were obtained for a reactor containing a control disk assembly versus one regulated by control rods. The analysis has shown a peak to average value of 1.175 for the disk assembly compared to 1.267 for the control rod regulated system. These values are for identical cores having the same Keff. It has been found that the axial profile can be tailored by varying the thickness of the control disks. In addition, the radial flux profile can be tailored by varying the size and location of the absorbing segments.

The Monte Carlo transport code MCNP was used to analyze the performance of the control disk assembly. This code was chosen due to the complex 3-D geometry required to arrive at an accurate solution. The default ENDF/B-V continuous energy neutron/photon cross section set was used with all materials at room temperature. Run times for the 81 cell model took approximately 70 minutes on an Apollo 10000 workstation. One million neutron histories were run for each case to ensure adequate statistics (relative error less than 0.005). Figure 3 shows two cross sectional views of the assembly as generated by the PLOT routine.

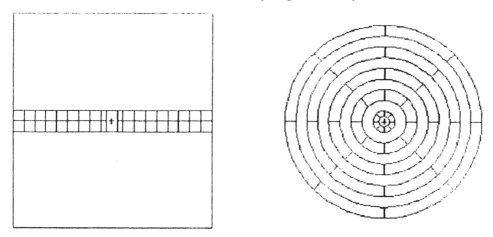

Figure 3. MCNP Model Geometry, Axial and Radial Profile.

RESULTS

Runs were performed in which the lower disk position was varied in 5° increments from 0° to 45°. The integral worth curve of the above system is shown in Figure 4. The total worth of the control disk from fully open to fully closed is $12.67. This result shows that the control disk concept has considerable promise. Assuming $7.00 reactivity is needed for shut-down margin and $7-10 excess reactivity is needed for achieving lifetime in a typical space reactor system, a control disk assembly should be able to meet these demands. This would make the need for separate shim and regulation control schemes unnecessary, as they could be combined into one disk assembly. A second safety system independent of the control disk may be required. The familiar "Godiva" type separation control scheme would lend itself to this purpose as due to the split geometry.

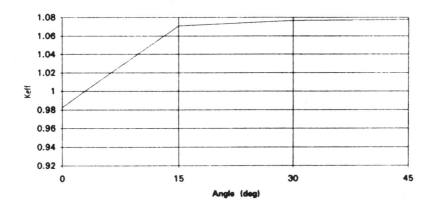

Figure 4. Control Disk Integral Worth Curve, 0°-45° Rotation.

Figure 5 shows the axial flux profiles obtained from the disk and control rod comparisons. As can be seen, the disk assembly provides a more even and symmetric profile as compared to the control rod regulated system. In addition to a more even fuel burnup, power distribution in the thermionic diodes would improve.

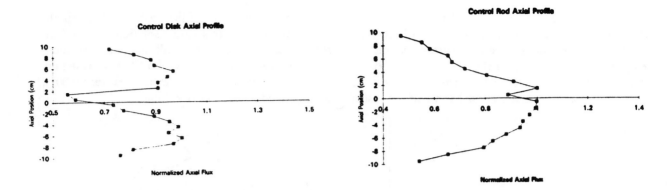

Figure 5. Axial Flux Profile, Control Disk vs. Control Rod.

Analysis of control disks in various applications has shown that variation of several design variables enable tailoring to a specific reactor configuration and desired control range. These design variables are listed below:

Materials - Any two materials possessing differing neutronic properties can be used to effect a change in reactivity for a given geometric configuration. Transparent, poison, fissile, or fertile materials can be used to provide differing worths. Burnable poisons or fertile isotopes can be used to enhance performance over the expected core lifetime.

Disk Thickness - Varying disk thickness provides a means of increasing or decreasing total worth for a given radial design without changing the behavior or shape of the worth curve.

Geometry - By changing the surface pattern of the different disk materials, reactivity worth can be varied. This also provides a means to tailor the radial power profile by allowing more or less inter-core transport where needed, thus enhancing the power profile.

In addition, the control disk assembly lends itself to combination with a separation or "Godiva" type control scheme. This would provide the two separate/redundant mechanisms required for reactor safety. A proposed thermionic system utilizing these schemes is shown in Figure 6. In this configuration, each core half is treated as a separate system, having individual heat rejection and power conversion systems. This also serves to increase the system survivability since multiple components could fail in one half without affecting the power output of the other part.

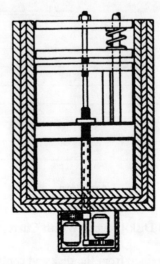

Figure 6. Hypothetical Control Disk Regulated Thermionic Space Reactor.

CONCLUSIONS

Control disks provide a novel control mechanism for highly compact thermionic space reactor systems. This control system has been shown to possess many advantages over conventional control schemes for very small cores. Analysis has shown that a control disk can be designed to compensate for built in excess reactivity while still providing enough worth to yield an adequate shut-down margin. Design variables enable the control system to be tailored to provide enhanced power profiles while extending core lifetime. The control disk concept can be combined with separation control to provide dual shutdown capability and a system less vulnerable to failure due to loss of key components.

Acknowledgments

This work was supported entirely by the Space & Nuclear Systems Division of the Babcock & Wilcox Company

References

Parlos, A., E. Khan, R. Frymire, S. Negron, J. Thomas, K. Peddicord (1991) "A Preliminary Feasibility Study of Passive In-Core Thermionic Reactors for Highly Compact Space Nuclear Power Systems," in Trans. 8th Symposium on Space Nuclear Power Systems, CONF-910116--Summs., held in Albuquerque, NM, 6-10 January 1991.

Negron, S.B., (1991) Nuclear Reactor Control Assembly, U.S. Patent No. 5 023 044, 11 June 1991.

Khan, E.U., and Parlos, A.G., "New Concepts for Compact Space Reactor Power Systems for Space Based Radar Applications: A Feasibility Study," BDM International, Inc. Report, Mclean, VA, December 1989.

Briesmeister, J., and J. Hendricks (1990) MCNP Version 4, Internal Memo X-6:JSH-90-153, Los Alamos National Laboratory, Los Alamos NM.

PERIOD-GENERATED CONTROL: A SPACE-SPINOFF TECHNOLOGY

John A. Bernard
Nuclear Reactor Laboratory
Massachusetts Institute of Technology
138 Albany Street
Cambridge, MA 02139
(617) 253-4202/4211
FAX (617) 253-7300

Abstract

Period-generated control is a method for tracking trajectories that are defined in terms of a demanded rate. It entails computing the control signal needed to achieve a specified trajectory by first using a form of proportional-integral-derivative feedback to generate a demanded inverse period (a velocity) and then substituting that inverse period into a system model. Terms that represent accelerations are included in the model and provide corrective action against deviations from the specified path. A characteristic feature of period-generated control is that rapid adjustments of the control signal are needed whenever a change is made in the demanded trajectory. Advantages to the technique are that it is readily implemented, that it is applicable non-linear systems, and that the resulting control laws may approach time-optimal behavior for the special case of rate-constrained processes. Period-generated control was developed for the automated operation of nuclear reactors and experiments have been performed to demonstrate its efficacy for that purpose. However, its potential is quite broad and examples are given of its use for the control of non-nuclear systems. In addition, the theory of period-generated control is presented together with an assessment of its major strengths.

INTRODUCTION

This paper describes a novel approach to the trajectory control of systems for which a demanded rate is to be observed. The technique, which has been designated as 'Period-Generated' control, was developed by the Massachusetts Institute of Technology (MIT) for the purpose of adjusting nuclear reactor power in a very rapid yet safe manner. The intended application is the control of the spacecraft reactors that will be used to propel manned expeditions to Mars. Experiments demonstrating that this control concept is capable of generating virtually any desired trajectory have been performed on both the 5-MWt MIT Research Reactor (MITR-II) and the Annular Core Research Reactor that is operated by Sandia National Laboratories (Bernard 1989, Bernard 1990, and Bernard et al. 1990). Although developed for the automated operation of nuclear reactors, period-generated control can be applied to many other systems and therefore constitutes a space-spinoff technology.

The specific objectives of this paper are to: (1) enumerate the theory of period-generated control, (2) give examples of its application to non-nuclear systems, and (3) provide an assessment of the technique.

THEORY OF PERIOD-GENERATED CONTROL

Period-generated control is a method for tracking trajectories that are defined in terms of a demanded rate. There are four major steps in its application. First, an error signal is defined by comparison of the observed process output with that which was specified. Second, a demanded inverse period (a velocity) is generated in terms of the error signal. Third, the demanded inverse period is processed through a system model to obtain the requisite control signal. Fourth, the control signal is applied to the actual system. Advantages to period-generated control are that it is readily implemented, that it is model-based and hence can be applied to non-linear systems, and that the resulting control laws may approach time-optimal behavior for the special case of rate-constrained processes.

For purposes of illustration, consider a second-order system of the type:

$$\dot{x}_1(t) = a(t)x_1(t) + \lambda x_2(t) \tag{1}$$

$$\dot{x}_2(t) = \gamma x_1(t) - \lambda x_2(t) \tag{2}$$

where x_1 and x_2 are the state variables and the other parameters are system coefficients. One of these is shown as time-dependent. For the general case, each could depend on both time and the state variables thus making the system non-linear. These equations might, for example, be used to describe population behavior. It is desired that the system output, $x_1(t)$, conform to a certain trajectory. Accordingly, it is convenient to define a measure of the system output's rate of rise. That quantity is the system's inverse period which is defined as:

$$\omega(t) \equiv \dot{x}_1(t)/x_1(t). \tag{3}$$

The inverse of $\omega(t)$ has physical meaning in that it is the 'e-folding' time or the time required for the quantity x_1 to change by the exponential factor.

The first step in applying period-generated control is to define an error signal, $e(t)$, such that:

$$e(t) = \ln(x_{1d}(t+j\Delta t)/x_1(t)) \tag{4}$$

where $x_{1d}(t)$ is the demanded trajectory, $x_1(t)$ is the observed trajectory, and j is a positive integer. A Taylor series expansion of this logarithmic expression reveals the rationale for selecting this particular arithmetic form for the error signal:

$$\begin{aligned} e(t) &= \ln(x_{1d}(t+j\Delta t)) - \ln(x_{1d}(t)) + \ln(x_{1d}(t)) - \ln(x_1(t)) \\ &\approx \ln(x_{1d}(t)) + j\Delta t \frac{d}{dt}(\ln(x_{1d}(t))) - \ln(x_{1d}(t)) + \ln(x_{1d}(t)) - \ln(x_1(t)) \\ &= j\Delta t \frac{d}{dt}(\ln(x_{1d}(t))) + \ln(x_{1d}(t)/x_1(t)) \\ &= j\Delta t\, \omega_d(t) + \ln(x_{1d}(t)/x_1(t)). \end{aligned} \tag{5}$$

Thus, the error signal used in period-generated control is the sum of a feedforward action from the demanded period and a proportional action from the quotient of the demanded and observed system outputs. The former defines the system path. The latter provides corrective action against deviations.

The second step in the application of period-generated control is to define a demanded period in terms of the error signal. Thus,

$$\omega_d(t) = \left(e(t) + (1/T_i)\int e(t)dt + T_d \dot{e}(t)\right)/j\Delta t \qquad (6)$$

where the parameters T_i and T_d correspond to the integral and derivative times in a conventional feedback expression.

The third step is to develop a system model that translates the demanded period into the requisite control signal. The process begins by substituting Equation (3) into Equation (1) and differentiating the result. So doing yields:

$$\omega(t)x_1(t) = a(t)x_1(t) + \lambda x_2(t) \qquad (7)$$

and, upon differentiating:

$$\dot{\omega}(t)x_1(t) + (\omega(t))^2 x_1(t) = \dot{a}(t)x_1(t) + a(t)\omega(t)x_1(t) + \lambda \dot{x}_2(t). \qquad (8)$$

The differentiation of Equation (7) has yielded two benefits. First, each parameter that can affect the tracking problem is now explicitly identified. For example, consider the term $a(t)x_1(t)$. Examination of Equation (1) gives the impression that only the magnitude of $a(t)$ is of importance. However, it is evident from Equation (8) that its rate of change, $\dot{a}(t)$, also bears on the solution of the tracking problem. In fact, for reasons that are explained below, the quantity $\dot{a}(t)$ is chosen as the control signal in the period-generated control law. The second benefit achieved from the differentiation of Equation (7) is that substitution can now be used to eliminate terms such as $\dot{x}_2(t)$ from the final control law. Such quantities may be difficult to measure or they may not be directly observable. Their elimination facilitates the practical realization of trajectory tracking schemes. Accordingly, Equation (2) is substituted into Equation (8) to eliminate $\dot{x}_2(t)$ Thus,

$$\dot{\omega}(t)x_1(t) + (\omega(t))^2 x_1(t) = \dot{a}(t)x_1(t) + a(t)\omega(t)x_1(t) + \lambda\gamma x_1(t) - \lambda^2 x_2(t). \qquad (9)$$

Next, x_2 is eliminated using Equation (1). So doing yields:

$$\dot{\omega}(t)x_1(t) + (\omega(t))^2 x_1(t) = \dot{a}(t)x_1(t) + a(t)\omega(t)x_1(t) + \lambda\gamma x_1(t) - \lambda^2(\omega(t)x_1(t) - a(t)x_1(t))/\lambda \qquad (10)$$

or

$$\dot{\omega}(t) + (\omega(t))^2 = \dot{a}(t) + a(t)\omega(t) + \lambda\gamma - \lambda\omega(t) + \lambda a(t). \qquad (11)$$

The next issue is the selection of the control signal. Both $\dot{a}(t)$ and $a(t)$ are possibilities because each is an externally-applied variable. The tracking of a specified trajectory requires that it be possible to alter the control signal on demand. This can be done with $\dot{a}(t)$ because it is itself a rate of change. In contrast, were $a(t)$ to be selected, the achievement of a sudden change of trajectory might necessitate a step change in that quantity's magnitude. This would be physically unrealistic for many systems. The quantity $\dot{a}(t)$ is therefore chosen as the control signal, and Equation (11) is rewritten as:

$$\dot{a}(t) = -a(t)\omega(t) - \lambda a(t) + \dot{\omega}(t) + (\omega(t))^2 + \lambda\omega(t) - \lambda\gamma. \qquad (12)$$

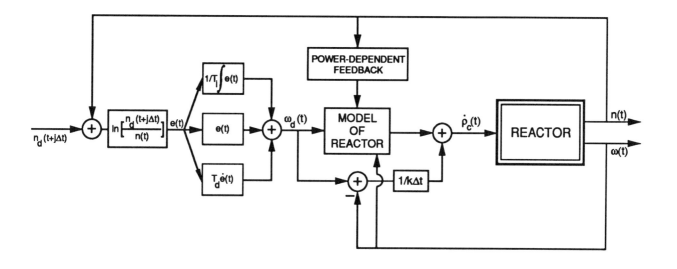

FIGURE 1. Period-Generated Control.

Equation (12) is referred to as an inverse model of the system dynamics because it gives the system input (the control signal) as a function of the system output (the inverse period).

It remains to address the term $\dot{\omega}(t)$ which represents the system acceleration. It is treated using the relation:

$$\dot{\omega}(t) = (\omega_d(t) - \omega(t))/k\Delta t \tag{13}$$

where $\omega_d(t)$ is the demanded inverse period, $\omega(t)$ is the observed inverse period, Δt is the time step, and k is the number of time steps over which it is desired that the system attain the specified trajectory. The quantity k should be chosen to be small because the objective of period-generated control is to cause the controlled parameter to begin rising (or falling) quickly at the demanded rate. For this to occur, the acceleration term must rapidly die out. However, as a practical matter, there is a lower limit to the value of k. Should it be made too small, $\dot{\omega}(t)$ will be quite large and an excessive rate of change in the control signal will be needed for transient initiation.

The desired control law can now be constructed by substituting $\omega_d(t)$ for $\omega(t)$ in Equation (12) and then substituting Equation (13) for $\dot{\omega}(t)$. Thus,

$$\dot{a}_c(t) = -\hat{a}(t)\omega_d(t) - \hat{\lambda}\hat{a}(t) + (\omega_d(t) - \omega(t))/k\Delta t + (\omega_d(t))^2 + \hat{\lambda}\omega_d(t) - \hat{\lambda}\hat{\gamma} \tag{14}$$

where the superscript (^) denotes an estimated quantity. The term $\dot{a}_c(t)$ is the control signal which, on application to the actual process, will cause the system output to track the demanded trajectory.

Figure 1 is a block diagram of the period-generated technique as applied to a nuclear reactor. The observed reactor power, n(t), is compared to that which is demanded, $n_d(t+j\Delta t)$. The difference between the two results in an error signal, e(t), which is used to generate the demanded inverse period, $\omega_d(t)$, that is needed to drive the reactor to the specified power level. This

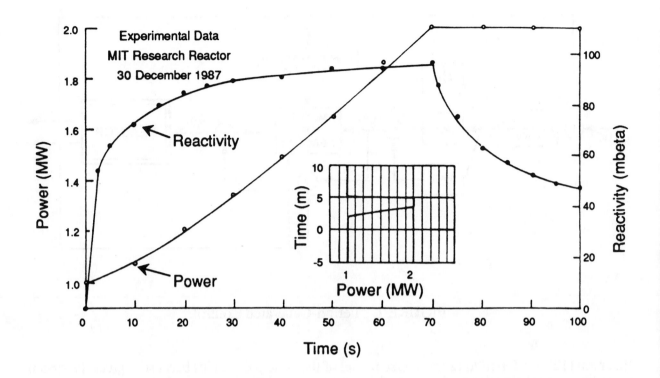

FIGURE 2. Closed-Loop Power Increase from 1 to 2 MW using Period-Generated Minimum Time Control Law on a 100-s Period.

demanded inverse period is processed through a system model to obtain the appropriate actuator signal which, in this case, is a rate of change of the reactivity, $\dot{\rho}_c(t)$. Shown in Figure 2 are the power and reactivity profiles obtained during a trial of period-generated control on the 5-MWt MIT Research Reactor. Also, the strip chart recording of this transient is shown as an inset. The reactor power was increased from 1 MW to 2 MW under conditions of closed-loop digital control using a variable speed stepper motor to adjust the net reactivity. The specified period was 100 s. The transient was completed at the expected time of 69 s and the shape of the power profile was exponential. Also of interest are the rapid changes in the reactivity during transient initiation and termination. These occurred because the reactor period was in effect being stepped from infinity (steady-state) to 100 s and, upon transient completion, from 100 s back to infinity. As is evident from the figure, the tracking performance achieved here was excellent. This result is typical of that attainable with period-generated control.

APPLICATIONS OF PERIOD-GENERATED CONTROL

Two applications of period-generated control are given here. They are fish-harvesting and recovery of a polluted lake.

Fish Harvesting

Suppose that it is desired to increase the yield from a fishery. This is to be accomplished by temporarily decreasing the harvest rate thereby allowing the fish population to rise. Upon achieving a new desired population, the harvest rate will be returned to its original value. The

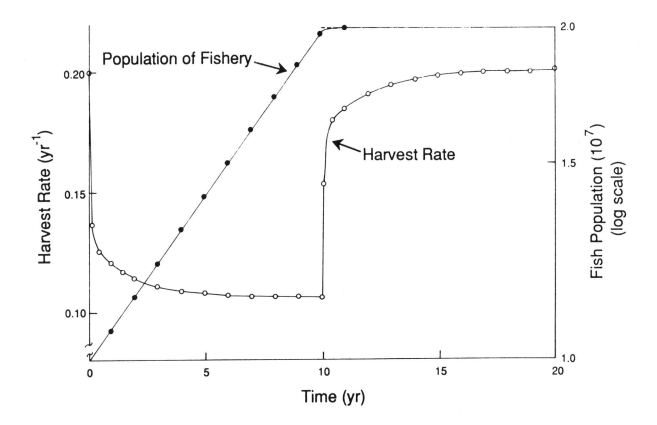

FIGURE 3. Adjustment of Harvest Rate needed to Double Fishery Population over Ten Years.

result should be both a stable fishery and a larger catch. A possible model of the fishery is:

$$\dot{F}(t) = -h(t)F(t) + \lambda J(t) \tag{15}$$

$$\dot{J}(t) = rF(t) - \lambda J(t) \tag{16}$$

where F is the population of fish, J is that of the fingerlings, h is the harvest rate, r is the reproduction rate, and λ is the rate at which fingerlings mature. It is assumed in this analysis that the maturation and reproduction rates are uniform rather than seasonal, that the normal attrition rate is small compared to the harvest rate, and that biological pressures such as the food supply are not limiting. Equations (15) and (16) are identical in form to those used in the previous section to explain period-generated control. Therefore, taking the rate of change of the harvest rate to be the control signal yields the following expression for the inverse dynamics model of the fishery:

$$\dot{h}(t) = -h(t)\omega(t) - \lambda h(t) - \dot{\omega}(t) - (\omega(t))^2 - \lambda\omega(t) + \lambda r . \tag{17}$$

Figure 3 shows the harvest rate and fishery population as a function of time for the case in which a doubling in population was demanded over an interval of ten years. An initial population of 10^7 fish was assumed together with reproduction and maturation rates of 20% and 50% per year respectively. It was also required that the specified trajectory be attained within 1.2 months. Thus, the parameter k in Equation (13) was 0.1 yr. The magnitude of the fishery is plotted on a logarithmic scale. If the demanded trajectory were tracked perfectly, then the slope of the resulting curve would be constant and the population would double in ten years. As is evident from the figure, the observed (simulated) trajectory is very close to the ideal result. Accomplishing this required that the harvest rate be manipulated so that the trajectory was driven within 1.2 months from steady-state to a period of 14.4 inverse years, held at that rate for ten years, and then stepped back to infinity. This in turn necessitated rapid changes in the harvest rate upon both transient initiation and termination. The minor deviations that do exist in the trajectory are the result of the finite time required for the acceleration term to die out. In situations where it is not possible to drive the acceleration term to zero quickly, the tracking performance will be poor.

Restoration of a Polluted Lake

The following model of the oxygen content of a polluted lake is taken from Beltrami (1987).

$$\dot{g}(t) = -kp(t) + k_1(g_s - g(t)) \tag{18}$$

$$\dot{p}(t) = \sigma(t) - kp(t) \tag{19}$$

where g is the dissolved oxygen gas level, g_s is the saturation level in water for dissolved oxygen, p is the mass of pollutant, σ is the rate of pollution inflow, k is the constant governing the rate at which the pollutant decomposes and thereby consumes oxygen, and k_1 is the rate constant governing oxygen uptake by the lake. Uniform mixing is assumed. It is desired to increase the lake's oxygen content by reducing the inflow of pollutants. Equations (18) and (19) differ in form from those previously considered and therefore indicate the versatility of period-generated control. The technique is applied by first defining $\omega(t)$ as:

$$\omega(t) \equiv \dot{g}(t)/g(t) . \tag{20}$$

Equation (18) therefore becomes:

$$\omega(t)g(t) = -kp(t) + k_1(g_s - g(t)) . \tag{21}$$

Differentiating and substituting to eliminate $\dot{p}(t)$ yields:

$$\dot{\omega}(t)g(t) + (\omega(t))^2 g(t) = -k\dot{p}(t) - k_1 g(t)\omega(t)$$

$$= -k\sigma(t) + k^2 p(t) - k_1 g(t)\omega(t) \tag{22}$$

with $\sigma(t)$ selected to become the control signal, the inverse dynamics model of the lake is:

$$\sigma(t) = \frac{-g(t)}{k}\left(k_1 \omega(t) + (\omega(t))^2 + \dot{\omega}(t)\right) + kp(t). \tag{23}$$

Feedback (proportional term only is shown) is generated using an expression of the form:

$$\omega_d(t) = \ln(g_d(t + j\Delta t)/g(t))/(j\Delta t) . \tag{24}$$

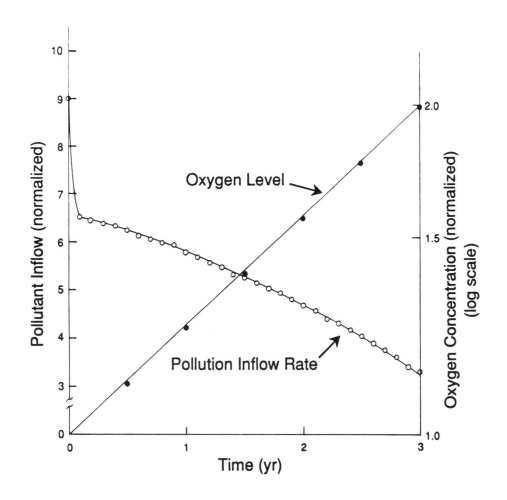

FIGURE 4. Reduction of Pollution Inflow needed to Double Oxygen Level in Three Years.

Figure 4 shows the reduction in the pollution inflow rate needed to achieve a doubling of the oxygen content in three years. The characteristic feature of period-generated control which is the rapid adjustment of the control signal upon transient initiation is again apparent. Note that an ever more severe decrease in the rate of pollutant inflow is needed to sustain the demanded rate of oxygenation.

ASSESSMENT OF PERIOD-GENERATED CONTROL

Major advantages to period-generated control are that it can treat non-linear systems and that, in the case of rate-constrained systems, the resulting control action approaches time-optimal behavior.

Non-Linear Control

The period-generated approach achieves proper control of non-linear systems through the use of a model of the process dynamics. Specifically, a feedback signal (the demanded inverse period) is computed from a comparison of the demanded and observed values of the system output. This signal is then input to an inverse dynamics model of the process that is being controlled. The solution is a form of feedforward control in the sense that the output of the inverse dynamics calculation is the control signal which, upon application to the actual process, will cause the system output to track the demanded trajectory (Lau et al. 1991). The merits of the approach are best illustrated by example. Specifically, a <u>functional</u> description of period-generated control can be written as:

$$e(t) = \ln(n_d(t + j\Delta t)/(n(t)) \tag{25}$$

$$\omega_d(t) = (e(t) + (1/T_i)\int e(t)dt + T_d \dot{e}(t))/j\Delta t \tag{26}$$

$$\dot{u}_c(t) = \hat{R}(t)\omega_d(t) - \hat{r}(t) + (\omega_d(t) - \omega(t))/k\Delta t \tag{27}$$

$$\omega(t) = (R(t))^{-1}(\dot{u}_c(t) + r(t) - \dot{\omega}(t)) \tag{28}$$

where the quantities R and r denote the system dynamics, $\dot{u}_c$ is the control signal, and where, for clarity of illustration, terms unimportant to the discussion have been omitted. The above equations show the basic steps in the implementation of period-generated control. Equation (25) gives the error signal in terms of the demanded and observed system outputs. Equation (26) defines the demanded inverse period as a function of that error signal. Equation (27) is the inverse model of the system dynamics that is used to translate the demanded inverse period into the appropriate control signal. Equation (28) is the actual system. It is instructive to substitute Equation (27) into Equation (28). So doing yields:

$$\begin{aligned}\omega(t) &= (R(t))^{-1}\left(\hat{R}(t)\omega_d(t) - \hat{r}(t) + (\omega_d(t) - \omega(t))/k\Delta t + r(t) - \dot{\omega}(t)\right) \\ &= (R(t))^{-1}\hat{R}(t)\omega_d(t) - (\hat{R}(t))^{-1}(\hat{r}(t) - r(t)) + (R(t))^{-1}\left((\omega_d(t) - \omega(t))/k\Delta t - \dot{\omega}(t)\right).\end{aligned} \tag{29}$$

If the quantities $\hat{R}$ and $\hat{r}$ are accurate, then the combination of the inverse dynamics calculation and the model-based feedforward action will result in the canceling of the system dynamics. Thus,

$$\omega(t) = \omega_d(t) + (R(t))^{-1}\left((\omega_d(t) - \omega(t))/k\Delta t - \dot{\omega}(t)\right). \tag{30}$$

It is evident from Equation (30) that once the acceleration term has been driven to zero, the actual and demanded inverse periods will be equal. This behavior is the strength of the period-generated approach and is of special importance for the trajectory control of non-linear systems. In particular, the result of the cancelation is that Equation (26), which is the conventional feedback expression, is the determining factor in the system's response. Its use here results in accurate tracking because the incorporation of a system model in the period-generated method causes the observed inverse period to equal that which is demanded once acceleration effects have died out. This will occur regardless of whether the process being controlled is linear or non-linear. In contrast, were that same feedback expression to be applied directly to a non-linear system with no use being made of a model, the tracking would not be accurate except for the specific trajectory for which the controller had been tuned.

Time-Optimal Response

Optimal control is normally achieved by applying techniques such as Bellman's dynamic programming or Pontryagin's maximum principle. These have the disadvantage of being computation-intensive. For example, application of the Pontryagin approach yields a set of partial differential equations with split boundary conditions. Such systems of equations must be solved iteratively. The result is that the time required to calculate the control action that corresponds to the optimal trajectory may exceed that available for implementing the associated control signal. Under such circumstances, the optimal control is calculated off-line and applied in an open-loop manner. No use of feedback is possible. Hence, if the system model is inaccurate or if a perturbation occurs, the resulting response will not be as desired.

A major advantage of the period-generated technique is that it results in closed-form control laws that can be implemented in real time and which may approach a time-optimal response. Specifically, for systems that are subject to a rate constraint, the time-optimal trajectory will be the one that moves the system along that constraint. Hence, rather than identify the optimal control by solving the system's describing equations subject to both the constraint and a performance index, it is more direct to define the physical conditions that correspond to system movement along that limiting constraint. This can be achieved using period-generated control by taking the demanded period to be that associated with the limiting constraint. For example, many nuclear research reactors are operated subject to limits on the power, temperature, coolant flow, and rate of rise of power. Suppose that the limit on the latter quantity for the MITR-II were a period of 100 s. In that case, the power and reactivity profiles shown in Figure 2 are those of the time-optimal trajectory.

The degree to which a period-generated control law approaches a time-optimal response depends on the treatment of the acceleration term. In the ideal case, the trajectory would be instantly switched to and from the limiting path. The presence of the acceleration term makes this scenario physically impossible. However, the impact of the acceleration term can be made quite small provided that the control signal can be rapidly changed and is not limited in magnitude. Under such circumstances, period-generated control laws can closely approximate time-optimal responses for rate-constrained systems.

CONCLUSIONS

Period-generated control is applicable to the tracking of trajectories that are specified in terms of a rate. It functions by requiring very rapid rates of change in a system's control signal. Application of these rates of change has the effect of 'stepping' the system from its initial condition to the desired path. Advantages to period-generated control are that it is model-based and can therefore address non-linearities, that the resulting control laws are readily implemented, and that the associated control action may approach time-optimal behavior in the special case of rate-constrained systems. Control laws formulated in terms of this concept have been successfully demonstrated on both the 5-MWt MIT Research Reactor and on the Annular Core Research Reactor that is operated by the Sandia National Laboratories. It is anticipated that these period-generated laws will be used to control power in one of the spacecraft reactors now under design as part of the space exploration program of the United States. In addition, period-generated control offers the promise of providing a general method for the tracking of rate-defined trajectories in a wide variety of non-nuclear systems.

Acknowledgments

Appreciation is expressed to Professor David D. Lanning (MIT), Professor Allan F. Henry (MIT), Mr. Francis J. Wyant (SNL), and Mr. Frank V. Thome (SNL) for their constructive comments. Mrs. Carolyn Hinds, Ms. Georgia Woodsworth, and Mr. Leonard Andexler assisted with paper preparation. Mr. Ara Sanentz performed the layout and prepared the graphics. This research was supported by the U.S. Department of Energy under Contract No. DE-FG07-90ER123.90.

References

Bernard, J. A. (1989) *Formulation and Experimental Evaluation of Closed-Form Control Laws for the Rapid Maneuvering of Reactor Neutronic Power*, MITNRL-030, Massachusetts Institute of Technology, Cambridge, MA, September 1989.

Bernard, J. A. (1990) *Startup and Control of Nuclear Reactors Characterized by Space-Independent Kinetics*, MITNRL-039, Massachusetts Institute of Technology, Cambridge, MA, May 1990.

Bernard, J. A., K. S. Kwok, P. T. Menadier, F. V. Thome, and F. J. Wyant (1990) "Experimental Evaluation of the MIT-SNL Period-Generated Minimum Time Control Laws for the Rapid Adjustment of Reactor Power," in *Space Nuclear Power Systems 1988*, M. S. El-Genk and M. D. Hoover, eds., Orbit Book Co., Malabar, FL, pp. 495-508.

Beltrami, E. (1987) *Mathematics for Dynamic Modeling*, Academic Press, Inc., Orlando, FL, p. 13.

Lau, S. H., J. A. Bernard, and D. D. Lanning (1991) "Experimental Evaluation of Feed-Forward Control for the Trajectory Tracking of Power in Nuclear Reactors," in *Proc. 26th Intersociety Energy Conversion Engineering Conference (IECEC)*, held in Boston, MA, August 1991, Vol. 5, pp. 13-19.

IRRADIATION TESTING OF A NIOBIUM-MOLYBDENUM DEVELOPMENTAL THERMOCOUPLE

R. Craig Knight
Westinghouse Hanford Company
P. O. Box 1970
Richland, Washington 99352
(509) 376-5419

David L. Greenslade
Westinghouse Hanford Company
P. O. Box 1970
Richland, Washington 99352
(509) 376-5601

Abstract

A need exists for a radiation-resistant thermocouple capable of monitoring temperatures in excess of the limits of the chromel/alumel system. Tungsten/rhenium and platinum/rhodium thermocouples have sufficient temperature capability but have proven to be unstable because of irradiation-induced decalibration. The niobium/molybdenum system is believed to hold great potential for nuclear applications at temperatures up to 2000 K. However, the fragility of pure niobium and fabrication problems with niobium/molybdenum alloys have limited development of this system. Utilizing the Fast Flux Test Facility, a developmental thermocouple with a thermoelement pair consisting of a pure molybdenum and a niobium-1%zirconium alloy wire was irradiated for 7200 hours at a temperature of 1070 K. The thermocouple performed flawlessly for the duration of the experiment and exhibited stability comparable to a companion chromel/alumel unit. A second thermocouple, operating at 1375 K, is currently being employed to monitor a fusion materials experiment in the Fast Flux Test Facility. This experiment, also scheduled for 7200 hours, will serve to further evaluate the potential of the niobium-1%zirconium/molybdenum thermoelement system.

INTRODUCTION

Accurate temperature measurement of nuclear reactor experiments requires specific capabilities. Thermocouples must be immune to irradiation damage that would alter the electro-chemical properties of the thermoelement wire pair. For low-to-moderate temperature applications, chromel/alumel (Type K) thermocouples have proven to be extremely reliable. However for extended service at temperatures in excess of approximately 1350 K, other systems must be considered.

Tungsten/rhenium (W/Re) and platinum/rhodium (Pt/Rh) thermocouples have higher temperature capability. However, these systems are subject to significant decalibration caused by irradiation effects including transmutation of the base materials (Browning and Miller 1962, and Heckelman and Kozar 1972) and second phase precipitation (Williams et al. 1983). Figure 1 shows the negative drift that occurred with a W/Re thermocouple during an irradiation experiment in the Fast Flux Test Facility (FFTF), a 400-mWt liquid metal cooled reactor located at the Hanford site near Richland, WA. This particular experiment operated at a temperature of 1350 K (Greenslade et al. 1990).

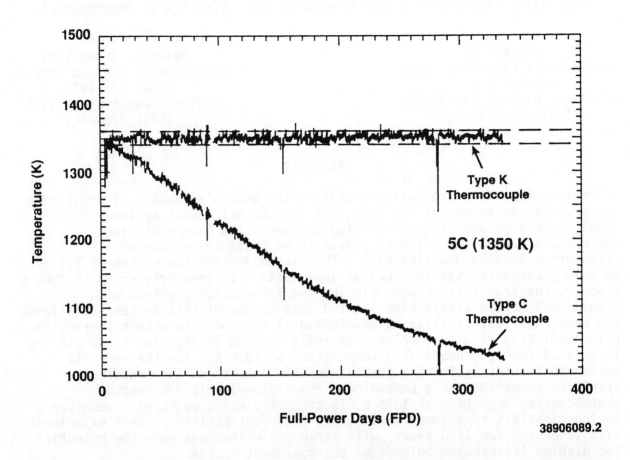

FIGURE 1. Thermocouple Performance in the FFTF Materials Open Test Assembly Canister 5C. Thermocouples were Exposed to a Fast Fluence (E > 0.1 MeV) of 4.0×10^{22} n/cm^2. Note: Type C is an Unofficial Designation for a W5%Re/W26%Re Thermocouple.

Niobium and molybdenum (Nb-Mo) alloys have previously been considered for thermocouple applications because these refractory metals have low neutron capture cross sections as well as high temperature capability. Significant work on Nb/Mo thermocouples occurred in France (Schley and Metauer 1982). Although the system showed significant potential, problems with alloy fabrication hampered this effort. Further work at the Idaho National Engineering Laboratory (INEL) also encountered fabricability problems (Wilkins 1988).

One reason that previous efforts focused on Nb/Mo alloys rather than pure metals was because the alloys exhibit better electromotive force (EMF) potential at temperatures above approximately 1750 K (Schley and Metauer 1982). However, 1750 K is considerably higher than the range of Type K thermocouples and encompasses the requirements for many nuclear experiments of interest. Because of this fact, as well as the availability of materials, the initial work at Westinghouse Hanford Company (WHC) concentrated on the pure metals. However, preliminary furnace tests with thermocouple probes utilizing 0.25 mm diameter thermoelement wires at temperatures up to 1675 K resulted in early failure attributed to excessive grain growth in the pure niobium wire.

As shown in Figure 2, the wire was transformed into a string of single crystals subject to transgranular failure caused by thermal stresses during temperature transients. The molybdenum wire did not exhibit such rapid grain growth, and no failures were experienced in this component.

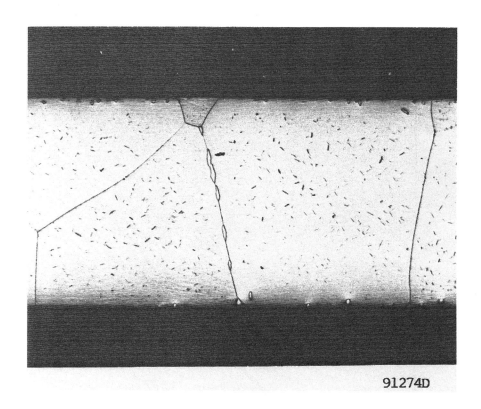

FIGURE 2. Longitudinal Photomicrograph of 0.25 mm Diameter Niobium Wire from Test Thermocouple, Furnace Aged for 260 Hours at 1475 K with a Brief Excursion to 1675 K (250X).

These results led to investigation of the relatively common Nb alloy, niobium-1% zirconium (Nb-1%Zr) used in conjunction with pure molybdenum to form the thermoelement wire pair.

EXPERIMENT DESIGN

Having limited resources, we decided to utilize materials on hand to devise a thermocouple experiment for testing in the FFTF. Both Nb-1%Zr and pure Mo wire of 0.25 mm diameter were available in our laboratory inventory. High purity (99.8%) alumina insulators, sized to fit the 0.25 mm wires, were also available. Unalloyed Mo sheath tubing, 1.8 mm OD x 1.2 mm ID, was fabricated from metal powder by a commercial supplier.

During testing of the pure metal system, a calibration test was run to a temperature of 1950 K utilizing a Type B (Pt/Rh) thermocouple for comparison. The EMF curve generated agreed closely with previous data (Wilkins 1988).

Upon incorporation of the Nb-1%Zr wire, we decided to perform a more precise calibration experiment. A thermocouple probe consisting of Nb-1%Zr/Mo wires, alumina insulators, and Mo sheath approximately 0.3 meters in length was assembled and inserted into a high temperature vacuum furnace. Pure copper lead wires were soldered to the thermoelement wires and run through an ice bath to a calibrated voltmeter. Millivolt readings were taken while the furnace temperature was cycled to 1450 K and back. The EMF curve shown in Figure 3 resulted. The curve was found to vary only slightly from that for the pure metals.

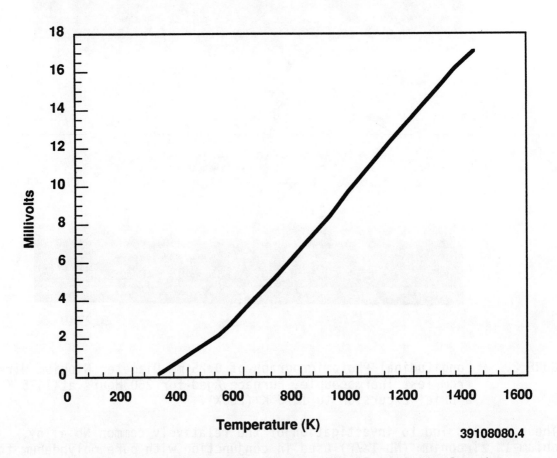

FIGURE 3.   Voltage Versus Temperature Curve for Alumina Insulated, Molybdenum Sheathed, Niobium-1%Zirconium/Molybdenum Thermocouple.

In order to obtain meaningful data in an irradiation experiment, it was necessary to find a way to generate a temperature gradient over the length of the Nb/Mo thermoelement wires. Since the system was experimental, we ruled out the use of a thermocouple to monitor an actual FFTF experiment. However, a method was devised to take advantage of nuclear heating of the molybdenum thermocouple sheath by enclosing the thermocouple probe in a vacuum chamber. The low thermal conductivity in vacuum would allow the probe to operate well above the temperature of the surrounding coolant. As shown in Figure 4, computer analysis disclosed that this method could be expected to generate a temperature of up to 1150 K, depending on the length of the thermocouple probe. Using this technique, all that remained was to find a vehicle to carry the test thermocouple into the core of the FFTF. A Materials Open Test Assembly (MOTA) (Greenslade et al. 1986) was selected.

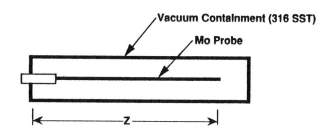

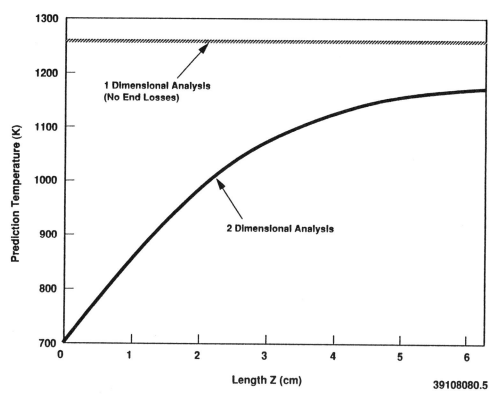

FIGURE 4. Computer Analysis Depicting Anticipated Performance of a Nb-1%Zr/Mo Test Thermocouple in the FFTF at Core Mid-Plane, Row 4 Position. Mo-Sheathed Probe is Enclosed in Stainless Steel Vacuum Containment.

FABRICATION AND INSTALLATION

Figure 5 shows the construction of the test thermocouples. In order to achieve the maximum operating temperature (see Figure 4), a probe length of 7 cm was selected. As noted in Figure 5, the Nb-1%Zr/Mo thermocouple wires were spliced to Type K wires adjacent to, but outside of, the vacuum containment. In this way, the dissimilar metal splices could be maintained at a constant (coolant) temperature--an important criteria for thermocouple accuracy. Standard MOTA Type K thermocouple cables were used. A vacuum of approximately 5×10^{-5} torr was achieved in the stainless steel enclosure by performing the final end cap weld in an electron beam welder. The thermocouple probe was backfilled at room temperature with one atmosphere of high purity helium.

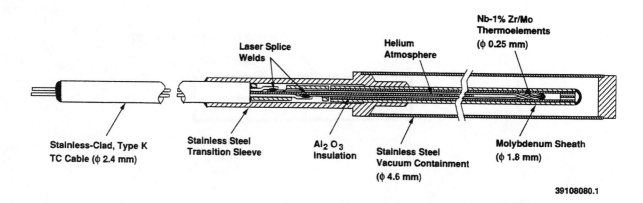

FIGURE 5.   Schematic Showing the Component Parts of the Test Thermocouples.

Two test thermocouples and one identically constructed Type K thermocouple were fabricated. The purpose of the Type K thermocouple was to provide comparative data (computer analyses showed that replacement of the Nb/Mo wires with Type K wires would have negligible effect on the temperature achieved by nuclear heating).

Due to space limitations, there was room for only two special thermocouples on the selected MOTA experiment. One Nb-1%Zr/Mo and one Type K thermocouple were affixed to the MOTA test train at the FFTF core mid-plane location. In addition, a standard MOTA Type K thermocouple was placed near the location of the dissimilar metal splices on the test thermocouple to accurately monitor the reactor coolant temperature at that point.

The corrected millivolt reading of the Nb-1%Zr/Mo test thermocouple assembly, including the Type K extension cable, was obtained using the following equation:

$$MV_c = MV_t - MV_{ks} + MV_{ns}$$

where MV_c is the corrected voltage, MV_t is the total indicated voltage, MV_{ks} is the voltage indicated by the Type K thermocouple at the splice, and MV_{ns} is the corresponding voltage that a Nb/Mo thermocouple located at the splice would indicate (see Figure 3).

During operation, the MOTA data acquisition system automatically performed the above calculation and converted the voltage outputs directly to temperature. Data points were taken every 5 minutes and logged once per hour for the duration of the test. The experiment began operation in the FFTF in January 1990 and concluded in March 1991, with 300 full power days (FPD) achieved.

RESULTS AND DISCUSSION

Both test thermocouples performed without failure for the duration of the MOTA irradiation. As shown in Figure 6, there was an initial period of instability before the two thermocouples began to indicate similar steady-state temperatures. The final temperature achieved averaged 1070 K, approximately 100 degrees less than predicted in Figure 4. We also observed that the Type K thermocouple indicated temperature continued to fluctuate by several degrees throughout the experiment, while the Nb-1%Zr/Mo test thermocouple was much more stable.

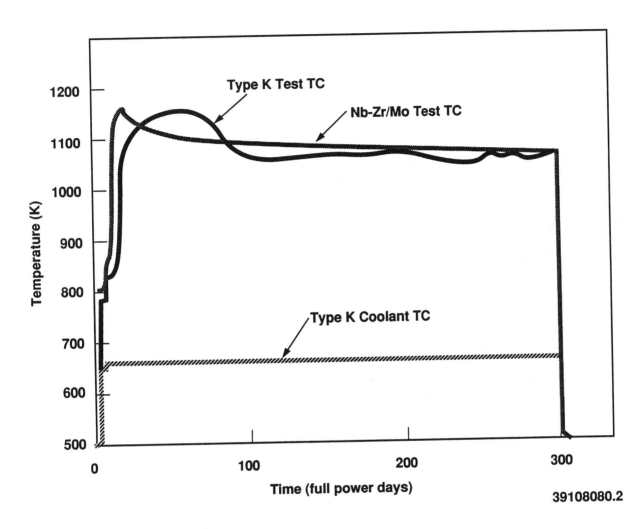

FIGURE 6. Performance of the Test Thermocouples in the FFTF During 300 Full Power Days Operating Cycle. Thermocouples were Exposed to a Fast Fluence of 7.5×10^{22} n/cm^2.

As noted in Figure 6, the average temperature of both thermocouples continued to decline slightly over the duration of the test. Over the final 100 FPD, the average temperature of the Nb-1%Zr/Mo thermocouple decreased from 1075 K to 1067 K--a total of 8 degrees. During this same period, the average temperature of the Type K thermocouple decreased by a similar amount, from 1070 K to 1061 K--a total decline of 9 degrees.

The causes of the instability at the start of irradiation, the continued oscillation of the Type K thermocouple, and the overall steady decrease in operating temperature are believed to include one or more of the following:

- A less-than-optimum vacuum atmosphere within the stainless steel outer enclosure;

- Radiation-caused changes in the emissivity of the component surfaces (OD of the Mo sheath and ID of the stainless steel vacuum containment tube); and

- Movement of the thermocouple probe within the vacuum containment tube caused by thermal stresses.

The degree of vacuum achieved in the containment tube by making the final end cap weld in an electron beam welder is questionable. Because of the tight fit-up of components, it is possible that some amount of air may have remained in the containment tube after welding. This would lead to conduction losses, oxidation, and surface emissivity changes during operation.

The temperature oscillations of the Type K thermocouple could be the result of movement of the probe within the vacuum containment. Although radiographic inspection showed excellent as-fabricated alignment, thermal stresses may have caused movement of the probe close to, or in intermittent contact with, the tube ID, resulting in substantial temperature variations.

CURRENT PROGRAMS

Because of the success of this experiment, we decided to use a Nb-1%Zr/Mo thermocouple to monitor a high-temperature experiment scheduled for irradiation in a follow-on FFTF MOTA.

The experiment, designed to evaluate candidate fusion reactor materials, required thermocouples capable of monitoring temperatures of up to 1475 K. Because of the potential to exceed the safe operating limit of Type K thermocouples, agreement was reached to place a Nb-1%Zr/Mo thermocouple with a Type K thermocouple in the MOTA test canister.

For this experiment, the Nb/Mo thermocouple consisted of the following materials:

- Thermoelement - Nb-1%Zr and unalloyed Mo, 0.25 mm diameter;

- Insulators - 99.8% purity polycrystalline alumina;

- Sheath - W22%Re, 1.6 mm OD x 1.2 mm ID; and

- Lead cable - 2.4 mm OD stainless steel sheathed Type K.

The Type K thermocouple was identically fabricated except for the thermoelement wires which were 0.25 mm OD chromel/alumel.

This experiment commenced operation in May 1991. Figure 7 shows the temperature history for the first operating cycle which totaled 53 FPD. As shown, the maximum experimental temperature achieved was approximately 1375 K. Both the Nb-1%Zr/Mo and Type K thermocouples operated flawlessly, with the latter averaging 42 K less than the experimental thermocouple because of the location of the thermocouples within the MOTA canister.

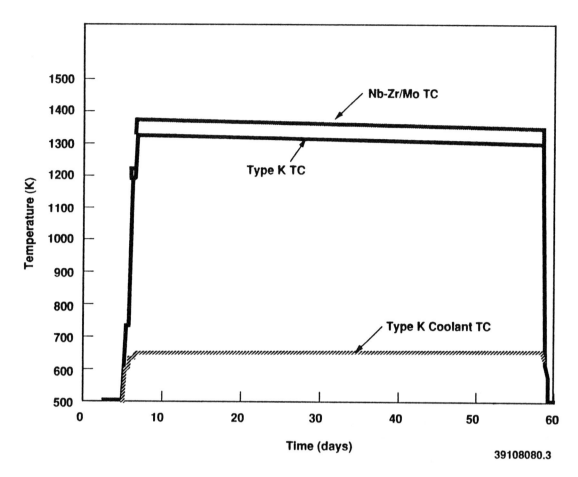

FIGURE 7. Comparison of the Performance to Date of a Test Nb-1%Zr/Mo Thermocouple with a Type K Thermocouple in a Fusion Materials Experiment in the FFTF.

Figure 7 shows a perceptible decline in the temperature indicated by both thermocouples. This is due to predicted burnup of the experimental materials and is not an indication of thermocouple decalibration. In fact, the actual difference between the two thermocouples varied only 3 degrees throughout the operating cycle--further proof that the Nb-1%Zr/Mo system is practically immune to radiation-induced decalibration.

The fusion materials experiment is scheduled to continue for a total irradiation cycle of 300 FPD.

CONCLUSIONS

The passive experiment demonstrated the stability of the Nb/Mo thermocouple system under severe irradiation conditions at an intermediate temperature. At core mid-plane in the FFTF, the thermoelement wires were subjected to a total fluence of 12.5×10^{22} n/cm^2 and a fast fluence of 7.5×10^{22} n/cm^2. The slight decrease in temperature noted as the irradiation progressed was concluded to be unrelated to nuclear transmutation or other irradiation effects as verified by the similar behavior of the Type K thermocouple.

Type K thermocouples have performed flawlessly in the FFTF under similar conditions of temperature and fluence in numerous irradiation tests.

The results also demonstrated the practicality of the experimental method. Although some refinement of the fabrication techniques is needed, the theory proved to be sound and the manufacturing cost was minimal. In addition, there was no risk to the operation of the plant or to other reactor experiments. While there are no current plans, the method could readily be used in the future to evaluate other developmental thermocouple systems.

The follow-on experiment will provide additional data regarding the viability of the Nb/Mo thermocouple system.

Acknowledgments

This work was performed by the Westinghouse Hanford Company under the Department of Energy Contract Number DE-AC06-87RL10930. The experiment was sponsored by the Fusion Materials Program.

References

Browning, W. E. and C. E. Miller, Jr., (1962) "Calculated Radiation-Induced Changes in Thermocouple Composition", Temperature: Its Measurement and Control in Science and Industry, 3(2): 271-276.

Greenslade, D. L., R. C. Knight, and A. M. Ermi, (1990) "The Design and Performance of High Temperature Irradiation Capsules", WHC-SA-0693-FP, Westinghouse Hanford Company, Richland, Washington.

Greenslade, D. L., R. J. Puigh, G. W. Hollenberg, and J. M. Grover (1986) "FFTF as an Irradiation Test Bed for Fusion Materials and Components", Journal of Nuclear Materials, 141-143: 1032-1038.

Heckelman, J. D. and R. P. Kozar (1972) "Measured Drift of Irradiated and Unirradiated W3%Re/W25%Re Thermocouples at a Nominal 2000 K", Temperature: Its Measurement and Control in Science and Industry, 4(3): 1935-1949.

Schley, R. and G. Metauer (1982) "Thermocouples for Measurements Under Conditions of High Temperature and Nuclear Irradiation", Temperature: Its Measurement and Control in Science and Industry, 5(2): 1109-1113.

Wilkins, S. C. (1988) "Low Cross-Section Mo-Nb Thermocouples for Nuclear Application: The State-of-the-Art", Space Nuclear Power Systems 1988, M. S. El-Genk and M. D. Hoover, eds. Orbit Book Co., Malabar, Florida, pp. 481-486.

Williams, R. K., F. W. Wiffen, J. Bentley, and J. O. Stiegler (1983) "Irradiation Induced Precipitation in Tungsten Based, W-Re Alloys", Metallurgical Transactions A, 14A: 655-666.

CONTROL ASPECTS OF HYDRIDE/BE MODERATED THERMIONIC SPACE REACTORS

Norman G. Gunther and Monte V. Davis
Space Power, Incorporated (SPI)
621 River Oaks Parkway
San Jose, CA 94134
(408)434-9500

Samit K. Bhattacharyya and Nelson A. Hanan
Argonne National Laboratory
9700 South Cass Ave.
Argonne, IL 60439

Abstract

Characteristics of small, moderated, incore thermionic space reactors have been investigated. Results are given for fuel Doppler, emitter Doppler, and collector Doppler. Reactivity coefficients have been calculated for moderator temperature, coolant temperature and core density effects. It is found that for a 44 kWe moderated system the moderator and coolant temperature coefficients can be made to be zero in magnitude, while fuel coefficients are slightly negative. This results in a system that, from any operating condition, can be scrammed to cold shutdown with only 1 or 2 of the 12 control-safety drums.

INTRODUCTION

Small, moderated, Incore Thermionic reactors have nuclear characteristics that can strongly affect their dynamic behavior. The Soviet TOPAZ, for instance, is reported to posses positive moderator reactivity temperature coefficient of about $+3.25 \times 10^{-5}/\Delta k/k$ per K isothermal from room temperature to operating temperature, see Figure 1 (Garin, 1990). The total coefficient is zero or slightly negative at power operating conditions due to the fuel and clad Doppler. The moderator temperature effect results from hardening of the thermal spectrum with temperature. This hardening increases thermal utilization since the highly enriched fuel is essentially black to all thermal neutrons.

The Authors have investigated some of the control, parameters of moderated incore thermionic reactors for power systems at 44 kWe and 14 kWe.

DOPPLER COEFFICIENTS

For safety and controllability the Doppler coefficient is the most important parameter affecting dynamic behavior for nuclear reactor. The Doppler coefficient is produced by the presence of shielded absorption resonances at relatively low energy. Usually, the coefficients are small, and usually there is only one, which is due to the fuel.

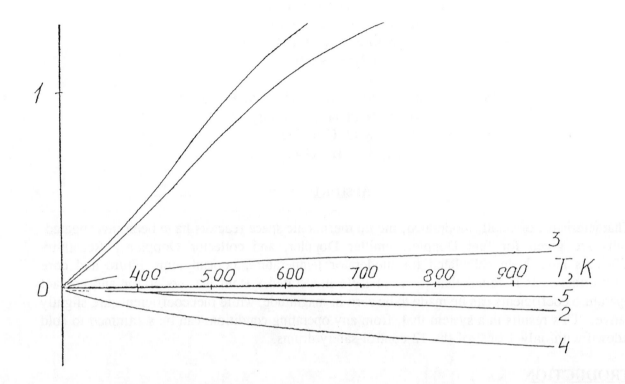

Figure 1. The Temperature Effect of Reactivity in the TCR with the Single-Element EGCs:

1 - Effect of the Moderator; 2 - Effect of the Tube Plates;
3 - Effect of the Reflector; 4 - Effect of the Fuel Elements;
5 - Effect of the Electrodes; 6 - Total Isothermic Temperature Effect.

Incore thermionic reactors have fuel elements which are clad in refractory materials. See Figure 2. Thus, there is a Doppler coefficient for both the fuel and the emitter. The reactors studied here also have molybdenum collectors so there is yet another coefficient due to this collector.

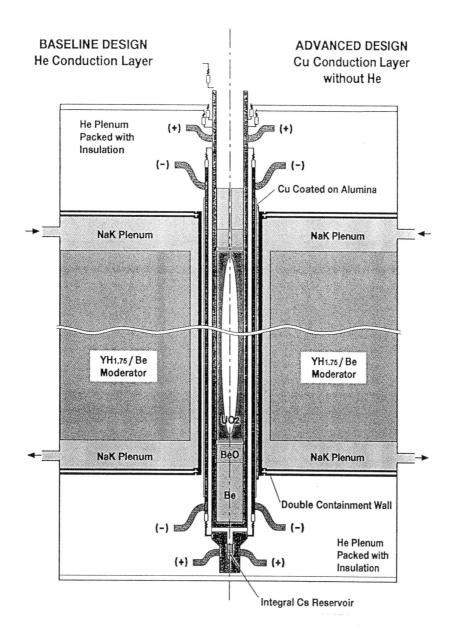

Figure 2. TFE and Core Structure - Baseline/Advanced Designs.

The emitter is in direct thermal contact with the fuel and is at the same temperature, whereas the collector is at the temperature of the moderator and is in thermal contact with the coolant. Consequently, the different coefficient produce different dynamic effects and are operative over different temperature ranges. Table 1 shows the calculated Doppler coefficient for a 44 kWe power system and for a 14 kWe system. The smaller system has driver fuel and so there is another set of coefficients associated with the driver fuel.

TABLE 1. Doppler Coefficients.

44 kWe System

Region	Doppler
Fuel	$-3.9 \times 10^{-6}/K$
Emitter	$-2.7 \times 10^{-6}/K$
Collector	$-2.2 \times 10^{-6}/K$

14 kWe System

Region	Doppler
TFE Fuel	$-4.5 \times 10^{-6}/K$
Emitter	$-2.3 \times 10^{-6}/K$
Collector	$-1.6 \times 10^{-6}/K$
Drum Fuel	$-2.5 \times 10^{-6}/K$
Clad	$-2.0 \times 10^{-6}/K$

MODERATOR TEMPERATURE COEFFICIENTS

The time constant for thermal response of the moderator to variations in reactor power is very long compared to the response of the fuel and emitter. Consequently, a positive moderator temperature coefficient such as in a heterogeneous $ZrH_{1.8}$ moderated core produces a slow reactivity effect that is almost independent of instantaneous controls (ie. like burn up). The TOPAZ reactors have relatively large positive moderator temperature coefficients, but their fuel coefficients are negative. Many Topaz reactors have been safely tested and controlled.

The cores which we have investigated in the study being reported herein, use moderators of metal hydride contained in beryllium. See Figure 3. As arranged in Figure 4, these cores have harder neutron spectrums than those moderated only with hydrides. The $ZrH_{1.8}$ + Be metal moderator results in temperature coefficients that are much smaller. Figure 5 gives results of

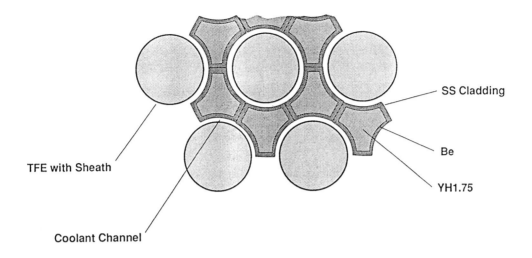

Figure 3.   Beryllium/Yttrium Hydride Moderator Schematic.

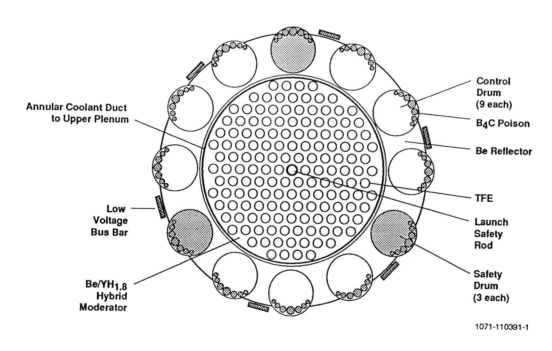

Figure 4.   Moderated Core Cross-Section.

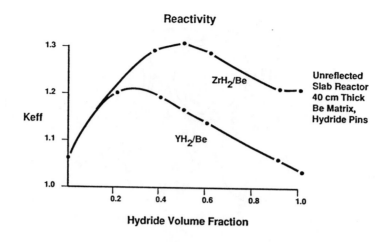

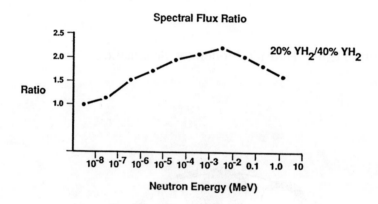

Figure 5.   Nuclear Design Hydride/Be Moderator.

a sensitivity study performed by SPI which shows the reactivity effect of varying the volume fraction of metal hydride in beryllium. This study shows that, over a sizable range of hydride volume fraction, the reactivity is insensitive to variation, whereas the neutron spectrum changes significantly. SPI has exploited this behavior to produce a core design with especially favorable control characterisitics. In the case of the yttrium hydride and Be metal moderated core at 44 kWe, the spectrum has been tailored to produce a zero moderator temperature coefficient.

Two other temperature coefficients are of importance. The first of these is the coolant temperature effect. This effect is due to the change in coolant density with temperature and its variation from core to core reflects the different weighing of leakage and parasitic absorption. The second is the core density effect. This effect was calculated separately from the coolant effect because the implications for dynamic behavior are different. Table 2 shows the calculated values of temperature coefficient for the various cores investigated. The calculations were done using the MCNP code (Briesmeister, 1990).

TABLE 2. Temperature Coefficients.

44 kWe Core YH_x Moderator	
Region	Doppler
Moderator	Zero
Coolant	Zero
Core	$-6.7 \times 10^{-6}/K$
14 kWe Core YH_x/Be Moderator	
Region	Coefficient
Moderator	$+2.4 \times 10^{-5}/K$
Coolant	$+4.2 \times 10^{-6}/K$
Core	$-6.7 \times 10^{-6}/K$
44 kWe Core $ZrHx$/Be Moderator	
Moderator	$+1.8 \times 10^{-5}/K$
6 kWe Core $ZrHx$ Moderator (TOPAZ)	
Moderator	$+3.25 \times 10^{-5}/K$

NEUTRON KINETICS PARAMETERS

Neutron kinetics parameters calculated for the 44 kWe YH_x moderated core are given in Table 3. These results do not account for the effect of photoneutrons produced in the beryllium moderator and reflector.

TABLE 3. Neutron Kinetics Parameters - 44 kWe YHx/Be Moderated Core.

Delayed Neutron Group	Beta	Lambda (sec^{-1})
1	2.6448×10^{-4}	1.2720×10^{-2}
2	1.4745×10^{-3}	3.1740×10^{-2}
3	1.3030×10^{-3}	1.1600×10^{-1}
4	2.8182×10^{-3}	3.1100×10^{-1}
5	8.8652×10^{-4}	1.4000
6	1.8003×10^{-4}	5.8700

Neutron Generation Time	1.267×10^{-5} sec
Prompt Lifetime	1.289×10^{-5} sec
Effective Delayed Neutron Lifetime	12.74 sec

Acknowledgments

This design study was conducted for the Phillips Laboratory of Albuquerque New Mexico. The authors are particularly grateful to Captain Jerry Fisher, Dr. Dan Mulder and Dr. Michael Schuller for their constructive assistance in the conduct of this study.

References

Briesmeister, J.F. ed.,"MCNP - A General Monte Carlo Code for Neutron and Photon Transport" LA-7396-M, 1990

Garin, V.P. et.al. "Development of a Mathematical Model of Neutron Physical Dynamics and Thermal Physical Processes in a SNPS with a Single-Element TIC". Thermionics Specialists Conference, Obninsk, USSR. May 1990.

DESIGN OF A PLANAR THERMIONIC CONVERTER TO MEASURE THE EFFECT OF DIFFUSION OF URANIUM OXIDE ON PERFORMANCE

Gabor Miskolczy and David P. Lieb
Thermo Electron Technologies
Waltham, MA 02154-9046
(617) 622-1357

G. Laurie Hatch
Rasor Associates, Inc.
Sunnyvale, CA 90489
(408) 734-1622)

Abstract

A plane parallel variable-spaced research converter was designed to investigate the effect of diffusion of uranium oxide fuel through the tungsten emitter. This work is part of the Thermionic Fuel Element (TFE) verification program. In the TFE program the cylindrical emitter is made of Chemical Vapor Deposited (CVD) tungsten from tungsten fluoride, which is over coated with CVD tungsten, from the chloride, resulting in a <110> crystal orientation. The planar converter has a similar emitter construction. This paper describes the design and preliminary testing of this converter.

INTRODUCTION

The effect of diffusion of the nuclear fuels, uranium oxide and uranium nitride, in thermionic fuel elements was studied by Shimada (1971), who built a cylindrical converter with uranium oxide fuel, and by Shimada and Cassell (1971) and Shimada (1972) who built planar converters with tungsten and rhenium emitters and with uranium nitride fuel. Shimada (1971) concluded that the oxygen and uranium both diffused through the rhenium emitter. The oxygen increased the uncesiated emitter work function, and the uranium continuously deposited on the collector. After 2400 hours of testing the cylindrical converter at 2000 K emitter temperature, the performance decreased by 15%.

Shimada (1972) tested a planar converter with a rhenium emitter fueled with uranium nitride at 1900 K which showed nearly 50% degradation in 2000 hours, while another converter with a tungsten emitter showed an increased performance up to 4300 hours. Then the performance totally degraded at 4700 hours. Shimada (1972) concluded that nitrogen and uranium both diffused through the emitter grain boundaries. The nitrogen reacted with the cesium and the uranium deposited on the collector degrading converter performance.

The current work is aimed toward testing the diffusion of uranium oxide fuel through tungsten emitters as shown in Table 1, which Shimada and co-workers did not investigate.

TABLE 1. Diffusion Study Matrix.

FUEL	EMITTER MATERIAL	
	TUNGSTEN	RHENIUM
URANIUM OXIDE	THIS WORK	SHIMADA 1971
URANIUM NITRIDE	SHIMADA 1972 SHIMADA AND CASSELL 1971	SHIMADA 1972 SHIMADA AND CASSELL 1971

In a related investigation von Bradke and Henne (1977) constructed an emitter of a cermet of uranium oxide and molybdenum. They observed beneficial effects of the oxygen.

The Thermionic Fuel Element (TFE) verification program proposes a test (General Atomics 1987) to investigate the diffusion of the oxide fuel through the emitter in a planar tungsten converter. The planar converter will be fabricated by chemical vapor deposition (CVD) in a manner similar to that in the TFE, namely the emissive surface will be deposited from tungsten chloride to yield a <110> crystal orientation and the substrate from tungsten fluoride. The thickness of the layers will be 0.3 mm and 0.7 mm respectively. This study will supplement the findings of earlier workers.

DESCRIPTION OF CONVERTER

The planar variable-spaced research converter shown in Figure 1 described by Hatch (1988) and Goodale et al. (1979), will be used for this test. It has an overall emitter diameter of 20 mm and an effective collector diameter of 11.3 mm, resulting in a collector area of 1 cm^2. The emitter is tungsten, with the emissive polycrystalline surface having a preferred <110> orientation derived through chemically vapor deposition from tungsten chloride. The collector is made of niobium which is ground flat. The emitter is supported on a thin sleeve made of a molybdenum-rhenium alloy and is electrically isolated from the collector by a high-purity aluminum oxide insulator. A niobium bellows between the emitter and collector allows the spacing to be changed from 0.5 mm to .125 mm. The cesium vapor for the converter is supplied from an external cesium-graphite reservoir which is the prototype for the in-pile converters. To make the experimental converter prototypic the graphite reservoir which will be charged with cesium through an isolation valve to the same loading which will be used in the TFE testing.

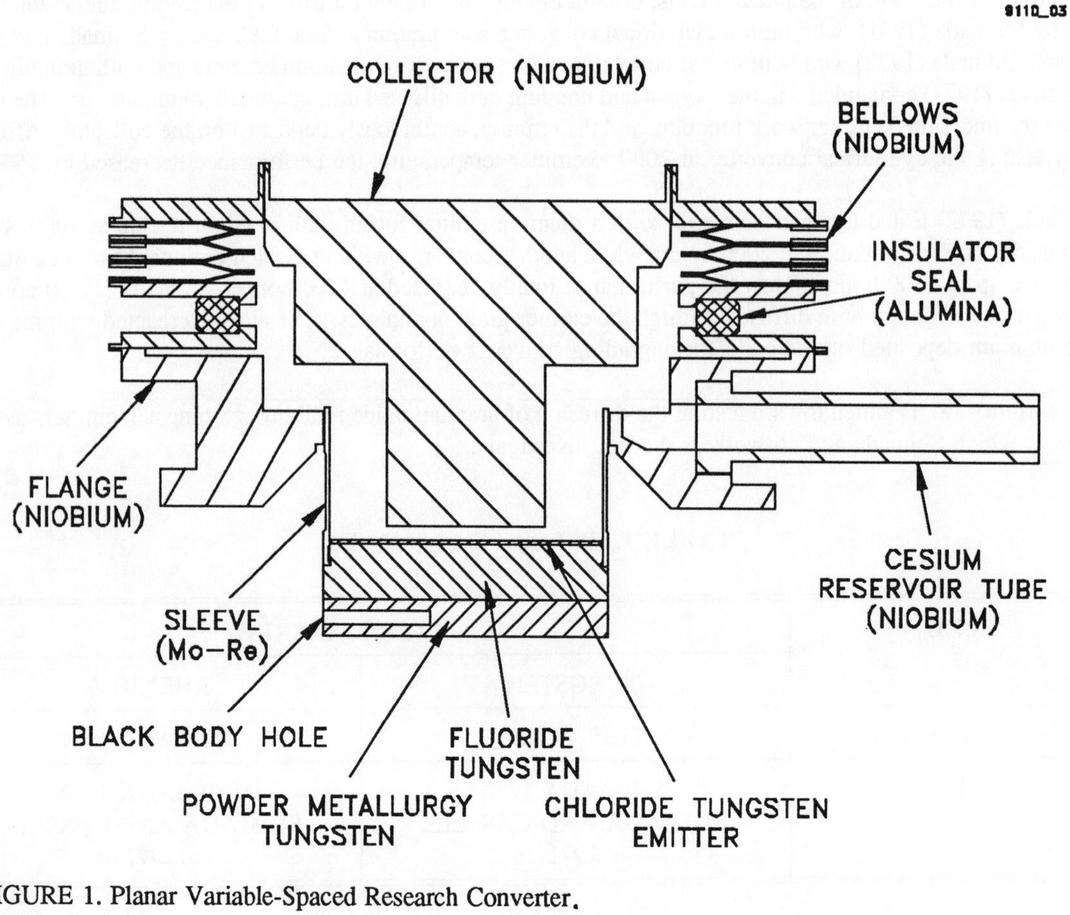

FIGURE 1. Planar Variable-Spaced Research Converter.

Initial experiments were carried out with an emitter with a single large, collector diameter (11.3 mm), cavity for the fuel. The first tests were run without the fuel pellet in place. The emitter surface adjacent to the fuel cavity distorted during this experiment, rendering the emitter unsuitable for further experiments. The distortion was probably caused by the thin unsupported central portion of the emitter.

A revised configuration of the emitter was proposed as shown in Figure 2 to minimize the emitter distortion. The new configuration will have four 4 mm diameter cavities instead of one large one for the fuel. While the new configuration will contain much less fuel than the old one, it should be less likely to distort when heated. The question of the thermal distortion of the emitter was studied using both a two-dimensional and a three-dimensional computer heat transfer model (Garrison 1991). This study confirmed that the multi-cavity design is much stiffer than that with a single large cavity. (The distortion was calculated to be .0008 mm).

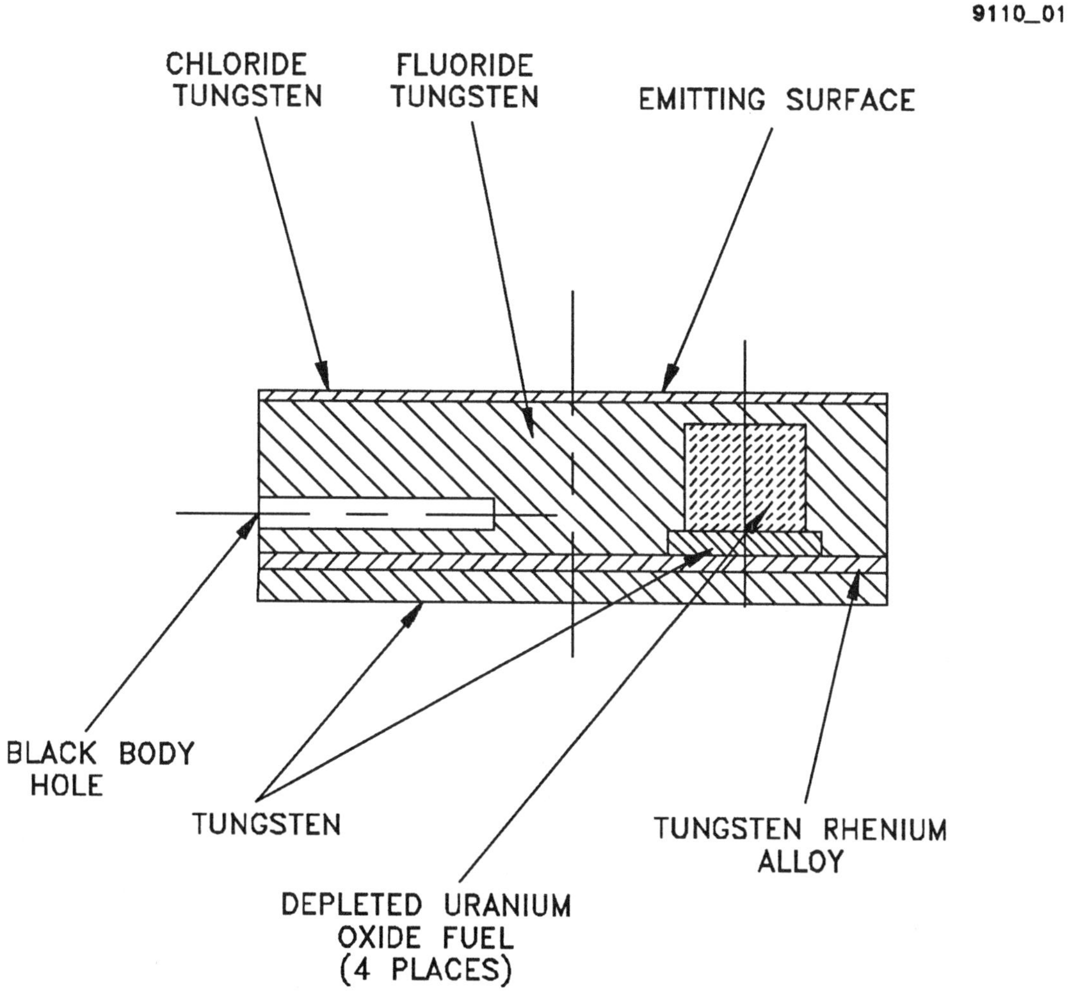

FIGURE 2. Fueled Emitter Configuration.

The emitter assembly will be fabricated as follows. First, the four cavities for the depleted uranium oxide fuel will be machined in a tungsten disc made by chemical vapor deposition from tungsten fluoride. Next, the fuel pellets will be inserted and two wafer lids, one of tungsten-25% rhenium alloy and one of tungsten will be assembled. The fuel pellets will be separated from the tungsten-25% rhenium lid by the tungsten wafers to minimize the possible interaction between the fuel and the tungsten-rhenium alloy. The entire assembly will be uniaxially pressure-bonded in vacuum. Next an emitting surface will be chemically vapor deposited from tungsten chloride, yielding a preferred <110> crystal orientation, when electro-polished.

The converter will be equipped with two reservoirs, one liquid cesium, the other cesium-graphite. After outgassing the converter, cesium will be distilled from the glass ampoules into a temporary receiver where residual gasses can be pumped out through a reflux column, and then distilled into the cesium reservoir. After this the cesium distillation apparatus will be pinched-off and removed, leaving a cesium and a cesium graphite reservoir as shown in Figure 3. A bakeable valve will be included to isolate the cesium reservoir from the rest of the converter.

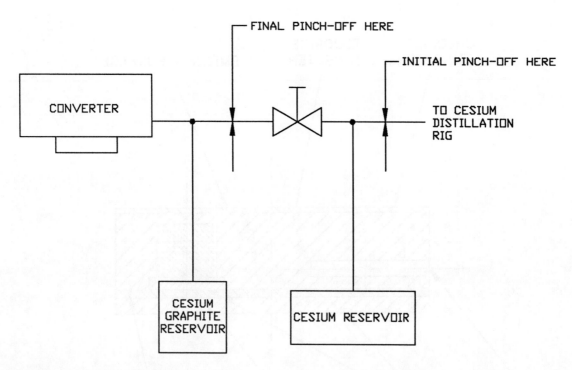

FIGURE 3. Reservoir Arrangement.

The converter will be tested with the cesium only reservoir active by heating the cesium-graphite reservoir to a high enough temperature to remove all the cesium. Then the converter can be tested with the cesium-graphite reservoir only by shutting off the valve. Following the establishment of the desired cesium loading of the graphite reservoir, the cesium reservoir will be pinched off. All subsequent testing will be performed with a single cesium-graphite reservoir.

TEST PROTOCOL

The converter will be characterized using the standard test conditions used by Hatch (1988), which are summarized in Table 2.

The test procedure is to fix the electrode temperatures and spacing. At each cesium pressure the J-V, (current-density versus voltage), curve is recorded while driving the converter with a 60 Hz alternating current sweep. Following the converter characterization with the liquid cesium reservoir, the graphite reservoir will be charged with cesium and the valve to the liquid cesium will be closed and the cesium reservoir will be pinched-off. Then the converter will be characterized again, adjusting the cesium-graphite reservoir temperature to yield the cesium pressures prescribed in Table 2.

TABLE 2. Converter Test Parameters.

	HIGH	LOW	INCREMENT
Emitter Temperature (K)	2000	1600	100
Collector Temperature (K)	1100	600	100
Interelectrode Spacing (mm)	.5	.125	.025
Cesium Pressure (Torr)	8	.5	Factor of 2

LIFE TEST

After full characterization, the converter will be life tested with an emitter temperature of 2000 K. The converter load point will be fixed at the optimum, and the performance will be monitored on a recorder. In addition J-V curves will be swept at a 1000 hour intervals up to 16000 hours or until a significant change of performance is observed.

If a significant change of performance is detected the converter will be completely characterized and life testing terminated. The converter will be disassembled and a complete metallurgical examination will be performed.

Acknowledgment

This work was supported by DOE and SDIO under the TFE Verification Program, General Atomics prime contractor, Department of Energy San Francisco contract No.DE-AC03-86SF16298.

References

Garrison, G. (1991) Personal Communication, MCR Associates, Sunnyvale, CA.

General Atomics (1987) *TFE Verification Program*, Semiannual Report, GA Project 3450, Contract DE-AC03-86SF16298, General Atomics, San Diego, CA.

Goodale, D. B., C. Lee, D. P. Lieb, and P.E. Oettinger, (1979), "Thermionic Converter Investigations" in *Proc. 14th IECEC*, Boston MA, p. 1862.

Hatch, G. L. (1988) *Thermionic Technology Program, Thermionic Converter Performance*, Report E-533-003-B-053188, Rasor Associates Inc., Sunnyvale, CA.

Shimada, K. (1971) "Out-of-Core Evaluations of a Nonfueled and a UO_2 Fueled Cylindrical Thermionic Converter," in *Proc. Thermionic Spec. Conf.*, San Diego, CA, p. 248.

Shimada, K. (1972) "Out-of-Core Evaluations of Uranium Nitride-Fueled Converters," in *Proc. 3rd Int. Conf. on Thermionic Power Generation*, Juelich, Germany, p. 1269.

Shimada, K. and P. L. Cassell, (1971) "Evaluation of Uranium-Nitride Fueled Converters," in *Proc. Thermionic Spec. Conf.*, San Diego, CA, p. 253.

von Bradke, M. and R. Henne, (1977) "Electrodes for Low Temperature Thermionic Energy Converter: Recent Results and Compatibility considerations", in *Proc. 12th IECEC*, Washington, DC, p. 1582.

ADVANCED THERMIONICS TECHNOLOGY PROGRAMS AT WRIGHT LABORATORY

Tom Lamp and Brian Donovan
Aerospace Power Division
Wright Laboratory, Wright-Patterson AFB, OH, 45433
513-255-6235

Abstract

This paper summarizes two thermionic technology programs being managed at Wright Laboratory: The Advanced Thermionics Initiative (ATI) Program, and The Thermionics Critical Technology (TCT) Investigation. These programs are supported by Strategic Defense Initiative Organization (SDIO) and United States Air Force, respectively. The described programs contain thermionics technology improvement efforts for both in-core and out-of-core reactor concepts; including, creep resistant emitter materials, cesium reservoirs, diamond film development, design analysis codes, and collector surface physics. This paper describes the current status of these programs, as well as, program goals for future evolutionary improvements in thermionic converter technology.

INTRODUCTION

The Air Force has had a long standing interest in thermionic space nuclear power (SNP) dating back to the early 1960s when a heat pipe cooled thermionic converter was demonstrated through work at the predecessor to Wright Laboratory (WL). With the exception of the short hiatus in the mid-70s, Air Force thermionics work at Wright Laboratory has continued to the present time with thermionic technology programs including the burst power thermionic phase change concepts, heat pipe cooled planar diodes, and advanced in-core concept developments such as composite materials, insulators and oxygenation.

Unique advantages offered by thermionic power systems include modularity, high waste heat rejection temperatures, and the stable electrical characteristics of the thermionic conversion process. The thermionic technology efforts at Wright Laboratory involve both in-core and out-of-core reactor concepts. Each of these reactor concepts is recognized for its specific virtues. The in-core is considered more scalable to the higher output powers while possessing more advanced and complicated technology elements. The out-of-core concept is simple, more straight forward, but is relegated to the lower power output levels. Both require some technology needs to reach fruition; for example, creep resistant emitters for in-core concepts and high temperature graphite nuclear fuel trays for out-of-core concepts.

The ATI Program was organized by SDIO/IST to integrate thermionic technology advances into a converter suitable for in-core reactor applications in the 10 to 40 KWe power range. The baseline design point is 30 KWe with the ability of scaling to 100 KWe as a desirable design feature.

The focus of the Air Forces' Thermionic Critical Technology Investigation is on an out-of-core, Romashka or Star-C type, reactor concept which can operate in the 5-40 KWe power range. Both of these concepts consist of a graphite moderated uranium carbide reactor, with out-of-core planar thermionic converters surrounding the perimeter. The converters are radiatively coupled to the reactor core.

THE ADVANCED THERMIONICS INITIATIVE

The ATI program organization consists of a number of contractors participating under several related programs sponsored by the Innovative Science and Technology office at SDIO. General Atomics (GA) is the primary contractor and is responsible for the ATI thermionic fuel element (TFE) overall design, TFE fabrication and in-core testing. Supporting contractors include Universal Energy Systems (UES), Space Power Incorporated (SPI), Rasor Associates Incorporated (RAI), Oregon Graduate Institute (OGI), Oregon State University (OSU), and Arizona State University (ASU). Beginning in late 1989 approximately fifteen programs have been initiated at Wright Laboratory to support the ATI effort. Some principal developments include: single-cell TFE design, improved emitters, simplified passive cesium reservoirs, diamond film coated collector electrodes, and a PC based SNP system design code.

The ATI program was originally established by SDIO/IST as a three year stand-alone effort; however, it is now envisioned as a program of continuing evolutionary improvement in thermionic converter technology. Presently, the ATI program has three technology stages identified as ATI-1, ATI-2 and ATI-3. Each program stage will yield testable TFE designs incorporating increasingly advanced thermionic technology concepts. Concepts targeted for development include metal matrix creep resistant emitters, single cell TFE designs, carbon and non-carbon intercalated cesium reservoirs, oxygenated converter electrodes, dual-vapor converters and diamond films. The ultimate goal, of course, is the development of a robust, long-life, high efficiency, simplified TFE which can be incorporated into an existing SNP in-core reactor program.

Core-Length Thermionic Fuel Element

At this stage, the program thrust is to adopt two design innovations: 1) reduce complexity by developing a more simple fuel element configuration, 2) increase life and efficiency by developing an emitter with improved creep resistance at 2000 K. The present design features a 25 cm long cylindrical fueled emitter which will extend the full length of the reactor core. This core-length emitter not only reduces complexity but it also makes possible the complete performance characterization of TFE's before loading with nuclear fuel. The core-length emitter is predictably lossy, recent analysis indicates about 20%, but the quality

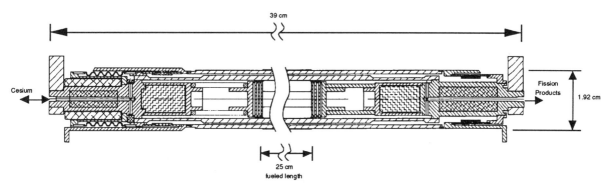

Figure 1. ATI Configuration.

assurance and reliability implications of pretesting the reactor system prior to loading are unquestionably positive. The present core-length emitter configuration can be seen in the engineering illustration contained in Figure 1. The emitter, which is being developed and supplied through a separate SBIR contract with SPI, will be reinforced to restrain fuel swelling by use of tungsten dispersion-strengthened with hafnium carbide. Some of the operating parameters of the ATI-TFE are listed in Table 1.

Table 1

Emitter Temp	2000 K
Collector Temp	880 K
Cesium Res Temp	600 K
Interelectrode Gap	0.05 cm
Current Density	>3.5 A/sq cm
Power Density	3.0 W/sq cm
Emitter Dimensions	1.5 cm diameter by 25 cm length

Creep Resistant Emitter

The emitter temperature of a thermionic converter has a significant effect on the electrical performance and mass of a thermionic SNP system. Studies have shown (Begg 1990) that increasing the emitter temperature from 1800 K to 2100 K yields a mass reduction on the order of 40%. Both electrical power density and thermal efficiency increase with increased emitter temperature. The net result is: 1) smaller reactor to produce the same electrical power; 2) smaller radiator because of the higher conversion efficiency; 3) smaller shield because of the smaller reactor core.

The improved emitter consists of a tungsten-hafnium-carbide wire reinforced emitter calculated to provide seven years life at a 2000 K operating temperature. This operating temperature is approximately 150°K above the operating temperatures of state-of-the-art (SOA) thermionic converters. Creep resistance of the ATI emitter is provided by tungsten-hafnium-carbide wire which has been over-coated with CVD tungsten. Figure 2 is a conceptual illustration of the wire-wrap CVD process. The superior creep properties of W-HfC is illustrated in Figure 3. W-Re-HfC has marginally superior creep performance; however, its larger thermal neutron cross-section drove the decision to use W-HfC. SPI has demonstrated void-free fabrication of wire-wrapped emitters with thoriated tungsten wire substituted for the W-HfC wire. Ten meters of W-HfC wire has been delivered to Wright Laboratory for physical evaluation which includes creep testing. Delivery of the full scale creep resistant emitters is scheduled for early 1992.

Other ATI Technology Support

Two cesium reservoir concepts have been selected for SBIR phase two funding as part of an overall effort to improve cesiation methods. Both concepts rely on passive cesiation techniques which will offer a large reduction in reactor core complexity by reducing the need for electrical control circuits as well as reducing the number of vacuum tight welds and/or brazes. Collector electrode sections have been fabricated with diamond coated insulators and efforts are presently underway to characterize diamond film insulators in high temperature nuclear environments. These diamond insulators will be incorporated in ATI-2 designs and are expected to improve thermal performance, reduce fabrication costs and eliminate insulator degradation caused by the reactor neutron environment. A PC based, user friendly SNP system design code has been developed which incorporates SOA thermionics, neutronics, thermal analysis modules. This design code will be made available for distribution throughout the SNP community. Oxygenation of thermionic converter electrodes significantly improves converter efficiency; however, reliable control of this additive within the operating converter has been very elusive. Work is presently underway on an oxygenation technique which promises controlled addition of oxygen to the converter and is expected to be a part of ATI-3 designs.

THERMIONIC CRITICAL TECHNOLOGY INVESTIGATION

The Thermionics Critical Technology Investigation was initiated by the United States Air Force to enrich the technology base for thermionic space nuclear power systems. This research effort includes four tasks,

Creep Resistant Emitter

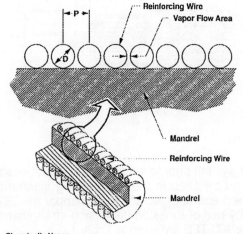
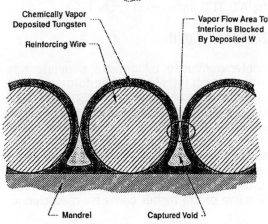

Figure 2

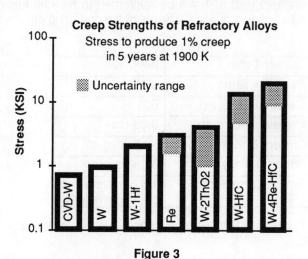

Figure 3

TABLE 2: Conceptual Thermionic Power System (UC_2 Graphite Reactor).

	Rasor's (1967)	STAR-C (1967)	LEOS (1968)
Reactor Power (KWt)	35	44.4	69.2
Electrical Output (KWe)	5	6	9.6
Thermionic Efficiency (%)	14.2	14.5	13.9
Core C/L Temperature (K)	2173	2290	2270
Core Surface Temp. (K)	2040*	2011*	1990*
Emitter Temperature (K)	1973	1863	1850
Collector Temperature (K)	1000	1016	990
Carnot Efficiency (%)	49.3	45.5	46.5
Core Diameter (cm)	27.9	25	25.5
Core Length (cm)	40.6	32	50

* Assume net emissivity exchange of 0.8.

parameters. Table 2 shows the results of tradeoff studies at the system level, and Table 3 lists the converter parameters that are configured to meet these system requirements. Figure 4 contains an illustration of the baseline excore converter configuration.

TABLE 3: Converter Design Parameters.

Emitter Temperature	1850 K
Collector Temperature	990 K
Cesium Reservoir Temp	600 K
Spacing	0.102 mm
Cesium Pressure	4 Torr
Output Voltage	
@ Max Power Dens. of 10.5 W/cm^2	0.57 V
@ Design Point (8.9 W/cm^2)	0.67 V
Converter Output per Cell	40 W (70 W Max)
Emitter Shoe Area	4 cm X 4 cm

spread over five years, and culminates in the demonstration of a series connected string of four advanced planar converters. This section discusses the results of the first two tasks: design, construction and preliminary test of two baseline converters. The first two converters have been delivered to Wright Laboratory where performance mapping and long-term life testing is being conducted. To date, converter performance tests have been conducted for emitter temperatures from 1850 K to 2000 K.

The focus of this research effort is on the converter which ultimately depends on the reactor system for its design The first converter was constructed with molybdenum emitter and collector electrodes. The second converter is based on CVD rhenium electrodes which have a bare work function of approximately 5.1eV. Previous work (Jacobsen and Hamerdinger 1969, and Campbell and Jacobsen 1968), illustrated in Figure 5, established that the use of rhenium as the collector electrode will yield a cesiated work function of about 1.4eV. The same work also established the optimum pressure-spacing product (pd) of 16 mil-torr. For the converter operating between 1800 K and 2000 K, the optimum spacing is between 0.102mm (0.004 inch) and 0.127mm (0.005 inch). The corresponding cesium pressure is 4 Torr, which is actively controlled by a boiling reservoir at 598 K.

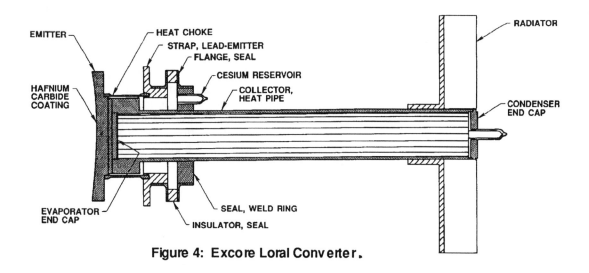

Figure 4: Excore Loral Converter.

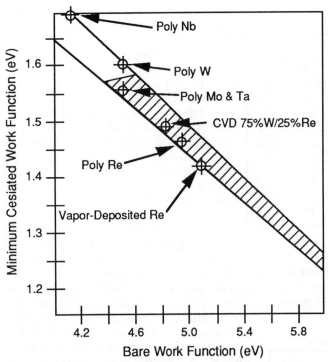

Figure 5: Cesiated & Bare Work Functions.

The collector of the converter is cooled by a 10 cm long sodium heat pipe. The sodium heat transports the waste heat through the neutron reflector which surrounds the reactor. The envelope of the heat pipe is constructed of niobium tube with 24 grooves on the ID. This wick design produces a heat transport capacity of 1000 watts. The small temperature differential afforded by the heat pipe allows the radiator to operate very close to the collector temperature which results in a minimum radiator area.

Performance Data

The emitter is designed to operate between 1850 K and 2050 K with the collector at 900 K to 1000 K. Peak converter efficiencies range between 11% and 14% at power outputs of 30 watts and 58 watts, respectively. The accumulated test time in the thermionics facility at Wright Laboratory has exceeded 600 hours. Figure 6 shows current density characteristics in the ignited mode for emitter temperatures between 1723 K and 2023 K. The collector and cesium reservoir are thermally coupled to the emitter; therefore, the collector temperature could not be varied independently.

The collector heat pipe was tested against gravity (evaporator up) under heat loads of 200 watts to 250 watts with operating temperatures of 923 K to 1100 K, respectively. The tests demonstrated an overall temperature differential within 10 K. The design load of 300 watts thermal represents approximately 27% of the transport capacity of this grooved heat pipe wick configuration. The only shortcoming observed in thermal performance concerned the radiator fin design. Poor conduction in the heat pipe/fin interface, and poor conduction along the fin resulted in poor radiation heat transfer performance. This issue is being addressed in the third task of the program with the incorporation of a variable conductance heat pipe (VCHP) concept for the radiator fin. The VCHP will control the temperature of the collector electrode independent of the thermal load from the emitter.

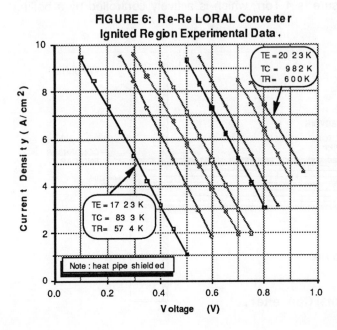

FIGURE 6: Re-Re LORAL Converter Ignited Region Experimental Data.

SUMMARY

The thermionics technology programs at Wright Laboratory offer excellent opportunities for synergism between the in-core and excore thermionic technologies. Such focusing can provide the national thermionic SNP program with viable in-core or out-of-core options for thermionics conversion. Out-of-core devices have been fabricated and are under test in Wright Laboratory's thermionics laboratory. In-core TFE devices are under construction. Testing of the in-core converters will commence in 1993; the electrically heated TFE tests will be conducted at Wright Laboratory, and the nuclear heated TFE tests will be performed in the TRIGA reactor at General Atomics.

The Thermionics Critical Technology Investigation has shown that a rhenium-rhenium converter operating in a close-spaced, ignited mode can achieve efficiencies of 14% with a cell potential of 0.7 volts. The test program at Wright Laboratory is expected to determine the long-term, as well as short-term, performance data needed for this type of thermionic energy converter.

The stated goal of the ATI program is the development of a robust, long-life, high efficiency, simplified TFE which can be incorporated into other SNP reactor programs. The first step in this process incorporates two innovations resulting in the "core-length, creep resistant emitter" TFE which is now approaching final design stages. This ATI-1 converter will become the test bed for further evolutionary and revolutionary improvements in thermionic converter technology. Successful incorporation of concepts such as dual vapor ionization and oxygenation is expected to boost efficiencies to the range of 15% to 25%. Enhanced reliability and producibility is expected to result through reduced complexity by incorporation of diamond films and simple, passive cesiation techniques.

Acknowledgements

Funding for the ATI program was provided by SDIO's Innovative Science and Technology Office under the direction of Dr. Len Caveny. Funding for SBIR's was supplied under the management of Mr. Carl Nelson, also of SDIO's Innovative Science and Technology Office. The Thermionics Critical Technology Investigation is funded by the United States Air Force through Exploratory Development Funds. The authors wish to thank William Homeyer of General Atomics, Charlie Jalichandra of LORAL Electro Optical Systems, and Dr Ramalingam of Universal Energy Systems for their generous assistance in making this paper possible.

References

AF Contract F33615-89-C-2948, First Interim Report,"Advanced Thermionics Initiative-TFE Program Support", General Atomics, San Diego, CA

AF Contract F33615-89-C-2018, Space Power Inc, San Jose, CA

AF Contract F33615-89-C-2950, Interim Report, "Advanced Thermionic Technology Initiative Program", Universal Energy Systems, Inc., February 1991.

AF Contract F33615-87-C-2706, Task 2 Final Report, "Thermionic Critical Technology Investigation", LORAL Electro Optical Systems, Pasadena, CA.

P. Jalichandra, R.W. Hamerdinger, E.A. Anderson, T.R. Lamp, and Brian Donovan (1991), "Thermionic Critical Technology Investigation", 26th Intersociety Energy Conversion Engineering Conference, Boston, MA..

Begg, L.L. (1990), "The Thermionic Fuel Clad/Emitter with Long Life Potential," 7th Symposium on Space Nuclear Power Systems, Albuquerque, NM.

BARIUM INTERACTION WITH PARTIALLY OXYGEN-COVERED Nb(110) SURFACES

Gerald G. Magera
Department of Applied Physics and Electrical Engineering
Oregon Graduate Institute of Science & Technology
19600 N.W. von Neumann Drive
Beaverton, OR 97006-1999
(503) 690-1130

and

Paul R. Davis
Linfield Research Institute
Linfield College
900 S. Baker Street
McMinnville, OR 97128-6894
(503) 472-4121

and

Thomas R. Lamp
Wright Laboratory Aerospace Power Division
Wright Patterson AFB, OH 45433-6523
(513) 255-6235

Abstract

We present results of investigations of the adsorption and desorption of Ba at the single crystal Nb(110) surface. The experiments were conducted under ultrahigh vacuum conditions, using the techniques of line-of-sight thermal desorption mass spectrometry, Auger electron spectroscopy and retarding potential work function measurement. All measurements were done on a stable oxygen layer on an Nb(110) surface that yielded a work function of 5.2 eV. Work functions as low as 2.47 eV were measured after Ba was deposited on the surface. Desorption of adsorbed Ba from the surface produces a Ba spectrum consisting of three peaks at temperatures of approximately 1110, 1225, and 1500 K, suggesting three different binding sites for Ba at the surface.

INTRODUCTION

Thermionic energy conversion (TEC) requires electrode materials with appropriate refractory character and electron emission or collection characteristics when operated in an environment of alkali or alkaline earth metal vapor. In in-core thermionic power applications, Nb is preferred as a collector material, since it is easy to fabricate, has a low neutron interaction cross section and exhibits a satisfactorily low work function with adsorbed Cs at the desired collector operating temperature at about 1000 K. While pure Cs has traditionally been used as the interelectrode vapor in these devices, recent converter experiments suggest that the use of dual vapor, Cs-Ba sources may provide significant improvements in device efficiency (Kalandarishvili et al. 1990).

As part of a program to investigate fundamental interactions of Cs and Ba at TEC electrode surfaces, we have been studying the adsorption of Ba onto partially oxygen-covered Nb(110) surfaces. Although some work on Ba adsorption onto Nb surfaces has been reported in the literature, very little, if any, work has been done on Nb surfaces with carefully determined levels of oxygen present. Since oxygen is difficult to remove completely from Nb surfaces, it seems likely that actual TEC collector surfaces operate with some oxygen present. The results reported here may therefore help improve our understanding of the interactions occurring at real collector surface in Cs-Ba TEC devices.

EXPERIMENTAL

The experimental setup was enclosed in a bakeable ultra high vacuum (UHV) stainless steel vacuum system with Ti ion pumps and turbomolecular pumping. Substrate and film studies were performed using Auger electron spectroscopy (AES), field emission retarding potential (FERP) method of work function measurement, and line-of-sight thermal desorption mass spectroscopy (TDS). The residual gas pressure in the vacuum chamber was measured by a Bayard-Alpert ion gauge and remained in the mid 10^{-10} torr range during work function and AES studies. The pressure rose to the low 10^{-8} torr range during TDS and flash cleaning. The substrate used in this investigation was a Nb(110) single crystal in the form of a disc of 10.0 mm diameter and 1.0 mm thickness which was cut from an electron beam zone refined single crystal. The surface was mechanically polished with diamond paste and was oriented within 2° of the <110> direction for the surface plane, as determined by the Laue X-ray diffraction technique. A tungsten-rhenium thermocouple (W-5% Re and W-26% Re) welded to the front edge of the crystal and calibrated by an optical pyrometer was used to monitor crystal temperatures. Estimated uncertainty in temperature values used is ± 50 K. Crystal heating was provided by means of electron bombardment.

Heated iron clad Ba getter wire was used as a source of metallic Ba. Auger electron spectroscopy (AES) measurements were made using a cylindrical mirror Auger electron optics system with a 3-0kV incident electron beam and an emission current of 10 μA. AES was used to measure initial substrate cleanliness and to monitor the surface during adsorption. Most AES data presented in this paper are in terms of peak heights or peak-to-peak ratios because of the well known uncertainties in determining percentage concentration values. Where percentage concentrations are given, they were found using relative sensitivities from published peak height data of the elements (Davis et al. 1976).

Work function measurements were taken using a prototype field emission retarding potential (FERP) gun. FERP is based on the basic principles of field-emission theory (Holscher 1966), and it provides a simple way to measure absolute work functions. A description of the FERP method is detailed elsewhere (Strayer et al. 1973). A quadrupole mass spectrometer was used to do line-of-sight thermal desorption mass spectroscopy (TDS) of Ba from the Nb(110) substrate. The desorption studies were done linearly by heating the Ba/Nb(110) surface in front of the quadrupole mass spectrometer.

RESULTS AND DISCUSSION

AES Measurements

In a typical experimental run the flashed Nb(110) substrate was exposed to a constant flux of Ba for a series of short time intervals. After each exposure the AES spectrum was recorded and then either work function or desorption data were taken. During the preparation of the Nb(110) substrate prior to Ba deposition, the Nb(110) was flashed to ~ 2300 K for 2 min. After this treatment, the surface was analyzed with AES, and showed ~ 6% oxygen, but the concentrations of other impurities such as carbon were $\leq$ 0.2%, that is, at or below the AES nominal detection sensitivity. AES measurements on the Nb(110) single crystal showed oxygen was always present. It has been observed by several others (Rieder 1980 and Haas 1966) that it is difficult to clean the Nb(110) single crystal face of oxygen, by heating and even by heavy Ar^+ sputtering. Hence, all the experiments described here were measurements of Ba on an Nb(110) single crystal with residual oxygen coverage. We made no attempt to remove the residual ~ 6% oxygen from the surface, but rather decided to study this surface as more representative of what would be used in an actual TEC device.

During the adsorption of Ba, the AES spectrum showed that the apparent oxygen concentration would typically increase very slightly, indicating that there may have been a small amount of BaO deposited on the surface. After each experimental run the Nb(110) substrate was flashed clean of Ba and AES was used to confirm that the oxygen and impurity concentrations remained constant.

Work Function Measurements

In this investigation we were interested in how the work function of the Nb(110) surface changed as a function of Ba coverage. Figure 1 shows how the work function changed as increasingly more Ba was adsorbed on the surface. The measure of coverage used in this figure is the Ba (57 eV)/Nb (167 eV) AES peak height ratio. The starting Nb(110) substrate had a work function of about 5.2 eV. It has been observed by others (Protopopov et al. 1968 and Lebanc et al. 1974), using thermionic emission methods that the work function for oxygen-free Nb(110) is 4.8-4.9 eV. The higher value we report is due to the presence of the residual oxygen on the surface. The adsorption of Ba reduced the work function of the Nb(110) substrate rather quickly, to a minimum of about 2.5 eV at an AES peak height ratio of ~ 0.027. As more Ba was adsorbed onto the Nb(110) surface, the work function would maximize at ~ 3.2 eV and remain fairly constant with increasing Ba coverage. The work functions measured on the Ba/Nb(110) system showed a repeatable pattern throughout the investigation.

TDS Measurements

A typical TDS curve of Ba from Nb(110) is shown in Figure 2. The starting Ba/Nb AES peak ratio was 0.38 for this particular curve. The peak that occurs at ~ 1495 K is a first order desorption peak and its position does not depend on coverage. The energy of activation for this peak was calculated to be 4.02 eV (Redhead 1962). No further desorption occurred above the temperature range indicated in Figure 2, and it was verified using AES that

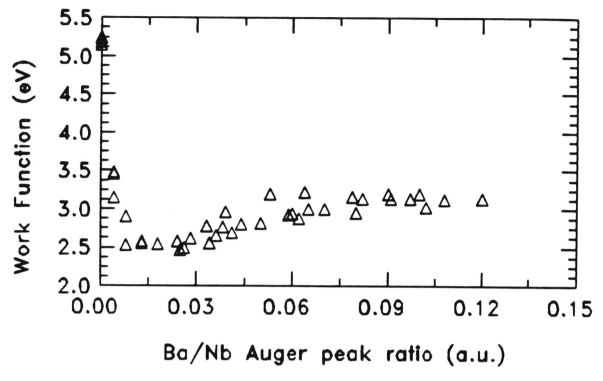

FIGURE 1. Illustration of the Change in the Work Function of the Nb(110) as Ba is Deposited at the Surface. The ratio is the Ba(57 eV) Auger peak to the Nb(167 eV) Auger peak. The minimum work function occurs at a ratio of 0.027.

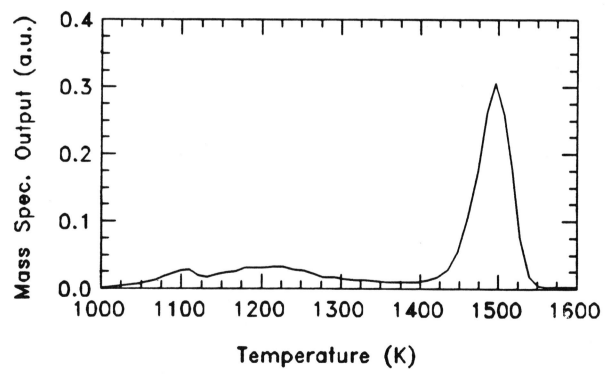

FIGURE 2. A Typical Desorption Curve from the Ba/Nb System. The Ba/Nb Auger peak ratio was 0.38 for this particular curve.

all the adsorbed Ba had been removed. The two lower temperature peaks have activation energies which depend on Ba coverage, and may be higher-order desorption peaks. In ongoing studies, we plan to determine how these peaks depend on Ba coverage.

CONCLUSION

The desorption spectrum shown in Figure 2 is different from typical results for electropositive atom adsorption onto metal surfaces. This spectrum, with its large high temperature peak, is very reminiscent of that reported by Desplat and Papageorgopoulos for Cs adsorption onto a thick film of WO_2 (Desplat 1980), which was explained by Cs diffusion into the oxide film. In the present case, we observe a work function minimum (Figure 1) at a very low Ba/Nb AES peak ratio, which Desplat and Papageorgopoulos did not observe in their experiments. Strangely, although the initial slope of our work function-coverage curve is steep, implying a large dipole moment per adsorbed Ba atom, the work function minimum is not as low as that reported for Ba adsorption onto the clean Nb(110) surface.

These findings, coupled with the additional information that Nb forms a stable oxide, and that oxygen is extremely difficult to remove completely from the Nb(110) surface, lead to some interesting possibilities. It may be possible that at least some of the surface of our sample consists of a relatively thick (more than one atomic layer) Nb oxide or substoichiometric oxide layer with an O-Nb structure extending into the bulk of the material. Further studies, including the adsorption of additional oxygen, the complete removal of residual oxygen, and the adsorption of Ba onto these carefully prepared surfaces, should provide us with a better understanding of the Ba/O/Nb(110) system.

Acknowledgments

Research activities are under UES, Inc. contract F33615-89-C-2950 sponsored by the Aero Propulsion and Power Directorate, Wright Laboratory, Aeronautical Systems Division (AFSC), United States Air Force, Wright-Patterson AFB, OH 45433-6563.

References

Davis, L. E., N. C. MacDonald, P. W. Palmberg, G. E. Riach, and R. E. Weber (1976) Handbook of Auger Electron Spectroscopy, 2nd Ed., Physical Electronics Industries, Edina, MN, pp. 11-15.

Desplat, J. L. and C. A. Papageorgopoulos (1980) "Interaction of Cesium and Oxygen on W(110), I. Cesium Adsorption on Oxygenated and Oxidized W(110)," Surf. Sci., 92: 97.

Haas, T. W. (1966) "A Study of the Niobium (110) Surface Using Low Energy Electron Diffraction Techniques," Surf. Sci., 5: 345-358.

Holscher, A. A. (1966) "A Field Emission Retarding Potential Method for Measuring Work Functions," Surf. Sci., 4: 89-102.

Kalandarishvili, V. G. Kashiya, and B. I. Ermilov (1990) "Experimental Studies on Thermionic Devices with Cs-Ba Fillings," in Proc. Conf. on Nuclear Power in Space, Obninsk, USSR.

Leblanc, R. P., B. C. Vandrugghe, and F. E. Girouard (1974) "Thermionic Emission from a Niobium Single Crystal," Canadian Journal of Physics, 52: 1589-1593.

Protopopov, O. D. and I. V. Strigushchenko (1968) "Emission Parameters of Faces of a Niobium Single Crystal," Sov. Phys-Solid State, 10: 747-8.

Redhead, P. A. (1962) "Thermal Desorption of Gases," Vacuum, 12: 203-210.

Rieder, K. H. (1980) "On the Interaction of Oxygen with Nb(110) and Nb(750)," Appl. of Surf. Sci., 4: 183-195.

Strayer, R. W., W. Mackie, and L. W. Swanson (1973) "Work Function Measurements by the Field Emission Retarding Potential Method," Surf. Sci., 34: 225-248.

SORPTION RESERVOIRS FOR THERMIONIC CONVERTERS

Kevin D. Horner-Richardson
Thermacore, Inc.
780 Eden Road
Lancaster, PA 17601
(717) 569-6551

Kwang Y. Kim
WL/POOC
Wright-Patterson Air Force Base
Dayton, OH 45433-6563
(513) 467-4780

Abstract

A metal matrix sorption reservoir has been designed and evaluated for use as an integral, self-regulating source of cesium for thermionic energy converters. The reservoir utilizes a sintered porous matrix of powder metal and graphite. The graphite acts as a sorption medium to store cesium as carbon-cesium compounds. The porous metal structure captures the graphite to prevent crumbling and splitting normally associated with graphite-based sorption reservoirs. Cesium vapor is supplied to the converter via high temperature carbon-cesium equilibrium reactions. Testing has demonstrated feasibility of the concept with graphite. Metal matrix reservoirs remained structurally intact under test conditions which caused pure graphite reservoirs to swell and split. Alternative sorption media have been identified to potentially replace graphite in the metal matrix. Metal matrix sorption reservoirs provide the potential for robust, internal, cesium reservoirs which naturally respond to changes in thermionic converter operating conditions to supply optimum cesium vapor pressure with no external power or control circuits.

INTRODUCTION

Thermionic energy conversion has been the subject of considerable investigation in the United States and abroad since the technology was first conceived in the mid 1950s. It has long been recognized that there is a need for practical methods for maintaining optimum cesium pressure within a thermionic converter. Cesiated thermionic converters used in research work most often use liquid cesium reservoirs because they present the simplest and most accurate method to control cesium pressure within the converter. A liquid cesium pool, connected to the converter through a tube, provides cesium vapor as the vapor pressure of the heated pool. As thermionic energy converters make their way out of the laboratory and into space reactors, these liquid reservoirs present at least two difficulties. First, the required cesium reservoir temperature typically ranges from 520 to 620 K (60-1300 Pa). In a thermionic reactor core, however, the minimum temperatures are typically 800-900 K so that liquid reservoirs must be located outside the reactor with separate electric heaters and control systems, and tubes to carry the cesium to the converters. Such a design constitutes a serious challenge to the practicality of any large scale thermionic system. Second, the reservoirs do not directly respond to temperature changes within the converters so that performance penalties and temperatures instabilities may result (Shock 1968).

An attractive solution to the problems associated with liquid cesium reservoirs is the use of an integral, solid sorption reservoir within each converter. These integral reservoirs can be designed to operate at some intermediate temperature between the emitter and the collector and thereby supply the required cesium vapor pressure to the converter without the need for auxiliary plumbing or heating and control circuits. High temperature sorption reservoirs have been pursued for 25 years. Harbaugh and Basiulis (1966) investigated sintered powdered metal reservoirs to store cesium by adsorption to the metal surface. These adsorption type reservoirs generally resulted in short lived converters due to the limited storage capabilities of the reservoirs. Yates and Fitzpatrick (1968) investigated cesium reservoirs based on carbon-cesium compounds. A series of compounds are formed from $C_{60}Cs$ to C_8Cs, when cesium intercalates into graphitic carbon. These compounds set up a two phase equilibrium in which cesium vapor exists at pressures from 60-1300 Pa (0.5-10 torr) over temperatures ranging from 700 K to 1200 K. Specific cesium pressure over the graphite depends only on the graphite temperature and the compounds involved. These graphite-cesium reservoirs overcome the low storage capacity limitations of metal adsorption reservoirs.

Extensive work in the area of graphite-cesium reservoirs has established the technology as the current state-of-the-art. The series of carbon-cesium compounds was identified and characterized by Salzano and Aronson (1965). Cesium loading parameters for specific carbon-cesium compounds were specified by Devin, et al. (1967). More recently, Smith et al. (1990) have investigated the effects of high neutron fluence on graphite-cesium reservoirs for thermionic fuel elements. Graphite-cesium reservoirs could provide several advantages over the liquid cesium reservoir approach including:

1. Optimum cesium pressure at temperatures which fall between the emitter and collector temperatures. The reservoirs eliminate the need for separate control of cesium temperatures.

2. Better controllability of cesium pressure. Graphite-cesium reservoirs are less sensitive to small temperature variations than liquid cesium reservoirs.

3. No liquid cesium present in thermionic converters for space applications.

4. The potential for direct temperature feedback from the converter to control cesium vapor pressure.

The primary difficulty encountered in using graphite cesium reservoirs has been the physical instability of the graphite. The graphite swells and crumbles as cesium intercalates through the interlaminar spaces. Investigators have focused on variations in graphite type and purity levels to minimize this structural instability, and each has experienced life-limiting swelling and crumbling (Yates et al. 1968, Salzano et al. 1965, and Devin et al. 1967). As a result, attempts to affix graphite-cesium reservoirs to locations within thermionic converters have been unsuccessful. The crumbling graphite-cesium material is redistributed to different locations within the converter, changing its cesium supply pressure and creating potential electrical short circuits.

This paper describes work conducted at Thermacore to develop a metal-matrix sorption reservoir (MMSR) to overcome the structural instability of graphite-cesium sorption reservoirs. The concept uses an open porous sintered powder metal to encapsulate a powdered sorption medium. The sintered powder metal gives structural support to the sorption medium and provides a mechanism to attach the reservoir to hot thermionic converter parts. Proof-of-concept was demonstrated by using powdered, high purity pyrolytic graphite as the sorption medium. Other potential storage media were investigated and proposed for further study. The metal matrix sorption reservoir resolves long standing structural problems associated with graphite-cesium intercalation compounds and makes possible the use of other powdered sorption media as alternatives to graphite.

EXPERIMENTAL PROCEDURE

The metal matrix sorption reservoir concept is being developed for use as an integral, self-regulating cesium reservoir for Thermionic Fuel Elements (TFEs). Initial investigations focused on establishing basic feasibility of the metal matrix concept with graphite as the sorption medium, and on identifying potential alternatives to graphite for use as cesium sorption media. Initial tests were performed to verify increased dimensional stability of the metal matrix reservoirs over pure graphite reservoirs. Feasibility testing consisted of 1000-hour life-tests of candidate reservoirs under cesium pressures and operating temperatures typical of thermionic operating conditions. This first phase study did not attempt to measure cesium vapor pressure above metal matrix sorption reservoirs. The basic pressure-temperature relationships for graphite-cesium compounds is well documented in the literature (Salzano et al. 1965 and Devin et al. 1967) and further verification was felt to be beyond the scope of this initial effort.

Two variations on the MMSR concept were investigated. The first consisted of a sintered matrix of nickel powder and pyrolytic graphite powder. The nickel powder was selected for its sinterability and compatibility with carbon and cesium. Pyrolytic graphite powder was selected as the most favorable graphite for cesium storage. The pyrolytic graphite is produced by vapor deposition followed by a high temperature heat treatment (3300 K). The result is a dense graphite with highly oriented lamellar planes. This form of graphite is reported to have well defined equilibrium transitions from C_8Cs up through the $C_{24}Cs \rightarrow C_{36}Cs$ reaction (Yates and Fitzpatrick 1968). The second MMSR design used a tungsten foam metal matrix fabricated by CVD. Tungsten was selected for its high temperature capability and its proven application to in-core TFEs. Graphite was formed in the pores of the foam by infiltration with carbon resin which was later graphitized by heating.

Life testing of the reservoirs was conducted under conditions which cause pure graphite reservoirs to swell and split. Four sample reservoir types were life tested. Details of these four reservoirs are given in Table 1.

TABLE 1. Cesium Reservoir Design Details.

Sample No.	Graphite Type, Weight & Source	Metal Matrix	Fabrication Type	Comments
1	High purity pyrolytic graphite powder, 100 mesh 0.66 g Pfizer, Inc.	Ni powder -100 +150 mesh 12.34 g	Sinter	• Good sinterability • Good matrix strength • Expected to provide greatest dimensional stability
2	Formed in stu by graphitizing carbon resin 0.37 g Ultramet, Inc.	Tungsten foam 1.03 g	CVD	• Open porous structure • Weaker strength than #1 • Expected to provide good dimensional stability
3	High purity pyrolytic graphite plate 11.0 g Pfizer, Inc.	-----	Machined from stock	• Highest graphitic carbon content • Expected to have poor dimensional stability
4	POCO graphite AXZ-5 4.0 g POCO Graphite, Inc.	-----	Machined from stock	• Expected to have better dimensional stability than pyrolytic graphite

Reservoir types 1 and 2 are the two variations on the MMSR concept discussed above. Reservoir types 3 and 4 are varieties of pure graphite currently being investigated as cesium reservoirs on other programs. The two pure graphite samples were included in the test matrix to confirm the previously reported dimensional instability of graphite reservoirs, and to gauge the degree to which testing simulated the conditions that cause graphite crumbling.

Three sets of the four test samples were fabricated for testing. Each reservoir was carefully cleaned, degassed and weighed using procedures developed in previous efforts (Yates and Fitzpatrick 1968). Table 2 gives the specific life test matrix used.

TABLE 2. Cesium Reservoir Life Test Matrix.

SAMPLE SET	GRAPHITE TEMPERATURE TG (K)	CESIUM TEMPERATURE TC, (K)	CESIUM LOADING EXPECTED (g)	COMPOUND EXPECTED (C_8cS)	TEST DURATION (HOURS)
Set #1 (control)	------	-----	0	------	0
Set #2	890	890	1.4 g Cs/1 g C	C_8Cs	1320
Set #3	890	590	1.1 g Cs/1 g C	$C_{10}Cs$	1320

Reservoirs from Set #1 were used as the control set and were not exposed to cesium. Reservoirs from Set #2 were each sealed in evacuated test capsules along with 5 grams of cesium. The test capsules were placed in a life test furnace and heated to 890 K to determine the effects of an unlimited supply of high pressure cesium. Reservoirs from Set #3 were placed in test cells and manifolded to a liquid cesium pool. These test cells were placed in a life test furnace and held at 890 K. The cesium pool was maintained at 590 K to supply a constant over-pressure of cesium to the reservoirs at 372 Pa. Testing of Set #3 was designed to closely simulate loading conditions typical of thermionic conditions. Figure 1 shows the life test apparatus for Set #3.

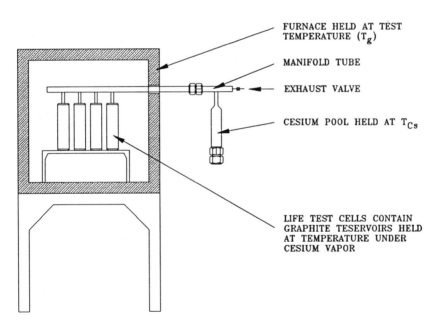

FIGURE 1. Reservoir Life Testing Apparatus.

TEST RESULTS

Life testing of sample reservoirs using the procedures described above was conducted for a period of 1320 hours. At the conclusion of the life test period, the test cells were cooled, removed from the test furnace and opened in an inert gas glove box for inspection. Each reservoir sample was removed from its test cell, photographed, and weighed. The post test reservoir weight was compared to the pretest weight to determine the quantity of cesium loaded into the reservoir. After inspection, each sample was potted in epoxy to stabilize the reactive carbon-cesium compounds. Potted samples were later sectioned and micrographed to observe changes in the graphite or metal structure.

A summary of life test results is given in Table 3. In general testing demonstrated that each sample reservoir absorbed cesium as expected. All of the pure graphite samples had crumbled or completely delaminated as expected. The metal matrix reservoir samples were essentially intact after life testing with the exception of one tungsten foam reservoir from set #2 which had crumbled. Reservoirs using sintered powder metal showed good dimensional stability and will be investigated further to optimize powder material and sintering procedures. Testing indicates that graphite can be captured and stabilized within metal matrix structures to prevent the splitting and crumbling normally associated with the formation of graphite-cesium compounds.

TABLE 3. Reservoir Life Test Results.

SAMPLE	PRETEST WEIGHTS		POST TEST RESULTS				COMMENTS
	A. SAMPLE WEIGHT (g)	GRAPHITE CONTENT (g)	B. SAMPLE WEIGHT (g)	CESIUM LOAD (B-A)(g)	MOLAR RATIO $c/_{Cs}$	PROBABLE COMPOUND FORMED	
Set #2							
2-1 Ni/graphite	13.11	0.67	16.20	3.09	2.4	C_8Cs	SAMPLE UNAFFECTED BY LIFE TESTING
2-2 W/graphite	1.58	0.42	2.80	1.22	3.8	C_8Cs	SAMPLE CRUMBLED DURING LIFE TESTING
2-3 Pyrolytic graphite	10.75	10.75	16.59	5.84	20.5	$C_{24}Cs$	SAMPLE SWELLED AND DELAMINATED
2-4 Poco graphite	3.89	3.89	9.40	5.51	7.7	C_8Cs	SAMPLE SWELLED AND CRUMBLED TO POWDER
Set #3							
3-1 Ni/graphite	12.89	0.66	13.45	0.56	13.1	$C_{10}Cs$	SAMPLE BROKE IN HALF-OTHERWISE UNAFFECTED
3-2 W/graphite	1.38	0.37	1.95	0.57	7.2	$C_{10}Cs$	SAMPLE UNAFFECTED BY LIFE TESTING
3-3 Pyrolytic graphite	9.93	9.93	20.62	10.69	10.3	$C_{10}Cs$	SAMPLE BROKE-UP SMALL PLATES
3-4 Poco graphite	3.90	3.90	7.52	3.62	12.1	$C_{10}Cs$	SAMPLE SWELLED AND CRUMBLED TO POWDER

ALTERNATIVE SORPTION MEDIA

Several alternatives to graphite as a sorption medium have been identified for use with the metal matrix sorption reservoir concept. One such alternative is the general class of materials known as dichalcogenides of metals. Dichalcogenides of metals, including molybdenum disulfide (MoS_2) and tungsten disulfide (WS_2) have bene found to intercalate cesium like graphite (Tsydinos) but have not been pursued by the thermionics industry, in part because they are typically only available in powder form. The metal matrix sorption reservoir concept provides the potential for exploiting dichalcogenides as cesium sorption reservoirs. Such a design would eliminate graphite from TFE systems, removing the associated risks of carburization of TFE parts and graphite migration causing electrical short circuits.

APPLICATION TO THERMIONICS

For application of sorption reservoirs to thermionic converters, it is necessary to know the total amount of cesium atoms which must be stored to maintain operating conditions over the converter life. Hernqvist and Levine (1966) proposed a method to calculate a total number of cesium atoms to be stored in a reservoir. During normal converter life, cesium is lost to adsorption on cool surfaces, small leaks, and oxidation. Therefore, for long converter life, the reservoir should hold 10 to 100 times the number of atoms required. The number of atoms of cesium required in a converter at any time is given by the equation.

$$N = nV + S\sigma_f \theta \qquad (1)$$

where

  N is the total number of atoms needed,
  n is the density of cesium in the converter volume,
  V is the converter volume,
  S is the total converter surface area for adsorption,
  σ_f is the density of a monatomic layer of cesium = 4.8×10^{14} (atoms/cm^2), and
  θ is the coverage fraction (assumed equal to 1).

Thus, the number of atoms to be stored in a reservoir holding 100 times the amount of cesium required is:

$$N_{reservoir} = 100[nV + S\sigma_f \theta] \qquad (2)$$

Using this requirement, the amount of sorption material needed can be established. For this study, graphite based metal matrix reservoirs were sized for the requirements of the U.S. Air Force Advanced Thermionic Initiative (ATI) TFE. Reservoir size is dependent on the particular equilibrium reaction used to provide cesium vapor, which in turn is dependent on the temperature of the location chosen for placement of the reservoir. Table 4 summarizes size requirements for potential operating temperature ranges of reservoirs in the TFE.

TABLE 4. Thermionic Reservoir Requirements.

TEMPERATURE OF RESERVOIR (K) (EQUILIBRIUM REACTION)	N_{REQ} # OF Cs ATOMS	GRAPHITE REQUIRED (g)	MATRIX REQUIRED (g)	MATRIX VOLUME REQUIRED (cm^3)
750 - 800 $[4C_{10}Cs\ (s) + Cs\ (g) = 5C_8Cs\ (s)]$	1×10^{19}	9.4×10^{-3}	0.17	4.3×10^{-2}
850 - 1000 $[5/7\ C_{24}Cs\ (s) + Cs\ (g) = 12/7\ C_{10}Cs\ (s)]$	1×10^{19}	4.0×10^{-3}	0.078	1.9×10^{-2}
1000-1150 $[2\ C_{36}Cs\ (s) + Cs\ (g) - 3C_{24}Cs\ (s)]$	1×10^{19}	1.7×10^{-2}	0.33	8.2×10^{-2}
1050 - 1250 $[3C_{48}Cs\ (s) + Cs\ (g) - 4C_{36}Cs\ (s)]$	1×10^{19}	3.4×10^{-2}	0.66	1.6×10^{-1}

By proper positioning of the integral cesium reservoir, the cesium pressure can be made to track automatically so that optimum pressure is maintained in the converter over a range of emitter temperatures. Gietzen (1969) analytically evaluated the influence of position and thermal coupling of integral graphite reservoirs on the performance characteristics of a thermionic converter. The study determined that to maintain optimum thermionic conditions at all power levels, the reservoir temperature must be strongly coupled to the emitter (that is, by conduction). For a cylindrical Thermionic Fuel Element with a metal matrix sorption reservoir, this condition can be achieved by attaching the reservoir to the cylindrical emitter lead. An optimized emitter lead produces a temperature gradient from the emitter temperature to the collector temperature. Proper placement of the MMSR by sintering on the emitter lead could provide an internal cesium sorption reservoir, operating at the optimum temperature, with feedback from the emitter required for self-regulation. Figure 2 shows the proposed placement of the sorption reservoir in a core length TFE. Analytical work is required to determine the optimum location on the emitter lead.

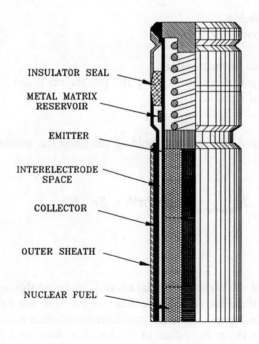

FIGURE 2. Internal Metal Matrix Sorption Reservoir for ATI-TFE.

CONCLUSIONS

Phase I research has successfully fabricated and tested metal matrix sorption reservoirs. Test results show that a sintered metal matrix can be made to support a powdered sorption medium. In Phase I, metal matrix sorption reservoirs using graphite were tested. These reservoirs demonstrated good dimensional stability under cesium pressures and temperatures which caused pure graphite reservoirs to swell and crumble.

Alternative sorption media were identified for use with the metal matrix sorption reservoir concept. These alternative materials may be found to have favorable cesium storage characteristics which would allow for the elimination of graphite from TFE systems.

Phase II work is underway to fully characterize the cesium pressure control capabilities of metal matrix sorption reservoirs. Mapping of the cesium pressure-temperature relationships is required for the graphite/matrix as well as for MoS_2, WS_2 and other alternative sorption media. Alternative powder metals such as tungsten will be investigated for improved strength and high temperature compatibility. A thermal model of potential locations within the TFE is required to determine optimum placement of the integral reservoir within the TFE. Phase II research is expected to result in stable, predictable, cesium sorption reservoirs for use as integral cesium reservoirs for Thermionic Fuel Elements.

Acknowledgments

This work was supported by the Innovative Science and Technology Branch of the Strategic Defense Initiative Organization. The program was managed by Wright Laboratory as a Phase I Small Business Innovative Research Program.

References

Devin, B., R. Lesueur, and R. Setton (1967) "Vapor Pressure of Cesium above Graphite Lamellar Compounds," in Proc. Thermionic Specialists Conference, held in Palo Alto, CA, pp. 277-286.

Gietzen, A.J. (1969) "Performance Evaluation of a Thermionic Converter a Graphite Integral Reservoir" in Proc. Thermionic Specialists Conference, held in Carmel, CA, pp. 535-541.

Harbaugh, W.E. and A. Basiulis (1966) "The Development of a High Temperature Reservoir for Automatic Control of Cesium Pressure", in Proc. Thermionic Specialists Conference, held in Houston, TX, pp. 243-245.

Hernqvist, K.G. and J.D. Levin (1966) "Adsorption Type Reservoirs for Gas Tubes," RCA Review, Vol. 27, No. 1, pp. 140-148, March, 1966.

Salzano, F.J. and S. Aronson (1965) "Thermodynamics of the Cesium-Graphite Lamellar Compounds," J. Chem. Phys., 43(1):149-154.

Shock, A. (1968) "Effect of Cesium Pressure on Thermionic Stability," in Proc. Second International Conference on Thermionic Electrical Power Generation, held in Stresa, Italy, May 1968.

Smith, J.N., T. Heffernan, T. Speidel, L. Lawrence, and S. Cannon (1990) "Evaluation of Poco Graphite as a Cesium Reservoir" in Proc. Seventh Symposium on Space Nuclear Power Systems, CONF 90019, M.S. El-Genk and M.D. Hoover, eds., University of New Mexico, Albuquerque, NM.

Tsydinos, G.A. The Properties and Structures of Inorganic Molybdenum-Sulfur Compounds," Bulletin Cdb-17 Ann Arbor, MI, Climax Molybdenum Company of Michigan.

Yates, M.K. and G.O. Fitzpatrick (1968) Cesium Sorption Materials for Thermionic Converters, U.S. A.E.C. Report A-8574, Gulf General Atomic, April 1968.

Yates, M.K. and G.O. Fitzpatrick (1968) "Cesium Sorption in Materials for Integral Reservoirs," in Proc. Thermionic Specialists Conference, held in Boston, MA, pp. 79-87.

COMPUTATION OF DIMENSIONAL CHANGES IN ISOTROPIC CESIUM-GRAPHITE RESERVOIRS

Joe N. Smith Jr. and Timothy Heffernan
General Atomics
PO Box 85608
San Diego, CA 92186
(619)455-3105

Abstract

Cs-graphite reservoirs have been utilized in many operating thermionic converters and TFEs, in both in-core and out-of-core tests. The vapor pressure of cesium over Cs-intercalated graphite is well documented for unirradiated reservoirs. The vapor pressure after irradiation is the subject of on-going study. Dimensional changes due to both intercalation and to neutron irradiation have been quantified only for highly oriented graphite. This paper describes extrapolation of the data for intercalated oriented graphite, to provide a qualitative description of the response of isotropic graphite to exposure to both cesium and neutrons.

INTRODUCTION

Cesium intercalation of highly oriented graphite is a process in which the galleries between carbon layer planes become occupied by cesium atoms. The separation between the carbon planes is increased from 3.35 Å to 5.95 Å by the insertion of cesium atoms. The change in thickness of a monolithic sample of oriented graphite can be calculated from this fact and a knowledge of the number of galleries that contain cesium. This feature of the intercalation compound is usually described by the "stage", or number of carbon planes between filled galleries. While the classical description of an intercalation compound suffices to describe this aspect of the response of graphite to intercalation, the pleated layer model shown in Figure 1 is more nearly descriptive of the compounds, when other aspects of the process are considered.

Note that every interlaminar space or gallery, contains Cs, but in localized regions (islands) only. In the perpendicular direction, these islands are separated by the same number of layers as in the classical model and the stoichiometry is still given by $C_{12n}Cs$ for $n \geq 2$. In this island model, transition from, for example, 2nd to 3rd stage is accomplished by a redistribution of Cs atoms within each occupied gallery. Initial intercalation, in the context of the island model, allows every layer to accept Cs and to retain it. Only the spatial distribution, or island size within the gallery is effected by the equilibrium state.

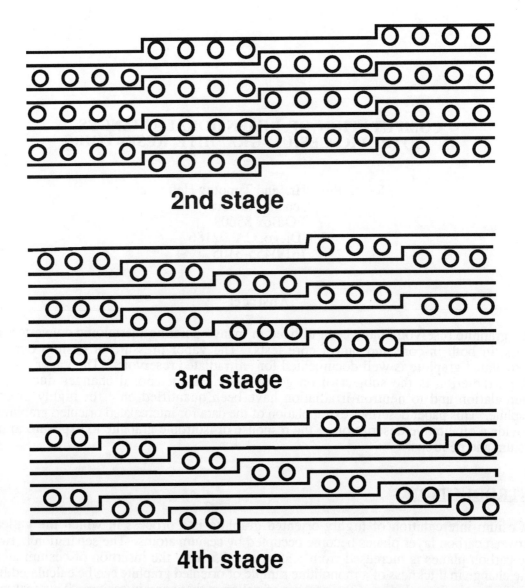

Figure 1. Pleated Layer Model for 2nd, 3rd and 4th Stage Cs-graphite Intercalation Compounds. For $C_{12n}Cs$ there are n carbon layers between Cs islands, in the vertical direction. Islands of Cs in the same interlaminar space, or gallery, are separated by a lateral distance equal to (n-1) islands. The solid lines depict the carbon layers and the open circles depict the cesium atoms.

The phenomenological model of Figure 1, in conjunction with earlier studies of neutron effects in pure graphite (Bokros et al. 1968), was used to describe the neutron induced swelling of HOPG graphite (Smith 1991). In that work, the graphite crystal was found to swell by nearly 400% in the direction perpendicular to the layer planes (that is, in the c-axis direction) and to shrink by 60% in the a-axis direction, due to exposure to 0.9×10^{22} nvt. These dimensional changes due to cesiation and to neutron exposure also occur in isotropic graphites, but the overall effects are very much diminished. The response of the isotropic material can be estimated by considering that it can be approximated by spatial averaging of the HOPG results. This approximation is performed below and its validity is borne out by a favorable comparison with experimental results on isotropic graphite.

ANALYSIS

An isotropic graphite may be considered as an isotropic distribution of oriented crystallites. If these individual oriented crystallites undergo c-axis swelling (or shrinking), ΔL_c, due to either Cs intercalation or neutron irradiation, the corresponding result for the isotropic material, ΔL_y, is obtained by spatially averaging the projection of ΔL_c in a convenient direction, say the y-axis:.

$$\Delta L_y = \int_0^{\frac{\pi}{2}} (\Delta L_c \cos\theta) \frac{(2\pi r \sin\theta\, rd\theta)}{4\pi r^2} = \frac{\Delta L_c}{4} \int_0^{\frac{\pi}{2}} \sin 2\theta\, d\theta = \frac{\Delta L_c}{4} \tag{1}$$

where $\Delta L_i = L_i - L_{io}$, the difference between the irradiated (intercalated) i^{th} dimension and the original i^{th} dimension. Table 1 gives the results of equation 1, for the first four stages of Cs-graphite.

TABLE 1. Computed swelling of Graphite due to Cs Intercalation.

Stage	Compound	Loading (mg Cs/gC)	L_c/L_{co} (oriented)	L_y/L_{yo} (isotropic)
1st	$C_{10}Cs$	1108	1.79	1.2
2nd	$C_{24}Cs$	461	1.40	1.1
3rd	$C_{36}Cs$	308	1.26	1.06
4th	$C_{48}Cs$	231	1.20	1.05

The values of L_y/L_{yo} for intermediate values of loading can be obtained by interpolation between the values for pure stage loading. For three values of loading, the comparison between the calculation and measured values is given in Table 2.

TABLE 2. Experimental and Computational Values of Swelling of Isotropic Graphite upon Intercalation.

Loading (mg Cs/gC)	L_y/L_{yo} (computed)	L_y/L_{yo} (experiment)
222	1.05	1.13
950	1.17	1.21
1130	1.20	1.18

The irradiation response of an intercalated isotropic graphite can be determined by averaging the results for HOPG in a similar fashion. In this case, however, the individual crystallites change dimension in both the c- and a-axis directions. If these two crystallographic directions are treated independently, the net result of the averaging process depends on the aspect ratio of the individual crystallites, that is, upon the ratio of crystallite diameter, d, to thickness, t, as shown in Table 3, where the results of Smith (1991) are used for c-axis swelling and a-axis contraction.

TABLE 3. Computed Neutron Induced Dimensional Changes in Cs Intercalated Isotropic Graphite for Different Crystallite Aspect Ratios; 0.9×10^{22} n/cm^2. (See Smith 1991)

Ratio of Crystallite diameter-to-thickness	Linear Dimension l_y/l_{y_0}	volume fraction V/V_0
1.0	1.77	5.5
3.0	1.46	3.1
5.0	1.15	1.52
7.0	0.84	0.59

In Table 3, note that if the aspect ratio of the individual crystallites is in the range of 6 to 7, the isotropic sample will shrink as a result of irradiation, in this approximation. In post-irradiation examination of an intercalated isotropic specimen that was in the UCA-2 series of experiments, it was found that a 14% shrinkage occurred during neutron irradiation (General Atomics 1989). Furthermore, preliminary indications are that an aspect ratio of 5 to 10 is to be expected in some isotropic graphites (Kopel 1991) These observations support the calculation leading to Table 3.

CONCLUSIONS

The reservoir volumes of TFEs must be designed with sufficient free volume to accommodate the expansion of the graphite during intercalation and also during in-core operation. The comparisons between the analytical and experimental results presented above show that the response of the isotropic graphite can be estimated from results obtained with oriented graphite. This is a useful finding, since the data with oriented graphites are generally easier to interpret. Furthermore, there is an indication that the irradiation response of isotropic graphites could be tailored by carefully monitoring the crystallite size of the material that is used.

Acknowledgments

This work was part of the Thermionic Fuel Element Verification Program (DE-AC03-86SF16298, GA Project 3450) which is jointly sponsored by the United States Department of Energy and Department of Defense under the technical oversight of the Los Alamos National Laboratory.

References

Bokros, J. C., G. L. Guthrie, and A. S. Schwartz (1968) "The Influence of Crystallite Size on the Dimensional Changes Induced in Carbonaceous Materials by High-Temperature Irradiation," Carbon, **6**:55-63.

General Atomics (1989) *TFE Verification Program semiannual report for the period ending April 30, 1989*. General Atomics report GA-A19666, Sept. 1989.

Kopel, J. (1991) POCO Graphite, Inc., private communication, Feb, 1991.

Price, R. J. (1974) "High-Temperature Neutron Irradiation of Highly Oriented Carbons and Graphites," Carbon, **12**:159-168.

Smith, J. N. Jr. (1991) "Neutron Irradiation-Induced Dimensional Changes in a Cesium-Highly Oriented Pyrolytic Graphite Intercalation Compound," Carbon, **29**(4/5):661-663.

DIAMOND FILM SHEATH INSULATOR FOR ADVANCED THERMIONIC FUEL ELEMENT

Steven F. Adams
Aerospace Power Division
Wright Laboratory
Wright Patterson AFB, OH 45433-6563
(513) 255-6235

Leonard H. Caveny
SDIO/IST
The Pentagon
Washington, DC 20301-7100
(703) 693-1671

Abstract

A continuing challenge to thermionic fuel elements is degradation of the sheath insulator in the high temperature and high neutron flux environment. Also, the present alumina insulator requires a barrier coating to isolate it from the alkali metals. The recent successes in applying polycrystalline films to metals led to the consideration of diamond as the ideal sheath insulator. Diamond possesses several of the intrinsic characteristics demanded of a sheath insulator, e.g., the highest combination of electrical resistivity and thermal conductivity at room temperature, compatibility to alkali metals, and high neutron radiation tolerance. However, a number of durability issues must be resolved before the practicality of diamond insulators can be assessed in prototype thermionic diodes. Also, since the fabrication techniques for thin diamond films are undergoing extraordinary advancement, the structural integrity and reproduceability are yet to be addressed in a systematic manner. Recent results increase the understanding of diamond as a sheath insulator. The paper quantifies the potential advantages, identifies the issues to be resolved, and summarizes the progress in resolving the issues.

INTRODUCTION

The application of thin diamond film as a high temperature electrical insulator is an intriguing goal for thermionic energy conversion researchers. The Advanced Thermionic Initiative (ATI), sponsored by SDIO's Innovative Science and Technology Directorate, is giving thermionic component developers the opportunity to evaluate concepts based on recent advances, such as diamond film. It has been proposed that chemical vapor deposited (CVD) diamond film could be applied as an advanced thermionic sheath insulator for a thermionic fuel element.

Historically, alumina (Al_2O_3) has been utilized as the sheath insulator material, but mechanical and electrical degradation of this ceramic in the harsh reactor environment of high temperature and high neutron fluences present the thermionic research community with difficult problems. A thermionic sheath insulator would typically be exposed to a temperature of 880K and a 1 MeV neutron fluence of 10^{22} n/cm^2. The alumina sheath hinders the conduction of heat away from the collector due to a relatively low thermal conductivity. The liquid alkali metal coolant, which flows around the thermionic fuel element (TFE), is not compatible with the alumina. An outer niobium sheath layer is, therefore, used in addition to the alumina both to protect the alumina and to enhance the thermal coupling to the coolant. This increases the complexity of the sheath fabrication process which already suffers from problems with cracking within the alumina layer.

The potential inertness of diamond film to the harsh reactor environment along with the prospect of easy fabrication prompted the investigation into applying diamond film sheath insulators for advanced thermionic reactors. Based on room temperature and atmospheric pressure resistivity measurements, previous research has shown diamond film to have great promise as an electrical insulator and heat conductor. Electrical resistivities of 10^{14} ohm-cm, electrical breakdown voltages of 3.5×10^6 V/cm and thermal conductivities of 18 W/cm-K have been measured on CVD diamond films under these normal environments. The values of these properties on CVD diamond is only Hxightly derated from natural diamond and may soon exceed those values.

The principal advantages of a diamond sheath insulator are: (1) simpler fabrication because the barrier sheath is not required to isolate the diamond from the alkali metal coolant, and (2) removes a failure mode if the diamond does not degrade as the alumina does. Other secondary advantages include: (1) thinner insulation layer reduces mass and volume of fuel element, and (2) reduced temperature drop across the insulation layer increases coolant temperature and thereby reduces radiator mass.

From its physical properties, diamond film is an ideal candidate for the thermionic sheath insulator, but several critical issues must be experimentally addressed before diamond film can be considered a viable alternative to the alumina sheath insulator. These issues are as follows:

1. The uniform and continuous deposition of diamond film onto the refractory metal outer surface of a cylindrical TFE;
2. The maintenance of sufficient electrical insulation and thermal conduction within the diamond film continuously operating at 880K;
3. The damage to the film by the neutron fluence of a thermionic reactor and the effects on the electrical and thermal properties of the film;
4. The degradation due to exposure of the diamond film to a flow of liquid alkali metal coolant at 880K within the thermionic reactor;

As research proceeds on these issues, the capability increases to quantify the gains achievable through the use of diamond films in the thermionic fuel element. The results of the diamond film studies conducted so far under the ATI program allow us to make certain quantified assessments of the advantages and disadvantages of the diamond sheath insulator.

ANALYTICAL PROCEDURE

To quantify the potential advantages of a diamond film sheath insulator, the results from several diamond film research efforts under the ATI program were considered. For systems level advantages, the present ATI fuel element design was used for comparison between the diamond film sheath insulator and the conventional alumina tri-layer sheath insulator. The present ATI fuel element design (25 cm in length) calls for a cylindrical niobium collector blanketed by a graded alumina sheath and an additional niobium/Nb1Zr outer sheath. The advanced conceptual design of interest in this study differed from the ATI by the inclusion of a molybdenum collector with a diamond thin film sheath and no outer sheath. Molybdenum as a collector material has superior compatibility with thin diamond film in both deposition and adherence.

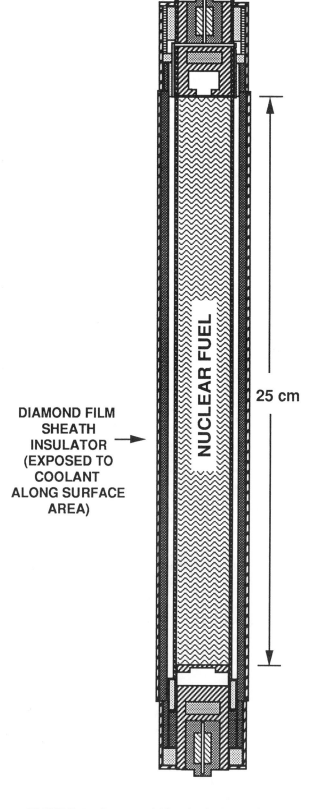

FIGURE 1. Conceptual Sketch of ATI Fuel Element with Diamond Film Sheath Insulator (Not To Scale).

RESULTS AND DISCUSSION

Electrical Resistivity

The fundamental function of the thermionic sheath insulator is to prevent the leakage of electrical current from the collector of the fuel element to other components within the thermionic reactor. A reactor which incorporates the ATI fuel elements will have a current of 390 A through each fuel element and an output voltage of 28 V. To insure that the current leakage across any 28 V drop is below 0.1% of the total current of the fuel element, it has been calculated that a minimum resistance of 72 ohms would be required from the diamond sheath insulator.

Recent high temperature electrical resistivity studies have provided data for the bulk resistivity of diamond film at the collector operating temperature of 880K. Electrical resistivities on the order of 10^8 ohm-cm have been measured for certain diamond film samples at this temperature. Using this data, a 10 μm thick diamond sheath insulator for the ATI fuel element with a surface area of 133.5cm, maintained at 880K produces a bulk resistance of 750 ohms.

This value is well within the acceptable range for the ATI application. There also exists the near term likelihood that thicker diamond films or films of improved quality will be produced which will improve the electrical performance of the diamond film insulator. Furthermore, it should be noted that the diamond structure would be very resistive to electrical breakdown due to ionic conduction, which is a concern with alumina within the thermionic reactor environment. Further evaluation of diamond film's insulating properties still need to be done, though, especially in the areas of neutron irradiation and long term performance degradation.

Thermal Conductivity

The thermal conduction of a thermionic sheath electrical insulator is of importance since the sheath is the path of heat rejection from the collector. The collector of the ATI fuel element is designed to be maintained at a temperature of 880K during operation. This temperature is maintained by a heat transfer system consisting of liquid alkali metal coolant flowing over the sheath insulator and eventually dissipating the heat to the space atmosphere through radiator fins. The rate of heat transfer through the insulator can be represented by the following equation,

$$q = -\lambda \left(\frac{A_s}{L}\right) \Delta T \qquad (1)$$

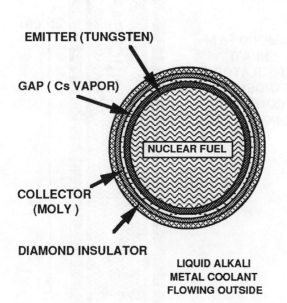

FIGURE 2. Cross Section of ATI Fuel Element With Diamond Film Insulator (Not To Scale).

where q is the heat transfer rate, λ is the thermal conductivity of the insulator, A_s is the surface area of the sheath, L is the thickness of the sheath, and ΔT is the temperature gradient across the insulator. In this analysis, the surface area used for both the diamond film and alumina insulators was 133.5 cm^2.

The thermal conductivity of alumina at 880K has been measured to be 0.093 W/cm-K. Recent data has shown the thermal conductivity of diamond to be approximately 6.2 W/cm-K at a temperature of 880K. In Case 1, the ATI fuel element with an 0.047 cm thick alumina sheath is designed to have a temperature gradient of 25 K across the sheath and a liquid metal coolant temperature of 855 K. Therefore, the rate of heat transfer from the collector surface is 7.81 kW.

In Case 2, if the same 7.81 kW of heat transfer was desired from an ATI fuel element with a 0.001 cm thick diamond sheath insulator, then the temperature gradient across the insulator is approximately 0.01K, which means the liquid coolant temperature is approximately 880K for Case 2. The advantage in Case 2 of this smaller ΔT is realized in the radiator area of the heat rejection system. The radiator area needed to dissipate the 7.81 kW of thermal energy was calculated according to the Stefan-Boltzmann Law,

$$q = \epsilon \sigma A_R (T_R^4 - T_o^4) \qquad (2)$$

where q is the heat dissipated through the radiator, ϵ is the emissivity of the radiator material, σ is the Stefan-Boltzmann constant, A_R is the radiator area, T_R is the temperature of the radiator, and T_o is the ambient space temperature (4K). Since q, ϵ, and σ remained constant in each Case, the ratio of the radiator areas in Case 1 and Case 2 can be found by,

$$\frac{A_{R2}}{A_{R1}} = \frac{(T_{R1}^4 - T_o^4)}{(T_{R2}^4 - T_o^4)} \qquad (3)$$

The heat rejection system of the fuel element with the diamond film sheath insulator, Case 2, was calculated to require 11% less area to dissipate the heat flux. This reduction in the radiator area requirement would result in a significant reduction in the mass of the thermionic reactor.

Cylindrical Surface Diamond Deposition

Diamond film with high electrical resistivity has been deposited uniformly and continuously onto the outer surface of molybdenum cylinders with the diameter of the ATI fuel element. This is an important development in the diamond film sheath program because parallel tests may now be run on diamond film and alumina sheath insulators.

The diamond coated molybdenum cylinders display bulk resistivities of 10^{13} ohm-cm at room temperature. This value shows that the cylindrical diamond films have electrical properties similar to diamond films deposited on planar silicon substrates, but still inferior to natural type IIa diamond.

In May 1991, a diamond coated molybdenum cylinder was placed in the EBR2 nuclear reactor test batch alongside several alumina tri-layer sheath insulators. This study will determine the neutron irradiation tolerance of the diamond film sheath as compared to the alumina tri-layer.

The deposition of diamond onto the molybdenum cylinders also proved that the diamond film sheath insulator, if selected to be integrated into an ATI fuel element, could be fabricated. Although the films grown to date have been on 7 cm long cylinders, the simple scale-up of the DC plasma enhanced chemical vapor deposition process to a 25 cm long cylinder is expected to be equally successful.

SUMMARY

The potential performance of a diamond film sheath insulator in an ATI fuel element is being quantified using data from recent ATI high temperature diamond film studies. The bulk electrical resistance of the diamond film sheath exposed to 880K has been calculated to be an acceptable 750 ohms. The excellent thermal conductivity of diamond film would payoff in a reduction of the reactor systems mass. The surface area of the reactor's thermal radiator will be reduced by 11% when a diamond film sheath insulator is used in the ATI fuel element as compared to the alumina tri-layer. High quality diamond film has been deposited onto the outer surface of a molybdenum collector. The reliability and performance of the diamond film sheath is presently being studied through parallel tests with the alumina tri-layer.

The projections of acceptable electrical performance and excellent thermal performance based on high temperature diamond film data is an indication that more advanced evaluation should be initiated. Diamond film's durability and reliability (or lack of) under the neutron environment of the reactor will determine its future as a sheath insulator material. A radiation hard diamond film insulator with a graceful degradation of performance would be an attractive option to the present alumina sheath design.

Acknowledgements

The authors thank Mysore Ramalingam and Bang Hong Tsao of Universal Energy Systems Inc, Jan Vandersande of Jet Propulsion Laboratory, Maurice Landstrass of Crystallume, and Elliot Kennel of Applied Sciences Inc for their valuable input into this study.

References

Adams, S.F., J.W. Vandersande, D. Zolton, B.R Stoner, and J.A. von Windheim (1991) "Electrical Conductivity of Diamond Films From Room Temperature to 1200C", Proceedings of the Second International Symposium on Diamond Materials, The Electrochemical Society.

Adams, S.F. and M.I. Landstrass (1991) "CVD Diamond Coated Cylinders for Advanced Thermionic Fuel Element", Application of Diamond Films and Related Materials, Elsevier, New York, pp371.

Herb, J.A., C. Bailey, K.V. Ravi, and P.A. Dennig (1989) "The Impact of Deposition Parameters on the Thermal Conductivity of CVD Diamond Films", Proceedings of The First International Symposium on Diamond and Diamond-Like Films, Proc. Vol. 89-12, pp366: The Electrochemical Society, Los Angeles.

Lake, M.L., J.K. Hickok, A.M. Lide (1990) "Diamond Thin Films for TFE Sheath Insulators", Final Report, Air Force Contract No. F33615-89-C-2960.

Lamp, T.R. and B.D. Donovan (1991) "The Advanced Thermionics Initiative Program", Proceedings of the 26th IECEC, Boston, MA, Proc. Vol. 91-3, pp132.

Landstrass, M.I. and K.V. Ravi (1989) "CVD Diamond: The Emerging Electronic Material", Proceedings of the Fourth International High Frequency Power Conversion Conference, Naples, Fl., pp103: Intertec Communications, Ventura.

Landstrass, M.I. and K.V. Ravi (1989) "The Resistivity of CVD Diamond Films", Appl. Phys. Lett., Vol. 55, pp975.

Landstrass, M.I., M.A. Plano, and D. Moyer (1991) "Diamond Coatings for Improved Thermionic Power Systems", Final Report, Air Force Contract No. F33615-90-C-2056.

Powell, R.W., C.Y. Ho, and P.E. Liley (1968) "Thermal Conductivity of Selected Materials- Part 2", National Standard Reference Data Series-NBS-16, Category 5, Figure 14.

Tsao, B.H., M.L. Ramalingam, S.F. Adams, and J.S. Cloyd (1991) "Adherence of Diamond Films on Refractory Metal Substrates for Thermionic Applications", Proceedings of the 26th IECEC, Boston, MA, Proc. Vol. 91-3, pp126.

Vandersande, J.W., C.B. Vining, and D. Zolton (1991) "Thermal Conductivity of Natural Type IIa Diamond Between 500K and 1250K", Proceedings of the Second International Symposium on Diamond Materials, The Electrochemical Society.

SAFETY QUESTIONS RELEVANT TO NUCLEAR THERMAL PROPULSION

David Buden
Idaho National Engineering Laboratory
P.O. Box 1625
Idaho Falls, ID 83415-1550
(208) 525-5626

Abstract

Nuclear propulsion is necessary for successful Mars exploration to enhance crew safety and reduce mission costs. Safety concerns are considered by some to be an impediment to the use of nuclear thermal rockets for these missions. Therefore, an assessment was made of the various types of possible accident conditions that might occur and whether design or operational solutions exist. With the previous work on the NERVA nuclear rocket, most of the issues have been addressed in some detail. Thus, a large data base exist to use in an assessment. The assessment includes evaluating both ground, launch, space operations and disposal conditions. The conclusion is that design and operational solutions do exist for the safe use of nuclear thermal rockets and that both the environment and crews can be protected against harmful radiation. Further, it is concluded that the use of nuclear thermal propulsion will reduce the radiation and mission risks to the Mars crews.

INTRODUCTION

Nuclear thermal propulsion (NTP) is critical for successful human Mars exploration (Synthesis Group Report, 1991). The safety of the crew is greatly enhanced by shorter trip times. This has the affect of reducing the crew exposure to high levels of galactic radiation, reducing the time that solar flares will be a problem, lowering psychological stresses of long periods in confined environments and reducing the time the crew is subjected to possible equipment malfunctions. The nuclear thermal propulsion rocket engine has many fewer moving parts then chemical rockets which it replaces and should, therefore, be more reliable. There is no need for a chemical rocket oxidizer system. Launch windows for departing Earth and for returning from Mars are significantly wider. Also, there are more opportunities to go to Mars, providing schedule flexibility and reducing the need for potentially hazardous decisions to meet limited Mars opportunities. In addition, with nuclear thermal rockets two to three times better performance than chemical rockets, less or no assembly is needed in Earth orbit. This makes the spacecraft more reliable, less costly and easier to meet schedule. In fact, the mass in low Earth orbit will be one-third to one-half of a chemical rocket mission configuration.

SAFETY QUESTIONS

A series of questions relevant to the use of nuclear thermal rockets is postulated in order to evaluate accident conditions. These cover ground, launch and space operations (See Table 1) and were designed to encompass the full range of accident conditions that a NTP will need to address to demonstrate safe operations. No particular rocket engine configuration is used as a model. Thus, the assessment is to determine whether generic design solutions exist. For the postulated accident conditions, the primary safety requirements were determined, design options examined, and the experience base reviewed. The results are given in terms of top level summary discussions. Once a particular design is selected for either unmanned scientific or exploration missions or for crew missions to Mars, detailed design and operational solutions will be needed. The important element here is to have examined the key questions in significant depth to show that solutions exist.

The questions related to safe ground operations in Table 1 will now be addressed.

TABLE 1.   Safety Questions Relevant to Nuclear Propulsion.

Ground Operations

What must be done to safely ground test nuclear rockets?

What special precautions will be needed at the launch pad?

How will radioactive material contamination at the launch site be avoided in rocket launch pad accidents?
 - Fires
 - Explosions

How will ground testing be handled so that there are not significant additions to the nuclear waste problem?

Who approves the launch of vehicles with nuclear rockets on-board?

Launch and Space Operations

How safe is the crew from reactor radiation?

How will inadvertent criticality be prevented for launch/ascent accidents?

If radioactive materials impact on land, what plans exist to clean up contaminated land areas?

If a reactor is started below a "Nuclear Safe Orbit" (NSO) or "Sufficient High Orbit" (SHO), how can reentry of a radioactive core be averted?

How is a "Nuclear Safe Orbit" determined?

Will nuclear engines release radioactive materials which contaminates near Earth space?

Will an operating nuclear rocket affect other satellites and experiments?

What are the plans for final disposal of nuclear engines in space?

Returning from Mars, how will a nuclear rocket be prevented from impacting the Earth?

Safe Ground Testing of Nuclear Rockets

Safety is the prime requirement in all testing and operational procedures. The established standards for radiation levels and radioactive releases levels must be met. Environmental Impact Statements will be needed before testing facilities can be constructed.

To meet environmental safety standards, radioactive material removal scrubbers will probably be needed to remove fission gases from the engine hydrogen exhaust and to catch any radioactive material releases. The basic technology has been demonstrated during Nuclear Furnace-1 testing in 1972 (see Figure 1). In addition, a scenario worst than what is considered the worst case credible scenario was intentionally tested in 1965 in Kiwi-TNT by building special rapid neutron control devices into the reactor. The result of the test showed that even in this extreme scenario that the reactor chemically exploded without significant nuclear contamination.

Special Precautions Needed At The Launch Pad

The requirement is to maintain the radiation dose levels below established health standards. Design options include not operating the reactor prior to launch (a zero power reactor test can be performed to verify the reactor physical assembly is correct). Acceptance testing at the launch facility will be needed

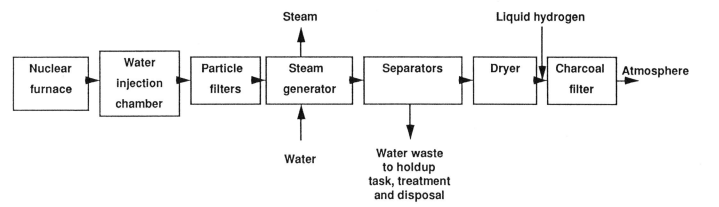

FIGURE 1. Nuclear Furnace Scrubber Concept.

to ensure that all components are functional prior to mating with the launch vehicle. This could include cold flow testing; that is, testing where hydrogen is run through the engine for short periods of time to demonstrate that all valves and the turbopump are operational while maintaining the reactor shutdown.

The U.S. has launched one space reactor. This reactor, SNAP-10A, demonstrated the capability to launch a reactor without special radiation handling at the launch site. Further, nuclear fuels and reactors are transported around the country using well established containers and procedures. Sufficient design and operational experience exist to avoid transportation criticality accidents.

Launch Site Contamination Accidents--Fires

The primary requirement is to maintain the reactor subcritical without releases of hazardous radiation or radioactive materials during a fire. Design options relate to choice of materials and physical layout on the launch vehicle. For example, in case of an accident, it is more desirable to have the nuclear rocket in-line with chemical boosters rather then along side of them.

A series of propellant fire tests were performed as part of a project called Pyro to investigate the temperatures and duration of liquid propellant fires (Project Pyro, 1968). Theoretical data showed a peak temperature of 2900 K for hydrogen-oxygen fires. The experimental data measured 2500K. This is below the melting temperatures of the nuclear rocket fuel, so that melting is not a problem. An analysis of the structural materials also indicate that melting is insufficient to cause a critical mass.

Solid propellant test show that they burn at approximately 3000 K, with some chucks burning for up to 10 minutes. Again, the fuel melting temperatures are above the fire temperature. Using evaluations of the Lincoln Laboratory Experimental satellites LES 8/9 that used a Titan III launch vehicle, the probability of an accident is 2-3 in a hundred. In a given accident, the probability of propellant chunks being in close proximity to the reactor is between one in a thousand and one in a million.

The conclusion is that the reactor can be designed not to melt or go critical in a launch pad fire. Detailed evaluations will be needed of particular nuclear thermal rocket and launch vehicle configurations.

Launch Site Contamination Accidents--Explosions

Here, the requirements are to prevent core compaction criticality and dispersal of radioactive materials. Design options are based on analysis from SP-100 where it was shown that the reactor would not go critical from the blast affects of launch vehicles. Similar design features can be built into nuclear rocket engines. Fragments may shear through the engine, but no fission fragment inventory exists within the core at this time. Therefore, no significant radiological risk from an explosion is projected. A major safety analytical and experimental program has shown that radioisotope generators are safe to launch (Bennett, 1990); NTPs, with their geometry and non-radioactive materials at launch, should be even less risk at launch.

Ground Nuclear Waste

The requirement is to minimize the amount of radioactive waste generated during the NTP program, especially long life waste. Detailed issues will be addressed as part of a programmatic Environmental Impact Statement. NTP characteristics tend to minimize nuclear waste because of the very short operating times, measured in hours. Reprocessing of the fuel and burning the actinides can minimize/eliminate nuclear waste. This was demonstrated when NERVA fuel was reprocessed and reused.

Launch Approval

It is required that a formal flight safety review be completed with the approval of the Office of the President before nuclear power systems can be launched in the United States. This process, shown in Figure 2, requires an independent review by the Interagency Nuclear Safety Review Panel that performs safety and risk evaluations (Sholtis and Nelson, 1990). The Panel provides the necessary independent risk evaluation that will be used by decision makers who must weigh the benefits of the mission against the potential risks. The agency

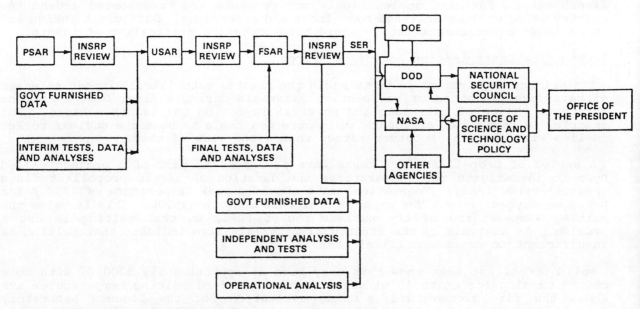

FIGURE 2. Flight Safety Review and Launch Approval Process for Space Nuclear Power Systems.

wanting to fly a nuclear powered payload than requests permission for flight to the Office of Science and Technology (OSTP). OSTP reviews the request and makes the launch decision; however, the Executive Office of the President makes the final decision if OSTP feels that is appropriate.

Turning now to launch and space operations, the questions in Table 1 will be addressed.

Crew Safety From Reactor Radiation

NASA crew dose guidelines for astronauts is 50 rem/year. Mars trips involve crew exposures to galactic cosmic radiation, Earth's radiation belts, solar radiation, and reactor radiation. Galactic cosmic radiation is continuous between 24 and 60 rem/year. Solar flares are stochastic short duration events with potentially high doses (>120 rem); the crew can be protected by a storm shelter for the limited duration of the events. Earth's belt radiation is minimized by limiting the amount of time spent there. The radiation to the crew from a NTP reactor is reduced by spacecraft geometry, local reactor shielding, hydrogen tanks and spacecraft shielding to levels of about one rem. For a typical NTP Mars trip, see Table 2, the radiation exposure levels for the crew is about 45 rem, of which the reactor contributes less than 3%.

TABLE 2. Typical Mars Mission.

Typical Mars Mission (rem)	
Galactic	34
Solar flares (with storm shelter)	7.7
Earth radiation belts	1.5
Nuclear rocket	<1.1
Mars (30 days)	<1
Total	45.3

Criticality Prevention During Launch/Ascent Accidents

Requirements are for the system to remain subcritical for all credible launch/ascent accidents and to have no power operations until the system achieves its intended orbit or flight path trajectory. Design options include a built-in redundant shutdown subsystems with sufficient design margins in each system to ensure shutdown in case of a failure within either subsystem. NERVA was designed, in addition to its control drums, with neutron absorption wires in the core through the nozzle to further protect against launch criticality. Configurations can include in-core safety/shutdown rods or wires with locking devices and weak links. Command destruction of the reactor can be provided to ensure that debris from an accident terminates in an ocean.

Plans To Clean Up Contaminated Land Areas

If radioactive debris is deposited on land areas, it will be necessary to remove the material to designated storage sites. The approach here is mainly a preventative one. If an abort occurs near the beginning of the mission, the vehicle will likely land in the Atlantic Ocean. Based on Titan and Shuttle data, one failure in 57 flights of the solid rockets have occurred; however, no land impacts have occurred on other continents. The footprint from aborts later in the flight profile can be controlled by command destruct mechanisms to cause debris to fall into an ocean. Also, the reactor contains no radioactive fission products at launch. In the unlikely event of land debris impact, standard clean up organizations and mechanisms are in place such as the NEST Team (Nuclear Emergency Search Team).

Operation Below "Nuclear Safe Orbit" or Sufficiently High Orbit"

Nuclear Safe Orbit (NSO) or Sufficiently High Orbit (SHO) refers to the acceptable reactor space storage location after use. The latter term, SHO, is now preferred. It means an orbital lifetime long enough to allow for sufficient decay of the fission products to approximately the activity of the actinides before reentry occurs. One design option is to initiate operations above the SHO for a given mission. However, for Mars missions and many others, it will be highly desirable to start below SHO. For these missions, provision can be made for on-board or external boost systems. Nuclear thermal propulsion stages can be throttled to ensure that the thrust vector is in an increasingly safe direction before accelerating to full propulsion power. The stage can be slowly rotated to average the thrust direction to safeguard against thrust nozzle misalignment failures. If fission gas retention is a problem at the higher temperatures, and correspondent higher rocket performance, the temperature can be reduced until the altitude is such that the fission gases are no longer a problem.

On-board devices have generally been used to boost low altitude satellites to higher orbits. This approach has been demonstrated on the USSR RORSAT satellites. However, these sometimes fail. An external capability is being evaluated under a project called SIREN (Search, Intercept, Recover, Expulsion Nuclear) for boosting radioactive materials to higher orbits (Lee, 1990).

Determination of "Nuclear Safe Orbit"

A NSO is a function of the geometry of the vehicle and operating history. Figure 3 shows the orbital decay time as a function of altitude in terms of mass, drag coefficient, and cross sectional area. Typically, an orbital lifetime of 300 years has been used as the time for the radioactive materials from nuclear power plants to decay to safe radioactive levels. This corresponds to a orbital altitude above 400-500 nmi.

Near-Earth Space Contamination

The requirements include no significant additions of radioactive materials to the near Earth environment and protection of crews from exposures that exceed safety limits. During flight operations, insignificant amounts of fission products are expected to be released. These should mostly be in the form of the fission gases. As part of the flight environmental impact statement, an assessment will need to establish acceptable fission gas release levels. If a sensitive environmental area is being traversed, power and temperature can be reduced to maintain releases to background levels.

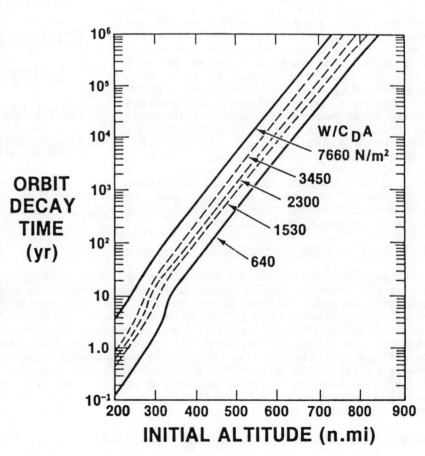

FIGURE 3. Orbital Decay Times.

Affects On Other Satellites and Experiments

It will be necessary to avoid/minimize affects on other satellites. This can be accomplished as part of particular mission planning. Operations should generally be well away from other satellites; the radiation level exposures at other Earth satellites is a function of distance, power level and duration. Just from the desire to avoid collisions these should be negligible; however, power can temporarily be reduced if necessary in the vicinity of other satellites. During the limited operating time while leaving the vicinity of Earth (about 90 minutes), radiation sensitive sensors on other satellites will probably record the nuclear radiation from the reactor.

Final Disposal of Nuclear Engines/Prevention Of Nuclear Rocket Earth Impact

Final disposal of nuclear engines must be such that there is negligible probability of intersecting or passing within the close proximity of Earth. From Mars, since NTR reuse is not planned by the Synthesis Group, it will be placed in a deep space orbit that will not intersect the Earth. The NTR stage can be ejected after the Mars burn or mid course correction used to return the manned spacecraft to Earth; the NTR is not planned to be used in spacecraft Earth capture or achieving Earth orbit. From the Moon, if reuse is not planned, the NTR can easily be placed in a deep space disposal orbit. For a nuclear tug, it will eventually be disposed of either above a Sufficiently High Orbit or in deep space, not back to Earth.

CONCLUSIONS

Nuclear thermal propulsion can be designed to operate safely if safety standards are defined at the initiation of any nuclear thermal propulsion program and continuously monitored for compliance. Design and operational solutions to meet these standards have been addressed in previous programs, such as NERVA. The solutions depend on particular concepts and their intended missions. However, after reviewing a wide range of questions related to safety, there were no questions that did not appear to have practical design/operational solutions.

Acknowledgments

This work was performed under the sponsorship of the U.S. Department of Energy, Idaho Field Office, DOE Contract #DE-AC07-76IDO1570.

References

Bennett, G. L., "Safety Status of Space Radioisotope and Reactor Power Sources," 25th Intersociety Energy Conversion Engineering Conference, Reno, Nevada, August 1, 1990.

Lee, J., et al, "Technology Requirements for the Disposal of Space Nuclear Power Sources and Implications for Space Debris Management Strategies," AIAA/NASA/DOD Orbital Debris Conference: Technical Issued and Future Directions, Paper No. AIAA 90-1368, Baltimore, Maryland, April 16-19, 1990.

Project Pyro, *Liquid Propellant Explosive Hazards, Final Report*, AFPRO-TR-68-92, December 1968.

Sholtis, J., Joyce, J.P., and Nelson, R. C., "U.S. Flight Safety Review/Approval Process For Nuclear-Powered Space Missions," Proceedings of the Seventh Symposium On Space Nuclear Power Systems, Albuquerque, New Mexico, January 1990, pp. 569-571.

Synthesis Group Report, *America At The Threshold, America's Space Exploration Initiative*, available from the Superintendent of Documents, U.S. Government Printing Office, Washington, D.C., 20402, June 1991.

RELIABILITY COMPARISON OF VARIOUS NUCLEAR PROPULSION CONFIGURATIONS FOR MARS MISSION

Donald R. Segna
Department of Energy
Field Office, Richland Washington
P.O. Box 550 M/S A5-90
Richland, WA 99352
(509) 376-8989

Jeffery E. Dagle
Pacific Northwest Laboratory
P.O. Box 999 M/S k5-19
Richland, WA 99352
(509) 375-3629

William F. Lyon, III
Westinghouse Hanford Company
P. O. Box 1970 M/S L5-02
Richland, WA 99352
(509) 376-8869

Abstract

Currently, trade-offs are being made among the various propulsion systems being considered for the Space Exploration Initiative (SEI) missions. It is necessary to investigate the reliability aspects as well as the efficiency, mass savings and experience characteristics of the various configurations. Reliability is a very important factor for the SEI missions because of the long duration and because problems will be fixed onboard. The propulsion options that were reviewed consist of nuclear thermal propulsion (NTP), nuclear electric propulsion (NEP) and various configurations of each system. There were four configurations developed for comparison with the NTP as baselined in the Synthesis (1991): 1) NEP, 2) hybrid NEP/NTP, 3) hybrid with power beaming, and 4) NTP upper stage on the heavy lift launch vehicle (HLLV). The comparisons were based more or less on a qualitative review of complexity, stress levels and operations for each of the four configurations. Each configuration included a pressurized NEP and an NTP ascent stage propulsion system for the Mars mission

INTRODUCTION

As indicated in the Synthesis Group Report nuclear thermal propulsion (NTP) for the Mars Transfer vehicle was selected on the basis of its substantially increased efficiency, safety, readiness, and its cost effectiveness in comparison to chemical propulsion. Reliability was only reviewed indirectly through safety, efficiency, etc. The intent of this paper is to extend the review of the various propulsion configurations that might be available for use on the SEI missions. Four configurations were developed. They include NEP, NTP/NEP hybrid, hybrid with laser power beaming, and finally using NTP as an upper stage on the HLLV.

A chemical propulsion comparison with NTP has been completed (Segna, 1991) and will not be repeated here. This comparison looked at those parameters such as complexity, stress levels, and operational difficulties that have a propensity of affecting reliability. Reliability as used in these comparisons relates to the ability to accomplish the mission and return the crew safely. While these parameters do not comprise a classic reliability study, they can be used effectively in the conceptual design phase when trade studies are being performed. It was the practice of Apollo and Shuttle program managers when selecting a propulsion system, that orbit mass and even cost was allowed to be compromised in order to enhance reliability. Obviously, a concept with more experience should have a high priority, however, systems with less experience but less complexity and with lower stress levels should also be considered. Higher reliability guided the decision to utilize storable, pressurized, and hypergolic propulsion systems for the Apollo space propulsion engine on the Command and Service Module, the descent and ascent engines on the Lunar Excursion Module, and the orbital maneuvering system on the Orbiter. These decisions were made even with a sacrifice in efficiency and mass savings that would have been afforded by the various cryogenic propulsion systems.

Importance of Propulsion System Reliability

For the Mars mission, there is probably no partial success for the main propulsion maneuvers. The propulsion maneuvers have to start on time and continue for the required duration. For instance, there is no practical backup to the Mars orbit insertion maneuver using an impulse maneuvering system. A free return, as was planned in the initial Apollo Missions in the event of a no-go at lunar orbit insertion, is not practical for Mars orbit insertion. The reason is that a short duration trip to Mars is necessary to minimize the galactic radiation effects and deconditioning during the lengthy, zero gravity period. The short duration trip times are not compatible with a free return because there is unreasonable excess velocity upon arriving at Mars. Even if free return, along with the longer trans Mars time were acceptable, the excessive duration for the return leg, two years or longer, is undesirable. In addition, proper operation of the propulsion system is required in other instances such as the trans-Earth injection maneuver leaving Mars and the braking maneuver just prior to entering the Earth's atmosphere. Even a partial burn in either situation could be a major problem. Therefore, a highly reliable propulsion system is imperative and will be a major factor in the final selection of the propulsion system.

Assumptions

This study makes the assumption that with increased complexity, stress levels and/or the operational requirements of a system, with all else equal, the less reliable that system becomes.

At the design phase, increased complexity and stress levels, in themselves, do not decrease reliability because designers can accommodate them. If NASA said, "design to 0.9999," that is what the designer would do, but in actuality the as-built and as-operated may have an entirely different reliability factor than that predicted from design. Many factors can compromise the predicted reliability such as how the organization makes decisions. How do they allow information to be communicated? How does the designer integrate changes made in one area to insure those changes do not affect other systems? Are the specifications right? How good is the suppliers' QA program? Is the fidelity of the test programs adequate? Is the training program complete, etc.?

Below are items that illustrate some of the consequences of these factors:

- The Challenger accident may be considered an organizational problem. The information "not to launch" was there, but was not communicated to the proper staff members. This case illustrates the point above, that complexity is not the whole story. The solid rocket booster is a simpler system than the SSME's and was expected to have a very high reliability -- after the fix it may be, but for that particular launch condition it was not. The problem could have been circumvented if the test program had been of higher fidelity, without compromising the predicted reliability. Simplifying the SRB would have circumvented the problem. This could have been accomplished by making the booster in one piece, thereby avoiding o-ring failure. An example of the reliability of a simple system is the RTG nuclear isotope power supply.

- On Apollo 13 the reason for the LOX tank explosion was traced to a checkout procedure problem. Procedures were misread and DC instead of AC voltage was applied to a circuit inside the oxygen tank overloading the circuit and causing it to short. When the circuit was energized in space the accident happened.

- The Hubble telescope didn't have high enough fidelity tests to see the error in the mirror grinding.

- Sometimes you get lucky. Apollo 8 POGO (propellant oscillations) was a design problem. While it was successful, if design margins hadn't added favorably, the thrust structure probably would have failed.

These examples are included to stress that if the system ignores good judgement, predicted reliability is worth no more than the paper the prediction was written on. Good judgement must prevail, and a thorough ground and flight test program must be conducted to screen any possible built-in problem(s).

So how is the comparison going to be valid relative to complexity? Simply, the more complex a system the higher probability that "Murphy" will cause a problem. Also, every part, component, and system has a finite time to failure. We may not know what that time is, but we can say that the probability of failure will increase with complexity. Again, the Challenger accident is a case in point. If the SRB had not been segmented, as mentioned above, there would not have been an o-ring to fail.

Equating increase in stresses on a system to lower reliability is not an easy relationship. One system can be designed for 1000 K and the other for 3000 K. As with complexity, either system can be designed to the same reliability. Let's look at factors where higher stresses might compromise reliability:

- Usually there is less experience at the higher stress level. Designers have to resort to using newer and more exotic materials as the stress levels are increased.

- The higher the system is stressed the greater the chance a failure will cause secondary damage. For example, failure of higher pressure systems creates higher speed projectiles; higher temperatures create more chemical reactivity, and higher temperatures increases heat transfer; likewise with higher voltage, etc.

- A specified safety factor for a higher stressed system usually has a higher penalty mass and/or cost than the same safety factor for a lower stressed system. Since mass and cost are always under scrutiny, it is a greater probability that the higher stressed system will have a lower safety factor thus the potential to be less reliable.

Chemical Vs. NTP Assumption

This comparison was previously completed under the same basis included in this paper. The result was the NTP has a very high potential for being more reliable than chemical, therefore, NTP as used in the Synthesis Report is the baseline for this study. Also, because of its higher reliability, NTP will be used in two other applications such as the ascent stage for Mars and a smaller version in conjunction with NEP in a hybrid mode. In addition, it was shown that a pressurized NTP is even simpler and is used in all the NTP configurations.

These configurations begin with the Synthesis Report NTP as the baseline. The first configuration developed for comparison replaced the NTP with a pressurized NEP and an NTP replaced the Mars ascent stage $H_2 O_2$ cryogenic propulsion system. Next, a relatively small NTP was added to the NEP configuration and is referred to as the hybrid followed by adding power beaming and finally adding pressurized NTP as the upper stage to the HLLV.

It did not appear that the NEP alone would be potentially more reliable than the NTP because of added complexity, so the hybrid and the power beaming were added for further analysis. The NTP upper stage was added separately and though not included in all the comparisons, but because of its much higher efficiency and simplicity it is considered for this review.

These four configurations are applied to a Mars mission and are summarized in Table 1. Each succeeding configuration builds on the previous configuration.

Comparison

The preceding configurations were then reviewed with respect to complexity, stress levels, and operations and then compared in Tables 2 through 5. Guidelines used for the comparisons are as follows:

- Each of the four configurations builds on the previous concepts.

- Benefits and adverse items will be listed compared to the previous concept. NTP will be considered the baseline and all the comparison items will be in the NEP column. For the NEP Plus Hybrid configuration, NEP will be the baseline and NEP Plus Hybrid will contain all the comparison items etc. This way the adverse items and the benefits are additive for each configuration change.

- Ascent stage will be NTP and pressurized (no propellant pumps). Also, reduces "g" build-up as propellants are used.

- NEP propellant tanks will be pressurized (some increase in mass but reduces complexity).

- Minor comparison items are ignored.

TABLE 1. Configuration Summary

MISSION PHASE	SYNTHESIS NTP	NEP	PLUS HYBRID	PLUS POWER BEAMING	PLUS HLLV PRESS. NTP UPPER STAGE
Basic Concept	Split/sprint	Same	Same	Same	Single mission
Launch	HLLV - 250MT	Same	Same	Same	HLLV - 350MT
LEO Ops					
unmanned	Assy for TMI	-Assy LEO -Spiral to escape	Same	-Dock components to OTVe & pwr/bm	NA
manned	Assy for TMI	-Assy. veh. & dock to OTVe -Crew board taxi	Same	-Dock components to OTVe & pwr/bm -Crew board taxi	Same
Staging					
unmanned	NA	NA	NA	-Beam OTV/veh. to L1 & assy.	NA
manned	NA	-OTV Spiral to L1 -Followed by crew taxi to L1	Same	-Beam OTV/veh to L1 & assy. -Followed by crew taxi to L1	Same
TMI					
unmanned	NTP	Spiral to escape	Same	Same	NA
manned	NTP	-Undock OTV -Taxi Dock -Crew xfer -Spiral escape	-Undock OTV -Taxi Dock -Crew xfer -Small NTP	Same	Same
Trans Mars	Coast	NEP	same	Same	Same
Mars Capture					
unmanned	NTP	Spiral to LMO	NEP spiral	Same	NA
manned	NTP	-Spiral to medium Mars orbit	-NEP braking -Small NTP capture HMO	Same	Same
Mars Orbital Operation					
unmanned	-Undock TEI propellant -Prepare for descent.	-Undock MTV/NEP & TEI propellant -Split NEP to be comm sat. -Prepare for des	-Same except -Spiral descent stage to LMO	-Same except -Split NEP to be comm & beam satellite -Beam OTV & des stg to LMO	NA
manned	-Dock w/MTV & TEI propellant -Prepare for descent.	-Dock w/MTV/NEP & TEI propellant -Prepare for des	-Des.to circ LMO with small NTP	Same	-Same except -Split redundant syst. to be comm & beam satellite.
Descent					
unmanned	-Descent by cryo	Same	Same	Same	NA
manned	-Descent by cryo	-Same except from medium orbit	-Same except from LMO	Same	Same
Surface Operation	-Short or long duration	Same	Same	-Same except surface power is beamed	Same
Ascent	-Ascent by cryo	-Same except to medium orbit by NTP (asc, stage)	-Same except to LMO	Same	Same
TEI	NTP from LMO	-NEP spiral to escape	Small NTP-2 burn	Same	Same
Trans Earth	Coast	NEP	Same	Same	Same
Earth Capture	Direct entry	-Same except from high L1 -Use taxi to return -Possible reuse	Same	Same	Same

TABLE 2. NEP SUMMARY

ADVERSE ITEMS	BENEFITS
COMPLEXITY	
• Very large radiators • Much smaller engines but more of them • More lines and valving • High current lines • Requires separate launch for crew and taxi vehicle • Large propellant tank required on the NTP ascent stage • Requires jettison of NTP on ascent stage	• Eliminates separate power supply • Smaller and/or fewer propellant tanks • Less radiation shielding requirements • Reduced structural requirements (less thrust) • Less precise reactor control mechanism • Eliminate need for separate comm sat. • Minimum storage requirements for photovoltaic power (standby or backup power) at L1 staging point • Reduces "g" loading due to propellant usage -- eliminates staging or wide range of throttling
STRESS LEVELS	
• Higher temperatures in engine • High electrical currents • Higher magnetic fields • Vehicle spends long time in Van Allen Belts • Increase radiation dosage from NTP ascent stage	• Much lower thrust levels • Lower tank and engine pressures • Less propellant slosh • Lower temperature in reactor • Much lower pressure levels in reactor • Reduced level debris field • Reduced mission duration • Less galactic ray exposure • Less probability to solar flare • Minimum atomic oxygen • Reduces refrigeration requirements over straight NTP (less fuel mass)
OPERATIONS	
• Large radiator deployment • Greater radiator shadowing/thermal control operations • Continuous propulsion on-time operations • Spiral out to escape at mass adds >20 days	• Minimum docking of propellant tanks • Larger launch windows from Earth and Mars • More abort capability • Reduced trip times • Longer Mars stay time for short duration stay missions • Less Mass in LEO less assy operations • Combining comm with MTV reduces number of separate satellites to manage • No nuclear operations near LEO • L1 staging reduces communication link complexity (reduces TDRS requirements) • No orbital decay • Back contamination can be accommodated at L1. Requires more time (~7 days longer + contamination clean up) • Further mass to LEO reduced by not carrying crew reentry vehicle to Mars and back

TABLE 3. NEP Plus Hybrid (Small NTP) Summary

ADVERSE ITEMS	BENEFITS
COMPLEXITY	
• Adds complexity of another propulsion system NTP	• The NTP would be non generic redundant propulsion system • Further reduction of mass to LEO

	STRESS LEVEL
	• Trip times further reduced • Further reduction in galactic ray and probability of solar flare exposure • Greatly reduced time in Van Allen Belts
	OPERATIONS
	• Less Mass in LEO further reduces assy operations • Mission duration further reduced • Two different types of propulsion systems to guard against generic problems

TABLE 4. NEP Plus Power Beaming Summary

ADVERSE ITEMS	BENEFITS
COMPLEXITY	
• Adds power beaming transmitters and receivers • Adds large radiator • Requires power beaming to OTV from Earth and Moon • Precise control for power beaming acquisition and tracking	• Further reduction of mass to LEO • Eliminates need for separate surface prime power • Redundancy is external to vehicle for some power and propulsion phases at Earth, Moon and Mars
STRESS LEVELS	
	• Trip times further reduced • Further reduction in galactic ray and probability of solar flare exposure
OPERATIONS	
• More radiator shadow/thermal control requirements	• Less Mass in LEO further reduces assy operations • Wider launch windows at Earth and Mars • L1 staging with beaming enhances reusable crew taxi which could be common with Mars ascent stage • Commonality with lunar operations • Enhances robotic operations because line-of-sight coverage from synchronous orbit is almost half the globe. • Crew excursions are also enhanced

TABLE 5. Plus NTP Upper Stage HLLV Summary

ADVERSE ITEMS	BENEFITS
COMPLEXITY	
	• Reduces total mass in orbit • Possible to reduce to single mission (non split/sprint)
STRESS LEVEL	
• If ocean launch salt water	• Trip times further reduced • Further reduction in galactic ray and probability of solar flare exposure
OPERATIONS	

- May launch from ocean
- Public opinion against nuclear

- Less assembly requirements in orbit
- Wider launch windows at Earth and Mars
- Can vary launch latitude and azimuth if ocean launch
- Less stringent launch abort requirements
  If ocean launch, non split/sprint capability if hybrid, power beam and L1 staging point used
- Non split/sprint, possible to use Mars excursion vehicle as life boat per Apollo 13
- Eliminates double missions (manned & unmanned) on succeeding launch opportunities

CONCLUSIONS

The NTP in a pressurized and somewhat reduced chamber pressure mode is a very simple propulsion system and has the potential of being very reliable. Replacing the cryogenic chemical ascent stage with an NTP system, using it in a hybrid mode with NEP and even as an upper stage of the launch vehicle, should increase the actual reliability of the Mars mission.

There are two major draw backs for the NEP configuration: very large radiators and long duration in the Van Allen belts for much of the spacecraft. To counter the drawbacks, there are many more benefits but of a lesser impact. Without a more detailed study, judging the potential reliability difference would be very subjective.

Adding the hybrid adds no measurable adverse items since a redundant propulsion system is required regardless. The different propulsion concepts are desirable and eliminate a generic problem making both propulsion systems inoperable. Power beaming does add complexity because of the transmitting antennas along with the large radiators needed to cool the antennas. However, the benefits of eliminating the surface prime, power equipment, the ability to be power rich, and to provide easily distributed power to exploring vehicles giving them unlimited range, will weigh heavily in the trade-off studies. The use of NTP for the upper stage of the HLLV may be precarious because of the public opinion against nuclear, but the reported (to the Synthesis Group) 60% increase in payload is extremely favorable. Also, the higher potential reliability over a cryogenic propulsion system, is most desirable. Especially since the Uranium 235 fuel is relatively safe (free of fission products and from transuranics) until the upper stage is started.

There are sufficient benefits when compared to adverse items that show NEP in a hybrid configuration with power beaming, should be included in the Mars technology trade-off studies. The added mission flexibility, power rich capability, and reduced trip times are very favorable. The blinders were on to take a qualitative look at the reliability comparison; therefore, the next effort should include the readiness level of the technology, the development schedule, and quantitative levels for both the adverse and benefit items. It is important to note the use of pressurized NTP for the hybrid and the Mars ascent stage propulsion systems and the use of L1 as the staging point for TEI and Mars return. It keeps from having to take the reentry vehicle to Mars and back; it also affords back contamination processing if required.

Acknowledgements

This work was performed at the Department of Energy's Field Office, Richland, Washington. I gratefully acknowledge the assistance of Joan Segna and Edward Coomes of the Pacific Northwest Laboratory, Richland, Washington, and the critique of Roy Rumery.

References

Segna, D. R., 1991. "Reliability Comparison of Nuclear with Chemical Propulsion for Mars Mission," AIAA 91-3629, American Institute of Aeronautics and Astronautics, New York, New York.

Synthesis Group. 1991. "America at the Threshold," Synthesis Group.

THE 1981 UNITED NATIONS REPORT: AN HISTORICAL CONSENSUS ON THE SAFE USE OF NUCLEAR POWER SOURCES IN SPACE

Gary L. Bennett
Propulsion, Power and Energy Division
National Aeronautics and Space Administration
Code RP
Washington, D. C. 20546
(202) 453-2856

Abstract

In 1981, after over three years of study and discussion, the technical specialists in the United Nations Committee on the Peaceful Uses of Outer Space agreed to a report which provided general safety criteria; a suggested format for notification for reentering space vehicles containing nuclear power sources; suggested improvements in orbit prediction; and recommendations relating to the search and recovery of a nuclear power source. The results of that first consensus on the use of nuclear power sources are summarized to provide an historical framework for developing international norms on the use of nuclear power sources in outer space.

INTRODUCTION

Following the 1978 reentry of the Soviet radar ocean reconnaissance satellite (RORSAT) known as Cosmos 954 over Canada the United Nations (U.N.) took up the question of the use of nuclear power sources (NPS) in outer space. Over the years most of the action has centered on the U.N. Committee on the Peaceful Uses of Outer Space (COPUOS) and its two standing subcommittees of the whole: the Scientific and Technical Subcommittee (STSC) and the Legal Subcommittee (LSC). A general logic was established that a technical Working Group on the Use of Nuclear Power Sources in Outer Space (WGNPS) would be established in the STSC to consider the technical aspects and safety measures relating to the use of NPS in outer space with the results of the WGNPS studies being passed through the STSC to the LSC for its consideration and follow-up action as appropriate (Bennett et al. 1988a).

By way of background it is worth noting that COPUOS and its two subcommittees operate on the consensus principle rather than voting. As stated in Bennett et al. 1988a: "In effect, this means that any member or group of members can prevent COPUOS or its subcommittees from taking action on that topic by formally objecting to such action, whether that action consists simply of adopting a report containing the conclusions of COPUOS or its subcommittees or proposed principles. In practice, the achievement of consensus takes time. This is especially true in regard to topics that are highly scientific or technical in nature and where the science and technology are continuing to evolve. . . On the other hand, the consensus principle provides a firm and uniform basis of support for any resulting agreement."

In the wake of Cosmos 954, many delegations to the 1978 STSC and LSC meetings expressed concern about safety measures for NPS and notification prior to reentry. Uncertainties over predictions of the impact point for Cosmos 954 led a number of delegations to suggest a need for improved orbit prediction models. Several delegations, in particular some from developing countries, expressed a need for assistance in search and recovery should an NPS impact their territories. These concerns and interests became the focus for a number of national studies completed during the period from 1978 to 1981. Using these studies and other papers the WGNPS met formally in three sessions (1979, 1980, and 1981) and, at the conclusion of its third session in 1981, issued a report that represents the first technical consensus on the technical and scientific aspects relating to the use of NPS in space. As noted in Bennett et al. 1988a: "The U.S. delegation worked actively with the Canadian delegation and the other WGNPS delegations to achieve consensus on the safety and technical aspects of the 1981 WGNPS report in an effort to maintain the momentum toward the LSC. Consequently, the U.S. delegation has consistently supported the text of the 1981 report." The report was adopted by the STSC and the STSC report was accepted by COPUOS. In view of the ongoing discussions in the U.N. on the use of NPS in outer space as summarized most recently in Hodgkins et al. 1991 it is worth reflecting back upon that first technical consensus in order to provide an historical framework for future discussions of the use of NPS in space.

The following sections summarize the consensus views of the 1981 WGNPS report with regard to various aspects of the use of NPS in space, including the types of NPS, safety measures, notification of reentry, orbit prediction, and search and recovery. One of the most important parts of the 1981 WGNPS report was its conclusion (U.N. 1981):

"The Working Group reaffirmed its previous conclusion that NPS can be used safely in outer space, provided that all necessary safety requirements are met."

TYPES OF NPS

The 1981 WGNPS report begins by noting ". . . that various types of power sources exist for use on spacecraft such as solar cells, fuel cells and chemical batteries, as well as nuclear systems. Selection of a suitable power source is a complex technical issue and in practice most space missions have used conventional power sources. The particular advantages of the use of NPS are their long life, compactness and ability to operate independently of solar radiation." (U.N. 1981).

The 1981 WGNPS report then stated that "For certain important space missions NPS have been the preferred technical choice. Provided the additional risks associated with NPS are maintained at an acceptably low level, *the Working Group considered that the basis of the decision to use NPS should be technical*." (Emphasis added.) The WGNPS report dealt with two kinds of NPS: radioisotope generators and nuclear reactors (U.N. 1981).

The WGNPS report defined radioisotope generators as consisting ". . . of radionuclide fuels surrounded by energy conversion systems. The radio-isotope decays spontaneously, emitting ionizing radiation which is absorbed as heat and can be converted into other forms of energy." The WGNPS report stated that nuclear reactors ". . . derive their thermal energy from the controlled fission of uranium 235. The reactor consists of an enriched uranium core with a reflector, producing heat for possible conversion to other forms of energy." (U.N. 1981). There was no intention in the 1981 WGNPS report to limit nuclear reactors to ^{235}U; rather, as stated in Bennett et al. 1988a: "Plutonium-239 reactors were not considered because there was no evidence that any nation was using them. A U. K. working paper, however, noted in comparison with ^{235}U that '. . . a plutonium fuelled reactor would be a somewhat greater risk, but it would take many times the five tonnes of plutonium dispersed in weapon tests before the hazard from fissile material could dominate.'" The source for the U.K statement may be found in U.N 1978.

The WGNPS report stated that "Both systems require that appropriate design and operational measures be taken, in order to protect the population and the environment for both normal and accidental conditions. Moreover the risks inherent in each particular application or project are to be assessed in terms both of the probability of failure or malfunction and the severity of its consequences." (U.N. 1981).

SAFETY MEASURES

The WGNPS report ". . . noted that the safety of radio-isotope systems was being assured by designing them to contain with a high probability of success the radio-isotope for normal and credible abnormal conditions." The report continued by stating that "The design should ensure minimal leakage of the radio-active contents with a reasonably high level of probability of success in all credible circumstances including launch accidents, re-entry into the atmosphere, impact and water immersion. The appropriate limits recommended by the International Commission on Radiological Protection (ICRP) should be met for normal operational conditions." (U.N. 1981).

Regarding nuclear reactors the WGNPS report stated that "The Working Group agreed that the safety of U-235 reactor systems did not present any difficulty when they were started and operated in orbits sufficiently high to give time for radio-active materials to decay to a safe level in space after the end of the mission. In this way the dose equivalents at the time of re-entry could be guaranteed in all circumstances to be within the limits recommended by ICRP for non-accident conditions. If reactors are intended for use in low orbits where the radio-active materials do not have sufficient time to decay to an acceptable level, safety depends on the start of the operation in orbit and the success of boosting NPS to a higher orbit after operation is completed. In the event of an unsuccessful boost into higher orbit the system should in all credible circumstances be capable of dispersing the radio-active material so that when the material reaches the earth the radiological situation conforms to the recommendations of ICRP when relevant. The

Working Group noted that ICRP publication 26 does not provide specific guidance for accidents and emergencies although it does address in general terms the circumstances in which remedial action might be taken." (U.N. 1981). Dispersal was agreed to in the 1981 report simply because that was the design approach being followed by both the U.S. and the U.S.S.R. at that time (Bennett and Buden 1983). Since 1981 the criterion for the SP-100 reactor has been changed to intact reentry so it is worth noting that in 1978 Canada had suggested intact reentry for reactors and in 1987 the Canadian delegation offered a working paper which stated that "Nuclear reactors shall be designed either to reenter to Earth's atmosphere and land while maintaining the functional integrity of the containment of radioactive materials, or to divide and disperse into fine particles the radioactive materials upon reentry into the Earth's atmosphere . . ." (Canada 1987 and U.N. 1978).

Throughout the deliberations leading up to the adoption of the 1981 WGNPS report there were discussions and working papers on applying various ICRP guidelines (including radiation dose limits) to NPS. The WGNPS recommended that the limits specified in ICRP publication 26 ". . . should not be exceed during normal phase of an NPS mission." However, the WGNPS was aware that in some accident cases the dose limits of ICRP publication 26 could be exceeded and so the WGNPS quoted ICRP publication 26 that "detailed guidance on the application of its recommendations, either in regulations or in codes of practice, should be elaborated by the various international and national bodies that are familiar with what is best for their needs." (U.N. 1981). Thus, the WGNPS report does not specify dose limits for accidents.

NOTIFICATION

The WGNPS report took note of the U.N. General Assembly (UNGA) resolution 33/16 ". . . which requests launching States to inform States concerned in the event that a space object with NPS on board is malfunctioning with a risk of re-entry of radio-active materials to the earth." (U.N. 1981). The report also states ". . . that States should be informed of a possible re-entry or malfunctioning of a spacecraft carrying an NPS so that those concerned might take necessary precautionary measures. The earliest possible notification of such an occurrence is deemed essential." (U.N. 1981). The Working Group felt that the early notification should be to the Secretary-General of the United Nations. The WGNPS ". . . agreed to the following format of notification for re-entering space vehicles containing NPS which may give rise to radiological hazards:" (U.N. 1981)

1. <u>System parameters</u>

    *1.1 Name of launching State or States including the address of the authority which may be contacted for additional information or assistance in case of accident

    *1.2 International designation

    *1.3 Date and territory or location of launch

1.4 Information required for best prediction of orbit lifetime, trajectory and impact region

*1.5 General function of spacecraft

2. <u>Information on the radiological risk of nuclear power source(s)</u>

2.1 Type of NPS: radio-isotopic/reactor

2.2 The probable physical form, amount and general radiological characteristics of the fuel and contaminated and/or activated components likely to reach the ground. The term "fuel" refers to the nuclear material used as the source of heat or power.

The asterisks denote the requirements contained in the Convention on Registration of Objects Launched into Outer Space. This convention, which entered into force on 15 September 1976, is generally referred to as the "Registration Convention". The above format has been adopted by COPUOS (Hodgkins et al. 1991).

ORBIT PREDICTION

The WGNPS was aware that reentry dates could only be predicted with an error of about 10 percent of the remaining lifetime and so ". . . noted that prediction of orbit lifetimes and re-entry paths of uncontrolled satellites remains at best an inexact science." (U.N. 1981). The WGNPS stated that "Accuracy could be improved by the implementation of additional degrees of control, further research and study and by extensive and co-operative use of tracking stations and communications lines." (U.N. 1981).

SEARCH AND RECOVERY

In the WGNPS report "The Working Group recommended that assistance in training be provided through appropriate international channels to personnel of States requesting training on hazard evaluation following re-entry of an NPS and on performing pertinent search and recovery and emergency planning operations." (U.N. 1981).

The WGNPS goes on to state that "The Working Group noted that the Agreement on the Rescue of Astronauts, the Return of Astronauts and the Return of Objects Launched into Outer Space and the Convention of International Liability for Damage caused by Space Objects are of direct relevance to search and recovery questions relating to NPS. It noted in particular that under the first agreement a launching State is obliged, at the request of a State affected, to eliminate possible damage or harm that might result from the return of a space object or its component parts which is of hazardous or deleterious nature, and that, under the second agreement, a launching State is obliged to examine the possibility of rendering appropriate and rapid assistance to a State suffering damage caused by a space object which presents a large-scale danger to human life or seriously interferes with the living conditions of the population or the functioning of vital centres, when that State so requests." (U.N. 1981). A summary of the relevance of these and other treaties to the use of NPS in space has been provided in Bennett 1988b.

CONCLUSIONS

As noted in Hodgkins et al. 1991, the U.N. is continuing its deliberations on the use of NPS in outer space. The 1981 WGNPS report represents the first technical consensus achieved on technical aspects and safety measures and, as such, provides a useful historical framework for the ongoing deliberations. In this connection, the conclusion of the 1981 WGNPS report that ". . . the Working group reaffirmed its previous conclusion that NPS can be used safely in outer space, provided that all necessary safety requirements are met" represents not only a consensus of international technical experts but a succinct statement of the U.S. position as well.

References

Bennett, G. L. and D. Buden (1983) "Use of Nuclear Reactors in Space," *The Nuclear Engineer*, Vol. 24, No. 4, pp. 108-117.

Bennett, G. L., J. A. Sholtis, Jr., and B. C. Rashkow (1988a) "United Nations Deliberations on the Use of Nuclear Power Sources in Space: 1978 - 1987," in *Space Nuclear Power Systems 1988*, M. S. El-Genk and M. D. Hoover, eds., Orbit Book Co., Malabar, FL, pp. 45-57.

Bennett, G. L. (1988b) "Proposed Principles on the Use of Nuclear Power Sources in Space," IECEC Paper No. 889027, in *Proceedings of the 23rd Intersociety Energy Conversion Engineering Conference*, held in Denver, Colorado, 31 July-5 August 1988.

Canada (1987) "The Elaboration of Draft Principles Relevant to the Use of Nuclear Power Sources in Outer Space," Working Paper submitted by Canada to the Legal Subcommittee of the U.N. Committee on the Peaceful Uses of Outer Space, U.N. Document A/AC.105/C.2/L.154/Rev. 1, United Nations, 12 March 1987.

Hodgkins, K. D., R. Lange, and B. C. Rashkow (1991) "United Nations Consideration of the Use of Space Nuclear Power," in *Proceedings of the Eighth Symposium on Space Nuclear Power Systems*, CONF-910116, M. S. El-Genk and M. D. Hoover, eds., American Institute of Physics, New York, NY, 1991.

United Nations (1978) "Question Relating to the Use of Nuclear Power Sources in Outer Space, Report of the Secretariat," U.N. Document A/AC.105/220, United Nations, 20 May 1978.

United Nations (1981) "Report of the Working Group on the Use of Nuclear Power Sources in Outer Space on the Work of its Third Session," Annex II of Report of the Scientific and Technical Subcommittee on the Work of its Eighteenth Session, U.N. Document A/AC.105/287, United Nations, 13 February 1981.

THE REGULATORY QUAGMIRE UNDERLYING THE TOPAZ II EXHIBITION

The Nuclear Regulatory Commission's Jurisdiction
Over the TOPAZ II Reactor System

John W. Lawrence
Winston & Strawn
1400 L Street, NW
Washington, DC 20005
(202) 371-5820

Abstract

At the 8th Symposium on Space Nuclear Power Systems, 6-10 January 1990, the Union of Soviet Socialist Republics displayed a TOPAZ II thermionic space nuclear reactor. Underlying that exhibition was a regulatory quagmire created by a decision of the Nuclear Regulatory Commission that an import license was required to bring the device into the United States, and that an amendment to their regulations governing exports was required to return the device to the Soviet Union later that summer. This paper briefly reviews the jurisdictional issue of how the Nuclear Regulatory Commission exerted its authority over the TOPAZ II reactor system, as well as the manner in which the import and export licensing actions were accomplished. In sum, the paper offers an independent interpretation of the applicable import and export regulations, and concludes that the Nuclear Regulatory Commission likely need not have exercised its import jurisdiction, and notwithstanding the initial assumption of jurisdiction, an export license likely could have been issued without an amendment of the then-existing regulations.

INTRODUCTION

Last year at the 8th Symposium on Space Nuclear Power Systems, the Union of Soviet Socialist Republics (USSR) displayed a TOPAZ II thermionic space nuclear reactor. Underlying that exhibition was a regulatory quagmire created by a decision of the Nuclear Regulatory Commission (NRC) that an import license was required to bring the device into the United States (US). Subsequent to the exhibition, plans for the purchase of the reactor system did not materialize and, accordingly, USSR officials requested the device to be returned. The NRC again exerted its jurisdiction by requiring an export license in order for the device to leave the US, but because an agreement between the US and the USSR covering exports of nuclear equipment did not exist, an export license could not be issued for the reactor system. Eventually the NRC amended its regulations to explicitly exempt the TOPAZ II reactor system from NRC export licensing requirements, thereby removing the NRC regulatory impediments that had blocked its return to the USSR. This paper briefly reviews the jurisdictional issue of how the NRC exerted its authority over the TOPAZ II reactor system, as well as the manner in which the import and export licensing actions were accomplished.

NRC JURISDICTION OVER THE IMPORT AND EXPORT OF NUCLEAR EQUIPMENT

Section 101 of the Atomic Energy Act of 1954, as amended, prohibits "any person within the United States to transfer or receive in interstate commerce, manufacture, produce, transfer, acquire, possess, use, import, or export any utilization or production facility except under and in accordance with a license issued by the [Nuclear Regulatory] Commission pursuant to section 103 or 104." 42 U.S.C. § 2131. Section 104 of the Act specifies several conditions which must be satisfied for the issuance of a license for research and development activities (Section 103 deals with commercial activities). For purposes of this paper, one particularly relevant condition associated with the issuance of an export license is the requirement for an "Agreement for Cooperation in the

Peaceful Uses of Atomic Energy" between the US and the country where the export of a production or utilization facility is bound. See 42 U.S.C. §§ 2134(d) and 2153.

With regard to both import and export licensing, Section 11(cc) of the Atomic Energy Act (AEA) defines a "utilization facility" as:

> any equipment or device, except an atomic weapon, determined by rule of the Commission to be capable of making use of special nuclear material . . . or peculiarly adapted for making use of atomic energy in such quantity as to be of significance to the common defense and security, or in such manner as to affect the health and safety of the public . . . .

For reasons reviewed in the next two sections, the NRC concluded that the TOPAZ II reactor system met the AEA definition of a utilization facility (and issued an import license to permit its entry into the US after reviewing information submitted regarding the device), but reached the exact opposite conclusion in determining that an NRC export license was not required to allow the device to return to the USSR.

WHY WAS AN NRC IMPORT LICENSE NEEDED FOR THE TOPAZ II REACTOR SYSTEM?

On 21 December 1990, the NRC Office of Governmental and Public Affairs (GPA) was notified in a meeting with representatives of the Strategic Defense Initiative Organization (SDIO) of the Department of Defense (DOD) that a TOPAZ II thermionic space nuclear reactor, owned by the USSR, was enroute to the US for exhibition at the 8th Symposium on Space Nuclear Power Systems in Albuquerque, New Mexico on 6-10 January 1991 (Hauber 1991). A subsequent letter from DOD/SDIO to NRC on 24 December 1990 formally requested "whatever prompt assistance" the NRC could provide to support the exhibition (O'Neill 1990a). The same day the NRC Office of General Counsel (OGC), responding to a request by GPA, stated that the decision as to whether or not the TOPAZ II reactor system could be made to produce or utilize special nuclear material (SNM), and thereby meet the definition of utilization facility, required a technical not legal analysis (Parler 1990). However, OGC noted that if the device was "a model and from [a] technical standpoint could not be made to produce or utilize SNM," it would not be a production or utilization facility subject to NRC licensing requirements (Parler 1990).

As part of the request for assistance, DOD/SDIO informed the NRC that the TOPAZ II reactor system to be exhibited in Albuquerque was "an inoperative model," and was in an "inoperable condition with no fuel or moderator or coolant installed" (O'Neill 1990a). Moreover, the moderator plates, which were essential to operation of the device but not included in the exhibit, were considered to be "unique" and capable of production "by a process known only in the Soviet Union, and are not available in the United States" (O'Neill 1990a). The NRC Staff considered these factors to be "outside the scope" of its import licensing review process (Shea 1990). Instead, the NRC Staff concluded that the TOPAZ II reactor system could become operable given the development of these missing components by US technology (Parler 1991). Accordingly, on 28 December 1990, the NRC informed DOD/SDIO that the TOPAZ II reactor system had been classified as a utilization facility, and that if the device was to be brought into the US, an NRC import license was required before its arrival (Hauber 1990).

It appears that the NRC's technical conclusion did not properly consider the latter half of the definition of utilization facility in AEA Section 11(cc). Specifically, the statute limits the definition of a utilization facility to equipment or devices that make use of special nuclear material or atomic energy "in such quantity as to be of significance to the common defense and security, or in such manner as to affect the health and safety of the public." The fuel, moderator, and coolant are physical attributes of the device that directly affect the quantity and manner of energy production, and hence should have been considered as one part of the NRC Staff's review of whether the TOPAZ II reactor system was a utilization facility. As another factor, the NRC was informed that this particular model had never been operated or tested as a reactor and was not radioactive (Hauber 1991, and O'Neill 1990a). Thus, the device exhibited was not inimical to the common defense and security and did not constitute an unreasonable risk to public health and safety, a conclusion eventually reached by the NRC upon issuance of the import license (Hauber 1991).

In addition, a utilization facility generally refers to nuclear reactors used to produce power or used in medical therapy, research, and testing (see 10 C.F.R. § 8.4(b)). In the case of the TOPAZ II reactor system, the stated purpose of the device was for exhibition. In fact, no TOPAZ II reactor system had ever been used to produce power in space flight (Leskov 1991). Moreover, since the USSR owned the device, even if the missing components could have been developed by US technology, it remains questionable as to whether the US government could have "married" the two technologies to create the potential for energy production. Thus, if the NRC Staff had considered these additional factors as part of its import licensing review, likely they could have concluded that the TOPAZ II reactor system was only a model. Then applying the OGC model exception (Parler 1990), the NRC could have concluded that the device was not a utilization facility and, hence, did not require an NRC import license.

Nevertheless, following notification that the TOPAZ II reactor system was a utilization facility and subject to NRC import jurisdiction, DOD/SDIO formally applied for an import license on 28 December 1990 (O'Neill 1990b), supplementing their earlier letter of 24 December 1990 (O'Neill 1990a). NRC also informed DOD/SDIO that any subsequent export of the TOPAZ II reactor system could not occur as the US and the USSR had not entered into an "Agreement for Cooperation" pursuant to Sections 104(d) and 123 of the Atomic Energy Act (Hauber 1990). Despite this forewarning, DOD/SDIO decided to import the device, apparently based on eventual plans to purchase it from the USSR (Britt 1991, Broad 1991, and Smith 1991a), and therefore acknowledged that any subsequent disposition of the TOPAZ II reactor system would involve advanced consultations and mutual agreement with the NRC (O'Neill 1990b).

While the NRC Staff was processing the application for an import license, another legal complication arose -- an import license only allows equipment or material of foreign origin to enter the US, it does not authorize the applicant to acquire, possess, receive, or use the equipment or material once in the US (see 10 C.F.R. § 110.50(a)(3)). In fact, 10 C.F.R. Part 50 requires another license for the acquisition, possession, or use of a utilization facility (see 10 C.F.R. § 50.10(a)). Although the TOPAZ II reactor system was to be used for exhibition purposes, the question of who would acquire and possess the device while in the US was raised by NRC Commissioner Curtiss on 3 January 1991 (Curtiss 1991a). DOD/SDIO responded to this latter complication the next day by designating the Sandia National Laboratories, a contractor to the Department of Energy, as the custodian of the TOPAZ II reactor system during the exhibition and while in storage immediately thereafter (Britt 1991, and Verga 1991).

As a result of this additional response, and because the Department of Energy and its contractors are exempt from the licensing requirements of Part 50 for utilization facilities (see 10 C.F.R. § 50.11(b)(2)), the NRC was able to issue Import License IR90002 for the TOPAZ II reactor system on 4 January 1991 (U.S. NRC 1991). The NRC concluded that the device did not constitute an unreasonable risk to public health and safety because the device was inoperable, did not contain fuel, moderator, or coolant, and was not radioactive (Hauber 1991). In addition, because of the involvement of DOD/SDIO and Sandia National Laboratories, the NRC concluded the import of the device would not be inimical to the common defense and security of the US (Hauber 1991). However, the NRC conditioned the license with the following two requirements:

1. The reactor system was subject to the custodial control of Sandia National Laboratories, Albuquerque, New Mexico; and

2. The license was effective only upon written acknowledgment by an authorized representative of the USSR that exportation of the device from the US was subject to the requirements of the Atomic Energy Act, namely the requirement for an Agreement for Cooperation in the Peaceful Uses of Atomic Energy between the US and the USSR.

HOW WAS THE TOPAZ II REACTOR SYSTEM RETURNED TO THE USSR?

Following the display at the 8th Symposium on Space Nuclear Power Systems, the USSR requested the return of the TOPAZ II reactor system when plans for the purchase of the device by the US government did not materialize (Smith 1991b). However, as noted by the NRC in their letter of 28 December 1990 to DOD/SDIO

(Hauber 1990), and as identified as a condition in Import License IR90002 (U.S. NRC 1991), the export of the TOPAZ II reactor system required an Agreement for Cooperation between the US and USSR governments before the device could be returned.

The statutory requirement for issuance of an export license, AEA Section 126, is codified in NRC regulations at 10 C.F.R. § 110.44 which states, in pertinent part (emphasis added):

> (a) The [Nuclear Regulatory] Commission will issue an export license if it has been notified by the State Department that it is the judgment of the Executive Branch that the proposed export will not be inimical to the common defense and security; and: (1) Finds, based upon a reasonable judgment of the assurances provided and other information available to the Federal government, that the applicable criteria in § 110.42, or their equivalent, are met.

10 C.F.R. § 110.42(a) identifies the eight criteria against which a license application for the export of a utilization facility is reviewed. Criterion 7 states that "[t]he proposed export of a facility . . . would be under the terms of an agreement for cooperation," which is defined in 10 C.F.R. § 110.2 as an agreement executed with another nation pursuant to AEA Section 123.

An Agreement for Cooperation pursuant to AEA Section 123 does not currently exist between the US and the USSR (U.S. DOS 1990), and thus the NRC Staff concluded that the TOPAZ II reactor system could not be exported to the USSR (Hauber 1990, and Shea 1990). However, this conclusion did not properly consider 10 C.F.R. § 110.44(a)(1) which conditions the issuance of an export license, in part, upon compliance with either the criteria of Section 110.42 "or their equivalent." The phrase "or their equivalent," also contained in AEA Section 126(a)(2), was included in the statute "to avoid technical disqualification of an export application simply because the phrasing of an assurance is not identical to that of a statutory criterion" (S. Rep. No. 467 (at 13), 95th Cong., 1st Sess. (1977), reprinted in 1978 U.S. Code Cong. & Admin. News 326, 338). The legislative history for Section 126 also states that "in most cases an export must be consistent with the terms of an existing agreement for cooperation," (S. Rep. No. 467, at 13, emphasis added), thereby leaving open the possibility that some NRC export actions may involve situations that do not meet the specific terms of an existing Agreement for Cooperation.

Thus, although an AEA Section 123 agreement does not exist, the NRC Staff could have considered the following bi-lateral agreements and multilateral treaties as "equivalents" for the purpose of exporting the TOPAZ II reactor system since, in sum, they address similar issues and because both the US and the USSR are signatory parties (U.S. DOS 1990, and Kavass 1991):

- Agreement on Scientific and Technical Cooperation in the Field of Peaceful Uses of Atomic Energy, 21 June 1973, US - USSR, 24 U.S.T. 1486, T.I.A.S. 7655, extended and amended, 5 July - 1 August 1983, T.I.A.S. 10757, extended, 21 November - 12 December 1989, KAV 1789;

- Agreement Concerning Cooperation in the Exploration and Use of Outer Space for Peaceful Purposes, 15 April 1987, US - USSR, KAV 1825, amended, 31 May 1988, KAV 1826, expanded, July 1991;

- Memorandum of Cooperation in the Field of Civilian Nuclear Reactor Safety, 26 April 1988, US - USSR, KAV 1786;

- Statute of the International Atomic Energy Agency, 26 October 1956, 8 U.S.T. 1093, T.I.A.S. 3873, 276 U.N.T.S. 3; and

- Treaty on the Non-Proliferation of Nuclear Weapons, 1 July 1968, 21 U.S.T. 483, T.I.A.S. 6839, 729 U.N.T.S. 161.

In addition, given the benign nature of the device, the NRC could have relied upon their earlier conclusion with regard to exports to the USSR in general. Specifically, "the Executive Branch and the Commission do not believe that the USSR poses any foreign policy concern or any proliferation risk with respect to any of the non-sensitive commodities under NRC's existing or proposed general licenses." 49 Fed. Reg. 47191, 47194 (1984) (Part 110 amendment expanding NRC authority to export non-sensitive nuclear equipment without specific licensing action). Moreover, the USSR already owned the device. With the Section 110.42(a)(7) "equivalents" in place, and considering these additional factors, the NRC likely could have issued an export license under the then-current regulatory structure.

However, after reviewing a variety of options (other than using the then-current regulatory structure) for returning the TOPAZ II reactor system to the USSR, the NRC Staff decided to amend their regulations to specifically exempt the device from the definition of "utilization facility," and hence eliminate the requirement for an NRC export license. Specifically, the NRC amended the definition of "utilization facility" in 10 C.F.R. § 110.2 by adding the following sentence:

> For purposes of export from the United States under the jurisdiction of the Nuclear Regulatory Commission, a utilization facility does not include the TOPAZ II Reactor System owned by the Union of Soviet Socialist Republics and imported into the United States pursuant to NRC License No. IR90002, issued January 4, 1991.

56 Fed. Reg. 24682, 24684 (May 31, 1991). The NRC Staff considered this approach the most practical manner of returning the TOPAZ II reactor system to the USSR in the most timely manner (Parler 1991).

As worded, the change to the NRC regulations only applies to the circumstances surrounding the export of this one particular TOPAZ II reactor system. Similar exports or imports from the USSR or other nations, and even an export or import of the same device at some future point in time, would not be covered by this amendment. In approving the amendment Commissioner Curtiss explicitly noted assurances provided by NRC/OGC that this amendment would have "no precedential ramifications, either domestic or international" (Curtiss 1991b). In addition, the NRC noted that this approach only affected the export controls contained in the AEA, and not any other export controls that may apply (Parler 1991). Of potential applicability are the State Department's International Traffic in Arms Regulations ("ITAR"), and the Commerce Department's regulations for exports of commodities exhibited at trade fairs (Liebman 1990). In fact, following the decision by the NRC to exempt the TOPAZ II reactor from the requirements of 10 C.F.R. Part 110, the State Department issued an export license for the TOPAZ II reactor system on 6 June 1991 (Blaha 1991a), and the device was shipped to the USSR on 24 July 1991 (Blaha 1991b).

CONCLUSION

The NRC is empowered, under AEA Section 11(cc), to define the term utilization facility in their regulations. Therefore, the NRC had the authority to revise the definition in 10 C.F.R. § 110.2 to specifically exempt the TOPAZ II reactor system from the NRC export licensing requirements. However, in order to conclude that such a change was warranted, the NRC confronted the same jurisdictional issues that arose during the import licensing review. For example, in a document preparing the change to 10 C.F.R. § 110.2 for the Commission's consideration (Parler 1991), the NRC Staff reasoned:

> Although the TOPAZ II reactor would have the capability to make use of SNM and is peculiarly adapted for making use of atomic energy, given the lack of fuel, coolant and moderator, the intended short stay and limited use as a model for exhibition purposes in the United States, and its return in the near future to the country of origin, the Commission may take all these circumstances into account and determine, in a connection with the export of the device, that it is not a "utilization facility."

If the NRC Staff had not considered these same factors as "outside the scope" of its import licensing review process, and if the NRC Staff had considered the additional factors noted in this paper, likely an NRC import license would not have been needed. Similarly, these factors could also have been applied in either concluding that an NRC export license was not needed. Alternatively, based on the existing "equivalent" agreements and the additional factors identified above, the NRC Staff likely could have issued an export license under the then-existing regulations. At bottom, the NRC Staff's exercise of jurisdiction with regard to the TOPAZ II reactor system created a regulatory quagmire for the movement of the device into and out of the US -- an unfortunate occurrence that in retrospect need not have occurred.

The complications encounter by the TOPAZ II reactor system in entering and exiting the US illustrate how even the most well-intentioned plan can become entangled in a regulatory quagmire. That is not to say that statutory requirements and their implementing regulations do not serve an important purpose, for as evidenced in this case the NRC import and export regulations have been developed to protect public health and safety and maintain the national defense and security. However, as shown in this paper, an independent review and assessment of a potential regulatory problem can often identify alternate solutions, which likely can be just as acceptable to the parties directly involved in the issue.

Acknowledgments

This paper was prepared in its entirety by Mr. John W. Lawrence (BSNE, 1983, Purdue Univ.; JD Candidate, 1993, Catholic Univ.). Mr. Lawrence has been employed in the nuclear power industry since 1983, principally working with electric utilities, and currently is a Technical Advisor and Legal Intern with the law firm of Winston & Strawn in Washington, D.C. However, the views and conclusions expressed in this paper are those of the author and do not necessarily reflect the policy or position of Winston & Strawn.

References

Blaha, J. L. (1991a) "Weekly Information Report - Week Ending June 7, 1991," U.S. Nuclear Regulatory Commission, Washington, DC, 14 June 1991, p. 2.

Blaha, J. L. (1991b) "Weekly Information Report - Week Ending July 26, 1991," U.S. Nuclear Regulatory Commission, Washington, DC, 1 August 1991, Enclosure E.

Britt, E. J. (1991) Letter to R. Hauber (U.S. NRC), Space Power, Inc., San Jose, CA, 4 January 1991.

Broad, W. J. (1991) "U.S. Ready to Buy Advanced Reactor from the Soviets," New York Times, New York, NY, 7 January 1991, pp. A1 and B8.

Curtiss, J. R. (1991a) "SECY-90-426 (Negative Consent) - Proposed Import of Soviet TOPAZ Reactor," U.S. Nuclear Regulatory Commission, Washington, DC, 3 January 1991.

Curtiss, J. R. (1991b) Affirmation Vote Response Sheet for SECY-91-136 (Return of TOPAZ II Reactor System to the Soviet Union), U.S. Nuclear Regulatory Commission, Washington, DC, 20 May 1991.

Hauber, R. D. (1990) Letter to M. O'Neill (U.S. DOD), U.S. Nuclear Regulatory Commission, Washington, DC, 28 December 1990.

Hauber, R. D. (1991) "Import of a Topaz II Reactor System from the U.S.S.R," U.S. Nuclear Regulatory Commission, Washington, DC, 4 January 1991.

Kavass, I. I. and A. Sprudzs (1991) "A Guide to the United States Treaties in Force, 1990 Edition, Part 1," William S. Hein Co., Buffalo, NY, p 555.

Leskov, S. (1991) "Soviet Reactor On U.S. Satellite?" News Article (translated) from IZVESTIYA, Moscow, USSR, 10 January 1991 (Union Edition), p. 6.

Liebman, J. R. and W. A. Root (1990) "United States Export Controls," Prentice Hall Law & Business, Englewood Cliffs, NJ, pp. 1-7/8, 1-17/18, and 3-5.

O'Neill, M. R. (1990a) Letter to R. Hauber (U.S. NRC), U.S. Department of Defense, Washington, DC, 24 December 1990.

O'Neill, M. R. (1990b) Letter to R. Hauber (U.S. NRC), U.S. Department of Defense, Washington, DC, 28 December 1990.

Parler, W. C. (1990) Internal Memorandum for S. Schwartz, U.S. Nuclear Regulatory Commission, Washington, DC, 24 December 1990.

Parler, W. C. (1991) "Return of TOPAZ II Reactor System to the Soviet Union," SECY-91-136, U.S. Nuclear Regulatory Commission, Washington, DC, 15 May 1991.

Shea, J. R. (1990) "Proposed Import of Soviet TOPAZ Reactor," SECY-90-426, U.S. Nuclear Regulatory Commission, Washington, DC, 31 December 1990.

Smith, R. J. (1991a) "U.S. Consortium to Buy Soviet Nuclear Reactor," Washington Post, Washington, DC, 8 January 1991, p. A6.

Smith, R. J. (1991b) "U.S. Won't Let Soviets Take Reactor Back Home," Washington Post, Washington, DC, 20 April 1991, p. A6.

Stratford, R. J. (1991) Letter to J. Shea (U.S. NRC), U.S. Department of State, Washington, DC, 9 May 1991.

U.S. DOS (1990) "Treaties in Force -- A List of Treaties and Other International Agreements of the United States in Force on January 1, 1990," Publication 9433, U.S. Department of State, Washington, DC, pp. 245, 284-86, and 357-58.

U.S. NRC (1991) Nuclear Equipment Import License No. IR90002, Docket No. 1104387, 4 January 1991.

Verga, R. L. (1991) Letter to R. Hauber (U.S. NRC), U.S. Department of Defense, Washington, DC, 4 January 1991.

AN INTEGRATED MISSION PLANNING APPROACH FOR THE SPACE EXPLORATION INITIATIVE

Edmund P. Coomes, Jeffery E. Dagle,
Judith A. Bamberger, and Kent E. Noffsinger
Pacific Northwest Laboratory
P.O. Box 999 M/S K5-21
Richland, WA 99352
(509) 375-2549

Abstract

A fully integrated energy-based approach to mission planning is needed if the Space Exploration Initiative (SEI) is to succeed. Such an approach would reduce the number of new systems and technologies requiring development. The resultant horizontal commonality of systems and hardware would reduce the direct economic impact of SEI and provide an economic benefit by greatly enhancing our international technical competitiveness through technology spin-offs and through the resulting early return on investment. Integrated planning and close interagency cooperation must occur if the SEI is to achieve its goal of expanding the human presence into the solar system and be an affordable endeavor. An energy-based mission planning approach gives each mission planner the needed power, yet preserves the individuality of mission requirements and objectives while reducing the concessions mission planners must make. This approach may even expand the mission options available and enhance mission activities.

INTRODUCTION

The direction of the American space program, as defined by President Bush, is to expand human presence into the solar system. Landing an American on Mars by the 50th anniversary of the Apollo 11 Lunar landing is the goal. This challenge has produced a level of excitement among young Americans not seen for nearly three decades. The exploration and settlement of the space frontier will occupy the creative thoughts and energies of generations of Americans well into the next century. The return of Americans to the moon and beyond must be viewed as a national effort with strong public support if it is to become a reality. Key to making this an actuality is the mission approach selected. Developing a permanent presence in space requires a continual stepping outward from Earth in a logical progressive manner. If we seriously plan to go and to stay, then not only must we plan what we are to do and how we are to do it, we must address the logistic support infrastructure that will allow us to stay there once we arrive.

A fully integrated approach to mission planning is needed if the Space Exploration Initiative (SEI) is to be successful. Only in this way can a permanent human presence in space be sustained. If SEI is to be affordable and acceptable, careful consideration must be given to such things as "return on investment" and "commercial product potential" of the technologies developed. An integrated power and propulsion infrastructure based on energy would reduce the number of new systems and technologies requiring development. The resultant horizontal commonality of systems and hardware would reduce the direct economic impact of SEI, while an early return on investment through technology spin-offs would be an economic benefit by greatly enhancing our international technical competitiveness. Such an approach will help win congressional support, help to secure financial backing, and ensure that human expansion into the solar system becomes a reality.

ENERGY TRANSMISSION

Space initiatives range from the development and exploitation of near-Earth space to missions to the outermost planets. The National Aeronautics and Space Administration's (NASA) mission to planet Earth will use orbiting satellites to collect data on planetary conditions in support of studies of global climate change. The Department of Defense (DOD) plans to increase its number of surveillance, navigation, and communications satellites in Earth orbit to support military missions and objectives. The success of all these efforts hinges on a single common need: the availability of power to operate the systems being proposed. Sufficient power must be available when and where it is needed. The power must be compatible with the various space systems being considered for deployment. Even more important, the method of power production and utilization must be acceptable from political, social, and economic perspectives and perceptions. These requirements all can be met by adopting a space power architecture based on energy transmission. The selection of energy transmission, commonly called power beaming, as the future space power approach would be a major step toward ensuring U.S. space program goals are achieved and the United States reestablishes and maintains its position of leadership in space.

ECONOMICS

Power beaming is a cost-effective approach to meeting space power needs. Building on existing technology development, base-load space power needs could be met for about the same cost as solar but with only half the on-orbit mass. Coupling power beaming with state-of-the-art electric propulsion technology, a beam-powered electric orbital transfer vehicle (EOTV) could deliver to geosynchronous Earth orbit (GEO) 80% of the mass initially placed in low Earth orbit (LEO) for about half the cost of a chemical OTV (Coomes, Johnson and Widrig 1990). Chemical upper stages today deliver only 20% to 25% of the initial mass in LEO to GEO.

By separating the power system from the end-use application, a standard power transmission satellite design can be adopted. A standard design would simplify technology development and system requirements. The power satellite design could be optimized for power production and transmission independent of the end-use mission requirements, thus reducing development costs. The more efficient dynamic systems would be more attractive because of their removal from mission platforms. Central-point power generation and distribution would reduce the total number of power systems needed and would greatly reduce or eliminate the need for onboard energy storage. The mass and volume savings on each satellite would allow the user to increase the satellite payload fraction, thus enhancing mission capabilities. This should increase the revenue-earning potential of each user satellite.

For the same power output, a laser energy receiver would be much smaller that a solar photovoltaic (PV) energy collector, thus reducing satellite drag. Decreased drag on the spacecraft would extend the useful life of onboard consumables needed for stationkeeping and attitude control, thereby extending the revenue-producing lifetime of a satellite. By trading drag reduction for increased power, highly fuel-efficient electric thrusters could be used for stationkeeping and/or attitude control, increasing a satellite's revenue-producing lifetime even further.

The commonality of power systems technology developed for power beaming would support military, commercial, and civilian space activities including advanced NASA missions to the Lunar and Mars surface. The long-life continuous nuclear power sources developed for beam-power satellites would also support nuclear electric propulsion for Lunar and Mars cargo transport spacecraft and OTVs. As the manned development of space progresses and a Lunar base is established, refurbishment and upgrading of beam-power satellites could become a Lunar-based endeavor, providing increased economic incentive for Lunar development.

Adopting power beaming does involve some economic uncertainties that must be addressed. As with any centralized system, the initial capital investment will be higher than that for smaller distributed systems. To keep power-user costs down, this higher initial capital cost must be spread over a longer operating time, thus increasing the need for long-life reliable systems. To ensure continuous power availability even in the event of a power satellite failure, the beam-power distribution grid must include excess generating capacity. Onboard limited backup and/or keep-alive power capabilities will be required to accommodate temporary power transmission loss.

Each beam-power satellite represents a significant portion of the power grid assets, and each unit is a high-value asset. This increases the economic consequences of a system failure or loss at launch or transfer from LEO to high Earth orbit, even though significantly fewer launches are required. Operation from high orbit implies additional expense for transportation to get the beam-power satellite on station. Because each power satellite is a high-value asset, periodic maintenance and/or repair may be used to extend beam-power satellite lifetime. This could imply additional transportation costs or higher repair and maintenance costs because the beam-power satellite is in high Earth orbit. These costs depend directly on the final operating scenario selected and must be factored into any deployment and operating analysis or evaluation that is performed.

Power beaming is a different method of providing power in space. As such, it will require a change by mission planners in their approach to meeting space power needs. The economic impact of transition from the one-on-one, onboard power approach to a central-point generation and distribution power approach could range from a minimum level (involving only change in configuration of existing solar power systems from rectangular to circular geometries) to a maximum level (requiring totally new receiver systems). A more complete understanding of the commonality between solar and laser-beam receiver technologies is necessary before transition impacts can properly be assessed.

MISSION APPLICATIONS.

Energy transmission can support a broad range of mission applications, because it is an integrated power approach that makes no distinction regarding the end use of the power delivered. Energy transmission provides two major mission benefits: increased power availability and easy transportability. Increased power availability alleviates the latent demand for power that exists with mission planning but is never directly faced. Power beaming is a space power utility approach separating production from use. It allows each mission planner the same access to power, yet preserves the individuality of mission requirements and objectives. Mission planning, an exercise in compromise, constantly trades off mission goals and objectives that can be achieved with a given payload against the amount of onboard power that can be packed into the finite satellite volume available. Power beaming provides more power with less volume of the satellite used and is a highly transportable and versatile method for supplying power as man expands his influence out into the solar system.

NASA's Office of Exploration has been examining the viability of putting a permanently manned base on the moon and on Mars. These bases will have a variety of power requirements for the base and for surface activities away from the base, such as mining or exploration. Power for transport vehicles or rovers will be needed to support these remote activities. The logistic support for these bases will require the use of electric propulsion on space transports for resupply of materials and equipment. A megawatt-class electric cargo transport with a long-life nuclear electric power source has new importance and greater value when coupled with an easily transported energy transmission system (Coomes and Dagle 1992).

Near-Earth Mission Applications.

Near-Earth mission applications include a mix of both military and civilian systems. Earth-orbiting satellites considered here include satellite operations from LEO to GEO and from equatorial to polar orbits. Low Earth orbiting satellites include remote sensing, intelligence gathering, and research satellites, while those in geosynchronous orbit consist primarily of communications satellites (Maral and Bousquet 1986), and in between are navigation satellites. All of these satellites typically use solar panels and lightweight batteries to provide the required power. Communications satellites constitute a major portion of the commercial satellites in Earth orbit while military satellites comprise the majority overall. In the near term, communications satellite operators are not likely to embrace power-beaming technology. However, the increased demand for power of direct broadcast and the anticipated market this would have may entice these users into adopting power beaming for future satellites.

For larger systems, power system area and mass become significant factors. The most notable large space system currently planned is the Space Station Freedom. Providing the power the station will require is a major issue at present. The initial station design was to have an average power of 75 kWe for housekeeping, attached payload requirements, and module payloads. The power was to be provided by eight large PV arrays covering a total of over 2200 m^2. Current station designs are now calling for 37.5 kWe, but this will still require over 1100 m^2 of solar panel area. With power beaming, a 7.5-m diameter laser energy receiver (45 m^2) could provide the station with 130 kWe. With three beam-power satellites orbiting at 20,000 km, the space station could be provided continuous power and the onboard batteries could be reserved for emergency power only. By pointing the laser energy receiver at the sun, 5 to 10 kWe of additional emergency power would be available. One approach for Space Station Freedom includes meeting growth power requirements with solar dynamic systems. These units would have solar collectors 50 ft in diameter and produce 25 kWe of power each. The use of solar collectors would be amenable to power beaming, as the collectors could also be used as laser energy collectors with little (4% or less) loss in efficiency--and this could be compensated for by increasing the laser intensity.

The Industrial Space Facility (ISF) is a commercial venture of Space Industries, Inc., to provide an Earth-orbital facility from which to conduct research and manufacturing operations. The ISF would use solar arrays to provide up to 10.8 kWe to its customers and require a panel area of approximately 250 m^2. With power beaming and 45 m^2 of receiver area, this system could also be provided with upwards of 130 kWe. With the ISF 120° out of phase with the station, it too could be provided continuous power from the same three beam-power satellites.

GEO and polar platforms, now being studied as part of the Space Station Program by the General Electric Company, would support such missions as Earth biological and geological observations, oceanographic and ice activity studies, Earth atmospheric monitoring and research, solar observations, and plasma physics measurements. By reducing the power feed to the station and the ISF to 90 kWe, these platforms could be supplied as multiple users of the same three beam-power satellites with 20 to 40 kWe.

The single most viable application is a beam-powered EOTV operating between LEO and GEO would justify the development and deployment of a beam-powered system. A fully operational EOTV (Dagle 1991) would use the same three beam-power satellites to ensure that continuous power for the EOTV. With this system, Dagle showed that a 6000 kg payload could be delivered to GEO in 100 days for an initial mass of 7221 kg delivered to LEO. A comparable chemical propulsion system using advanced engine technology would require a gross weight of over 22,500 kg to deploy the same payload. Although the trip time for the chemical system is only hours, the beam-powered OTV cost per kilogram is only one-half that of the chemical system. In addition, the lower thrust of electric propulsion would allow the satellite to be transported fully deployed. In this way, satellites could be operationally checked out in LEO and repaired or returned to Earth if necessary, saving millions of dollars worth of hardware.

Lunar Mission Applications.

The technology base developed to support near-Earth energy transmission would eliminate the need to land large power sources on the Lunar surface. Lunar surface power could be available almost immediately upon arrival if electric propulsion is used for the Lunar cargo transport craft. Consider the scenario where energy transmission has been implemented in support of near-Earth operations and mankind is ready to expand outward by establishing a manned Lunar base. By then the first generation power transmission satellites in Earth orbit should be reaching the end of their operational life. Systematically, each old power satellite is being replaced with a second-generation power satellites with upwards of 10-MWe capability. The power and propulsion technologies needed to establish and provide logistics support for a permanently manned Lunar base have been developed and operationally tested in support of near-Earth energy transmission. The cargo transport vehicle, needed to initially deliver the materials and equipment from Earth orbit to Lunar orbit and later to provide continued logistical support to the manned base, would be a straightforward engineering effort based on the beam-powered EOTV and power satellite technologies. By adding an energy transmitter to the cargo manifest, a Lunar beam-power transmission system would readily be available. The Lunar transporter would deliver its cargo to the proper Lunar parking orbit and then move out and take up position at the L1 libration point 35,000 miles above the Lunar surface. From here, depending on the design of the power transmission system, a 10-MW nuclear electric power system could provide upwards of 5-MW of electrical power for use on the Lunar surface, greatly enhancing current planned activities and enabling new mission options not previously considered possible because of power limitations.

Mars Mission Applications.

The development of a manned Mars base would be a natural progression and extrapolation of the power infrastructure developed for the Lunar base. The power infrastructure established on Mars would be a seasoned technology, having been fully tested during Lunar development. Energy transmission systems can be transported to Mars and implemented directly, again eliminating the need to land large power sources on the Mars surface and making surface power available almost immediately upon arrival.

A logical approach to the development of a manned Mars outpost is the proposed split Mars mission scenario. An unmanned nuclear electric cargo vehicle would be sent to Mars, carrying all the supplies and equipment needed by the manned mission that would follow later, using a chemical propulsion system. The use of a nuclear electric-propelled cargo vehicle offers several unique mission options not possible with a chemically-powered cargo vehicle (Coomes et al. 1988). Because the cargo vehicle would be sent in advance, it could carry the fuel needed for the manned return flight. This option is also possible with a chemically-powered vehicle but requires tradeoffs of other mission objectives to accommodate the added mass. The mass savings of electric propulsion would reduce the impact of this option on the overall mission. With the cargo vehicle already in orbit around Mars, a backup return vehicle is available for the crew, should the primary manned space craft become disabled. By proper modular design and common interfacing of systems of the cargo vehicle and the manned Mars spacecraft, the crew could reconfigure the two spacecraft and use the propulsion system of the cargo vehicle to propel them home. Taking advantage of the fuel savings offered by electric propulsion and with no need for the life support, habitat, or additional shielding systems a manned spacecraft would require, a cargo transport vehicle could carry enough supplies in a single flight to support several manned flights to Mars. By configuring the spacecraft to take advantage of aerobraking at Mars, even more propellant could be saved and additional supplies could be carried. If the nuclear electric power system is designed for long-life operation and includes a beam-power transmission system, electric power can be supplied from space to support a broader range of activities on the Mars surface. Placing the power and propulsion unit in geosynchronous Mars orbit (GMO) above the selected landing site, 20% to 60% of the onboard electric power capacity would be available at the Mars surface to support mission activities, both manned and unmanned. Automated systems deployed from the cargo vehicle or left operating when the crew returns to Earth could convert raw materials on the Mars surface into usable form, available to the manned flights that follow.

Upon arrival at Mars, the transport vehicle would deliver its cargo to the desired parking orbit and move outward, taking a position in GMO above the desired landing site. From there it would beam power to the Mars surface. Initially, this electric power would be used to operate fully automated systems released earlier and, having landed on the Mars surface, would prepare the landing site for the Mars crew arriving later. This same system would later provide the surface power needed to support crew activities after their arrival. With megawatt levels of electric power available on the Mars surface, mission activities could be expanded. The power-beaming option can provide the energy needed to support a manned Mars base that would allow a Mars colony to develop and grow.

CONCLUSIONS

Central-point space power generation and distribution by energy beams--power beaming--is technically feasible and should be included in the U.S. space power program as an option that warrants serious consideration as our future baseline approach to meeting this nation's space power requirements. Power beaming is a true integrated approach to providing power in space. It can provide significant benefit to this nation's overall space efforts and offers an opportunity to greatly enhance U.S. space leadership. Power beaming is a viable concept that allows power to be generated at one location and transmitted to a remote user at another location by energy beams. Power beaming can provide power users with up to 10 times more power than a solar PV system for the same collector area or, conversely, power beaming could provide the user with the same power as a solar PV system with only 10% of the collector area. A beam-powered electric OTV would reduce the cost of satellite transfer from LEO to GEO to half that for a chemical-based OTV and would increase the payload capability delivered to GEO for existing launch systems lift capability to LEO. A space power architecture based on power beaming would support both military and civilian space operations and would be mission-enhancing, if not enabling, for manned activities planned for beyond-Earth orbit. Power beaming would increase the acceptability and use of nuclear power in space by mission planners and satellite designers because it isolates the end-use satellite from the nuclear power source while providing significantly more power than a solar source. The SP-100 Program can serve as the baseline power system technology for the first-generation beam-power satellites. SP-100 technology-based beam-power satellites of 500- to 1000-kWe output could be developed and deployed in the 2000 to 2010 time frame. Second-generation beam-power satellites could use multimegawatt power system technology. These would be phased in as replacements for first-generation systems, with beam-power satellite output capability reaching levels of 10 MWe to 20 MWe by 2030. Compared with solar PV with energy storage or onboard nuclear reactors, power beaming results in the lowest total power system mass placed in orbit.

Acknowledgments

This work was performed at the Pacific Northwest Laboratory and supported by the U.S. Department of Energy under Contract DE-AC06-76RLO 1830.

References

Coomes, E.P., B.M. Johnson, and R.D. Widrig (1990) Space Power Generation and Distribution (SPGD) Program Basis Document: Volume 1: Concept Description. PNL-7162, Vol. 1, Pacific Northwest Laboratory, Richland, Washington.

Coomes, E. P. and J.E. Dagle (1992) "A Low-Alpha Nuclear Electric Propulsion System for Lunar and Mars Missions" in Proceedings 9th Space Nuclear Power Symposium, American Institute of Physics, New York, New York.

Coomes, E. P., L. A. McCauley, J. L. Christian, M. A. Gomez, and W. A. Wong (1988) "Unique Mission Options Available with a Megawatt-Class Nuclear Electric Propulsion System." AIAA-88-2389, American Institute of Aeronautics and Astronautics, New York, New York.

Dagle, J.E. (1991) "Performance Enhancement Using Power Beaming for Electric Propulsion Earth Orbital Transporters" in Proceedings: 26th Intersociety Energy Conversion Engineering Conference, American Nuclear Society, La Grange Park, Illinois.

Maral, G., and M. Bousquet (1986) Satellite Communications Systems, John Wiley and Sons, Chichester, Great Britain.

FUELS AND MATERIALS DEVELOPMENT FOR SPACE NUCLEAR PROPULSION

S.K. Bhattacharyya
Argonne National Laboratory
9700 S. Cass Avenue
Argonne, IL 60439
(708) 252-3293

R.H. Cooper
Oak Ridge National Laboratory
P.O. Box 2008
Oak Ridge, TN 37831-6079
(615) 574-4470

R.B. Matthews
Los Alamos National Laboratory
Department of Energy
SN-GA155, Forrestal Building
1000 Independence Avenue
Washington, DC 20585
(202) 586-4754

C.S. Olsen
Idaho National Engineering Laboratory
P.O. Box 1625
Idaho Falls, ID 83415-3511
(208) 526-9094

C.E. Walter
Lawrence Livermore Nat'l. Lab.
7000 East Avenue, L-144
Livermore, CA 94550
(415) 422-1777

R.H. Titran
NASA Lewis Research Center
21000 Brookpark Road
MS 49-1
Cleveland, OH 44135
(216) 433-3198

Abstract

Nuclear propulsion has been identified as an enabling technology in meeting the missions of the Space Exploration Initiative (SEI). Both Nuclear Thermal Propulsion (NTP) and Nuclear Electric Propulsion (NEP) have roles to play in the initiative. Of the numerous specific development items that need to be undertaken for these technologies, nuclear fuels and materials are considered by experts as the most challenging and therefore requiring early attention. One of the six panels organized by the NASA/DOE/DoD Steering Committee to help plan the nuclear propulsion development tasks was dedicated to nuclear fuels, materials and related technologies. Considering only the solid core concepts presented in the 1990 workshops on NTP and NEP, the panel concluded that there were classes of fuels and materials that had the potential to meet the demanding temperature and lifetime requirements for SEI in the timeframe of interest. Plans for the development of fuels and materials were prepared. A full development plan is going to involve the construction of several major new facilities and the modification of many existing ones. For the fuels, areas of commonality have been explored to allow for effective early activity while the downselection process to focus the development takes place. For materials, the lack of definition of candidate selection by the concept proposers makes this somewhat more difficult.

Careful coordination of the work will be essential to keep the development and characterization of fuels and materials on schedule and costs within bounds.

INTRODUCTION

The use of nuclear energy for space propulsion has been under consideration from the early days of the development of nuclear power. Recent activity in the Space Exploration Initiative has rekindled interest in nuclear power and propulsion for use in space missions. The interest in nuclear propulsion stems primarily from the promise of reduced trip times to Mars for manned missions relative to chemical propulsion. This is a major advantage because effects of radiation (cosmic and solar), microgravity, and long-term isolation are anticipated to be major obstacles to overcome for human astronavigation. Any reduction in trip times will result in enhanced performance. In a broader sense, nuclear power and propulsion systems will enable a number of activities in space commerce, defense and exploration areas.

Two classes of nuclear propulsion systems are under consideration. Nuclear Thermal Propulsion (NTP) systems are characterized by very high temperature exhaust gas from the reactors (~3000K) and relatively short operating times (hours). Nuclear Electric Propulsion (NEP) systems are characterized by lower coolant temperatures (1500-2000K) but considerably longer operating times (years). The NEP reactor systems bear a close relationship to surface power systems and could imply a shared development strategy. The NEP and NTP systems, however, involve different development paths and will generally involve different fuels and materials. The testing required to develop and qualify the fuels and materials are considerably different.

Since it was clear from earlier development activities that reactor fuels and materials and related technologies would be major technology drivers, a panel consisting of DOE, DoD and NASA experts was established in December 1990 to address the principal issues and prepare a development plan for these technologies. The membership of the panel represents a broad participation of the DOE laboratories in addition to the DoD and NASA representatives. In addition to the actual membership, the panel benefitted greatly from the active participation of a number of industry experts.

This panel (like the five other companion panels that together addressed the Nuclear Propulsion development issues) was composed of volunteers. The modus operandi was to meet as a group approximately once a month starting January, 1991 to discuss the major issues. Considerable activity was undertaken by separate subgroups. The panel took full advantage of the large body of systematic preparations and evaluations of development plans for the SDI multi-megawatt program. In addition, the ongoing DOE-supported work at Idaho National Engineering Laboratory (INEL) and Los Alamos National Laboratory (LANL) in the fuels development area provided a large amount of information.

REQUIREMENTS AND DEVELOPMENTS NEEDS

Requirements

The operating requirements for fuels and materials are derived from the large amount of mission analyses performed for potential SEI missions by NASA analysts and others. From these, a specific impulse (I_{sp}) value of 925 seconds has generally been accepted as baseline. This value corresponds approximately to a chamber temperature of 2750K, from extrapolation of NERVA data. The fuel temperature will exceed the chamber temperature by an amount determined by safety margins and temperature drops. These are dependent upon specific designs; a nominal 10% increase has been assumed. Thus, a baseline temperature is approximately 3000K. These and other requirements are listed in Table 1. Since the basic mission requirements can be achieved in a number of ways, it is appropriate to list a range of requirements. The burn time and number of cycles take account of the fact that re-usability of the engines will be a consideration. The criteria for reliability and fission product release have not been quantitatively established, but it is clear that the former is as high as possible and the latter is As Low As Reasonably Achievable (ALARA). Finally, the TRL5 number for NEP signifies the fact that no integral ground test of the whole system is planned. The first integral test will be the first space test.

State-of-the-Art

If one assumes that past demonstrations constitute state-of-the-art, the technology status is defined by the NERVA/ROVER program for NTP and the latest results on SP-100 for NEP. State-of-the-art needs to be understood as a combination of temperature, operating time (in appropriate environment) and fuel/materials integrity (ALARA fission product release). The state-of-the-art conditions are shown in Table 2.

Note that the NTP case represents a full system while the SP-100 case represents individual fuel pin irradiations. Also, note that the fission product release considerations imply a greater degree of technology development requirement than is evident from the data.

Recent information from the Soviets (Goldin et al. 1991) suggest that they have attained temperatures (exit gas?) of 3000K in their test rocket engines. While details were not available, there were references to solid solution carbide fuels (UC-ZrC, UC-ZrC-NbC) and heterophase composition (ZrC-UC+C). Information on the lifetimes achieved at these temperatures was unclear although a one hour burn was implied. There are also references to the production of very high temperature carbide solid solution fuel--but indications are that these are highly unstable and would not be able to reach the lifetime goals of SEI missions.

The materials state-of-the-art is derivable from the results of a number of earlier and currently ongoing space and advanced aeronautics programs. These include the SNAP, ROVER/NERVA, SP-100, MMW and NASP programs. An assessment of the hydrogen-oxygen rocket engine program was also considered. Because of the large number of materials involved and the wide range of data considered, it is not possible to make a concise statement of the "state-of-the-art." These programs, in general, establish the potential feasibility of a number of materials (refractory alloys, carbon-carbon and other composites and ceramics) for operation at elevated temperatures, some in the H_2 or liquid metal environments. However, significant amount of work will be needed to develop these materials for SEI/nuclear propulsion use.

Summary of Concepts

References 2 and 3 present capsule summaries of the NTP concepts and the NEP concepts presented at the workshops. Several points are noted.

- For NTP, the temperatures proposed range from 2500K to 3600K for solid core reactors.

- For NTP, H_2 is the coolant/propellant of choice.

- For NEP, there are two approaches--liquid metal (Li) cooled reactors or gas cooled (He-Xe) reactors.

- The information on materials is sparse. For fuels there was some definition, but generally with no basis for the temperatures projected.

The commonality of coolants and of the basic building blocks in fuels for solid core systems allowed for a possible generic initial phase of fuels development. The solid core reactors were judged to be the only ones capable of meeting the TRL6 date of 2006.

Approach

Given the above NTP and NEP requirements and the descriptions of the concepts proposed, the Panel approach was to focus primarily on the solid core concepts and prepare lists of the technical issues for the development of fuels and materials. These issues included fabrication of the fuels and materials components. In order to obtain inputs from the various concept focal points, a detailed questionnaire was sent to each of them. Their responses, received in a numerical format, provided good corroboration of the panel's judgment. From these lists were derived testing requirements and a corresponding set of facility requirements. The development plans so produced were for individual classes of fuels and materials. The actual program plans for nuclear propulsion will involve an initial screening effort to focus on the "best" candidate fuels and materials in order to keep development costs within bounds. These candidates will be developed and "qualified" for the SEI missions. Final selections will be based upon systems and mission considerations but the demonstrated performance capabilities of fuels and materials will undoubtedly provide a significant input. The development plans for the individual classes of fuels and materials serve as our data base and contribute in the selection of the early screening tests and in selecting the later development tasks.

FUELS AND MATERIALS DEVELOPMENT

Fuels Development

A general view of fuel lifetime behavior as a function of temperature can be stated as follows. At relatively low temperatures the fuel can survive a long time, the life limiting conditions being imposed by burnup and associated problems. As fuel temperature increases the various phenomena depicted in the figure come into play and lifetime is shortened. At the limit, fuel melting occurs. It should be noted that the practical limit of fuel usage is sufficiently below the melting point to allow structural stability and allowances for uncertainties. In addition, other failure mechanisms can come into play well before the melting point is reached.

From the various classes of fuels that have the potential of meeting the requirements listed in the time frame of interest, the following were evaluated in detail. They essentially spanned the solid core reactor classes of interest.

NTP: Hydrogen cooled reactors
    NERVA derivative (carbide fuels)
    Particle bed (carbide fuels)
    Cermet (UO_2)

NEP: Liquid-metal cooled reactors
    UN pellets
    UN/W-Re cermet
    UO_2 thermionics

Gas-cooled reactors
    NERVA derivative (carbide fuels)
    Particle bed (carbide fuels)

The development items for the fuels covered the following areas:

- Fuel "element" development and characterization
- Fabrication process development
- Non-nuclear tests to establish capability, lifetime
- Nuclear capsule tests (small fuel samples)
- Nuclear loop tests (near prototypic size fuel)
- Nuclear furnace tests (prototypic fuel in prototypic operating conditions)

Tables were prepared which listed the specific issues in each class of fuels. The information was used to produce requirements for fabrication and test facilities as well as detailed test plans for development and "qualification" of fuels.

The overall plan so depicted consists of the entire sequence of development fabrication and testing. The general purpose of testing is to provide data to resolve technical feasibility issues (screening), provide data for design and specifications, provide reliability data (all tests), and establish safety margins (tests to failure).

In actual practice, the development plan will probably proceed so as to include several options early on. Initial work will be started on several different classes of fuel and the results of screening tests will feed into the down selection process. Only the most promising candidates will be developed through to the ground engine test.

The balanced development program involves the use of a number of in-reactor capsule and loop tests (in existing reactors) and tests in new "Element Test" reactors before the ground test of the engine. This is an expensive proposition. All of the steps are necessary to minimize risk to the program. If a higher level of risk is made necessary by cost constraints, some of the inter-mediate steps may have to be dropped. At this point, there is no consensus of opinion on how this should be done, and very careful planning will be required to ensure acceptable development and qualification.

Materials Development

Tables 3 and 4 show a list of candidate materials for use in the SEI applications for both NTP and NEP. There are clearly a number of plausible choices and an early task will be to select lead candidates for detailed development. The general development tasks consist of the following items:

- Procurement of candidate material for testing (long lead time items): this includes the preparation of detailed specifications.

- Mechanical and physical property determinations: this includes the usual creep, fatigue fracture and tensile strength determinations along with specialized studies of thermal cycling. The thermal properties need to be determined as well.

- Compatibility studies: this needs to be performed under prototypical (temperature, environment, etc.) conditions. For NEP the long term exposure to liquid metal coolants or to impurities in the He/Xe gas are major issues. For NTP the erosion and corrosion due to hot H_2 are principal issues.

- Irradiation tolerance: the radiation behavior of candidate materials needs to be determined as early as possible. The magnitude of irradiation induced swelling, property degradation and phase stability has to be determined under prototypic neutron fluences and spectra.

- Processing and component fabrication: since the NP systems will involve complex mechanical shapes, the fabricability of the materials is of major concern. This will involve development and demonstration of joining and forming methods. For coatings, methods to produce high quality, uniform coatings have to be developed.

SUMMARY

The development of high temperature fuels and materials for SEI nuclear propulsion applications represent significant advances from current state-of-the-art and therefore significant development challenges. The best technical assessments of the specific development needs suggest that the required fuels and materials can be developed, but the costs will necessarily be large. In order to keep the development costs and risks within bounds, it is essential that:

- Commonality in fuel development be exploited.

- Work be focussed to relevant concepts via downselections.

- Maximum use be made of parallel work on other projects.

- Maximum use be made of existing DOE, NASA, DoD and industry facilities and capabilities for testing and evaluation.

- Careful coordination of the development work be maintained.

References

Goldin, A.Y. et al. (1991) *"Development of Nuclear Rocket Engines in the USSR,"* Presented at the AIAA Propulsion Meeting.

Barnett, J.W. (1991) *"NEP Technologies: Overview of the NASA/DOE/DOD NEP Workshop,"* Proceedings 8th Symposium on Space Nuclear Power Systems, American Institute of Physics CONF 910116, Part 2, p. 511.

Clark, J.S. (1991) *"A Comparison of NTP Concepts: Results of a Workshop,"* Proceedings 8th Symposium on Space Nuclear Power Systems, American Institute of Physics CONF 910116, Part 2, p. 740.

TABLE 1. Requirements Derived from NTP and NEP Panels.

Parameter	NTP	NEP
Exit Coolant Temperature (K)	3000 (2500 - 3200)*	- 1350 TE - 1350 - 1700 Rankine - 1700 - 1900 Brayton ~ > 2000 Thermionic**
Lifetime	- Maximum single burn: 1 hour - Total run time under 10 hours	7 year full power
Number of Cycles	6 per mission (maximum 24)	Multiple startups/shutdowns
Safety	Minimize fission product release (ALARA)	Withstand high burnups
Reliability	TBD (as high as reasonably possible)***	TBD
Power Level	1-2 GWt (75,000 lb. thrust)	5 MWe - 10 MWe
Anticipated Ramp Rates (startup)	TBD†	TBD†
Readiness Dates	- TRL6 2006 - Lab scale 2700°K capability early	TRL5 @ < 50 1998 TRL5 @ < 20 2001 TRL5 @ < 5 2005

* A temperature of 3600K was presented by one concept. This was based upon the reported melting point of HfC. The actual value will be considerably lower.

** Concept emitter temperature presented was 2400K.

*** NERVA value was 0.995 at 90% confidence level.

† Determine acceptable ramp rates. These are not expected to be major problems.

TABLE 2. Anticipated Operating Conditions for Major NEP Subsystems.

Subsystem	Nominal Temperature (K)	Environment	Radiation Fluence (n/cm^2)	Lifetime Goal (h)
Nuclear Heat Source	1350 to 1900 (to 2000)[a]	Liquid metal or inert gas	10_{22}	6×10^4
Shield	500 to 1000	Major components (i.e., hydride and structural materials)	10^{15} to 10^{20}	6×10^4
Power Conversion	1350 to 1900	Liquid metal or inert gas	10^{12}	6×10^4
Thermal Management	900 to 1200	Liquid metal	10^9	6×10^4
Electric Thruster	TBD	Hydrogen helium or lithium	TBD	1×10^4

[a] In core thermionic emitter temperature may approach 200 K.

TABLE 3. Summary of Currently Identified Candidate Materials Organized by Major NTP Subsystems and Operating Conditions.

Subsystem	Major Component	Operating Temperature Range (K)	Environment	Candidate Materials			
				Alloy Metal Matrix Composites	Ceramic and Ceramic Composites	Graphites, Carbon/ Polymer Matrix, and C/C Composites	Coatings
Propellant Tank	Tank	4 to TBD	Cryogenic hydrogen	Aℓ alloys Aℓ-Li alloys		C/polymeric matrix composites	
Turbo Pump	Pump	4 to TBD	Cryogenic hydrogen	Fe alloys Aℓ alloys Ti alloys			
	Turbine	900 to 1300	Hot hydrogen	Ni alloys		H_2 protected C/C composites	ZrC, NbC, TaC, HfC
Radiation Shield	External shield	TBD	Radiation shield components	W-Ni-Fe Aℓ Alloys (structure)	LiH Borated ZrH BATH lead		
Nuclear Heat Source	Fuel support	20-3000	Cold hydrogen to hot hydrogen	Mo alloys W-25 Re alloys In-718	TaC ZrC	Various graphites	ZrC NbC TaC HfC
	Reactivity control	20-3000	Cold hydrogen to hot hydrogen	Be B-Cu	B_4C ZrH		Tribological
	Core support	20-3000	Cold hydrogen to hot hydrogen	Mo alloys			
	Pressure vessel	20-3000	Cold hydrogen to hot hydrogen	Aℓ alloys Ti alloys Ni alloys Fe alloys			
Nozzle	Exhause nozzle		Hot hydrogen			H_2 protected C/C composites	

- Neutron fluence at the components expected to be small.

TABLE 4. Summary of Currently Identified Candidate Materials Organized by Major NEP Subsystem and Operating Conditions.

Subsystem	Nominal Temperature (K)	Environment	Neutron Fluence n/cm^2	Major Component	Candidate Materials			
					Alloy Metal-Matrix Composites	Ceramic and Ceramic Composites	Graphitics and C/C Composites	Coatings
Nuclear Heat Source	1350-1900 (up to 2000K)[a]	Liquid metal or inert gas	10^{22}[b]	Fuel cladding and reactor structure	W-based Mo-based Ta-based Nb-based W•HfC/Nb•1Zr		C/C composites	
				Reactivity control	Be	B$_4$C, BeO Boral		Tribological
Power Conversion	1350 to 1900	Liquid metal or inert gas	10^{12}	Turbine, shell, and valves	See refractory alloys above	BeO	C/C Turbine	
Radiation Shield	500 to 1000	Major components (i.e., hydride and structural materials)	10^{15} - 10^{20}	Neutron shield	LiH (Other)			
				Gamma shield				
				Structural				
Thermal Management	900 to 1200		10^9	Pumps Radiators				
Thruster	TBD		TBD	Ejector				

a In core thermionic emitter temperature may approach 2000K.
b Lifetime goal of all components is 6 x 10^4 hours. The electric thruster is expected to last 1 x 10^4 hours.

NUCLEAR THERMAL PROPULSION TEST
FACILITY REQUIREMENTS AND DEVELOPMENT STRATEGY

George C. Allen
Sandia National Laboratories
P.O. Box 5800
Albuquerque, NM 87185
(505)845-7015

John S. Clark
NASA/Lewis Research Center
MS 3-8
2100 Brook Park Road
Cleveland, OH 44135
(216)977-7090

John Warren
Division of Defense Energy Projects
Office of Nuclear Energy
U.S. Dept. of Energy, NE-52
Washington, DC 20545
(301)353-6491

David R. Perkins
OL/AC/Phillips Laboratory
PL/RAS
Edwards AFB, CA 93523-5000
(805)275-5640

John Martinell
Idaho National Engineering Laboratory
EG&G Idaho, Inc.
P.O. Box 1625
Idaho Falls, ID 33415
(208)526-8593

Abstract

The Nuclear Thermal Propulsion (NTP) subpanel of the Space Nuclear Propulsion Test Facilities Panel evaluated facility requirements and strategies for nuclear thermal propulsion systems development. High pressure, solid core concepts were considered as the baseline for the evaluation, with low pressure concepts an alternative. The work of the NTP subpanel revealed that a wealth of facilities already exists to support NTP development, and that only a few new facilities must be constructed. Some modifications to existing facilities will be required. Present funding emphasis should be on long-lead-time items for the major new ground test facility complex and on facilities supporting nuclear fuel development, hot hydrogen flow test facilities, and low power critical facilities.

INTRODUCTION

The United States' Space Exploration Initiative (SEI) has as one of its goals a manned mission to Mars by the year 2019. While it will enable a number of space missions, nuclear thermal propulsion (NTP) has been specifically identified as a critical technology for reaching Mars. The National Aeronautics and Space Administration (NASA) has begun to study NTP for this purpose. The NASA Lewis Research Center, the Department of Energy (DOE), and the Department of Defense (DoD) sponsored a workshop on Nuclear Thermal Propulsion in July of 1990 (Clark 1991). In the fall of 1990, a group of six interagency technology panels was formed to evaluate a number of issues related to nuclear propulsion. One of these panels was the Space Nuclear Propulsion Test Facilities Panel, whose purpose was to evaluate test facility needs and considerations for supporting the development of nuclear propulsion systems.

The Space Nuclear Propulsion Test Facilities Panel was divided into two subpanels: One subpanel focused on nuclear thermal propulsion (NTP) facilities and the other on nuclear electric propulsion (NEP) facilities. The Nuclear Thermal Propulsion Facilities Subpanel evaluated facility issues related to nuclear thermal propulsion development. The work of the NTP Facilities Subpanel is the focus of this paper.

NTP FACILITIES SUBPANEL OBJECTIVES

The NTP Facilities Subpanel consisted of volunteer representatives from NASA, DOE, DoD, NASA centers, DOE and DoD laboratories, and private industry, who held monthly meetings during government fiscal year 1991 to evaluate NTP facility requirements and strategies.

The specific objectives of the NTP Facilities Subpanel were to:

1. Define NTP test facility needs based on NTP technology development requirements;
2. Evaluate existing facility capabilities that meet these requirements;
3. Identify new facility development or existing facility modification needs;
4. Identify critical path facility development requirements;
5. Recommend facility development strategies; and
6. Comment on frequently asked questions related to NTP facilities.

In addition to its own expertise, the subpanel interacted frequently with other NASA/DOE/DoD panels that were addressing nuclear thermal propulsion technology needs. Specifically, input from the NTP Technology, NTP Fuel and Materials, and NTP Safety panels was key in developing facility requirements. The NTP Facilities subpanel also solicited information from owners of existing facilities. Data on more than 200 facilities were compiled by Sverdrup, Inc. for NASA Lewis Research Center (see Baldwin 1991). Additionally, the subpanel visited several potential facility sites.

The subpanel compared NTP facility requirements against the capabilities of existing facilities, and discussed and debated development strategy, critical paths, and facility issues. However, no funding was provided to allow a detailed analysis to verify the NTP Facility Subpanel positions.

SCOPE OF EVALUATION

Because high pressure propulsion systems were the only concepts judged to be capable of completing full system ground testing (TRL-6) by 2006, high pressure systems were considered as the baseline, with low pressure concepts considered as an alternative. The NTP Subpanel, therefore, focused on facilities for developing both nuclear and non-nuclear components and systems for solid core concepts such as Nuclear Engine for Rocket Vehicle Application (NERVA) derivatives, particle bed, wire core, cermet, pellet bed, and Dumbo (see Clark 1991). Facilities for open cycle liquid or gas core systems were not specifically discussed by the subpanel, although some information on early proof-of-principle test facility needs for highly innovative concepts is included in the subpanel report.

The major working assumptions of the NTP Subpanel were:

- A NASA/DOE/DoD Memoranda of Agreement will exist for coordinating nuclear propulsion activities;
- Technical feasibility, schedule times, and cost envelopes will be success-oriented;
- Evolving "innovative" technologies such as open cycle, gas core engines cannot be developed in the near-term, while mainline solid core concepts probably can;
- The current environmental, safety, and health requirements may evolve but will not undergo quantum changes;
- Nuclear tests will be conducted at DOE facilities;
- An open cycle effluent treatment system will work and will be environmentally acceptable;
- Full-scale reactor/engine tests to failure will not be conducted at ground test sites;
- Engines will not be tested at power in clusters at the ground test facility;
- Full expansion-ratio nozzle tests will not be conducted at the ground test facility;
- Reactor assembly and low power critical tests will not be required at the launch site; and
- Unmanned demonstration flights will be conducted in space prior to manned flight.

NTP DEVELOPMENT TEST LOGIC

The NTP Subpanel, based on its own discussions and on input from other NASA/DOE/DoD Nuclear Propulsion panels, developed the summary test logic shown in Figure 1.

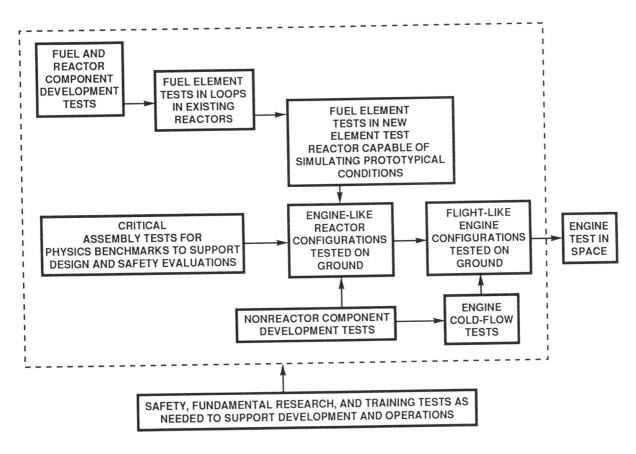

FIGURE 1. Summary Test Logic for NTP Development.

The other NASA/DOE/DoD panels provided extensive input to the facility requirements. Figure 2 shows the NTP Facilities Subpanel interaction with other panels.

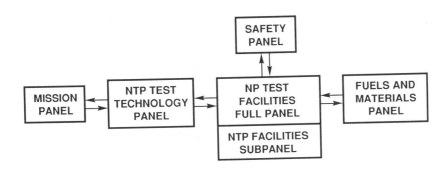

FIGURE 2. NTP Facilities Subpanel Interactions with Other NASA/DOE/DoD Panels.

The subpanel established 19 categories of test facilities which it used to guide data collection on test locations and to evaluate current capabilities. These categories were:

1. Fuel Fabrication Facilities;
2. Unirradiated Fuel Materials Test Facilities;
3. Unirradiated Materials Test Facilities;
4. Hot-hydrogen Flow Test Facilities;
5. Fuel Irradiation Test Facilities;
6. Material Irradiation Test Facilities;
7. Fuel Element Loops in Existing Reactors;
8. Low-power Critical Facilities;
9. Prototypic Fuel Element Test Reactor;
10. Reactor Test Cell;
11. Engine Ground Test Cell;
12. Remote Inspection/Post-irradiation Examination Facilities;
13. Component Test Facilities without Hot-hydrogen or Irradiation Environments;
14. Control System Test Facilities;
15. Component Safety Test Facilities;
16. Training and Simulator Test Facilities;
17. Engine Integration Test Facility;
18. Flight Test Facilities; and
19. System-level Safety Test Facilities.

The Hot-hydrogen Flow Test Facilities category was further divided into:

a. Fuels and Materials Hot-hydrogen Flow Test Facilities;
b. Hot-hydrogen Flow Test Facilities for Turbopump Development;
c. Hot-hydrogen Flow Test Facilities for Nozzle Development; and
d. Hot-hydrogen Flow Altitude Simulation Facility for Nozzle Demonstration.

For each of the 19 categories, the NTP Facilities Subpanel identified test objectives, top-level facility requirements, details of facility capability needs, and potentially available existing facilities.

FACILITY ISSUES

The NTP Facilities Subpanel discussed a number of issues that affect facilities development. The following paragraphs summarize some of the topics discussed. Environmental, safety, and health considerations were the top priority of the subpanel discussions.

Test Issues

The scope of an appropriate flight qualification ground testing program was considered a key issue in defining the requirements for a new ground test complex. The number of test cells required, test cell throughput requirements, potential source terms for environmental impact assessments, and posttest hardware handling and storage requirements depends on the amount of testing needed for flight qualification. This concern over the amount of testing extended to both full engine and fuel element testing. Multiple test cells are recommended, but the subpanel did not suggest an exact number.

The subpanel evaluated the impact of bypassing fuel element testing. Fuel element testing at lower powers, lower power densities, shorter test durations, and lower fuel temperatures is possible in several existing test reactors. A few of these reactors (such as the Idaho National Engineering Laboratory's Advanced Test Reactor) approach some nominal nuclear rocket operating conditions, but have significant shortfalls that will leave unanswered key fuel element development questions for some baseline concepts. The uncertainty is even greater for advanced innovative reactor concepts. Such questions will require prototypic fuel element test reactors or full engine tests for answers. The subpanel recommended including the element test reactor in the test program.

The subpanel did not take a position on qualification testing of large area ratio nozzles (nozzles with ratios as high as 500 to 1). Due to large physical dimensions coupled with low nozzle exit plane pressures, ground testing may not be feasible.

The subpanel evaluated the need to test a complete stage on the ground and determined that only a close-coupled representative section of the tank bottom would be required. Any portion of the nozzle that is regeneratively cooled should be included in the ground test unit.

The subpanel did not identify the need for a specialized facility for safety testing at the full systems level. This position was consistent with the conclusions of the safety panel. However, the subpanel recommended that vibration tests simulating dynamic flight environments be performed on subcritical systems.

Test Reactor/Engine Issues

A facility to accommodate the needs of all possible reactor configurations would be prohibitively expensive. Because solid core concepts were considered to be the only concepts capable of producing near-term results, the subpanel defined a minimum set of facilities needed to develop a solid core nuclear rocket. The subpanel recommended that the reactor/engine test facilities should be designed for single engine tests at power. Multiple reactors would only be ground tested in clusters in low-power critical experiments.

The NTP Facilities Subpanel did not resolve the question of whether a driver reactor or a self-driven element test reactor would be needed. The advantages of a driver core include greater integrity for a larger portion of the core, separation of experiment coolant from driver elements, a potentially smaller effluent treatment system, and potentially lower long-term test costs. The advantages of a self-driven core include the potential for testing more experimental fuel elements at the same time, more flexibility to test different experimental fuel element designs, and the elimination of the development cost of driver reactor fuel. In addition, there would be no reactor to maintain when the experiment is removed and no large permanent build-up of long-lived fissions products in a driver core. The subpanel recommended that the driver/self-driven element test reactor decision be left to the facility designer.

The element test reactor should be designed to permit the evaluation of different fuel concepts, as the cost for multiple element test facilities is prohibitive. The fuel test facility should be developed modularly.

Site Requirements Issues

Some facilities could take as long as ten years from the start of the program to full development. If an early flight test becomes a requirement, the program would be forced to accept significantly higher risks in the first space flights since full system ground testing could not be conducted.

Low pressure rocket engine test facilities are complicated by the very low nozzle exhaust pressures of such engines. In past and present test facility concepts, the exhaust pressure serves as the driving pressure for an effluent treatment system. The proper method for maintaining high standards of effluent scrubbing of fission fragments with low rocket exhaust pressure is not clear, but may require a fundamentally different effluent treatment system design. Much work remains to identify low pressure effluent cleanup options.

The rocket development hardware tested on the ground must be retained and stored after all posttest evaluations are completed, because the requirement for very high rocket reliability demands that development hardware be available to resolve subsequent development, qualification, and/or flight anomalies.

Fuel loading and zero-power critical testing of flight reactors must be conducted at facilities qualified for nuclear operations. The subpanel recommended that this be done at the reactor manufacturing and assembly location.

Site Location Issues

Planning for the fuel element test reactor facility and full system ground testing must begin as early as possible in the project. The environmental and safety concerns for such a facility are significant, and a highly sophisticated test site with an effluent treatment system that minimizes radionuclide release is essential. The site will also require expensive, long-lead-time, special order equipment. Years will be required for facility design, synthesis, and approval before equipment can be assembled and actual site construction completed.

Ground tests will generate waste from three principal sources: (1) the filters used for effluent treatment systems, (2) radioactive fuel, and (3) non-nuclear hardware. In addition, at the conclusion of the development program the ground test site and its equipment will have to be decontaminated and decommissioned. The NTP Facilities Subpanel recommended that the program minimize waste and that the test site be colocated with a site having low-level waste disposal capability.

RESULTS

The NTP subpanel study revealed that the United States has a wealth of test facilities available for supporting NTP technology development. While some modifications will be required to support specific NTP development actions, there is a solid base of existing facilities available to satisfy a large majority of test needs.

Facilities Status

The subpanel found that NTP facilities could be divided into four major groups: (1) those that do not exist; (2) those that exist but need modification or equipment purchases; (3) those that exist and can be used as is, but may need later modification or equipment purchases; and (4) those that exist and for which no modifications are anticipated to be needed. Table 1 divides the 19 categories into their respective groups.

TABLE 1. Current Status of NTP Facilities.

No current facilities	Exist but need modifications or equipment purchases
Prototypic Fuel Element Test Reactor	Fuel Fabrication Facilities
Reactor Test Cell	Unirradiated Fuel Materials Test Facilities
Engine Ground Test Cell	Hot-hydrogen Flow Test Facilities
Flight Test Facilities	Fuel Element Loops in Existing Reactors
System-level Safety Test Facilities	Remote Inspection/Post-irradiation Examination Facilities
Training and Simulator Test Facilities	Low-power Critical Test Facilities
Exist but may need eventual modification or equipment purchases	**Exist and need no modifications**
Fuel Irradiation Test Facilities	Unirradiated Materials Test Facilities
Material Irradiation Test Facilities	Component Test Facilities without Hot-hydrogen or Irradiation Environments
Engine Integration Test Facility	Control System Test Facilities
	Component Safety Test Facilities

As Table 1 shows, the only facility categories that do not currently exist are the Prototypic Fuel Element Test Reactor, Reactor Test Cell, Engine Ground Test Cell, Flight Test Facilities, System-level Safety Test Facilities, and Training and Simulator Test Facilities. Of these six, three (Flight Test Facilities, System-level Safety Test Facilities, Training and Simulator Test Facilities) are anticipated to not be needed or could be incorporated into other categories. Modifications to existing facilities also could be made to accommodate these categories. Therefore, only the Prototypic Fuel Element Test Reactor, Reactor Test Cell, and Engine Ground Test Cell would be new constructions.

Tables 2 and 3 show examples of the top-level requirements for the Prototypic Fuel Element Test Reactor and the Engine Ground Test Cell. The NTP Subpanel report (Allen 1991) contains such detailed requirement listings on all 19 of the test facilities categories.

It should be noted that the lead-time and cost of facility construction or modification are very dependent on the test capabilities required. In the case of the Prototypic Element Test Reactor and the Reactor/Engine Test Complex the key drivers are:

- Environmental and safety regulations;
- Total reactor power or thrust level;
- Test run time;
- Reactor power density;
- Exhaust temperature;
- Exhaust backpressure; and
- Tests to performance margins that include potential fuel failure. (This is primarily an issue for the element test reactor.)

Colocation of Similar Test Functions

The NTP Subpanel members agreed that the reactor and engine ground test facilities, which generate neutrons and large amounts of energy, should be colocated on the same site. The subpanel also agreed that the Element Test Reactor could be located with the Reactor Test Cell and Engine Ground Test Cell, forming a single element/reactor/engine test site. This test complex would be located on an existing DOE site or reservation and could use the existing permits, environmental assessments, infrastructure, and waste management/fuel processing facilities as much as possible. Such an approach would save time, effort, and cost. Multiple cells and/or other physical separations should be included in the test site complex to allow work on different test articles to proceed in parallel.

Nevada Research and Development Area (NRDA)

During the early 1960s, NASA tested a nuclear thermal propulsion reactor and engine system (NERVA) at the Nuclear Reactor Development Station in Nevada. The project was stopped, but the facilities still remain in a test complex renamed as the Nevada Research and Development Area (NRDA). The NTP Facilities Subpanel visited the NRDA to determine if the site could be reused for current NTP development. The NRDA test cell facilities would require extensive refurbishment and modification to be useful for current nuclear rocket development. The effects of long dormancy coupled with the requirements of much more restrictive environmental standards probably makes the existing NRDA facilities unviable. Additionally, much of the equipment at NRDA has been scavenged and some of it is currently being used by other programs.

TABLE 2. Example of Top-Level Facility Requirements - Prototypic Fuel Element Test Reactor.

1. Test reactor configuration capable of simulating desired prototypical operating and transient conditions to fuel element(s) being tested.

   Operating Assumptions
   Total Power: >50MW
   Power Density: Prototypic value for given concept (2 to 20 MW/ℓ)

   Test Environment: Hydrogen
   Exhaust temperature: 1000-3500 K
   Pressure: 15-1500 psia
   Duration/Cycles: Sufficient to test elements beyond design basis of engine test article (up to 2 h single burn, 4.5 h cumulative burn, up to 24 cycles).

2. Reactor has capability to test alternate fuel concepts with maximum reuse of components feasible.

3. Facility is capable of fast turnaround of element tests.

4. Reactor complex will comply with all environmental and safety regulations. This includes being able to subject fuel to be tested up to and through failure thresholds as a planned, normal operational event.

5. Facility can supply process fluids as required for both operations and posttest decay heat removal according to specification.

6. Facility can maintain effluent releases within regulatory limits and as-low-as reasonably achievable.

7. Facility has robust instrumentation capability for meeting both operational requirements and experiment data acquisition needs.

8. Facility has capability to test nonfuel components (for example, electronics, valves) in NTP environment (that is, radiation, hot H_2).

9. On-site posttest examination and handling capability is the baseline with off-site inspection/examination capabilities an option, provided the associated handling, packaging, transportation, and posttest examination can be accomplished effectively and in a manner which does not perturb the test articles/assemblies or invalidate the test results.

10. Facility lifetime and reusability should be sufficient for the entire NTP ground test program.

11. Facility should be kept as simple as possible to reduce test costs.

12. Facility accommodates interim storage of test articles.

13. Facility accommodates efficient decontamination, decommissioning, and disposal of waste.

14. Facility complies with applicable security and safeguards requirements.

15. Facility has capability for recovery and reuse after major fuel element failure event.

RECOMMENDATIONS AND CONCLUSIONS

The subpanel concluded that while upgrades and modifications may be made to many existing facilities to support NTP development, only the prototypic element test reactor and the reactor/engine test facilities need to be constructed from the ground up. However, this positive finding must be tempered with the realization that a significant amount of program funding will still be required for new facility development, existing facility modifications, and test operations.

Safety and protection of the environment will be the highest priority of nuclear thermal propulsion technology development. These issues were foremost in the subpanel's considerations. While always considering safety goals, the NTP Facilities Subpanel recommended that NASA, DOE, and DoD:

1. Focus first on facilities needed for fuel development and new facilities with long lead-times. The need to perform fuel element testing under fully prototypical conditions and to evaluate reactor/engine systems on the ground is anticipated to make the prototypic element test reactor and the reactor/engine test facilities fall on the NTP critical path. Major new test facilities of these types will probably take seven to ten years to develop and, therefore, development of these facilities and high-priority facility modifications should receive high funding priority.

TABLE 3. Example of Top-Level Facility Requirements - NTP Engine Ground Test Cell.

1. NTP Engine Ground Test Facility will be colocated on same site or facility as Reactor Test Facility with maximum efficient use of same support infrastructure. Multiple test cells are anticipated for redundancy and to prevent scheduling conflicts.

2. Test cells capable of supporting operations meeting capability requirements for engine system verification and engine flight qualification.

   Operating Assumptions
   Single Engine Tests with a total power up to 2000 MW
   Maximum Allowable Normal Operating Exhaust Pressure at Nozzle Exit: TBD
   Thrust Vector control Operation: 0 to 5%
   Exhaust Chamber Pressure: 15 to 1500 psia
   Mixed Mean Exhaust Temperature Range: 1000 to 3500 K
   Coolant Supply: Liquid or slush H_2
       20 - 40 K
       25 to 100 psia
   Topping or Bleed Cycle for Turbopump
   Maximum Single Burn: 1-2 h
   Cumulative Reactor Run Times: 1.5 to 4.5 h
   Restarts/Cycles: Up to 24

3. Test cells can test alternative solid-core concepts.

4. Test cells can simulate or accommodate close coupling of lower portion of propellant tank.

5. Test complex will comply with all environmental and safety regulations.

6. Test complex will supply process fluids as required for both operations and posttest decay heat removal according to specification, and will have ability to handle slush hydrogen.

7. Facility will maintain effluent releases within regulatory limits and as-low-as reasonably achievable. Flaring of exhaust hydrogen is baseline.

8. Facility has robust instrumentation capability for meeting both operational requirements as well as experiment data acquisition needs. (~1000 channels of experimental data anticipated).

9. On-site posttest examination and handling capability is the baseline with off-site inspection/examination capabilities an option, provided the associated handling, packaging, transportation, and posttest examination can be accomplished effectively and in a manner which does not perturb the test articles/assemblies or invalidate the test results.

10. Facility accommodates interim storage of test articles accommodated.

11. Facility accommodates efficient decontamination, decommissioning, and disposal of waste.

12. Facility complies with applicable security and safeguards requirements.

13. Facility has capability for recovery and reuse after major fuel element failure event.

---

2. Start now on some essential near-term activities such as the National Environmental Policy Act (NEPA) process, NTP technology and facility development plans, conceptual design studies of high-priority items, formal site and facility evaluations, evaluations of impacts of testing different fuel forms in key facilities, major system acquisition/construction project documentation, and high-priority modifications to existing facilities.

3. Develop facilities intelligently and modestly, emphasizing modular expansion capability and multi-user/multi-use facilities with possible applications beyond NTP activities.

4. Use existing facilities and related program resources wisely. The SP-100 and the National Aerospace Plane (NASP) programs might have some synergy with NTP development; multiple use of facilities currently under development by related or parallel programs will have major benefits.

5. Develop a minimum number of facilities/sites where capabilities do not presently exist.

Based upon its reviews and its assessment of NTP development requirements, the subpanel presently recommends the funding priority for facility development shown in Table 4. Facilities required for fuel development have highest priority, followed by hot-hydrogen flow test facilities, and then low power critical facilities. The prototypic element test reactor and reactor/engine facilities are high on the list, not because they would be used first, but because they are long-lead-time items.

TABLE 4. Present Facility Development Funding Priority.

Priority	New	Existing	
Highest	Prototypic Element Test Reactor	Fuel Fabrication Facilities	Unirradiated Fuel Test Facilities
	Reactor/Engine Ground Test Facility	Hot-hydrogen Test Facilities Fuel Element Test Loops	Low power Critical Facilities Post-irradiation Examination Facilities
Medium		Fuel Irradiation Test Facilities Material Irradiation Test Facilities Engine Integration Test Facility	Unirradiated Material Test Facilities Component Safety Test Facilites
Low	Flight Test Support Facility	Control System Test Facilities	
	Training and Simulator Facilities	Non-irradiation/Non-hydrogen Component Test Facilities	
	System-level Safety Test Facilities		

Certainly, as NTP development activities evolve, this priority list will change. But, at the present time, funding emphasis should be on facilities required to support nuclear fuel development and long lead-time facilities such as the prototypic element test reactor and reactor/engine test facilities.

The approach suggested by the NTP Facilities Subpanel will make maximum use of the many existing facilities in the United States and the facilities' experienced staffs in developing space nuclear thermal propulsion. At the same time, our approach requires a minimum of new construction. Wise choices, careful planning, and sufficient funding will ensure that the NTP program attains its goal of completing full system ground testing by 2006.

Acknowledgments

This work was sponsored by NASA, DOE, and DoD, who paid for the efforts of the individual agency or national laboratory contributors. The authors acknowledge the contributions of other subpanel members including Tom Byrd, National Aeronautics and Space Administration/Marshall Space Flight Center; Dallas Evans, National Aeronautics and Space Administration/Johnson Space Center; Sam Bhattacharyya, Argonne National Laboratory; Bill Kirk, Los Alamos National Laboratory; Walt Kato, Brookhaven National Laboratory; Roger Pressentin, Department of Energy; and Keven Freese, Arnold Engineering Development Center. Thanks also go to Darryl Baldwin of Sverdrup, who compiled the facility data base and all of the numerous industry, laboratory, and government personnel who made presentations, supplied data, or participated in the subpanel meetings. Special thanks go to Daryl Isbell of Tech Reps, Inc. in Albuquerque, who provided extensive editorial support to both this paper and the NTP Facilities Subpanel report.

This work was supported by the United States Department of Energy under Contract DE-AC04-76DP00789.

References

Allen, G. C., ed. (1991) *Space Exploration Initiative Nuclear Thermal Propulsion Facilities Subpanel Report - Results of an Interagency Evaluation* (Draft), Sandia National Laboratories, Albuquerque, NM.

Baldwin, D. H., ed. (1991) *Space Exploration Initiative Candidate Nuclear Propulsion Test Facilities*, Sverdrup Technology, Inc., Lewis Research Center, Cleveland, OH.

Clark, J. S., ed. (1991) *Nuclear Thermal Propulsion - Proceedings of a Workshop*, NASA-CP-10079, held in Cleveland, OH, July 1990.

THE NASA/DOE SPACE NUCLEAR PROPULSION PROJECT PLAN - FY 1991 STATUS

John S. Clark
NASA Lewis Research Center
Cleveland, OH 44135
(216) 977-7090
FTS 297-7090

Abstract

NASA and the DOE have initiated critical technology development for nuclear rocket propulsion systems for U.S. Space Exploration Initiative (SEI) human and robotic missions to the Moon and to Mars. This paper summarizes the activities of the interagency project planning team in FY 1990 and 1991, and summarizes the project plan. The project plan includes evolutionary technology development for both nuclear electric propulsion (NEP) systems and nuclear thermal propulsion (NTP) systems. The NEP development expands upon the SP-100 technology, being developed for lunar and Mars surface nuclear power, as a foundation from which to develop more powerful, lighter systems for interplanetary probes, lunar and Mars cargo vehicles, and lunar and Mars human-piloted vehicles. Similarly, the NTP technology development plan expands upon the substantial NERVA data base developed in the 1960s and early 1970s for solid core thermal reactors, by improving nuclear fuels to permit higher operating temperatures (with appropriate safety and reliability margins), and ultimately, to upgrade the system to liquid or gaseous core concepts. Thus, a logical, step-wise program is planned that will have systems ready for initial unmanned flights by about 2008, will include robotic lunar missions to gain operational experience prior to manned flight, and will then proceed to a manned lunar mission that will simulate a full-up Mars mission. Mars robotic missions are planned to begin in about 2011-2012, with the first manned Mars mission in about 2014. System upgrades are expected to evolve that will result in shorter trip times, improved payload capabilities, or enhanced safety and reliability.

INTRODUCTION

The Lewis Research Center was selected to lead the nuclear propulsion technology development project for NASA. Also participating in the project are the Marshall Spaceflight Center and the Jet Propulsion Laboratory. The Department of Energy will provide nuclear technology for the systems. The project includes both nuclear electric propulsion (NEP) and nuclear thermal propulsion (NTP) technology development.

The technology development project will be guided by Space Exploration Initiative (SEI) mission requirements as they are defined by the agency SEI planning team. These mission requirements will probably remain a "moving target" for some time as SEI studies continue and the mission architecture is selected. The project will include an iterative, parallel systems engineering and enabling technology development phase, followed by extensive system testing to verify technology readiness. This project will develop the technology to "technology readiness level 6" - (TRL-6), full system ground testing by 2006. It should be noted that flight hardware system development and testing and space qualification and testing are not included in this project, but also must be conducted before a nuclear rocket system will be ready for operational status. Innovative technology development will also be included in the project because of the potential for significantly higher performance and hence, reduced trip times and lower cost.

Workshops were held in 1990 on Nuclear Electric Propulsion and Nuclear Thermal Propulsion technologies (Clark 1991a, Miller et al. 1991, and Clark 1991b). Panels of technical experts were assembled to assess the concepts and technologies presented based on:

- Mission benefit (performance),
- Safety,

- Technology plans and risk,
- Development schedule and cost.

The technical panel recommendations formed the foundation for the initial project planning.

In fiscal year 1991, six interagency technical panels were formed to address key issues identified at the workshops and to continue to refine the technology project plans. Each of these technical panels are preparing a final report to be published in the fall of 1991. Each of the panel chairmen presented a summary of their final reports at the AIAA/NASA/OAI Conference on Advanced SEI Technologies in Cleveland in September. The panel results will be summarized in this paper, and the resulting nuclear technology project plan will be summarized (Clark et al. 1991c).

FY 1991 INTERAGENCY PANEL ACTIVITIES

Mission Analysis Panel

The mission analysis panel was chartered to provide consistent mission performance data and studies to provide a fair comparison between the various nuclear propulsion concepts proposed. Reference missions were selected early in FY 1991 to focus the technology requirements and enable the early definition of facility requirements. These major test facility requirements are summarized in Table 1. For both NEP and NTP systems an "all-up" manned mission to Mars in 2016 was assumed. Subsequently, the Stafford Synthesis Group Report (Synthesis Report 1991) has recommended "split-sprint" manned missions to Mars starting in 2014.
This mission has been studied extensively by the members of the panel, including the effect of engine-out on mission abort scenarios (Wickenheiser 1991). The panel also studied performance tradeoffs associated with:

- Specific impulse and thrust:weight ratios,
- Crew size,
- Opposition-class missions and conjunction-class missions,
- Nuclear heating of system components,
- Cooldown propellant requirements, and
- Engine clustering effects.

Quantified Figures-of-Merit were identified and a structure developed to evaluate and compare competing systems; much work remains to complete this evaluation, however.

Nuclear Safety Policy Working Group

A joint interagency Nuclear Safety Policy Working Group was formed to develop a policy on nuclear propulsion safety (Marshall 1991). A recommended policy has been proposed:

"Ensuring safety is a paramount objective of the Space Exploration Initiative nuclear propulsion program; all program activities shall be conducted in a manner to achieve this objective. Stringent design and operational safety requirements shall be established and met for all program activities to ensure the protection of individuals and the environment. These requirements shall be based on applicable regulations, standards, and research. The fundamental program safety philosophy shall be to reduce risk to levels as low as reasonably achievable.

A comprehensive safety program shall be established. It shall include continual monitoring and evaluation of safety performance and shall provide for independent safety oversight. Clear lines of authority, responsibility, and communication shall be established and maintained. Furthermore, program management shall foster a safety consciousness among all program participants and throughout all aspects of the nuclear propulsion program."

Several program safety requirements are proposed, including: (1) reactor subcriticality prior to the system achieving earth orbit (except for zero power testing on the ground), (2) risk identification and reduction efforts, and

TABLE 1.   Reference Missions Led to Facility Requirements.

NTP	NEP
- "all-up" mission	- "all-up" manned mission
- (3) perigee burn	- multiple missions (reusable) refurbish thrusters, replace propellant
- (3) engine cluster 75,000 lb thrust each	- 10 MWe power
- 1 1/2 hrs. total burn time	- 10 kg/kWe
- 1/2 hr. max. single burn	- 400 mt IMLEO (2016)
- (8) Restarts/mission	- 250-300 days one-way transit
- 2700 K exhaust temp. with appropriate safety & reliability margins	- 500-600 days opposition-class mission
- 925 seconds Isp	- "split" missions → 400 days R.T.
	- cargo → minimum energy

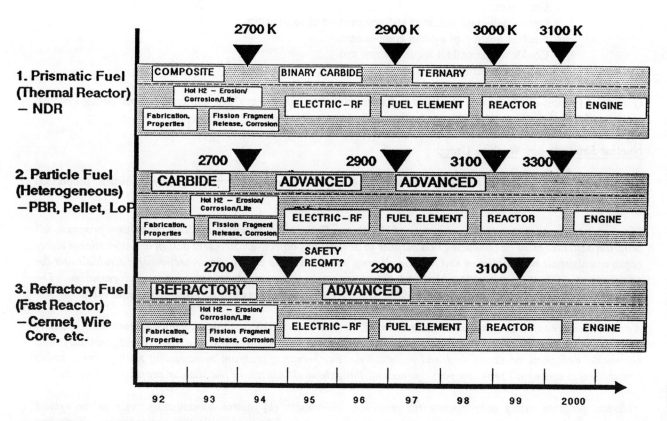

FIGURE 1.   NTP Evolutionary Fuel Strategy.

probabilistic goal demonstration through testing and analysis, (3) inadvertent entry shall not be a planned mission event, and the probability of entry shall be made as low as practical (The reactor must be designed to remain subcritical during any accidental entry and impact), and (4) adequate disposal of spent reactor subsystems shall be explicitly included in the mission planning and design activities.

The working group also made several recommendations to further enhance mission safety. These included specific recommendations regarding:

- Radioactive release, normal operation;
- Radioactive release, accidental;
- Safety validation testing;
- Launch safety;
- Powered flight safety;
- Ground test and equipment safety; and
- Special nuclear materials safeguards.

Nuclear Fuels and Materials Panel

This panel was chartered to define a fuels and materials technology program for both NTP and NEP reactor systems (Bhattacharyya 1991). An early output from this panel described facility requirements for nuclear fuel testing. Detailed test objectives were defined for a wide range of possible fuel types (reactor types). The initial test program includes fuel fabrication and production development, measurement of fuel properties for design and evaluation, fuel concept screening in relevant temperature, fluid and nuclear environments, and development of adequate safety and reliability data.

An assessment was made of the potential commonality between NTP and NEP fuel types. The assumed requirements for NTP reactors included:

- Operating temperatures, 2500 to 3600 K;
- Mission firing time less than 10 hours;
- Low fuel burnup (low fission inventory); and
- Reactor power from 1000 to 5000 MW_{th}.

For NEP reactors the assumed requirements included:

- Temperatures from 1350 to 2000 K,
- Reactor firing time up to 7 years,
- High fuel burnup (high fission fragment inventories), and
- Reactor power 25 to 100 MW_{th} (5 to 10 MW_e).

Thus, NTP and NEP fuel commonality is relatively limited. Gas cooled NEP reactors could utilize prismatic or particle fuels developed for NTP, provided adequate provision is included for the longer life and higher fission fragment buildup in the fuels.

Fuels Strategy

A joint subpanel of members of the Fuels/Materials Panel, NEP Technology Panel and NTP Technology Panel recommended the following evolutionary fuels strategy (See Figure 1 for NTP and Figure 2 for NEP). The strategy includes three difference fuel classes, ranked in priority order.

NTP: For NTP, a composite prismatic fuel for a thermal reactor (typical of the NERVA/ROVER concepts tested in the 1960s and early 1970s) is the top priority because of the extensive database that exists for this concept and the fuels tested. Particle fuels offer performance potential, but must be verified. Similarly, cermet fuels offer

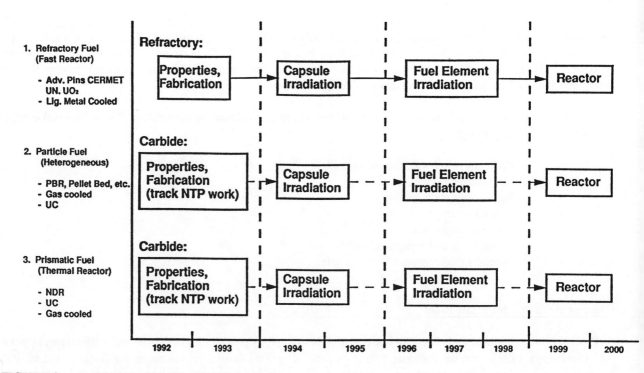

FIGURE 2. NEP Evolutionary Fuel Strategy.

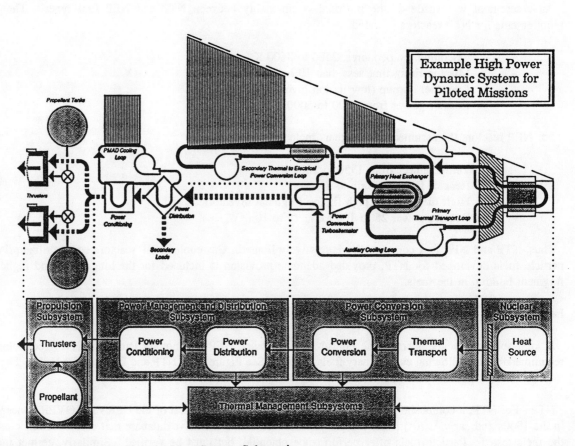

FIGURE 3. Nuclear Electric Propulsion System Schematic.

possible fission fragment retention advantages, and may also have high performance (Polansky 1991).

The initial target exit gas temperature for the NTP system will be about 2700 K (with appropriate safety and design margins), with composite fuels. As higher temperature fuels are developed and validated, they will be incorporated into reference system designs.

Each of the fuel classes will follow a similar test sequence as indicated on figure 1. Initial efforts will focus on fabrication techniques and property measurements. Hot hydrogen coupon tests will be conducted to measure erosion and corrosion rates in a non-nuclear test. Next, a nuclear test with hot hydrogen will be conducted to establish a database for fission fragment release, life and corrosion rate as a function of temperature. This data should permit a selection of fuel form or forms to meet fission fragment release requirements.

Next, the fuels will be made into (sub-scale) fuel element configurations, for electric heating tests (coating integrity, cooling performance, and so on). When a prototypic fuel element tester is built, full fuel elements will be tested in a nuclear, hot hydrogen environment. Finally, full scale reactor and engine tests are planned to verify technology readiness.

1. <u>Prismatic Thermal Reactor Fuels</u>

UC-ZrC-C composite fuel is currently the top priority for SEI missions. Appropriate coatings will be required for prismatic fuel elements initially to about 2700 K (a.k.a. NERVA-derived, NDR), evolving to binary carbide and/or ternary carbide fuels as they are developed (2900-3100 K). There is a substantial NERVA database (detailed system design and full system tests completed (with duplex fuel), and system improvements have been identified. Also, composite fuels were tested in nuclear furnace tests in the 1970's with significantly improved corrosion resistance compared to duplex bead fuel. NDR concept development would be the lowest technical risk, lowest cost, and shortest schedule to TRL-6 (Clark 1991a). The concept can evolve to higher performance; higher performance will require binary/ternary carbide fuel development (with more uncertainty).

2. <u>UC-ZrC Particle Fuels</u>

Particle fuels are applicable to several concepts (particle bed reactors -PBR, Pellet bed, and Low-Pressure reactors), and either thermal or heterogeneous reactors (and possibly even fast reactors). Possible higher fuel temperature capability is claimed, but must be verified; binary/ternary carbide fuel development is higher risk than composite fuel. Since there are essentially no structural loads on the fuel particles, the high strength outer sphere may contain fission products, and the fuel kernel can operate near it's melting temperature.

Very large surface area to volume ratio maximizes the heat transfer area around each sphere, and the tiny particles have a very short heat transfer path, so the fuel kernel temperature and the sphere surface temperature can be maximized. The high surface to volume ratio may also promote higher erosion rates and thus, shorter reactor life at a given temperature, however.

Very high system power density should lead to high system thrust/weight. A more detailed conceptual design will be required for an astronaut-rated SEI mission to verify this potential. There is currently no consensus that high power density or high thrust/weight is required for SEI missions. There may be some advantages in clustered engines, however.

Proof-of-concept testing will be required to verify:

- Mass loss (lifetime) versus temperature at prototypic power generation rates and cooling flow rates,
- Flow distribution and control,
- Laminar flow stability, and
- Cold frit/hot frit porosity/flow control.

No existing experimental reactor is capable of the very high power densities required to test these fuel elements.

3. <u>Refractory Cermet Fuels, Fast Reactor</u>

Some concept design work was done in the 1960s, and fuel processing/fabrication techniques were studied extensively for the nuclear airplane program (Clark 1991a). Refractory metal structural integrity may result in improved fission fragment retention by the cermet fuel compared to other concepts; this must be verified early in project. The rugged construction may offer improved shock loading. Thus, the concept may provide additional safety margins for compaction criticality and immersion criticality.

Cermet fuel development may have application for both NTP and nuclear electric propulsion (NEP) systems. High temperature performance (to 3100 K) has been claimed for cermet fuels (Clark 1991a). System thrust/weight may be lower than other concepts because of higher mass required for fast reactors, thus, performance may be lower. However, if a requirement for very low release of fission fragments is imposed, this concept could be the only way of meeting the requirement. An important effort early in the project will be to evaluate fuel lifetime versus temperature versus fission fragment release for each fuel type in an actual nuclear, hot hydrogen environment. There may be a temperature penalty for low fission product release, however. Also, grain growth at high temperature may make for brittleness and loss of fission fragments.

Figure 2 shows a similar fuels strategy for nuclear electric reactor systems, also in priority order. It should be noted that the cermet fuels, for a fast reactor with liquid metal coolant, are the top priority for NEP. Particle fuels and prismatic fuels are also included for gas cooled reactors; some synergy with the NTP fuels development may be possible in this area.

<u>Nuclear Electric Technology Panel</u>

The NEP Technology Panel was chartered to characterize NEP system options, including integrated reactor/thruster interactions, using common, consistent ground rules and assumptions, and to develop an NEP technology development plan. A schematic of an NEP system is shown in Figure 3. Five major sub-systems usually make up an NEP system: reactor, power conversion, thermal management, power management and distribution (PMAD), and thruster. Many component options have been identified that could be used to make up the NEP system, but an "optimized" system has not been designed, nor is the technology in hand for a manned NEP mission to Mars. A conceptual design study is proposed early in the program to focus on the optimum design and to help focus the technology development activities.

The technology development plan is evolutionary, in that it contains interim milestones and missions to verify the technology readiness in lower power, interplanetary mission applications first, and then progressing to the more challenging Mars cargo and piloted missions (Doherty 1991). The plan is highly integrated with the existing SP-100 space reactor program and the SP-100 technology is an effective "jumping off" point from which the NEP systems can be developed.

The major NEP technical challenges include:

- High temperature reactors, turbines and radiators;
- High burnup fuels and reactor designs;
- Efficient, high temperature power conditioning; and
- High efficiency, long life thrusters.

The high system specific impulse, however, makes the system ideally suited for reusable, long missions where minimum propellant usage is critical.

<u>Nuclear Thermal Propulsion Technology Panel</u>

The NTP Technology Panel was chartered to evaluate nuclear thermal propulsion (NTP) concepts on a consistent basis, and to continue technology development project planning for a joint project in nuclear propulsion for Space Exploration Initiative (SEI) (Clark et al. 1991d). Concepts were categorized based on probable technology readiness

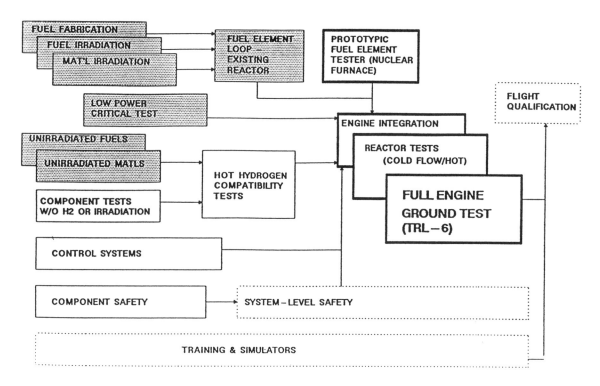

FIGURE 4.   NTP Summary Test Plan.

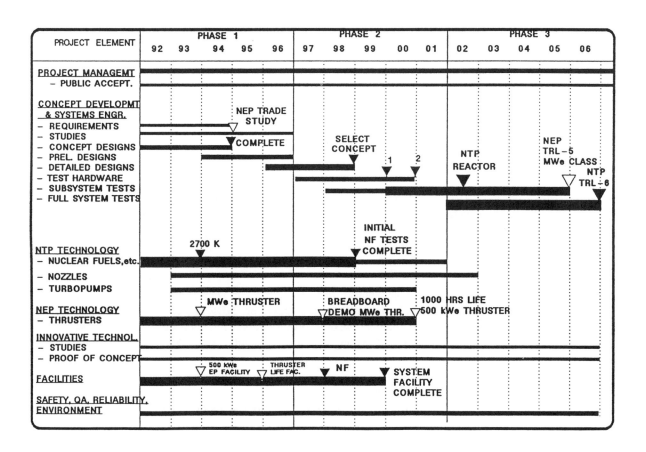

FIGURE 5.   Nuclear Propulsion Project Plan Overview.

date, and innovative concept "proof-of-concept" tests and analyses were defined. Further studies will be required to provide a consistent comparison of all of the NTP concepts (Polansky 1991).

The panel agreed that the highest priority technology development efforts should be (1) high temperature fuels and materials development, (2) long lead time facilities design and construction for technology validation testing, and (3) conceptual design contract studies to help focus the technology development efforts.

Instrumentation development, neutronics, controls, and diagnostics system integration will also be very important and should be included in the project from the beginning. A test plan for the NTP technology development was developed jointly with the Facilities Panel and the Nuclear Fuels/Materials Panel (see Figure 4), (Clark et al, 1991d).

The concepts presented at the NTP workshop were reviewed in detail by the NTP panel (Clark 1991a). Of the solid core concepts presented, it was the consensus that any of the three reactor types (and hence, fuel types), could be developed through full system ground test completion (technology readiness level 6: TRL-6) by 2006, provided adequate funding is provided:

- Prismatic fuel, thermal reactor (NDR),
- Particle fuel, heterogeneous fuel (PBR), and
- Cermet fuel, fast reactor (cermet, wire core).

In addition, it is believed that several other concepts could also be developed to TRL-6 by 2006, provided (1) they overcome relevant "proof-of-concept" issues, and (2) adequate funding is provided; development of these concepts will almost certainly be higher cost. These concepts are:

- The low pressure solid core concept,
- The closed cycle gas core "Nuclear Light Bulb" concept, and
- The closed cycle vapor core NDR.

Proof-of-concept testing and analysis of these concepts will be a high priority in the project as "Innovative Technology."

Innovative Concepts are concepts that are probably second- or third-generation SEI systems, that offer significant performance advantages (shorter transit time) compared to first generation systems. A number of the concepts presented at the NTP workshop are in this category, and critical tests were identified for these concepts (Clark 1991a).

Low-Pressure Concept: The concept is claimed to take advantage of significant hydrogen dissociation and recombination to enable specific impulse as high as 1300-1500 seconds. Studies are underway at Lewis and the Idaho National Engineering Laboratory (INEL), to evaluate the potential of this phenomenon, and experimental proof-of-concept testing is planned. Several other concepts also would benefit from hydrogen dissociation-recombination, if the effect can be realized.

Closed Cycle Gas Core Concept: The nuclear light bulb concept was studied extensively by the United Technologies Research Center (UTRC), in the 1960s, and a vapor core NDR has been proposed by the University of Florida (Clark 1991a). Critical tests include window opacity versus radiation dose, and window/fuel erosion studies. Radiation transport, gas seeding and cooling of critical structural components remain key issues.

Open cycle Gas Core Concepts: The open cycle gas core has the highest potential performance of the NTP concepts, with significantly lower transit times. Key proof-of concept issues include hydrodynamic containment, gas seeding and radiation transport, and plasma/propellant mixing. Analytical studies are underway to update computer programs and to define appropriate initial experiments. Critical proof-of-concept issues remain in:

- Neutronics,
- Materials,

- Fluid dynamics/thermal hydraulics, and
- Facilities.

Facilities Panel

The Test Facilities Panel was chartered to identify nuclear propulsion test facility test requirements and options early in the panel deliberations, since major facilities are known to be long-lead-time elements of a test program (Martinell et al. 1991). Reference mission requirements were provided by the Mission Analysis Panel, and test requirements were provided by the NTP, NEP and Fuels/Materials Panels. A number of potential test sites were visited, and a significant database was established of potential facilities that may be used in the technology development project.

Major new test facilities include a MW-class NEP thruster life facility, an NTP fuel element tester (nuclear furnace) capable of testing a wide range of element concepts, an NTP system ground test facility, with multiple cells for reactor and engine tests, and flight system engine qualification tests. The NTP test facilities will include full effluent cooldown and cleanup to ensure environmental compliance. It is estimated that the earliest that a nuclear furnace facility could be completed is 1997, and the full system ground test could be available in 1999.

TECHNOLOGY PROJECT PLAN SUMMARY

A summary of the nuclear propulsion technology development project with major milestones, is shown in Figure 5. The plan includes both NTP and NEP technologies. The solid milestone symbols are for NTP, and the open symbols are for NEP. Project activities are indicated in public acceptance, innovative technology development, and safety, quality assurance, reliability, and environmental compliance.

Facilities

Major activities early in the project are focussed on designing and building (or modifying existing) facilities to perform the system and subsystem tests. Agency and Department approvals, safety analyses and reporting, and environmental documentation requirements will contribute to the lead times indicated for these major facilities.

Technology Development

Fuels technology development will be the primary focus in the early stages of the project for both NTP and NEP. Long life, high power thruster technology will also be developed early for NEP. Other technologies will be included as funding becomes available.

Systems Engineering

Initial trade studies for NEP and conceptual design studies for NTP will help to provide a consistent comparison of concept options, and will help to guide the technology development activities. Hardware design activities will lead to initial concept selection for system testing in the 1999 time period, which should provide systems validated to TRL-6 by the 2006 target date.

CONCLUSION

This paper has summarized the status of planning activities for the Nuclear Propulsion Office at the NASA Lewis Research Center for a technology development project in nuclear propulsion systems. Early workshops in 1990 "cast a wide net" to identify a wide range of concepts and their associated technology needs. Interagency panel activities in 1991 have helped to clarify a number of issues, recommended technical approaches, and evaluated a number of options. A broad base of government and industry support for the nuclear propulsion option for the SEI missions has been developed. Final reports of each of the panels will be published soon and will provide a very comprehensive database for the status of nuclear propulsion technology, and future plans.

Acknowledgments

The author wishes to acknowledge the support and guidance provided by the participants in the workshops and the Interagency technical panels, on a non-reimbursed basis. Their efforts have contributed in a very direct way to the quality of the technology development plans, and their enthusiasm for the technology has been noted both within the technical community and by the national policy makers.

References

Bhattacharyya, S. (1991) "Development of Nuclear Fuels and Materials for Propulsion Systems for SEI," AIAA 91-3452, AIAA/NASA/OAI Conference on Advanced SEI Technologies, 4-6 September 1991, Cleveland, OH.

Clark, J. S., Editor (1991a), Proceedings of the NASA/DOE/DOD Nuclear Thermal Propulsion Workshop, NASA CP-10079, held in Cleveland, OH, 10-12 July, 1990.

Clark, J.S. (1991b) "A Comparison of Nuclear Thermal Propulsion Concepts: Results of a Workshop," in Proc. Eighth Symposium on Space Nuclear Power Systems, M.S. El-Genk and M.D. Hoover, eds., American Institute of Physics, New York, Part II, pp. 740-748, Jan. 1991.

Clark, J.S. and T.J. Miller, (1991c) "The NASA/DOE/DOD Nuclear Rocket Propulsion Project: FY 1991 Status," AIAA 91-3413, AIAA/NASA/OAI Conference on Advanced SEI Technologies, 4-6 September 1991, Cleveland, OH.

Clark, J.S., P. McDaniel, S. Howe, and M. Stanley (1991) "Nuclear Thermal Propulsion Technology: Summary of FY 1991 Interagency Panel Planning," AIAA 91-3631, AIAA/NASA/OAI Conference on Advanced SEI Technologies, 4-6 September 1991, Cleveland, OH.

Doherty, M. (1991) "Blazing the Trailway: Nuclear Electric Propulsion and it's Technology Development Plans," AIAA 91-3441, AIAA/NASA/OAI Conference on Advanced SEI Technologies, 4-6 September 1991, Cleveland, OH.

Marshall, A. (1991) "A Recommended Interagency Nuclear Propulsion Safety Policy," AIAA 91-3630, AIAA/NASA/OAI Conference on Advanced SEI Technologies, 4-6 September 1991, Cleveland, OH.

Martinell, J., J. Warren, and J. Clark (1991) "NASA/DOE/DOD Nuclear Rocket Propulsion Major Facility Requirements," AIAA 91-3414, AIAA/NASA/OAI Conference on Advanced SEI Technologies, 4-6 September 1991, Cleveland, OH.

Miller, T.J., J. S. Clark, and J.W. Barnett (1991) "Nuclear Propulsion Project Workshop Summary," in Proc. Eighth Symposium on Space Nuclear Power Systems, M.S. El-Genk and M.D. Hoover, eds., American Institute of Physics, New York, Part I, pp. 84-92.

Polansky, G. (1991) "Comparison of Selected Nuclear Thermal Propulsion Concepts: Preliminary Results," AIAA 91-3505, AIAA/NASA/OAI Conference on Advanced SEI Technologies, 4-6 September 1991, Cleveland, OH.

Synthesis Group Report (1991), America at the Threshhold - America's Space Exploration Initiative, Available from the Superintendent of Documents, U.S. Government Printing Office, Washington, DC 20402, June 1991.

Wickenheiser, T., K. Gessner, and S. Alexander (1991) "Performance Impact of NTR Propulsion on Manned Mars Missions with Short Transit Times," AIAA 91-3401, AIAA/NASA/OAI Conference on Advanced SEI Technologies, 4-6 September 1991, Cleveland, OH.

A UNIQUE NUCLEAR THERMAL ROCKET ENGINE USING A PARTICLE BED REACTOR

Donald W. Culver
Dept. 5252, Bldg. 2019
(916) 355-2083

Wayne B. Dahl
Dept. 5154, Bldg. 2019
(916) 355-3956

Melvin C. McIlwain
Dept. 5154, Bldg. 2019
(916) 355-6057

Aerojet Propulsion Division
P.O. Box 13222
Sacramento, CA 95813-6000

Abstract

Aerojet Propulsion Division (APD) studied 75-klb thrust Nuclear Thermal Rocket Engines (NTRE) with particle bed reactors (PBR) for application to NASA's manned Mars mission and prepared a conceptual design description of a unique engine that best satisfied mission-defined propulsion requirements and customer criteria. This paper describes the selection of a sprint-type Mars transfer mission and its impact on propulsion system design and operation. It shows how our NTRE concept was developed from this information. The resulting, unusual engine design is short, lightweight, and capable of high specific impulse operation, all factors that decrease Earth to orbit launch costs. Many unusual features of the NTRE are discussed, including nozzle area ratio variation and nozzle closure for closed loop after cooling. Mission performance calculations reveal that other well known engine options do not support this mission.

INTRODUCTION

It is now widely recognized that manned transfer to and from the planet Mars must be done quickly, so that deleterious effects of galactic cosmic rays on the astronauts are minimized. This paper describes development of a NTRE specifically for this purpose, using modern nuclear and rocket technology to achieve low program cost.

ENGINE DEFINITION

Mission Selection

We selected the most performance stressing manned Mars mission to drive out limit propulsion requirements. It is a sprint type mission, beginning and ending in low earth orbit (LEO) with either a Phobos or circular orbit rendezvous at Mars. The NTRE propelled Mars transfer vehicle (MTV) carries a Mars lander/ascent vehicle as outbound payload and returns without it. Propulsion performance studies show that staging is necessary; propellant tanks must be dropped after each of the first three of four major transfer burns of the mission and at least half of the MTV engines must be dropped before the last burn (Earth capture). Dropped NTRE are left on Phobos along with empty tankage and lander vehicle as emergency MTV equipment for future missions, or they and their tankage are returned as a stage to a high Earth orbit partway through the Earth escape burn. In either case staged, reusable NTRE could be resupplied for a subsequent mission by a slow, unmanned propellant tanker mission at Phobos or with an earth to orbit (ETO) launch at Earth orbit.

We selected the direct earth escape mission scenario and we fixed the vehicle's initial thrust/weight ratio to yield a nearly optimum value of 0.30-G throughout our study. The MTV propulsion ideal burnt velocity increments we used for approximately 100 day outbound and 150 day inbound transit times are: Earth escape 9.75 km/s, Mars capture 6.10 km/s, Mars escape 4.57 km/s, and Earth capture 6.10 km/s.

We did not select aerobraking for Mars or Earth capture maneuvers because of flight path and attitude accuracy demands for success and the consequences of failure with manned payload and previously operated NTRE aboard. Additionally, aerobrake hardware appears to be difficult to transport to LEO whole or to assemble in LEO.

Propulsion System Design and Operation

We selected a propulsion system with four identical engines, a fixed hydrogen run tank, and three sets of drop tanks. The mission begins with all engines burning in parallel for earth escape. One million pound initial weight in LEO (IMLEO) was assumed for all flights; four engines of 75-klb thrust accelerate a one million pound vehicle at 0.30-G. Two engines are used during the subsequent Mars capture maneuver. An engine failure during Earth escape may abort the mission safely, while a failure during Mars capture is tolerable, because the system has engine-out capability throughout the mission after Earth escape. When activities at Mars are completed, MTV propulsion returns the habitat payload to LEO with its fixed tankage, Mars escape drop tanks, and two engines.

Engine Concept Development

Fundamental to engine design is the propellant feed concept and its operating pressure and temperature. Tank pressure fed engines operating at high pressures are not practical for space exploration on a propulsion system performance basis, where low density hydrogen propellant causes large, heavy tankage and pressurization and heavy MTV weight. This increases launch costs to LEO for any mission and payload. We studied low pressure, pressure-fed systems and found that they did not pass our payload performance, operational, and safety/reliability screens. Such systems having low engine pressure suffer from large nozzle diameters. Those with small, high pressure run tanks require a large array of valves that have excessive cycle life requirements. Therefore, pressure-fed NTRE were screened out during our study. A high pressure, pump-fed engine was selected for the concept, because no significant engine performance benefit is predicted for low pressure NTRE with near term fuel (2800 K reactor outlet gas temperature).

Engine cycle is a major driver in development complexity and overall engine efficiency, because it affects how the energy is produced to drive the turbopump in the propellant feed system. In our study, four different engine cycles were evaluated: (1) expander bleed, (2) mixed bleed, (3) expander, and (4) augmented expander. Bleed cycles, such as were used in the NERVA ground test engine built by Aerojet, are inherently inefficient, because turbine drive exhaust gas is vented overboard without recovering full propulsive efficiency (losses are typically 1 to 4%). Conventional expander cycles, such as the cycle selected for the NERVA flight engine design, are sensitive to heat extraction efficiencies in regeneratively cooled components which can limit power available to drive the turbopump. These uncertainties can complicate and drive up development cost because of iterating redesigns of multiple components. An augmented expander cycle, where regeneratively heated turbine drive fluid is further heated by special "augmenter fuel elements" in the reactor core, provides excess power for turbine drive and avoids the necessity for expensive design iteration to eliminate power limitations. For this reason the augmented expander cycle was chosen for our concept.

Limited trades studies were conducted on other critical engine characteristics to support design point selection. These results are summarized in Table 1.

TABLE 1. Design Decisions Based on Performance and Weight.

Study	Results
1. Working Fluid	LH2 with active refrigeration for IMLEO advantages of 20:1 versus NH3.
2. Propellant Feed System	Feed system advantage >2:1 versus pressurized storage tank propulsion IMLEO.
3. Reactor Outlet Pressure	Use high pressure designs with PBR mixed mean outlet temps to 2800 K.
4. Reactor Outlet Temperature	Isp is controlled by outlet temp and it has the largest impact on IMLEO. A 10% change returns a 50% change in IMLEO. It also controls reactor life.
5. NTRE Thrust/Weight	Shielded NTRE thrust/weight has a major impact on IMLEO at low values. Need ~10:1 technology.

ENGINE DESIGN

Engine Description

Figure 1 is a layout drawing of our selected NTRE with weight breakdowns and Table 2 summarizes its design data. Major engine components are turbopump, shadow shield, short L/D expansion deflection (ED) rocket nozzle with annular sonic throat, and PBR nuclear reactor heat source. The reactor is suspended within the ED nozzle bell to further reduce engine length. Thrust vector control is provided in conjunction with non-gimbaled engine hardware by use of a semi-flexible, radiation cooled, coated carbon-carbon (C-C) nozzle and four shield-mounted actuators.

During operation, the turbopump feeds liquid hydrogen through itself and its shadow shield on the way to reactor vessel coolant passages and a two-pass reactor core. Hydrogen flows through cooling passages in a hollow, central engine support column and in the reactor vessel wall to the smaller of two axial core heater element arrays. This aftmost array heats the gas flow to supply the needed turbine drive power. Hydrogen at less than 500 K flows forward through the central engine support column to the turbine inlet manifold. This yields efficient turbine drive at temperature within present materials capability. Turbine exhaust gas is returned down the annulus around the central support column and reheated in the larger, upper axial core array. Fully heated, 2800-K gas exits the reactor forward into an annular plenum and, by turning 90 degrees subsonically, into an annular, sonic rocket nozzle throat. Expanding nozzle gases flow radially outward, and aft through the forced deflection supersonic rocket nozzle.

The ED nozzle (also called a forced deflection (FD) nozzle) is flexible and strong, because it is made of coated C-C composite, which has near maximum strength at reactor outlet gas

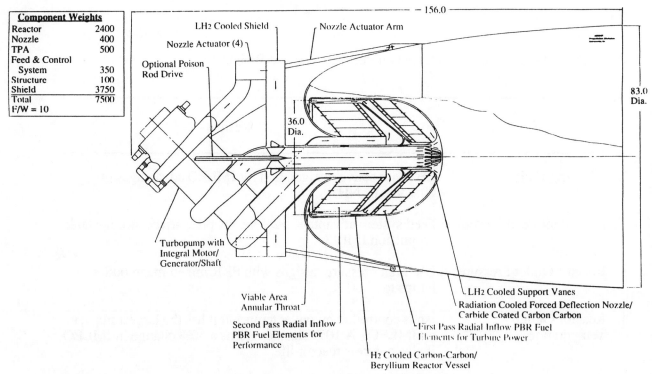

FIGURE 1. NTRE is Short, Lightweight, and Delivers High Specific Impulse.

TABLE 2. Selected NTRE Design Data.

Vacuum Thrust	75,000 lbf	Reactor Outlet Temp	2750 K to 3000 K
Propellant	Liquid Hydrogen	Feed System Type	Turbopump
Reactor Outlet Press	750 psia	Reactor Flow Geometry	2-Pass, Axially
Feed System Power Cycle	Augmented Expander	Reactor Power Control	Bistem
Reactor & Shield Cooling	Regen, LH2	Nozzle Type	Forced Deflection
Nozzle Cooling	Regen & Radiation	Specific Impulse	915 to 940 s
Nozzle Area Ratio	Variable: 100 to >400	Thrust Vector Sys.	Nozzle Deflection in Pitch, Yaw Planes
Engine Flow Rate	82 lb/s max.	Engine Mounting	Rigid/Non-Gimbal-ling
Flexible Lines	None	Nozzle Material	Coated C-C
Reactor Vessel & Structure	Be & C-C	Coolant Ducts	Al
Turbine Press Ratio @ Inlet Temp	1.3 @ 600 K	Throat Width	24.1 mm
Heat Source	PBR	Throat Diameter	914 mm

temperatures. It is expected that a thin coating of rhenium or possibly zirconium carbide or boron carbide will be used to protect the C-C from chemical attack by hot hydrogen. The rocket nozzle's position and throat area are controlled by four graphite composite actuation rods that connect to electromechanical ballscrew actuators attached to the radiation shield. Spherical actuator mounts allow differential rod motions in the two pitch and yaw planes to cant the nozzle and skew the annular throat area and propellant flow to any side, providing large side loads for thrust vector control of long, NTRE powered, manned MTV.

Engine Features

Our ED nozzle design can deliver high specific impulse. Figure 2 (Mockenhaupt et al. 1981), shows that high turning angle ED nozzles attain high nozzle efficiencies at high area ratios with much shorter contours than do conical or bell contoured DeLaval nozzles. Thus, a significant Earth launch packaging feature of this nozzle design is that it shortens the engine length by about eight feet compared to a NERVA-type design. Also, weight and equivalent gimbal loads are reduced by shortening the nozzle.

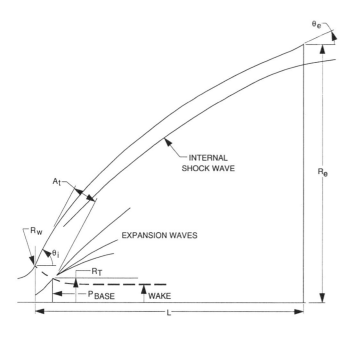
Forced Deflection Nozzle Geometry & Flow Field

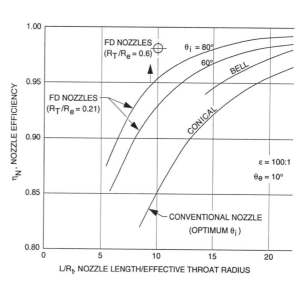
Effects of Nozzle Length on Calculated Efficiencies of Forced Deflection Nozzles

FIGURE 2. Short Forced Deflection Nozzles with Large Turning Angles have High Efficiencies.

A large part of the eight feet engine length reduction accrues from locating the nozzle throat forward of the reactor, so the reactor length is not a part of the engine length. This concept is practical, because the reactor's location is remote from the majority of supersonic gas that expands off the nozzle wall, having been directed there by the annular throat. Small diameter reactor designs are always sought for NTRE, because they minimize shadow shield weight, and in our design they also minimize any nozzle centerbody impact. If necessary, centerbody streamlining can be supplied by a small gas bleed from the center of the reactor vessel's aft dome or by a small conical C-C extension.

Engine pressure and nozzle area ratio control is available with this design by varying nozzle throat area. Thus, engine operation at low thrust can provide higher specific impulse than at full thrust. The specific impulse benefit is more than 3% (nearly 30 seconds Ispvac) at 40% of maximum thrust (30 k-lbf) and nearly twice that improvement at 10% thrust. Nozzle throat area changes are made by operating all four actuators in unison to increase or decrease the annular nozzle throat plane gap width. After engine shut down waste heat may be removed with closed loop coolant flow to power active Hydrogen tank refrigeration systems by closing the nozzle throat.

Mission Performance

Payload performance maps were calculated using fundamental mission relationships to provide a method of comparing engine concepts and of evaluating how improvements in engine thrust-to-weight ratios (F/W) and specific impulse (Isp) would benefit specific missions. The manned portion of a split sprint, previously described, is depicted in Figure 3 and a more leisurely Venus fly-by mission in 2016 is shown in Figure 4. It can be seen in Figure 3 that the more conventional engines, for example, NERVA and its derivative, can not perform this 100-day outbound mission.

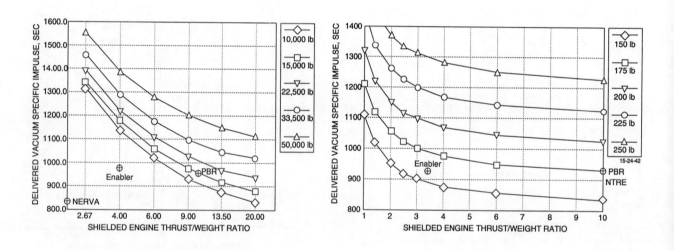

FIGURE 3.   Mars Mission Payload Weight Returned - Split Sprint Mission.

FIGURE 4.   Mars Mission Payload Weight Returned - All Propulsive Venus Fly-by Mission.

However, on the slower mission, the difference in performance between the PBR concept described and the NERVA/NERVA derivatives are less. The question is can the astronauts really perform without assistance on arrival at Mars after 160 days of zero-G and galactic radiation.

The flatness of the two curves show that engine F/W is not terribly important above 10 on the fast mission or above 5 on the slower transfer. However, Isp can significantly improve performance at any F/W, but being more important at the high F/W. If goals of F/W>5 and Isp>900 are established, the slower mission appears feasible with margins. Otherwise, F/W approximately equals 10 is needed.

RECOMMENDATIONS

This study has revealed possibilities for developing a compact NTRE that enables more rapid, safe, manned missions to the planet Mars and heavy payloads to the Earth's moon at potentially lower cost than with prior NTRE concepts. More work is needed in the areas of system design, materials, controls, and reactor development and concurrent engineering leading toward full scale NTRE development.

Acknowledgments

NTRE definition and design results reported in this paper were performed at Aerojet Propulsion Division, Sacramento, CA, funded in part by Boeing Aerospace & Electronics (Culver, et al.) under the direction of Gordon Woodcock and Benjamin Donahue.

References

Culver, D.W. et al., (1991) Nuclear Thermal Rocket Engine Study, Aerojet Propulsion Division, contract HM 9618, for Boeing Aerospace and Electronics, Huntsville, AL, and Prime contract NAS8-37857, with George C. Marshall Space Flight Center, 24 April 1991.

Mockenhaupt, H.D. and G.J. Felix, (1981) "Cold Flow Tests of Forced Deflection Nozzles for Integrated Stage Applications," AIAA-811420, in Proc. AIAA/SAE/ASME 17th Joint Propulsion Conference, Colorado Springs, CO, 27-29 July 1981.

ASSESSMENT OF THE USE OF H_2, CH_4, NH_3, AND CO_2 AS NTR PROPELLANTS

Elizabeth C. Selcow, Richard E. Davis, Kenneth R. Perkins, Hans Ludewig, and Ralph J. Cerbone
Brookhaven National Laboratory
Upton, NY 11973
(516) 282-2624

Abstract

In this paper the effect of changing from the traditional NTP coolant, hydrogen, to several alternative coolant is studied. Hydrogen is generally chosen as an NTP coolant, since its use maximizes the specific impulse for a given operating temperature. However, there are situations in which it may not be available as optional. The alternative coolant which were considered are ammonia, urethane, carbon dioxide and carbon monoxide. A particle bed reactor (PBR) generating 200 MW and coolant by hydrogen was used as the baseline against which all the comparisons were made. Both 19 and 37 element cases were considered and the large number of elements was found to be necessary in the case of the carbon monoxide. The coolant reactivity worth was found to be directly proportional to the hydrogen coolant content. It was found that due to differences in the thermophysical proportions of the coolant that it would not be possible to use one reactor for all the coolants. The reactor would have to be constructed specifically for a coolant type.

INTRODUCTION

Nuclear Thermal Rocket (NTR) engines are traditionally thought of as using liquid hydrogen as a propellant. This is the preferred propellant, since, for any given exhaust temperature, it results in the highest possible specific impulse, approximately twice that of the best chemical rocket. However, there are situations when other propellant types are more desirable:

1. Non-cryogenic storage and handling,
2. Availability,
3. Safety, and
4. Mission optimization.

Furthermore, since an NTR does not require a combustible mixture, the possible propellant types are large and can include such diverse fluids as water, octane, ammonia, methane, and carbon monoxide. In the following study, four propellant types will be studied. This study will consider the physics, fluid dynamics, heat transfer, and materials implications of these propellants. Changes to the baseline design are suggested, and where possible, solutions are suggested. The four propellants chosen for this study are hydrogen (H_2), methane (CH_4), carbon dioxide (CO_2), and ammonia (NH_3). These propellants cover the range from reducing (H_2) to oxidizing environments (CO_2), high hydrogen content (NH_3 and CH_4) to no hydrogen content (CO_2) and finally the last three all dissociated in the core. In this manner, the problems and suggested solution should cover other propellants with similar chemical and physical properties.

A basic requirement of an NTR is that its coolant is initially liquid and then changes to a gas at the exhaust. This implies that the phase of the coolant changes as it passes through the system. It is important that this phase change does not take place within the reactor since it would lead to major control problems. In order to satisfy this requirement, the reactors are operated at pressures that are high enough to ensure that the reactor outlet pressure is beyond the critical pressure for the coolant. In this way, the fluid from the pump outlet to the reactor outlet is all a single phase and no sudden phase change takes place in either the moderator or core. Physical properties for the four coolants to be considered in this study are given below on Table 1. It should be noted that most of the properties are specific to a Particle Bed Reactor (PBR) based design, with properties appropriate to the inlet of the cold frit.

In the following sections, the reactor model will be described. This will be followed by three sections in which preliminary thoughts on physics, fluid dynamics, and heat transfer and material compatibility issues will be discussed.

TABLE 1. Coolant Properties at Cold Frit Inlet.

Property	Coolant Type			
	H_2	NH_3	CO_2	CH_4
Critical Pressure (MPa)[a]	1.3	11.5	7.4	4.65
Critical Temperature (K)[a]	33.2	405.5	304	190.0
Temperature (K)	100.0	340.0	300.0	220.0
Pressure (MPa)	7.0	15.0	10.0	10.0
Enthalpy (kJ/kg)	1314	1323	288	-410
Viscosity (Kg/m-s)	4.2(-6)	1.20(-5)	4.5(-5)	N/A
Density (kg/m^3)	17.0	90.4	176.2	88.2
Number Densities (atm/b-cm)	H- 1.0(-2) ----	H- 9.6(-3) N- 3.2(-3)	C- 2.4(-3) O- 4.8(-3)	C- 3.3(-3) H- 1.3(-2)
Thermal Absorption Cross Section (cm^{-1})	3.3(-3)	9.11(-3)	8.0(-6)	4.3(-3)
Thermal Scattering[b] Cross Section (cm^{-1})	2.1(-1)	.57(-1)	3.0(-2)	7.2(-1)
Slowing Down Power ($\xi\Sigma_s$)	2.1(-1)	1.9(-1)	9.0(-3)	2.6(-1)

[a]Not at cold frit inlet.
[b]Approximate values since scattering law data does not exist for these molecules.

MODEL

The reactor design to be used in this study is based on the one used in the 200 MW orbital transfer vehicle NTR. This NTR was based on the well known PBR concept described in Powell et al. (1987). Two reactor designs based on the 200-MW power level were considered in this study. The design differed primarily in the number of fuel elements. Shown in Table 2 are the basic reactor parameters for the two designs.

TABLE 2. Reactor Design Parameters.[a]

Parameter	Design 1	Design 2
Power (MW)	200	200
Bed Power Density (MW/l)	10	10
Number Of Elements	19	37
Chamber Temperature (K)	3000	3000
Chamber Pressure (MPa)	7.0	7.0
Hot Frit ID (cm)	1.111	1.568
Fuel Bed ID (cm)	1.411	2.168
Fuel Bed OD (cm)	2.928	3.058
Cold Frit OD (cm)	3.128	4.258
Plenum OD (cm)	3.428	4.858
Element Pitch (cm)	11.125	9.647
Core Diameter (cm)	55.63	67.53
Core Height (cm)	58.56	67.53
Radial Reflector Thickness (cm)	10.0	10.0
Top Axial Reflector (cm)	3.0	3.0
Bottom Axial Reflector (cm)	5.0	5.0
Fuel Particle OD (μm)	500	500

[a]The materials of construction for both reactors are assumed to be the same and are given in Table 3.

TABLE 3. Reactor Materials.

Component	Material
Hot Frit	Coated Carbon/Carbon
Fuel	ZrC Coated (U,Zr) Particles
Cold Frit	Al
Moderator	Be (80% Dense) + Coolant
Radial Reflectors	Be (90% Dense) + Coolant
Upper Axial Reflector	Be
Lower Axial Reflector	C

PHYSICS ANALYSIS

Physics analyses were carried out to determine the multiplication factor for the two reactor designs described above. These analyses were carried out with and without propellant in order to estimate the propellant worth. Furthermore, since the propellants being studied here are not customarily found in reactors, various approximations were used to represent the molecular binding. In one case (CO_2), no approximate model could be used and all the calculations were carried out using a free gas scattering model. Finally, since the worth of CO_2 was so low, several designs are proposed for the reactor cooled by CO_2.

These determinations were carried out using the MCNP Monte Carlo code. The use of this code makes it possible to explicitly represent all the geometric detail necessary to make an accurate determination of the multiplication factor. Figure 1 shows the schematic representation of a fuel particle, a fuel element bedded in the moderator and both a radial and axial section through a PBR based NTR. In these analyses, the heterogenous nature of this reactor is preserved and all the leakage paths are explicitly represented. Neutron spectral shifts which occur between the moderator and fuel zones are accounted for accurately in this analysis technique. The MCNP code uses point cross sections data from the ENDF/B files and thus avoids the inaccuracies implied by group averaged cross sections.

Table 4 shows the multiplication factor (k_{eff}) for various configurations, propellants and scattering models. The average standard deviation in k_{eff} is .002.

TABLE 4. Multiplication Factors (k_{eff}).

Case	Propellant	Number of Elements	Radial Reflector Thickness	k_{eff}
1	--	19	10	.925
2	H_2	19	10	1.024
3	NH_3	19	10	1.120
4[a]	NH_3	19	10	1.100
5	CO_2	19	10	.944
6	CO_2	19	20	1.014
7	--	37	10	1.079
8	CO_2	37	10	1.098
9	CH_4	19	10	1.083
10[a]	CH_4	19	10	1.067

[a]Hydrogen bound in Benzene molecule.

From these results, it is clear that a critical reactor in the size of interest (200 MW) can be designed for all propellants considered in the study. In all cases, except the CO_2 cooled reactor, a 19-element arrangement with a 20-cm thick radial reflector is required. However, in the case of CO_2, the poor moderating properties of the coolant ($\xi\Sigma_s$) results in a system with a very small coolant worth. Thus, a core which has a sufficiently large value of k_{eff}, without coolant in it, must be designed. This increase in k_{eff}, while maintaining the same total power, power density and fissile loading can be achieved by using more fuel elements. An increase in the

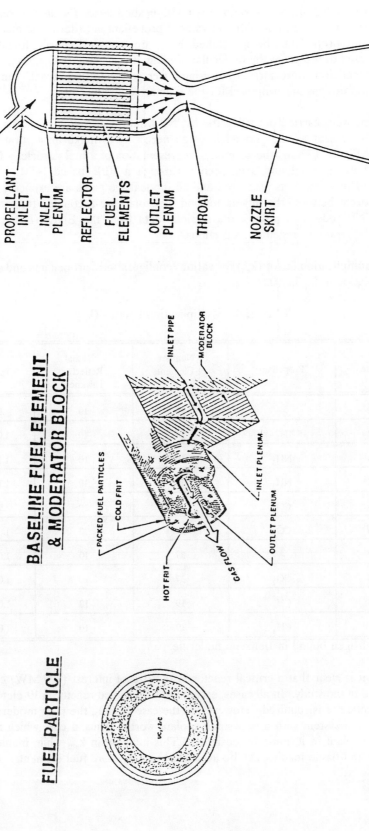

Figure 1. Schematic Representation of a Particle Bed Reactor Based Rocket Concept.

number of fuel elements, while maintaining all mission performance parameters, makes the core more homogeneous, and thus increases the value of k_{eff}. Hence, a 37-element core was designed for the CO_2 cooled reactor, and it can be seen that a satisfactory value of k_{eff} results.

It should be noted that the worth of the coolant is largest for NH_3 and CH_4 ($.15 < \Delta k < .2$), intermediate value ($\Delta k = .1$) for H_2 and the smallest value for CO_2 ($\Delta k = .02$). This variation in coolant worth will impact the start up and control scenarios for the various reactor designs. Furthermore, the concept of using a single reactor design to operate on all the four coolant types is unrealistic. Finally, calculations were carried out to estimate the effect of binding on the hydrogen scattering model. Since neither CH_4 or NH_3 kernels exist, these were approximated by hydrogen as bound in benzene (C_6H_6) in the cooler parts of the reactor and unbound (free gas) in the outlet duct. It can be seen that the change in k_{eff} is approximately .02 for this change in scattering kernel.

MATERIAL COMPATIBILITY

At high temperature, all the gases (CH_4, CO_2, NH_3) are unstable at a pressure of approximately 2 MPa.

$CH_4 = C(s) + 2 H_2$ quantitative by ~ 2000 K,
$CO_2 = CO + 1/2 O_2$ 5 to 30% over 2400 to 3000 K, and
$NH_3 = 1/2 N_2 + 3/2 H_2$ quantitative by ~ 1000 K.

The use of methane, a common feedstock for pyrocarbon deposition, as propellant is unlikely. It would appear that kinetic hinderance of carbon deposition might prevent this system from clogging or insulating the fuel bed/element.

Carbon dioxide is a possibility if fuel, frit, and ancillary hardware are fabricated from materials compatible with hot $CO_2/CO/O_2$. Refractory oxides are an obvious possibility. Information on HTGR particle fuel kernels of UO_x and $UThO_x$ exist. Ceramic frits of ThO_2 (melting point (mp) = 3370 C), Y_2O_3 (mp = 2704 C), MgO (mp = 2825 C), and others are possibilities. However, these materials are brittle and their thermal shock resistance is fair to poor.

In the case of ammonia, a "new" materials effort, nitrides for instance, would require examination of the full range of issues: similar to efforts in the CO_2 system. However, nitrides are generally less stable and have lower melting temperature than their oxide or carbide counterparts. While oxides are most likely unstable in the hot N_2/H_2 (H_2 being the concern), certain metal carbides might resist hot N_2/H_2; for example, TaC. Thus, alternatively, an assessment of the stability of refractory metal carbides in hot H_2/N_2 should be pursued.

Finally preliminary thermochemical calculations indicate TaC stability in hot CO, however, compatibility must be shown to exist with other materials, and the potential for carbon deposition in the nonisothermal reactor environment must be explored.

FLUID DYNAMICS AND HEAT TRANSFER

Detailed fluid dynamics and heat transfer calculations of these reactor designs have not been carried out at this stage. However, the following preliminary insights into potential design changes can be made.

1. The high value of specific heat for hydrogen compared to the other coolants indicates a much lower mass flow rate for hydrogen to extract the same power. This lower mass flow will result in a lower thrust rocket, despite the higher value of specific impulse.

2. The lower value for viscosity for hydrogen has implications in pressure drop determinations. This is particularly true in the case of the cold frit, whose pressure drop is a strong function of coolant viscosity. This is the primary fluid dynamic reason why one reactor cannot be used with different propellant types.

3. The much higher density for NH_3, CO_2, and CH_4 somewhat off-sets the lower value of specific heat compared to hydrogen. Thus, the higher mass flow rates required for these coolants may be accomplished through similarly sized coolant passages, for example, moderator, and pressure vessel.

CONCLUSIONS

The following conclusions can be drawn from this preliminary study.

1. The use of alternative coolants in a PBR based NTR does not pose any insurmountable physics problems. Those coolants with the highest hydrogen content will require care in design in order to minimize the reactivity swing associated with startup.

2. The excellent heat transfer characteristics typical of a PBR ensure that the fuel particles will always be coolable, even in the cases where the thermophysical properties are less desirable than those corresponding to hydrogen.

3. Chemical and material compatibility problems would seem to imply the largest amount of effort in order to make the use of alternative propellants practical.

4. Finally, due to its high power density and the small size, the PBR would be the ideal system to design especially using a coolant available at the mission destination. Such a system could be carried along as part of the payload and could then be used at this destination.

References

Powell, J. R. et al. (1987) "Nuclear Propulsion Systems for Orbit Transfer Based on the Particle Bed Reactor," in *Space Nuclear Power Systems 1987*, M. S. El-Genk and M. D. Hoover, eds., Orbit Book Co., Malabar, FL, pp. 185-198.

ENABLER-II: A HIGH PERFORMANCE PRISMATIC FUEL NUCLEAR ROCKET ENGINE

Lyman J. Petrosky
Westinghouse Advanced Energy Systems
P.O. Box 158
Madison, PA 15663
(412) 722-5110

Abstract

The Rover/NERVA program demonstrated graphite/carbide based prismatic fuel nuclear thermal rockets (NTR). The fuel performance increased steadily throughout the program, but data received late in the program did not have an opportunity to be incorporated into the engine designs. This article investigates NTR engines derived from the Rover/NERVA program database, utilizing all the available data developed in the program. This work does not assume any material advancements; the engine designs are based on demonstrated fuel performance levels and utilize established design margins. The Rover/NERVA database guarantees a minimum achievable performance, but has considerable growth potential which has yet to be realized. This article discusses the design issues and presents the performance figures for the ENABLER-I and ENABLER-II engines designs, which represent a low-risk first stage embodiment of this growth potential.

INTRODUCTION

The Rover/NERVA program (WANL 1972 and Koenig 1986) demonstrated graphite/carbide based prismatic fueled nuclear thermal rockets (NTR). The fuel performance increased steadily throughout the program, but data received late in the program did not have an opportunity to be incorporated into the engine designs. This article investigates NTR engines derived from the Rover/NERVA program database, utilizing all the available data developed in the program. This work does not assume any material advancements, i.e., the engine designs are based on demonstrated fuel performance levels and utilize established design margins. First we briefly review the Rover/NERVA Program and the past ENABLER engines. Then there is a discussion of the design issues, and finally the current ENABLER engine designs are presented up through ENABLER-II.

Review of Prior Programs

Extensive work has been done in the past on the design and testing of nuclear rocket engines. The Rover/NERVA program concentrated on the design and demonstration of cylindrical core reactors with axial hydrogen coolant flow through prismatic fuel elements. This technology was directed toward the objective of demonstrating an in-flight test of a nuclear rocket engine sized at 1100 MW with a thrust of 245 kN (55 thousand pounds, 55 klb) and with a specific impulse (I_{sp}) of over 7460 m/s (760 seconds).

Twenty reactor and engine system tests were conducted in the KIWI, Phoebus, NERVA, and Pewee series, and advanced fuels were tested in Nuclear Furnace 1 (NF-1). The NRX series tests began in 1964 and culminated with the successful one hour full power run of the NRX-A6 in December 1967. The integration of the reactor, pressure vessel, nozzle, turbopump, and control system which had been tested on the NRX/EST were combined into a full prototype system test, the NRX-XE', which was tested at simulated altitude conditions and achieved a total of 28 startups. Detailed flight engine specifications and candidate designs were completed just prior to program termination in 1973.

The Rover/NERVA technology base provides a performance level which enables manned Mars missions even if no further improvements in performance are contemplated. The program covered a period of over fifteen years and was very aggressive by todays standards. Over 20,000 fuel element were tested during the program. A multitude of reactor tests allowed the engineers to characterize the performance levels of the materials and components and proved the robustness of the reactor design. The Rover/NERVA database guarantees a minimum achievable performance, but has considerable growth potential which has yet to be realized. The ENABLER and ENABLER-II represent a low-risk first stage embodiment of this potential. Advanced prismatic cores are also under consideration and will be presented in future publications.

The ENABLER NTR Engine

The ENABLER class of NTR engines (Livingston and Pierce 1991) is based on the nuclear rocket technology developed in the Rover/NERVA programs. The ENABLER designs incorporate NERVA type fuel elements which are 19-mm (0.75-inch) hexagonal extrusions of graphite based fuel with a 19-coolant channel array within the element. The fuel elements are fabricated from the (U,Zr)C-Graphite composite material developed late in the Rover/NERVA program, which exhibits improved corrosion resistance and allows higher operating temperatures (Lyon 1973 and Taub 1975). Zirconium-hydride moderator is placed in the core support elements (demonstrated in the Pewee reactor) to increase the neutronic reactivity and thereby decrease the required uranium fuel loading. All ENABLER engines which use the 19-mm fuel have the ENABLER-I designation. The ENABLER-II is a more advanced engine family that is discussed later in the article.

The technology used in the ENABLER reactor can be sized for thrusts from 65 kN (15 klb) to over 1100 kN (250 klb). The rating of the baseline ENABLER reactor design is 1613 MW, 335 kN (75 klb) thrust, with a specific impulse (I_{sp}) of 9080 m/s (925 s). The reactor runs an expander cycle with chamber conditions of 2700 K (4860 R) and 7 MPa (1000 psia). The ENABLER engine design has been evolving over the past few years. The thrust to weight ratio (F/W) of the engine (inclusive of the nozzle, thrust structure, turbopump, and associated hardware) is 59 N/kg (6.0 lbf/lbm) without internal shield, and 50 N/kg (5.0 lbf/lbm) with shield.

NTR DESIGN CONSIDERATIONS

NTR must meet a multitude of design requirements to be useful for a manned Mars mission. This section provides a brief overview of the design considerations that must be addressed. The discussion is broken into four categories: fuel issues, core issues, engine issues, and other considerations.

Fuel Issues

To achieve good engine performance, the fuel materials for a NTR must sustain high temperatures (>2500 K) in a hydrogen environment while transferring several MW/L of fission heat to the coolant. The choice of materials is very limited in the domain of interest, being primarily metal carbides, carbide-graphite composites, and tungsten cermets.

Fuel performance is limited by its melting temperature. The best performance fuel which has been tested in the prismatic form is the carbide-graphite composites. Any fuel with carbide in contact with graphite is a carbide-graphite composite material system. The melting point of the fuel is limited by the solidus line for the (U,Zr)C - C interaction (Lyon 1973). The melting point is strongly dependent on the uranium loading of the fuel and ranges from 2900 to 3100 K (5200 to 5600 R) for typical fuel loadings. There are two basic ways to increase the fuel melting temperature, reduce the uranium loading or eliminate free carbon in the matrix. The former is limited by the need to maintain core criticality and is therefore a possibility only in large or well moderated cores. The latter case results in the pure carbide fuel type, which has been investigated in the past. Carbide fuel has a much higher melting point, but is difficult to fabricate and has a relatively low thermal stress limit. It is not considered fully demonstrated based on the Rover/NERVA data base.

The fuel melting point places an absolute limit on the fuel operating temperature. The temperature to which the coolant gas may be heated must be less than the fuel melting point. The required difference is dictated by the system film drops, peaking factors, and hot channel factors. For a well designed prismatic core the peak-to-average coolant channel power is in the range of 1.1 to 1.3. The required hot channel factors relate to the ability to match the flow to power within the fuel element. The flow in every single coolant channel of a prismatic core is controlled with a precision orifice at the channel inlet. Under these conditions the hot channel factor can be as low 5 percent.

The allowable fuel power density is limited thermal stress in the fuel material. The limit is determined by the properties of the fuel material and is temperature dependent. The thermal stress limit when combined with the thermal conductivity and expansion coefficient determines the allowable power density for a given fuel geometry assuming that fuel fracture is to be avoided. It should be noted that in prismatic fuels fuel fracture does not constitute fuel failure, but rather it is a reasonable design limit. Prismatic fuel has proven to be very robust in that the elements continue to function even with significant damage. For composite fuel in the element geometry used in the NF-1 test, the allowable power density at the core midband temperature is in the range of 4.5 to 6.0 MW/L of fuel matrix. The composition of the composite fuel may be altered to enhance the power density limit. This possibility was explored extensively near the end of the Rover/NERVA program and resulted in the higher of the two values quoted above. It is therefore unlikely that further significant improvements will be accomplished via this route. A second method of improving the fuel power density (without actually changing the thermal stress limits) is to change the fuel geometry, primarily by fuel scaling (Petrosky 1991a). This fuel scaling has proven to be a potent method of effecting the fuel power density limit without requiring improved material properties. Optimization of the fuel scale is the basis for the ENABLER-II engine discussed later.

The fuel life is limited by the slow ablation of the surface during operation. The ablation is a combination of evaporation and hydrogen corrosion. The core coolant, hydrogen, will react with free carbon on the fuel element surface to produce hydrocarbons which are exhausted from the core. This reaction is inhibited by coating the coolant channels with barrier materials, typically zirconium carbide, ZrC, but at high operating temperatures the ZrC slowly evaporates (Storms et al. 1991). The carbon mass loss from the fuel reduces core criticality, changes the core power distribution, and reduces the structural strength of the fuel elements.

Core Issues

The core as a whole must meet numerous design criteria. The first of these is criticality. Weight is an important driver in space applications, therefore the cores in NTR engines tend to be small. Small size enhances neutron leakage, which must be offset by the use of core reflectors and moderators resulting in increased reactor mass. The fuel loading, reflectors, and moderators must be adjusted throughout the size range that a given NTR design is to span. In NERVA derivative reactors such as ENABLER the core support structures can incorporate various amounts of zirconium hydride (ZrH) to adjust the moderator to the required level throughout the range of engine sizes.

Related to core criticality is the core reactivity control span. The core must have sufficient reactivity control to go from a hard shutdown to full power throughout the life of the engine. The control span must be sufficient to overcome poisoning, burnup, and mass loss due to fuel ablation. As the reactivity control is varied the core power distribution is altered, an effect which must be factored into the fuel peaking factors (see above).

To control the reactor power level requires that the core have a negative thermal coefficient of reactivity. The thermal reactivity coefficient is a combination of Doppler effects, coolant density, and thermal expansion. The reactors in the Rover/NERVA program were very well behaved in their reactivity feedback and would self regulate to match power with coolant flow.

The core support scheme is a critical element of any reactor design. The differential expansion of the fuel and supports must be accommodated. The dimensions of the hot components may expand by as much as 2 percent (2 cm per meter). This expansion cannot be permitted to alter the flow balance of the core and must not result in thermal ratcheting or the engine will not be restartable. In addition, the support system should not allow unrestrained fuel motion at any time. The NERVA type reactors have special provisions in the vertical and lateral support structures to repeatedly allow core expansion and contraction without damage.

Engine Issues

All NTRs require a turbopump to force the coolant through the core (except certain very low thrust, low pressure concepts). The pump is driven from a turbine powered by heated hydrogen. The amount of heat required to drive the turbine can range as high as 15 percent of the total reactor power. There are two basic types of engine cycles available, bleed and expander. The bleed cycle taps heated hydrogen from the engine, runs it through the turbine, and exhausts the turbine to space. This cycle is simple, but the exhausted hydrogen reduces the net I_{sp} of the engine. The expander cycle taps high pressure hydrogen from heated core peripheral and support structures, runs it through the turbine, and exhausts the turbine to the core inlet, thus the full specific impulse capability of the engine is maintained. Few engine designs other than the NERVA derivative engines are able to use the expander cycle.

Engine restartability will likely be required on any major space mission. Restartability imposes several requirements on the reactor. First, the engine must not be altered during operation and cooldown. This relates in part to the fuel restraint scheme mentioned above. Second, the decay heat must be removed after engine shutdown. This is accomplished initially using some of the propellant, but the engine cooling should be switched to a closed system as soon as possible to reduce the loss of propellant which reduces the net I_{sp}

for the engine run. Thirdly, the temperature distribution in the reactor changes significantly after shutdown and the non-fuel core periphery generally has materials which melt or distort at moderate temperatures; therefore, the temperatures of all these materials must be kept within their respective limits during the entire cooldown period. Finally, most of the engine will be exposed to a hard vacuum after engine shutdown. Many materials, and hydrogenous moderators in particular, have some volatility at moderate temperatures. There is a critical period after shutdown when many reactor components experience moderate temperatures in combination with vacuum. This is of particular concern for moderating materials located within the core, where hydrogen loss is likely and could prevent engine restart. Most of the above problems has been solved for the case of prismatic core engines. The NERVA NRX-XE' engine demonstrated 28 restarts. The ENABLER engines all have the capability of closed cycle post-shutdown core cooling, and feature hydrogen overpressure of the moderator during the cooldown period.

Shielding of the reactor is generally required based on three criteria: protection of engine components, propellant heating in the tank, and crew protection for manned vehicles. The major shielding mass is in the direction of the propellant tank and crew. Shielding generally adds dead weight to the engine system, except for the reflector which is required for criticality and control. The optimal distribution of shielding depends on how the direct and scattered radiation compares with the three criteria. Multiple engines generally require a disproportionate increase in the total shielding because of the forward scattering of radiation from each engine off another's nozzle, thrust structure, and tankage. To attenuate the scattered radiation requires a shadow shield much larger than would be required for a single engine.

Reliability and robustness of the engine is a major concern. The engine must tolerate abnormal conditions. The Rover/NERVA type engines have a sufficiently long thermal time constant (low power density and high thermal inertia) to tolerate fluctuations in the coolant flow rate. The Rover/NERVA program included not only the testing of full size reactors, but also the testing of over 20,000 fuel elements, and the manufacture and operation of all subsystem components, control devices, instrumentation, and other non-nuclear and servicing components of the reactor system. Proof testing of the system reliability took place in the major full scale nuclear engine tests, in particular the NRX-A1 through A6 (which demonstrated continuous operation for 60 minutes at full power) and the NRX-XE' which underwent 28 full startups and shutdowns while accumulating 228 minutes power operation. The NERVA NRX series engines demonstrated continued operation despite fuel cracks and damage. It is very difficult to completely prevent a NERVA or ENABLER type engine from providing thrust, and thus this engine type has good operating capacity beyond normal design limits.

Other Issues

Safety requires that their be a low risk to the workers and the public. Unused NTRs have a low level of radioactivity and must be handled in accordance with established procedures. A greater concern is that the reactor be kept subcritical while on earth under all conceivable circumstances. Safety was a critical issue during the Rover/NERVA program as it is today. The design, analyses, and testing were all performed in accordance with the nuclear safety standards that were in effect at the time. Nuclear safety considerations included design for avoidance of inadvertent criticality during assembly and transport of fully loaded reactors from the manufacturing site in Pittsburgh, handling and launch activities, and analysis of reentry conditions.

Some level of redundancy in propulsive capability will be required for manned missions. The mix of engine redundancy versus component redundancy must be determined based on mission abort analyses and reliability data of the systems. As a result of the design, analyses, and test activities accomplished during the Rover/NERVA program, a reasonable, although not complete, data base has been established to be able to project the reliability of a NERVA or ENABLER type engine.

ENGINE PERFORMANCE OPTIMIZATION

Once an engine design is specified that can meet all the requirements given in the preceding section, then the engine performance may be considered. All of the necessary requirements are satisfied by the NERVA or ENABLER engine designs. Beyond this, it is desirable to provide the best possible performance from the engine. The performance of an NTR is measured primarily by its I_{sp}, its thrust to weight (F/W), and its operating life t_L. The best possible engine would have the maximum I_{sp}, F/W, and t_L; however, in the range of interest these parameters are coupled, being determined by the fuel selection, chamber temperature, reactor design, and thrust level. The optimal choice of engine design parameters can only be determined based on a complex set of constraints and trade-off factors defined by the proposed mission. This trade space is not well defined at this time and therefore only limited engine optimization may be performed. The dominant engine trade-offs are discussed below.

The engine I_{sp} is dependent on the allowable fuel operating temperature and any parasitic hydrogen effluents (bleed cycle exhaust or other hydrogen flows not passed through the fuel). The allowable fuel operating temperature is limited by either fuel melting or fuel ablation over the life of the engine (Petrosky 1991b and Storms et al. 1991). The ENABLER engines utilize the composite fuel elements developed late in the Rover/NERVA program. These elements exhibited superior corrosion resistance and thus the reactor exhaust temperature could be increased to nominal 2700 K, which allowed an increase in I_{sp} to 9080 m/s (925 s). Further increases in temperature and I_{sp} are possible if better corrosion resistant coatings are developed or the elements are fabricated from pure carbide. Both of these options are beyond the Rover/NERVA database and therefore involve some program risk.

The engine F/W is dependent on the reactor design and fuel power density. The ENABLER-I engines have gone through three design iterations (A, B, and C). ENABLER-I versions A and B are minor variations on the NERVA R-1 reactor design. The B design engine has F/W at the 335 kN (75 klb) thrust level of 47 N/kg (4.8 lbf/lbm) without internal shield and 41 N/kg (4.2 lbf/lbm) with internal shield. A notable feature of the weight breakdown of ENABLER-I B is that the fuel accounts for less than 20 percent of the engine weight. This low fuel weight fraction prompted a reevaluation of the reactor design, that revealed that the design had many poorly chosen features and was not weight optimized. The ENABLER-I C design is the first attempt at reducing the engine weight using appropriate selection of materials and design features in the peripheral regions of the core. The reactor weight is also driven by the fuel volume. The fuel volume depends on the thrust level and power density of the fuel. The fuel volume strongly effects the reactor weight because it defines the size of the reactor peripheral components. The ENABLER-I engines all use fuel power densities consistent with the Rover/NERVA database; however, fuel scaling (Petrosky 1991a) has been found to be a way of changing allowable fuel power density through the geometry of the element while not altering the required material properties. ENABLER-II is the designation used to denote an engine using a fuel that is scaled. Both the ENABLER-I C design and the ENABLER-II are described in the following paragraphs.

ENABLER-I C Engine Design

The design features of the ENABLER-I C engine (shown in Figure 1) were selected to minimize the engine weight. Analysis of the performance parameters and weight schedules for an array of design options yielded a combination of features that resulted in an improvement in both I_{sp} and F/W. The desirable features are:

- Expander Cycle Powered by Tie Tubes,
- High Specific Strength Pressure Vessel,
- Low Heat Loss State-of-the-art Nozzle, and
- Weight Optimized Thrust Structure.

The reactor cycle is designed such that the heated hydrogen required to drive the turbopump is derived from flow through the tie tubes (core support structure) only, and the reflector coolant is feed directly to the core. Previous reactor cycles utilized hydrogen heated in the tie tubes and core reflector area to drive the turbopump. The significance of the change is that in an expander cycle the hydrogen circuit used to drive the turbopump operates at a much higher pressure (typically 1.8 to 2.2 times the core inlet pressure) than a circuit that feeds the core and the design pressure of the reactor pressure vessel is determined by the hydrogen pressure in the core reflector area. This change in the hydrogen flow circuits thus reduces the pressure seen by the pressure vessel by a factor of 1.8, and also eliminates a massive flow baffle located between the reflector and core inlet. These changes eliminate several hundred kg of mass.

The second feature involves the use of a high specific strength material for the pressure vessel. The NERVA and earlier ENABLER designs use an aluminum pressure vessel. ENABLER-I C uses a titanium pressure vessel to save weight. Additional saving may be possible if a carbon filament pressure vessel can be used. It is not currently known if a carbon filament or other composite material can endure the environment around the reactor.

The low heat loss nozzle is designed with an insulating layer on the cooled surfaces. NTRs are different than chemical engines in that the heat transferred to the nozzle surfaces reduces the engine I_{sp} even if the nozzle coolant is fed back into the core. This results from the NTR peculiarity that the chamber temperature is limited by the core fuel capabilities and not the enthalpy rise of the propellant. The addition of insulation in the nozzle adds minimal weight, but increases the engine I_{sp} by 5 seconds relative to a conventional nozzle.

Finally, the turbine and thrust structures have been optimized to reduce weight. Extensive work at Rocketdyne using present day technology resulted in a redefined thrust structure which weighs less than half that designed by the old criteria. Turbopump weights were revised based on the technology level used in the space shuttle engines.

The net result of these design changes is an engine with 5 seconds more I_{sp} and with a F/W at the 335 kN (75 klb) thrust level (inclusive of the nozzle, thrust structure, turbopump, and associated hardware) of 59 N/kg (6.0 lbf/lbm) without internal shield, and 50 N/kg (5.0 lbf/lbm) with shield.

ENABLER-II NEXT STEP ENGINE

The ENABLER-II engine is the next step toward increased performance. The reactor design is the same as ENABLER-I C, but the fuel size is scaled to achieve optimal fuel

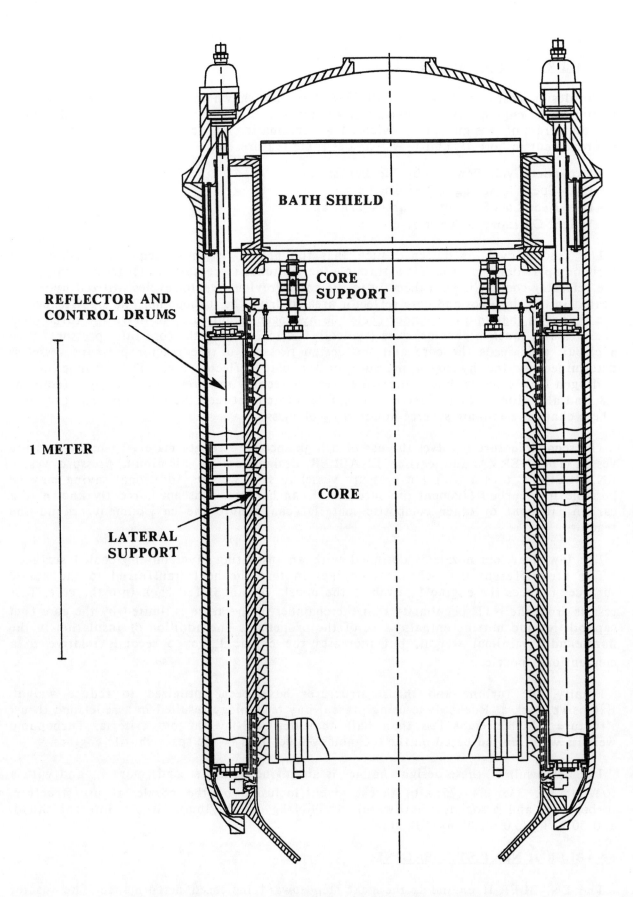

FIGURE 1. ENABLER-I Reactor Design Layout C.

performance (Petrosky 1991b). Fuel scaling is a uniform (photographic) reduction in all the dimensions of a fuel element. Channel size, channel spacing, and exterior element dimensions are all changed from a baseline element by multiplying the dimension by a scale factor α. An α less than one is an element that is reduced in size.

Prior work (Petrosky 1991a) has shown that scaling maintains similitude in the thermal and hydraulic profiles of fuel elements and of the core as a whole. A key result of scaling is that the allowable power density for an element increases by $1/\alpha^2$ while the thermal stress levels remain unchanged. Therefore, scaling maintains a strong linkage with the existing Rover/NERVA database while allowing departures from the previous fuel power density limitations. The higher allowable power density translates into a smaller core and lighter engine at a given thrust level. This results in an increase in the F/W performance measure.

Based on the analysis of fuel scale optimization (Petrosky 1991b), it is apparent that for typical mission exchange factors, the standard Rover/NERVA fuel is oversized. This is partially due to the success the program had in reducing the corrosion rates. Over the life of the Rover/NERVA program, fuel corrosion rates were decreased by more than an order of magnitude while the fuel size remained unchanged. Since lower corrosion rates push the optimal toward smaller α, the fuel which was near optimal early in the program became suboptimal as the corrosion problem was reduced.

Although the precise optimal value cannot be determined until the mission is better defined, the optimal scale factor α will most likely lie in the range of 0.5 to 0.75. Based on this observation, a design point of $\alpha = 0.667$, 12.7 mm (0.5 inch) hexagonal fuel, was chosen to characterize an engine utilizing this fuel scaling. At an $\alpha = 0.667$ the corresponding core length is 89 cm (35 inch). The core diameter varies with thrust level. Moderator (ZrH in the tie tube core supports) will be required for thrust levels less than 667 kN (150 klb) to maintain criticality.

Table 1 compares the F/W at various thrust levels of the ENABLER-I C design with the ENABLER-II, both designed for a 2 hour operating life. The comparison also includes the effects of the design features listed above. The ENABLER-I F/Ws are substantially better than the old NERVA designs and are well into the acceptable range even with shield mass included. The ENABLER-II with scaled fuel elements offers a further increase in F/W.

TABLE 1. F/W Comparison for ENABLER I & II.

Thrust Level kN (klb)	F/W N/kg (lbf/lbm)			
	ENABLER-I		ENABLER-II	
	w/shield	w/o shield	w/shield	w/o shield
130 (30)	34 (3.5)	41 (4.2)	48 (4.9)	64 (6.5)
220 (50)	40 (4.1)	48 (4.9)	58 (5.9)	75 (7.7)
330 (75)	50 (5.0)	59 (6.0)	69 (7.0)	90 (9.2)
440 (100)	53 (5.4)	63 (6.4)	74 (7.5)	96 (9.8)
560 (125)	55 (5.6)	66 (6.7)	80 (8.2)	105 (10.7)
670 (150)	57 (5.8)	68 (6.9)	83 (8.5)	109 (11.1)

CONCLUSIONS

The database provided by the Rover/NERVA program forms a basis for a low risk approach to NTR development. The guaranteed performance level of NERVA is sufficient for a manned Mars mission, but the improved performance achievable in the ENABLER engines begins to realize the potential possible with this technology.

Acknowledgments

This work was performed under Westinghouse Electric Corporation internal funding. Rocketdyne Division of Rockwell International provided support with their internal funding. The design of the non-nuclear components was headed by S. Gunn.

References

Durham, F.P. (1972) *Nuclear Engine Definition Study Preliminary Report Volume 1 - Engine Description (U)*, LA-5044-MS, Vol. 1, Los Alamos National Laboratory, Los Alamos, NM.

Koenig, D.R. (1986) *Experience Gained from the Space Nuclear Rocket Program (Rover)*, LA-10062-H, Los Alamos National Laboratory, Los Alamos, NM.

Livingston, J.M. and B.L. Pierce (1991) "The ENABLER - Based on Proven NERVA Technology" in *Proceedings of the Eighth Symposium on Space Nuclear Power Systems*, CONF-910116, M.S. El-Genk and M.D. Hoover, eds, American Institute of Physics, New York, Vol 2, pp 598-602.

Lyon, L.L. (1973) *Performance of (U,Zr)C-Graphite (Composite) and of (U,Zr)C (Carbide) Fuel Elements in the Nuclear Furnace 1 Test Reactor*, LA-5398-MS, Los Alamos National Laboratory, Los Alamos, NM.

Petrosky, L.J. (1991a) "Scaling Laws of Prismatic Fuel Nuclear Rocket Engines", Paper AIAA-91-2337, AIAA/SAE/ASME/ASEE 27th Joint Propulsion Conference, June 1991, available from the American Institute of Aeronautics and Astronautics, 370 L'Enfant Promenade S.W., Washington, DC.

Petrosky, L.J. (1991b) "Optimal Scaling of Prismatic Fuel Elements in Nuclear Rocket Engines", Paper AIAA-91-3506, AIAA/NASA/OAI Conference on Advanced SEI Technologies, September 1991, available from the American Institute of Aeronautics and Astronautics, 370 L'Enfant Promenade S.W., Washington, DC.

Storms, E.K., D. Hanson, W. Kirk, and P. Goldman (1991) "Effect of Fuel Geometry on the Life Time-Temperature Performance of Advanced Nuclear Propulsion Reactors", Paper AIAA-91-3454, AIAA/NASA/OAI Conference on Advanced SEI Technologies, September 1991, available from the American Institute of Aeronautics and Astronautics, 370 L'Enfant Promenade S.W., Washington, DC.

Taub, J.M. (1975) *A Review of Fuel Element Development for Nuclear Rocket Engines*, LA-5931, Los Alamos National Laboratory, Los Alamos, NM.

WANL (1972) *Technical Summary Report of NERVA Program*, TNR-230, Westinghouse Astronuclear Laboratory, available from Westinghouse Advanced Energy Systems, Madison, PA.

A PROTOTYPE ON-LINE WORK PROCEDURE SYSTEM FOR RTG PRODUCTION

Gary R. Kiebel
Westinghouse Hanford Company
Mail Stop N1-42, Box 1970
Richland, WA 99352
(509) 376-4995

Abstract

An on-line system to manage work procedures is being developed to support radioisotope thermoelectric generator (RTG) assembly and testing in a new production facility. This system implements production work procedures as interactive electronic documents executed at the work site with no intermediate printed form. It provides good control of the creation and application of work procedures and provides active assistance to the worker in performing them and in documenting the results. An extensive prototype of this system is being evaluated to ensure that it will have all the necessary features and that it will fit the user's needs and expectations. This effort has involved the Radioisotope Power Systems Facility (RPSF) operations organization and technology transfer between Westinghouse Hanford Company (Westinghouse Hanford) and EG&G Mound Applied Technologies Inc. (Mound) at the U.S. Department of Energy (DOE) Mound Site.

INTRODUCTION

The Procedure Manager (PM) is a major component of the Production Integration and Certification System (PICS), which provides on-line information support for the assembly and testing of space and terrestrial RTGs. The PICS system is part of a new facility, the Radioisotope Power Systems Facility (RPSF), which currently is being developed at the Hanford Site in southeastern Washington State. The PICS is a distributed computer-based system with redundant server computers and portable field workstations. A local area network (LAN) provides communications between workstations and servers (Kiebel 1991).

The PM implements RPSF production work procedures as interactive electronic documents. Authors prepare or revise the originals on word processors and transfer them to the PICS system, which translates them into a format that is directly usable by the system. Procedures are used directly from the field workstation; instructions are read from the screen, and results are entered via the keyboard. A record copy is printed when the procedure is complete.

DESCRIPTION

Benefits of Procedure Manager

Manufacturing and handling of RTGs produces a large volume of quality records (paperwork). Record copies of the work procedures make up much of this volume. The RTG production work must be performed correctly and good records must be kept in order to certify each RTG. The computer-based PM handles this information as electronic data more efficiently and more accurately than a paper-based system and can apply active checking and verification at many places throughout the process. Correct and legible quality records are produced with all the required information in the proper place. Work flow is controlled to ensure that work steps are not inadvertently skipped or performed in the wrong order. The correct version of a procedure is available instantly wherever it is needed on the process floor.

The PM also interfaces directly with other on-line systems within PICS. It communicates with the material tracking system and ensures that the correct material is used by a particular procedure and that all operations on the material are entered automatically. The PICS also maintains a list of training records for PICS workers and a list of training requirements for signature authority. The PM electronic signatures are cleared through this central authority, which ensures that work is authenticated only by properly qualified people whose training is current.

Evolution of Prototype

It was recognized during the conceptual design that the PM would be developed in-house because no commercial products could be found.

A successful development approach should answer several needs. Detailed and correct requirements are needed to make (and certify) good software.
Application experience is needed, especially by end users, to know whether requirements are correct and complete. The people with the most appropriate production experience (Mound) are not using an on-line computer based system. The Westinghouse Hanford operations people are not experienced yet with RTG production or on-line computer systems. The people with computer expertise at Westinghouse Hanford had no significant operations experience.

The PM prototype was undertaken as a way to bring the right knowledge and people together. Prototype software simulates important features of the software being developed and provides users and other affected parties with some 'hands on' experience. They then may make educated comments and ensure that the requirements and design for the proposed software properly address their needs. Prototype software also is useful for assessing the human factors provisions of the user interface. Prototype software is implemented quickly and cheaply. It does need not need to be complete, or durable, or maintained beyond its initial use.

The prototype fit well into an existing technology transfer program of Mound production expertise to Westinghouse Hanford.

Features of Procedure Manager

The PM is modeled on Mound procedures and leavened with Westinghouse Hanford nuclear fuel fabrication experience, which is the closest relevant local equivalent. The RPSF production procedures also must comply with Westinghouse Hanford Quality Assurance and other company policies.

A procedure is identified uniquely by its procedure identification (ID) and its version number. Only one master copy of a procedure with a given procedure ID and version number is allowed, but successive generations of a procedure with different version numbers all can be on the system at once because they are considered to be completely separate entities. The PM carries master copies of a procedure both in the original text form, as written by the author, and in the translated form used by the computer. When a procedure is to be used for actual work, a specific copy is created from the master. This copy carries its own serial number and is called a Procedure Record (PR). Many PRs may be made from a single master.

The master can be thought of as the 'generic' form of a work procedure; it deals with the elements of the procedure in abstract terms (e.g., kinds of material, types of people, types of events, etc.) as opposed to specific terms. The PR, on the other hand, deals with actual elements. It provides for recording the serial numbers of material items, instrument readings actually taken, names of people performing and authenticating work, actual time and date of events, and so forth.

Because the procedure must be understood by the computer, it must be written in a special format. In effect, it is a computer program but looks more like a typical work procedure than a typical computer program.

A procedure is divided logically into the following sections.

Procedure Identification Section specifies the procedure ID, its version number, and a title.

Configuration Control List Section lists the documents that control the procedure. Each PR records the specific revision and engineering change notice (ECN) level that pertain to it.

Materials List Section specifies the part numbers of all material to be used by the procedure. The serial or lot number of the actual materials used by a particular instance of the procedure is recorded in the PR.

Location List Section specifies all the locations in which the material defined in the materials list section will reside during the execution of the procedure. This is needed because the PM interfaces with the material tracking system to record material movements.

Data Definition Section specifies the data items to be taken during the performance of the procedure. The actual data values are recorded in the PR. Each data item is given a unique identifier by which it is referenced from other sections of the procedure.

Signature Definition Section lists the signature categories that are valid for the procedure. The PM uses 'electronic signatures,' which are entered by workers to authenticate work steps and to close out the procedure when completed. These electronic signatures are defined by category. Each signature category corresponds to a body of qualifications that a person must have on record in order to be allowed to sign.

Step Definition Section describes the required work and permitted activities for a single work step. Each step definition section is composed of the following subsections.

Opening Constraints Subsection defines the conditions that must be satisfied before the step is allowed to be opened for work. These conditions are defined by expressions that test the state of other work steps or data. Opening constraints typically require that some combination of previous steps be completed or that specific data be available.

Drawings and Specifications Subsection lists which of the drawings and specifications listed in the configuration control list apply to this particular step.

Worker Cautions Subsection describes any special precautions that the worker or workers must be aware of for that step.

Worker Instructions Subsection describes the work that is to be performed by the worker or workers in order to complete that step.

Step Operations Subsection specifies a set of operations that the PM will perform for the worker for that step. These operations are presented in menu format, and the worker can select any operation. A typical step operation prompts the worker to enter a specific reading or observation that is recorded in a data item. Math functions on data and interaction with the material tracking system also are done in this subsection.

Signature Subsection lists the signature categories that are required to authenticate the work done by this step.

Closing Constraints Subsection defines the conditions that must be satisfied before the step is allowed to be closed, that is, marked as completed. These conditions are similar to the ones for the opening constraint subsection.

As may be inferred from the sections and subsections described above, an author must do the following things in order to write a procedure acceptable to the PM:

- Identify and name the work procedure,
- Specify the documents that control the work or product,
- Break the work down into manageable work steps,
- Define the materials to be used during the procedure,
- Define the locations involved in material operations,
- Define the data to be recorded during the procedure,
- Define valid signature categories for the procedure,
- Detail the work to be done at each work step,
- Define the work flow rules for each work step, and
- Define signature categories for each work step.

The PM cannot apply judgment or make loose interpretations of the rules and conditions that are specified in the work procedures that it manages; it enforces these rules rigorously. Therefore, a burden will fall upon the authors of RPSF work procedures to think them through thoroughly and remove inconsistencies and errors, particularly in the overall organization. Also, procedures will need to be verified by dry runs before committing actual production work to them. This is the perception of how the process works at Mound, and it is in keeping with the high level of quality required for RTG production. The PM supports and controls procedures that are in a provisional status (while undergoing verification) and an approved status (when released for production operations).

Corrective Procedures

What happens when things go wrong with production operations and the procedure cannot be carried forward as written? The procedure cannot be changed simply on the spot with the differences penciled in the margins because changes to the master would create an entirely new revision of the procedure. There would be no way to connect the PR that contained the work already done to such a new revision. Somehow, the work necessary to correct the operation must be captured in a procedure that the worker can use, the affected parts of the original PR marked, and a connection between the two made.

The mechanism that the PM provides is the corrective procedure. This is a work procedure that can designate an existing procedure as its target and 'reach into' a PR made from that procedure to mark (but not change) the affected parts. The corrective procedure itself describes all the work necessary to correct the problem that the target procedure encountered. It will mark as 'superseded' those work steps and data items in the target PR that are replaced by the corrective work. This is a 'lab notebook' model, where everything is written in ink and mistakes can be crossed out but not erased. In some cases, the original PR will be able to resume work beyond the superseded areas, and in some cases it will not. Cross-references between the corrective PR and target PR are maintained.

Evaluation Process

The main thrust of the evaluation process has been to have RPSF operations people use the PM prototype to prepare, verify, and perform work procedures. These procedures are part of mock-up testing of the process equipment (predominately assembly equipment for hot-cell operations). A draft user's manual was prepared when the effort began. The software development staff is consulted whenever a problem is encountered by the operations staff. Resolution of the problem may require the modification of the prototype, revision of the user's manual, or changes to the PICS software requirements specification.

The prototype also is demonstrated frequently to various individuals or groups whose opinions or comments are wanted (quality assurance is an important example).

CONCLUSIONS

The prototype largely has accomplished its goals. It has been an excellent vehicle for soliciting comments. A number of design oversights have been identified and corrected at relatively low cost. The general human factors of the user interface also have improved (such as better screen layout, more menu selections, and more understandable commands). There also has been a pervasive improvement in the design of the product, which might be described as having the operator's point of view become more dominant in the organization and behavior.

Applications for transfer of the technology to other Hanford Site facilities are being considered as a result of the many prototype demonstrations.

Acknowledgments

The Mound RTG production staff is acknowledged for the tremendous body of knowledge represented by the Mound operating procedures that have served as an important basis for the PM concept. Terry Ward and Frank

Moore of the RPSF Operations Integrations Organization are acknowledged for their efforts as the principal prototype evaluators.

The project under which the system described in this paper was developed was funded by the U.S. Department of Energy.

References

Kiebel, G. R. (1991) "An On-Line Information System for Radioisotope Thermal Generator Production," in Proc., *8th Symposium On Space Nuclear Power Systems*, CONF-910116, M. S. El-Genk and M. D. Hoover, eds., American Institute of Physics, New York, NY, pp. 21-25.

APPLICATION OF PbTe/TAGS CPA THERMOELECTRIC MODULE TECHNOLOGY IN RTGs FOR MARS SURFACE MISSIONS

Wayne M. Brittain
Teledyne Energy Systems
110 W. Timonium Road
Timonium, MD 21093
(301) 252-8220

Abstract

This paper presents the status of PbTe/TAGS close-pack-array (CPA) thermoelectric module technology developed by Teledyne Energy Systems for use in terrestrial radioisotope thermoelectric generators (RTGs), and discusses application of this technology to RTGs for Mars surface missions. Low-power RTGs with power levels ranging from 3 to 20 We have been identified as candidate power sources for various scientific packages to be landed on the Martian surface. This power range is consistent with the use of PbTe/TAGS CPA technology. Modifications required for the current CPA technology to increase mechanical shock loading capability and to permit operation at the higher temperatures required for efficient operation in the Martian planetary environment are identified. A conceptual 15-We space RTG design incorporating CPA technology is presented. CPA technology permits a significant improvement in the specific power of PbTe/TAGS RTG units, even at these relatively low power levels. Specific power is in the 3.7 We/kg (1.7 We/lb) range for RTGs employing CPA technology compared with approximately 2.2 We/kg (1 We/lb) for the RTGs employed in the NASA Pioneer Jupiter Fly-by and Viking Mars Lander missions in the 1970s.

INTRODUCTION

Effort on the DOE-sponsored Special Applications RTG Technology Program and Two-Watt RTG Program has focused on PbTe/TAGS CPA technology for terrestrial applications where the RTG will be exposed to relatively low temperature environments, for example subsea. Thus, effort has been oriented towards design optimization at cold junction temperatures in the 283 to 366 K (50 to 200 F) range. However, for other more severe design environments (such as space applications where a high heat rejection radiator temperature in the 450 to 477 K (350 to 400 F) range is required to minimize RTG size and weight, and high shock/vibration capability is necessary) a modified CPA thermoelectric module design is dictated. In order to minimize the RTG system size and weight, and to increase the mechanical strength of the thermoelectric module to withstand increased dynamic loads, a monolithic PbTe/TAGS CPA module configuration is desirable. A monolithic CPA module is especially attractive for applications where severe structural loads will be imposed, such as "hard" planetary landers and penetrators.

CPA DESIGN MODIFICATION

The technical approach for modifying the terrestrial CPA thermoelectric module to expand its applicability to planetary space missions is to increase the operating temperature capability of the current design as well as reconfigure the minicouple array into a more densely packed arrangement with the objective of increasing the module dynamic loading capability while minimizing size and weight. The techniques and processes developed for fabricating small-cross section minicouples (shown in Figure 1) serve as the foundations for the CPA module development program. Minicouples are currently grouped to form a thermoelectric module as shown in Figure 2 which depicts the terrestrial Two-Watt RTG module. In this configuration the minicouples are relatively widely spaced within a 8.9-cm (3.5-in) diameter circle, and the void between minicouples filled with powdered Min-K insulation. The objective of the modified CPA module development program is to group the minicouples into a compact array similar to that shown in Figure 3 which depicts an early PbTe/TAGS-type CPA development module fabricated on the Special Applications RTG Technology Program. Use of modified CPA technology will result in approximately a factor of five reduction in the area of the thermoelectric module cross section from the current minicouple packing density exemplified by the Two-Watt RTG module.

Increasing Temperature Capability

To optimize the current Two-Watt CPA thermoelectric module design for increased temperature capability, the following design modifications will be implemented:

o   The maximum cold junction temperature limit of the Two-Watt thermoelectric module for long-term operation is 422 K (300 F) due to the fiberglass printed circuit board used for cold end electrical connections between the minicouples. To permit operation at a higher temperature, the printed circuit board material will be changed to one with increased temperature capability; alumina and Kapton are prime candidates.

o   The maximum hot junction operating temperature of the Two-Watt thermoelectric module is dictated by the Special Applications Terrestrial 3-layer fuel capsule technology which has a temperature limitation of 873 K (1112 F) on the inner liner for long-term operation. For space applications the use of the GPHS high temperature heat source technology permits increasing the hot junction temperature to the 811 K (1000 F) level.

o   In the Two-Watt CPA thermoelectric module (refer to Figure 2) Johns-Manville powdered Min-K provides for both thermal insulation between the minicouples and sublimation suppression of the thermoelectric materials at the hot end. A major effort of the modified CPA development program will be to identify and demonstrate an effective material (or materials) to provide both electrical and thermal insulation between the minicouples within the CPA module. Materials such as Kapton have been used at lower temperatures, and are candidates for insulation at the module cold side for a "staged" insulation system. However, a higher temperature insulating material is required at the hot side of the high temperature CPA module. Ideally this hot side insulation will serve both as an electrical and thermal insulator between the minicouples, and provide a degree of sublimation suppression for the thermoelectric materials.

FIGURE 1. Typical Special Applications Minicouple Configuration.

FIGURE 2. Two-Watt RTG Minicouple Module.

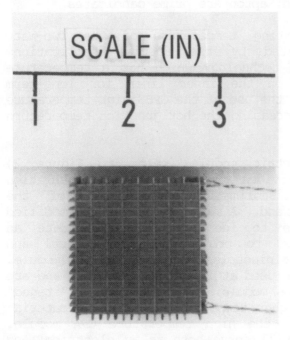

FIGURE 3. Early Special Applications CPA Module

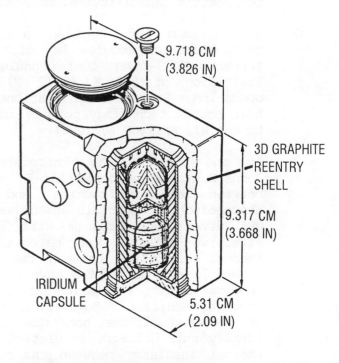

FIGURE 4. GPHS 250-Wt Module.

15-We RTG CONCEPTUAL DESIGN

The conceptual design of a 15-We RTG for powering Martian surface planetary exploration instrumentation packages has been generated as an illustration for using CPA thermoelectric module technology. Assumptions for establishing the conceptual design are based on experience from the design and integration effort for the PbTe/TAGS RTGs used on the two NASA Viking Mars Landers as well as preliminary information available for planned near-term Mars surface missions. The following is a synopsis of the assumptions:

o   Modified Two-Watt RTG CPA thermoelectric module technology;
o   MgTh alloy housing with 0.236 cm (0.093 in) wall thickness (same as Viking RTGs);
o   Electrical receptacle with redundant Viton O-ring seals that permit controlled release of the He gas generated by the Pu-238 decay to the Martian atmosphere (same as Pioneer and Viking RTGs);
o   311 K (100 F) totally absorbing radiation sink on the Mars surface (same as Viking RTG thermal design under Lander wind covers);
o   General Purpose Heat Source (GPHS) technology (only current Pu-238 heat source technology approved for application in space RTGs); and
o   Maximum landing shock load in the 40 to 100G range.

Figure 4 shows the 250-Wt GPHS heat source module which is the current "building block" for space RTGs. The heat source has four fueled clad (FC) assemblies, each containing nominally 62.5 Wt; two FCs are contained in a graphite impact shell (GIS) assembly; two GIS assemblies are in turn assembled into the graphite reentry shell. The GPHS module weighs 1.44 kg (3.16 lb).

Figure 5 shows a cross section of the 15-We RTG concept. The dimensions of the basic housing are 16.0 cm (6.3 in) diameter by 16.0 cm height. The diameter of the conical fin is 24.9 cm (9.8 in). Figure 6 is an expanded view of the RTG. A spring arrangement preloads the heat source and thermoelectric modules to withstand vibration and shock dynamic loads. Four CPA thermoelectric modules, electrically connected in series, provide the electrical power circuit. Circuit redundancy within each module is provided by a series/parallel electrical interconnection of the minicouples. Each thermoelectric module contains 82 minicouples, approximately the same number of couples as used in the terrestrial Two-Watt RTG module; the minicouple N and P element cross sections (0.259 cm (0.102 in) square) are the same as those of the Two-Watt RTG.

Table 1 summarizes the RTG performance and design characteristics. Power output at beginning-of-life (BOL) is nominally 15 We at 12V. Hot and cold junction temperatures are 783 K (950 F) and 477 K (400 F), respectively. The unit weighs 3.9 kg (8.7 lb) and overall system conversion efficiency is 6.0%. Table 2 presents the RTG weight statement.

CONCLUSIONS

RTGs employing PbTe/TAGS CPA thermoelectric module technology are attractive power source candidates for the Mars surface exploration missions anticipated for the late 1990s. The employment of CPA module technology results in a significant improvement in the specific power of PbTe/TAGS RTGs. The performance of the SNAP-19 Viking RTGs on the Martian surface, which operated

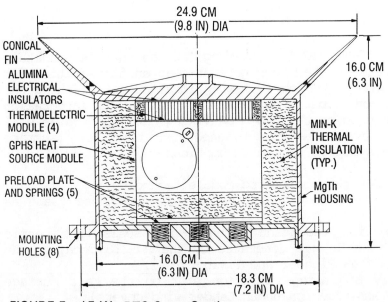

FIGURE 5. 15-We RTG Cross Section.

TABLE 1. 15-We RTG Design Summary.

BOL Performance	
Power Output (We)	15
Load Voltage (V)	12
Hot Junction Temperature (K)	783 (950 F)
Cold Junction Temperature (K)	477 (400 F)
Average Housing Temperature (K)	450 (350 F)
Thermoelectric Efficiency (%)	6.9
System Efficiency (%)	6.0
Thermoelectric Converter	
Materials	
N-element	PbTe
P-element	TAGS/PbSnTe segment
Size	
N-element (cm)	0.259 (0.102 in) sq x 1.27 (0.500 in) lg
P-element (cm)	0.259 (0.102 in) sq x 1.27 (0.500 in) lg
Number of Couples	328
Circuit Connections	Parallel/Series
Technology Base	Pioneer, Viking and Two-Watt T/E Generators
Module Configuration	Close-Pack-Array (CPA)
Heat Source	
Thermal Inventory (Wt)	250
Technology Base	GPHS Module
Overall Dimensions	
Basic Housing (cm)	16.0 (6.3 in)
Dia. across Conical Fin (cm)	24.9 (9.8 in)
Overall Height (cm)	16.0 (6.3 in)
Weight (kg)	3.9 (8.7 lb)
Power-to-Weight Ratio (We/kg)	3.7 (1.7 We/lb)

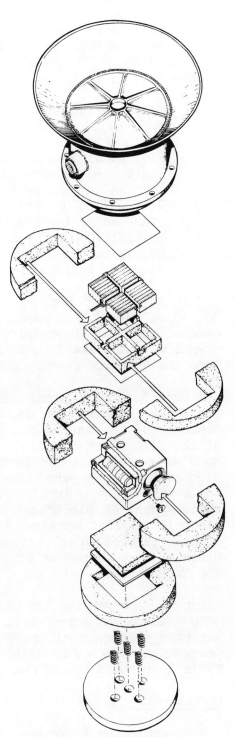

FIGURE 6. Expanded View of 15-We RTG.

for over 6 years before the last Lander was shut down, has demonstrated the long-term reliability of RTG power for such planetary missions.

TABLE 2. 15-Watt RTG Weight Summary.

Item		Weight	
		kg	lb
GPHS Heat Source Module		1.44	3.16
MgTh Housing/Fin Assembly		1.20	2.65
Min-K Thermal Insulation		0.36	0.79
Thermoelectric Modules (4)		0.65	1.43
Hot/Cold Alumina Insulators		0.06	0.13
Preload Plate and Springs		0.12	0.27
Receptacle/O-Rings/Wires/Crimp, and other		0.09	0.20
	Total	3.92	8.63

Acknowledgments

The Two-Watt Special Applications RTG effort is being sponsored by the United States Department of Energy under contracts DE-AC01-83NE32115 and DE-AC01-88NE32142. The conceptual design effort for the 15-We RTG was company-funded by Teledyne Energy Systems.

References

Brittain, W. M., M. F. McKittrick, W. H. Hayes, and K. Weide (1991) "Production of PbTe/TAGS/Bi_2Te_3 Minicouple Modules," in Proceedings of the 8th Symposium on Space Nuclear Power Systems, CONF-910116, M.S. El-Genk and M.D. Hoover, eds., American Institute of Physics, New York, NY, pp. 1309-1314.

Brittain, W. M. (1991) "Close-Packed-Array (CPA) Thermoelectric Module Development Status," in Proceedings of the 8th Symposium on Space Nuclear Power Systems, CONF-910116, M.S. El-Genk and M.D. Hoover, eds., American Institute of Physics, New York, NY, pp. 923-927.

Dick, P. J., W. M. Brittain and R. C. Brouns (1985) "Special Applications RTG Technology Program," in Proceedings of the 20th IECEC, Miami Beach, FL, August 1985.

Dick, P. J. and W. M. Brittain (1987) "Special Applications RTG Performance Testing," in Proceedings of the 22nd IECEC, Philadelphia, PA, August 1987.

Trimmer, D. S. (1986) "Testing of Thermoelectrics for the Pu-238 Special Applications RTG," in Proceedings of the 21st IECEC, San Diego, CA, August 1986.

DEVELOPMENT OF A RADIOISOTOPE HEAT SOURCE FOR THE TWO-WATT RADIOISOTOPE THERMOELECTRIC GENERATOR

Edwin I. Howell, Dennis C. McNeil, and Wayne R. Amos
EG&G Mound Applied Technologies
P. O. Box 3000
Miamisburg, OH 45343-3000
(513) 865-4163

Abstract

Described is a radioisotope heat source for the Two-Watt Radioisotope Thermoelectric Generator (RTG) which is being considered for possible application by the U.S. Navy and for other Department of Defense applications. The heat source thermal energy (75 Wt) is produced from the alpha decay of plutonium-238 which is in the form of high-fired plutonium dioxide. The capsule is non-vented and consists of three domed cylindrical components each closed with a corresponding sealed end cap. Surrounding the fuel is the liner component, which is fabricated from a tantalum-based alloy, T-111. Also fabricated from T-111 is the next component, the strength member, which serves to meet pressure and impact criteria. The outermost component, or clad, is the oxidation- and corrosion-resistant nickel-based alloy, Hastelloy S. This paper defines the design considerations, details the hardware fabrication and welding processes, discusses the addition of yttrium to the fuel to reduce liner embrittlement, and describes the testing that has been conducted or is planned to assure that there is fuel containment not only during the heat source operational life, but also in case of an accident environment.

INTRODUCTION

Since the early 1960s, Mound (presently operated by EG&G Mound Applied Technologies, but operated by Monsanto Research Corporation prior to October 1988) has worked with Teledyne Energy Systems (TES) of Timonium, Maryland in the production of radioisotope heat sources that are employed in radioisotope thermoelectric generators (RTGs) for both space and terrestrial applications. The space program RTGs have provided electrical power for the Pioneer spacecraft and the two Viking missions that landed on the surface of Mars. Terrestrial applications have included the Five Watt RTGs and the High Performance Generator (HPG) MOD-3 RTGs.

Recently, the U.S. Navy has shown an interest in a Two-Watt RTG (Figure 1), which will generate 4.9 We at the beginning of life (or 2.6 We at the end of its 10 year life). Since 1989, EG&G Mound has been actively involved with TES in the development of a radioisotope heat source to support the Two-Watt RTG program. Much of the design philosophy has been established from studies performed at TES (Dick 1988).

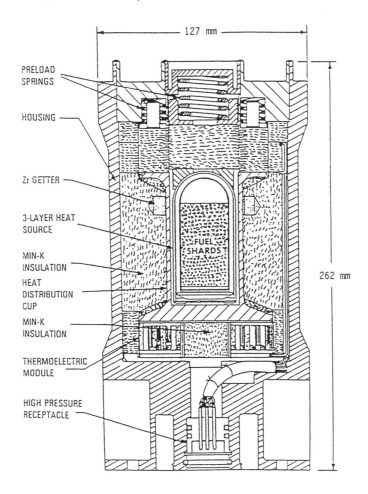

FIGURE 1. Two-Watt RTG.

DESIGN CONSIDERATIONS

Plutonium-238 has always been attractive as a heat generator for RTGs because it has a relatively long half-life (87.79 years) and it is predominately an alpha emitter rather than an emitter of more penetrating radiation. However, plutonium is extremely toxic, which necessitates designing a total containment heat source that is capable of withstanding high temperatures and internal pressures, especially during potential accident conditions such as fire and impact.

To allow the design to be established early and to eliminate the need to duplicate many studies already performed in the areas of compatibility and safety, existing materials technology was utilized. Therefore, the design chosen for the Two-Watt RTG heat source was essentially a scale-up of the Milliwatt Generator (MWG) heat source.

The MWG heat source, manufactured from 1972 until 1990, consisted of a three-layered, non-vented capsule. The liner (which contained the plutonium dioxide) and the strength member, were T-111, a tantalum based alloy (nominally 90% Ta, 8% W, and 2% Hf). The clad, or the outermost component, was the oxidation- and corrosion-resistant Hastelloy C, a nickel based alloy (nominally 56% Ni, 16% Mo, and 15.5% Cr). The plutonia (4.0 Wt) was in the form of high-fired (to 1873 K) shards which were 53-500 µm in size. The Savannah River Plant provided the plutonia which had been oxygen-16 exchanged to reduce the neutron radiation due to the alpha-neutron reaction occuring with natural oxygen. The plutonium-238 isotopic content was approximately 80%. Interspersed with the plutonia were yttrium chips, 1.6 mm by 1.6 mm by 0.25 mm thick. The purpose of the yttrium was to reduce the PuO_2 to $PuO_{1.75}$, thus lowering the partial pressure of oxygen within the capsule and reducing the amount of oxygen available to react with the liner. The success of this yttrium reduction has been experimentally proven (David 1980). Without the yttrium the plutonia would become substoichiometric by reacting with the T-111 liner and would cause liner embrittlement. It was learned that pretreatment of the welded liner - strength member assembly at 1573-1673 K for one hour would force this reaction in-situ and significantly reduce T-111 oxidation (embrittlement) from either capsule aging or a fire situation (Jones et al. 1975 and 1979). The cap to body joints for each liner, strength member, and clad were gas tungsten arc (GTA) welded.

RESULTS

With only minor variations, the Two-Watt heat source is simply a larger MWG capsule. There was a change in the clad material from Hastelloy C, as Hastelloy C did not possess all of the favorable characteristics required for Two-Watt RTG application. Five alloys were considered: Hastelloy C-276, Hastelloy S, Haynes 230, Inconel 600, and Inconel 625. The corrosion resistance, creep strength, thermal stability, formability, weldability, and availability of each alloy was assessed, and Hastelloy S was selected. With the chemical and physical property similarities of Hastelloy C and Hastelloy S (Hastelloy S is nominally 68% Ni, 15% Mo, and 15% Cr), relatively little work will be required to verify that Hastelloy S is compatible with the T-111 strength member at the operating temperatures. During MWG compatibility tests at Mound, no Hastelloy C/T-111 reactions were observed at or below 1173 K for up to two years. Also, Inconel 600/T-111 compatibility studies (Teaney 1979) and information obtained from literature searches involving constituents in nickel and tantalum alloy systems indicate that good compatibility would be expected for Hastelloy S and T-111 at the operating temperature (~800 K) and life (10 years) of the RTG. A compatibility study is underway to confirm this.

The fuel and the yttrium getter for the Two-Watt were identical to the MWG unit with the exception of the larger Two-Watt quantities (75 Wt versus 4.0 Wt, 180 g of plutonia versus 10 g, and 9.8 g of yttrium versus 0.54 g), and the plutonium-238 isotope content was increased from 80% to 83%. The pretreatment operation (yttrium reduction of the plutonia) was the same as previously described for the MWG unit.

A sketch of the Two-Watt heat source is shown in Figure 2 and a photograph of the actual hardware is shown in Figure 3. The purpose of the liner shim was to deflect the heat away from the fuel during the body-cap welding operation, thereby minimizing fuel outgassing. The hardware dimensions are given in Table 1, and a brief description of the hardware fabrication process for the cylindrical (body) components is provided in Table 2. All body forming operations utilized punch and die tooling manufactured from Ampco Bronze 25 material. This material was selected over the standard hardened steel tooling as it resulted in reduced galling. Another standard throughout the forming operations was the use of "Cimflo 10" (Cincinnati Milacron) lubricant. The forming rates ranged from 0.8 to 2.5 mm/s and all draws were performed at room temperature.

The three caps were all manufactured by conventional machining techniques. The 0.25 mm thick T-111 shim was blanked directly from sheet stock.

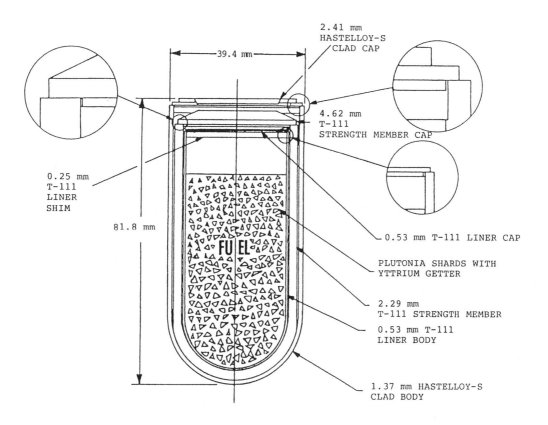

FIGURE 2. Two-Watt RTG Radioisotope Heat Source. (Weld joint details shown in encircled enlargements.)

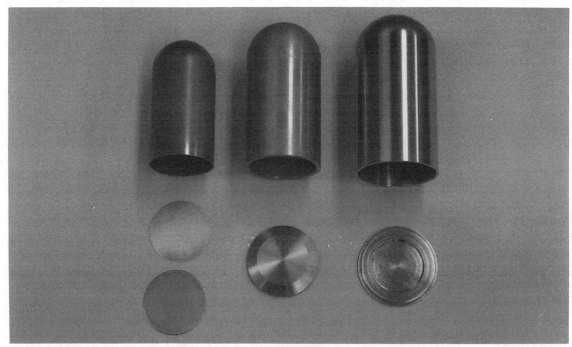

FIGURE 3. Two-Watt RTG Hardware.

TABLE 1. Heat Source Component Dimensions (in mm).

Component	Minimum Dome Wall Thickness	Outside Diameter At Open End	Inside Length
Liner Body	0.41	31.67	69.19
Strength Member Body	2.21	36.91	71.12
Clad Body	1.30	40.16	78.31

Component	Major Diameter	Thickness
Liner Shim	30.61	0.25
Liner Cap	31.19	0.53
Strength Member Cap	34.67	4.62
Clad Cap	38.84	2.41

TABLE 2. Variables and Processing Flow for the Cylindrical Components of the Two-Watt Heat Source.

Operation/Variable	Liner	Strength Member	Clad
BLANK CHARACTERISTICS:			
Material	T-111	T-111	Hastelloy S
Thickness	0.58 mm	3.18 mm	1.57 mm
Diameter	99.06 mm	102.36 mm	114.30 mm
MANUFACTURING FLOW:			
Blanking of Diameter	Wire EDM [a]	Wire EDM [a]	Laser
First Draw Reduction	46%	48%	53%
Rough Machine	Length	Length/Diameter	Length/Diameter
Clean	Defined Below [b]	Defined Below [b]	Defined Below [c]
Vacuum Heat Treat	1700 K/1 h	1700 K/1 h	1353 K/1 h
Second Draw Reduction	26%	26%	26%
Rough Machine	Length	Length/Diameter	Length/Diameter
Clean	Defined Below [b]	Defined Below [b]	Defined Below [c]
Vacuum Heat Treat	1700 K/1 h	1700 K/1 h	1353 K/1 h
Third Draw Reduction	22%	18%	5%
Clean	Defined Below [b]	Not Applicable	Not Applicable
Vacuum Heat Treat	1700 K/1 h	Not Applicable	Not Applicable
Sizing Draw Reduction	1%	Not Applicable	Not Applicble
Establish Final Configuration	Machine	Machine	Machine
Clean	Defined Below [b]	Defined Below [b]	Defined Below [c]
Dye Penetrant Check	Defined Below [d]	Defined Below [d]	Defined Below [d]
Grit Blast/Oxidation	Not Applicable	Not Applicable	Defined in Text
Dimensional/Visual Check	Inspect	Inspect	Inspect
Clean	Defined Below [b]	Defined Below [b]	Defined Below [c]
Vacuum Heat Treat	1700 K/1 h	1700 K/1 h	1353 K/1 h

[a] Electrical discharge machining.

[b] T-111 cleaning procedure: methylene chloride; acetone; solution of 2 parts hydrofluoric acid, 2 parts nitric acid, and 5 parts demineralized water; tap water and demineralized water rinse; and ethanol.

[c] Hastelloy S cleaning procedure: methylene chloride; acetone; 1:1 nitric acid; tap water and demineralized water rinse; and ethanol.

[d] Dye penetrant is Sherwin HM608, developer is Sherwin D-90G; inspect for defects, cracks, pores, and inclusions.

Another variation from the MWG design involved an emissivity coating that was applied to the Two-Watt clad body to enhance the heat flow from the heat source to the RTG thermoelectrics. An emissivity of 0.8 to 0.9 was achieved by grit blasting the clad body (except within 2.5 mm of the open end where welding would occur) with -65/+150 mesh tungsten carbide powder to a 0.08 to 3.2 AA surface finish followed by a one hour air oxidation at 1311 K.

The welding equipment for the GTA joining of the three caps to the bodies consisted of computer controls which permitted all welding variables (such as weld current, pulse, and travel speed) to be entered through a keyboard. Welding current was supplied by a 300 A direct current transistor inverter power supply. A slow pulse mode (6 pulses per second) was incorporated into the welding parameters to reduce zone refinement (Zielinski et al. 1977). The welding fixture consisted of a chuck equipped with nickel plated Be-Cu collets. The collets used for the liner and strength member had tantalum inserts to withstand the high temperature (~3400 K) required to melt the T-111.

As noted in Figure 2, a small step was machined on each of the caps to provide proper cap-to-body configuration. The joint was also designed to provide a weld cross section equal to the body wall thickness without excessive weld bead build-up on the outside of the capsule. Welding parameters were developed to produce complete joint penetration with an inside fillet of approximately 0.76 mm. As the T-111 near the weld (heat-affected zone) was prone to embrittlement when melted in an atmosphere containing minor amounts of oxygen and nitrogen, all welding was performed in a helium atmosphere with oxygen and nitrogen maintained at less than 5 parts per million.

After development of the initial processes, a series of capsules were welded to verify that the parameters were satisfactory. The liners were loaded with 200 g of fuel simulant (55 wt % Mo and 45 wt % W) which had been vacuum outgassed at 873 K at ≤ 1.3 µPa. Each welded member received a helium leak test, visual inspection, dye penetrant evaluation, and radiographic examination. After these assessments were completed, the welds were cross-sectioned and metallographically examined. Once all the data indicated the welds were consistently defect-free, the welding parameters were considered acceptable for test capsule fabrication.

The testing that is planned for the Two-Watt RTG heat source hardware will parallel that performed on the MWG units with minor differences. Emphasis will be placed on potential fire and impact situations. A test matrix was developed for stress-rupture testing (Table 3) based on preliminary fire response studies and the need to expand the stress-rupture data base for T-111 for a biaxial state of stress. The fire environment implies a time-dependent increase in stress as a result of gas expansion.

As it is assumed that the thinner liner component will be compromised during its 10 year life, the strength member must remain integral during this fire test. A multi-axial mechanical test will demonstrate this integrity. By means of high pressure tantalum tubing, a strength member will be filled with helium of a known amount at ambient temperature, then the pressure will be monitored when the specimen is heated to the test temperature in a vacuum furnace. If the helium leak rate (monitored within the furnace) increases to a value in excess of 90 pmol/s, the test specimen is considered a failure. The test procedure was described in detail by Zielenski et al. (1977).

TABLE 3. Stress-Rupture Stress Matrix.

Test Pressure (MPa)	Equivalent Stress (MPa)	Temperature (K)	Expected Nominal Time to Rupture (Hours)
69	486	1277	10.3
59	414	1277	24.2
49	345	1277	63.4
49	345	1444	3.1
44	310	1444	5.4
39	276	1444	10.1
29	207	1444	45.8
24	172	1444	119.9

The impact tests will be comparable to those performed on MWG hardware (Teaney et al. 1982). Capsules, filled with Mo-W powder fuel simulant, will be heated to operating temperature (873 K) and then impacted against a hardened steel surface by a gas-fired impact gun.

The desired impact angle is achieved by placing the capsule to be impacted in a specially designed nesting device. Emphasis will be focused on impacting the units at the most vulnerable angle (expected to be 45° on the cap end) and then determining the failure velocity (expected to be greater than 130 m/s) and failure mechanism. To date, one impact test has been performed at 75 m/s (45°, cap end), with no breach of the liner or strength member assemblies occurring.

Another completed test series subjected two capsules to high external hydrostatic pressure. These capsules were filled with the Mo-W simulant and welded in a helium atmosphere. The tests showed that although hardware deformation occurred, there was no release of helium at hydrostatic pressures of 100 MPa.

The hardware fabrication and testing operations were controlled by standard quality engineering processes in an attempt to detect, prevent, or minimize product defects and variability, and to assure conformance of the product to design requirements.

CONCLUSIONS

The Two-Watt RTG heat source hardware fabrication and welding processes have been successfully developed. To date, seventeen complete sets of hardware have been fabricated. Impact testing has been initiated on capsules filled with simulant fuel, two capsules have been hydrostatically tested, and fire tests are planned in the near future on representative strength member hardware. Because the testing results to date have been favorable and because comparable safety and compatibility testing has been completed on MWG hardware, there are no problems expected to surface that will alter the present established design.

With the Two-Watt heat source hardware design and production techniques established, a production campaign can be initiated at any time.

Acknowledgments

Mound is operated by EG&G Applied Technologies for the U.S. Department of Energy under Contract DE-AC04-88DP43495. The authors wish to acknowledge the technical assistance of P. E. Teaney of EG&G Mound and R. W. Saylor of Westinghouse Savannah River Company and the heat source design contributions from Teledyne Energy Systems. Also acknowledged is the capsule welding support provided by C. E. Burgan of EG&G Mound.

References

David, D. J. (1980) *A Thermodynamic Study of MWG System/Components and Measurement of the Oxygen Partial Pressure in the Heat Source Capsule*, MLM-2719, Monsanto Research Corporation, Miamisburg, Ohio.

Dick, P. J. (1988) *Heat Source Development Summary Report for the Special Applications RTG Technology Program*, TES-32115-185, Teledyne Energy Systems, Timonium, Maryland.

Jones, G. L., J. E. Selle, and P. E. Teaney (1975) *Plutonium-238 Dioxide/T-111 Compatibility Studies*, MLM-2209, Monsanto Research Corporation, Miamisburg, Ohio.

Jones, G. L., J. E. Selle, and P. E. Teaney (1975) *Radioisotopic Heat Source*, U.S. Patent No. 3,909,617, 30 September 1975.

Teaney, P. E. (1979) "Inconel-600/T-111 Alloy Compatibility Studies," *Microstructural Science*, 7: 187-191.

Teaney, P. E., W. B. Cartmill, and R. L. Wise (1982) *Testing and Evaluation of Doubly Impacted Simulant Fueled Milliwatt Generator Heat Sources*, MLM-MU-82-64-0001, Monsanto Research Corporation, Miamisburg, Ohio.

Zielinski, R. E., E. Stacy, and C. E. Burgan (1977) "Failure Analysis of Radioisotopic Heat Source Capsules Tested Under Multi-Axial Conditions," *Microstructural Science*, 5: 443-453.

RE-ESTABLISHMENT OF RTG UNICOUPLE PRODUCTION

Kermit D. Kuhl   and   James Braun
215-354-3505   215-354-3437
General Electric Company
P.O. Box 8555
Philadelphia, PA 19101

Abstract

The manufacturing process that was utilized to start up fabrication of the thermoelectric unicouple devices for the CRAF(Comet Rendezvous Asteroid Flyby)/Cassini RTG (Radioisotope Thermoelectric Generator) program are described in this paper. The overall approach taken to re-establish a qualified production line is discussed. Key elements involved in this effort were: engineering review of specifications; training of operators; manufacturing product verification runs; and management review of results. Appropriately, issues involved in activating a fabrication process that has been idle for nearly a decade, such as upgrading equipment, adhering to updated environmental, health, and safety requirements, or approving new vendors, are also addressed. The cumulative results of the startup activities have verified that a production line for this type of device can be reopened successfully.

INTRODUCTION

The unicouple is a thermoelectric device employed in Radioisotope Thermoelectric Generators (RTGs) where the heat from the radioisotope fuel is converted by the unicouple directly into electricity. Unicouples were last manufactured in 1983 under the GPHS-RTG (General Purpose Heat Source) program for use on the Galileo and Ulysses spacecraft. The base thermoelectric material used for the unicouples is silicon-germanium (SiGe). These SiGe unicouples for the GPHS-RTG program were identical in design to those manufactured in the MHW-RTG (Multi-Hundred Watt) program for the LES 8 and 9 and Voyager 1 and 2 space missions. All totaled to date, RTGs employing SiGe unicouples have demonstrated over one half million hours of successful operation in space. Continuing to rely on this exceptional performance, the CRAF/Cassini program selected the same GPHS-RTG SiGe unicouple and RTG design for its power system. Figure 1 provides a detailed view of the unicouple.

Vacuum casting, powder crushing and blending, hot pressing, diffusion bonding, chemical vapor deposition, mechanical assembly, hydrogen brazing, and insulation wrapping are the major manufacturing processes used to fabricate these unicouples. An overview of these fabrication steps is shown in Figure 2. The manufacture of the RTG and unicouple for the CRAF/Cassini program is being performed by the General Electric Astro Space Division (GE-ASD) located in Valley Forge, Pennsylvania. The CRAF/Cassini program requires the fabrication of over 3000 flight quality unicouples following the re-establishment of the production line.

RE-ESTABLISHMENT OF THE UNICOUPLE PRODUCTION LINE

The overall approach implemented to re-establish the unicouple manufacturing process is outlined in Figure 3. Key aspects of this approach with respect to manufacturing are further discussed in Reference [1]. The initial startup effort consisted of four major tasks performed in parallel:

1. Implementation of an EMQ (Engineering, Manufacturing and Quality) team to review all the process documentation from the former program and to control all proposed changes to the former process methods;

2. Training and certification of manufacturing operators;
3. Tooling and fixture readiness; and
4. Update of facilities and equipment.

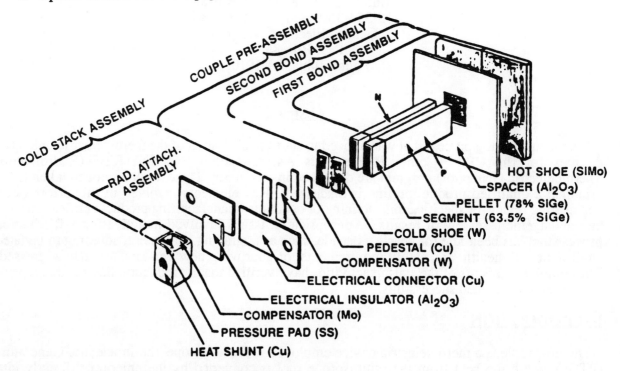

FIGURE 1. Exploded View of SiGe Unicouple.

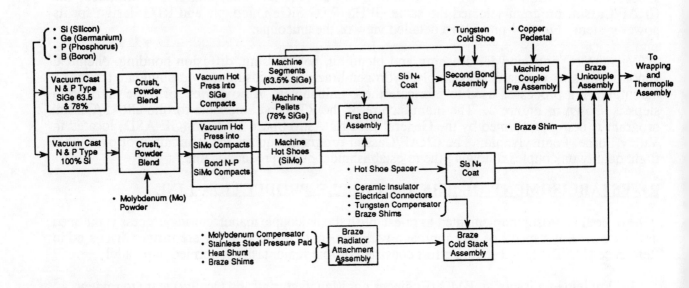

FIGURE 2. Manufacturing Process Flow Diagram - Unicouple Fabrication.

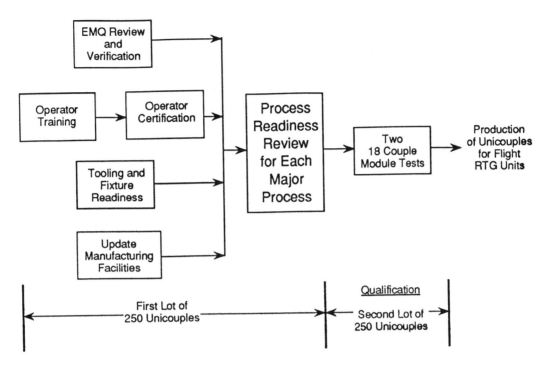

FIGURE 3. Overview of Manufacturing Approach to Re-Establish Unicouple Manufacturing Process.

EMQ Review and Verification Efforts

Since the goal of the CRAF/Cassini RTG program is to produce unicouples of the same design as used on the former GPHS-RTG program, the general manufacturing philosophy consistent with this goal was to use the exact same materials and fabrication techniques. There were, however, several areas identified where changes were required for the following reasons: (1) manufacturing technology advancements in several processes developed for the MOD-RTG multicouple showed significant cost savings by using up-to-date equipment; (2) identical replacements for old equipment were not available; (3) some of the qualified vendors that had supplied raw materials or piece parts were no longer in business and new suppliers had to be found and qualified. To control which changes were incorporated into the CRAF/Cassini unicouple fabrication process, a manufacturing verification effort was established to identify and resolve each of the previously listed items.

The EMQ team, with representatives from Engineering, Manufacturing, and Quality was established to perform a detailed review of all the engineering drawings and specifications, manufacturing planning, and tooling, and quality inspection documentation. The focus of the EMQ team was centered on resolving issues where a potential change to the former process was needed. A management steering committee (including a customer representative) was created to review the findings and recommendations of the EMQ team. Approval of the steering committee was required before any changes to the original GPHS documentation or process could be made. Where recommended changes were approved, the EMQ team and steering committee agreed in advance to a verification test program to assure that the product resulting from the recommended change was equivalent to the original GPHS product. As each step of the manufacturing process flow review was completed, a "Red-Lined" documentation set was created reflecting the approved changes. This documentation was put in the shop as soon as it was ready and used for initial operator training. The initial hardware fabricated from this documentation was used for the verification tests. Upon completion of the verification tests, the steering committee reviewed the process and inspection data to decide if final approval of the recommended change was appropriate. If approval was granted, the involved engineering drawings or specifications were changed through configuration control procedures. An overview of the EMQ review and verification process is provided in Figure 4. In addition to its involvement in verifying potential process changes, the EMQ team also examined product and corresponding processes which remained unchanged from the original GPHS-RTG method. The purpose of this examination was to ensure that new CRAF/Cassini product matched the former GPHS product. The summary result of the EMQ effort, despite proposed changes and new vendor sources, produced few changes in the unicouple manufacturing methods and the materials and overall construction of the unicouple duplicated that of the GPHS-RTG.

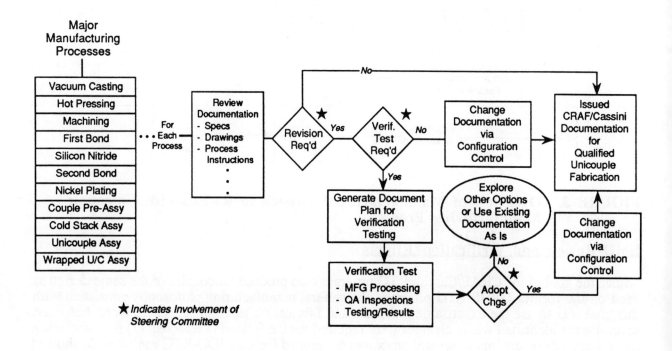

FIGURE 4. Overview of EMQ Review and Verification Process.

Operator Training and Re-Certification

With the lengthy break in production, virtually all the shop operators were new and complete training was necessary for all operators in the required skills for silicon-germanium unicouple manufacturing. The EMQ team members were intimately involved as trainers to ensure that the recommended changes were incorporated correctly.

As indicated in Table 1, key unicouple fabrication processes were identified as requiring certified operators. This involved intense training and discipline in all aspects of performing each manufacturing operation to specific instructions. As part of the introduction to the manufacturing facility, operators also received the appropriate safety and hazardous material handling training. After each operator was trained, an oral examination had to be passed regarding knowledge of the proper processes and the discipline of following the documentation. In addition, a specified number of hardware samples had to be made and be accepted by Quality Assurance to the drawing/specification requirements. Only after the operator had passed the certification requirements was he/she permitted to produce hardware for flight use.

TABLE 1. Key Processes Requiring Operator Certification and Process Readiness.

Manufacturing Process	Operator Certification	Process Readiness Approval
Vacuum Casting	Yes	Yes
Powder Blending	Yes	Yes
Hot Pressing	Yes	Yes
Machining	Yes	Yes
First Bond	Yes	Yes
Silicon Nitride Coating	Yes	Yes
Second Bond	Yes	Yes
Nickel Plating	Yes	Yes
Couple Preassembly	No	Yes
Cold Stack Assembly	No	Yes
Unicouple Assembly	No	Yes
Wrapped U/C Assembly	No	Yes

Tooling and Fixture Readiness

The processes involved in unicouple fabrication require specialized tooling and fixtures in many areas. A considerable amount of these resources was still available from the former GPHS-RTG program, but these existing items had to be cleaned, inspected, and entered into the tool control system prior to use on the CRAF/Cassini RTG program. Manufacturing Engineering reviewed the process documentation for the required tooling against the existing inventory, and determined the items that could be acceptably reworked along with the quantities and types of new items to make/buy. There were several instances where the former processing specifications were not clear with regard to tooling description. In these occurrences, the direct involvement of the EMQ team was employed to perform historical record checks and conduct trial operations. As deviations in documentation related to tooling and fixtures were resolved, the EMQ team also updated the controlled documentation.

Update of Facilities and Equipment

The majority of the processing equipment from the former GPHS-RTG program was still in place and in operating condition. Much of this equipment, however, was over ten years old and subject to frequent downtime for repairs. Manufacturing planning had determined that without modification, the existing equipment situation would not be adequate for the CRAF/Cassini production schedule. Specific pieces of equipment were identified to be replaced, upgraded/refurbished or supplemented with new units. In cases where upgraded or new equipment resulted in a functional change from the GPHS-RTG processing equipment, verification tests under the framework of the EMQ team review process were conducted to ensure that the product was not impacted. Re-training and re-certification of operators was also performed on new processing equipment. In addition to equipment improvements, the manufacturing facility was also upgraded. Major areas of improvement were: addition of a recirculating cooling water loop, and separation of CRAF/Cassini unicouple production areas from other programs. The end result provided a manufacturing center for CRAF/Cassini unicouple production and future thermoelectric programs.

Process Readiness Reviews

As shown by Figure 3, the culmination of the start-up activities resulted in process readiness reviews of each major manufacturing process. The process readiness review was a Quality Assurance led procedure where again, an EMQ team reviewed in detail all the paperwork, tooling, procedures, equipment, maintenance, and operator capability to determine if everything is in place to successfully produce acceptable flight hardware on a repetitive basis. Table 1 shows that all unicouple fabrication processes were subject to process readiness reviews. Reference [2] provides a full description of the process readiness review plan. A specified quantity of acceptable hardware had to be produced by certified operators before the process was considered approved for production.

Qualification

The initial startup phase of unicouple production required the production of 250 unicouples. Following the successful completion of the readiness reviews, a second lot of 250 unicouples is produced. This second lot is planned as a pre-production run of qualification unicouples. From this lot, two eighteen couple test modules are assembled and put on test to gather performance data. Also during this pre-production run, data is gathered to assess process yields. This information is used during later production runs to monitor process yield data and identify where process engineering attention is required to improve yields and reduce costs.

CONCLUSIONS

Re-establishment of the CRAF/Cassini unicouple production line has been successfully accomplished. Starting up the production of a discontinued item can be accomplished, although it must be recognized in advance that changes are inevitable since time and technology do not stand still. A review and verification of all proposed changes must be performed before they are implemented, but an emphasis should be placed on maintaining the original fabrication techniques. Training or retraining of engineers and operators requires a significant effort and a comprehensive review of old documentation and data is required. This effort will likely be an order of magnitude larger than expected by those who are still around and remember "how it was done" the last time.

Acknowledgments

The authors wish to acknowledge the support of the CRAF/Cassini manufacturing team and participants on the EMQ teams for information contained in this paper and for review of its contents. Unicouple production for the CRAF/Cassini RTG Program is being performed under DOE contract.

References

1. GESP-7219, "Fabrication Plan," General Electric Company, Astro-Space Division, Philadelphia, PA, 12 April 1991.

2. GESP-7233, "Process Readiness Review Plan," General Electric Company, Astro-Space Division, Philadelphia, PA, 23 August 1991.

TRANSPORT AND HANDLING OF RADIOISOTOPE THERMOELECTRIC GENERATORS AT WESTINGHOUSE HANFORD COMPANY

Carol J. Alderman
Westinghouse Hanford Company
P. O. Box 1970
Richland, WA 99352
(509) 376-0275

Abstract

Westinghouse Hanford Company is converting part of the Fuels and Materials Examination Facility, located at the Hanford Site in southeastern Washington State, to fuel and test radioisotope thermoelectric generators. The facility is referred to as the Radioisotope Power Systems Facility (RPSF). Historically, this activity has been performed at EG&G Mound Applied Technologies Inc. in Miamisburg, Ohio. Transportation and handling of the generator within the RPSF will be facilitated by the re-design of the Mound system to reduce the number of steps and therefore reduce operator dose. The overall handling concept has been retained and, wherever possible, the original design has been maintained. Changes reflect a simplified design, reducing the number of parts and a consistent interface between the generator to both the testing equipment and the transport cart. The new system design also improves operator efficiency and reduces operating costs.

INTRODUCTION

Westinghouse Hanford Company (Westinghouse Hanford) is installing the Radioisotope Power Systems Facility (RPSF) in the Fuels and Materials Examination Facility (FMEF) at the Hanford Site for assembling and testing radioisotope thermoelectric generators (RTGs) (Alderman 1990 and Eschenbaum and Wiemers 1990). The assembly process is based on one developed at Mound Laboratory (currently operated by EG&G Mound Applied Technologies) (Amos and Goebel 1990). Westinghouse Hanford is designing the process to ensure the heat source assemblies are produced in a manner that is consistent with today's environmental and operational safety standards. The facility will be ready to support future National Aeronautics and Space Administration (NASA) missions for radioisotope power systems.

General Electric (GE) Astro Space Division, King of Prussia, Pennsylvania, designed the present RTG and has supplied the generators for the fueling and testing of the latest NASA missions. The fuel is a plutonium-oxide powder that is pressed into ceramic fuel pellets approximately 2.5 cm high, 2.5 cm in diameter and clad in iridium. Each fueled clad generates 62.5 W, and a fully fueled generator contains 72 fueled clads. This, then, produces 4500 W and a dose rate of approximately 1.2 mSv/h (120 mR/h) at 0.61 m. The fixturing originally developed at Mound and GE has been modified to improve handling efficiency and minimize operator doses by reducing the time operators must spend in close proximity to the fueled generator.

HANDLING REQUIREMENTS

Following fuel loading, the generator must be placed in various configurations (inboard end up, inboard end down, and the generator in a horizontal configuration) for testing and evaluation. Mound devised a system that operated effectively; however, with the ever-increasing awareness of the potentially harmful effects of radiation, the RPSF is making every effort to provide shielding, minimize handling of the radioactive components, and have improved efficiency when direct contact with the components is necessary. In fact, all operations at the RPSF are designed to a neutron quality factor of at least 20.

The RPSF is divided into two primary, interrelated engineering organizations—GPHS Engineering and RTG Engineering. General purpose heat source (GPHS) modules consist of four fueled clads assembled into a graphite block. The GPHS Engineering group is responsible for the design of RPSF processes involved in receiving, inspecting, preparing, and assembling the fuel into the GPHS modules. The RTG Engineering designs the systems for loading the GPHS modules into the generator and testing the fueled generator. Although the transport and storage system is part of GPHS Engineering, it supports both organizations. It is transport and storage that is responsible for handling fuel, GPHS modules, and RTGs.

Figure 1 shows the basic RTG transport cart configuration. The cart base provides support for the rotation frame. The generator (approximately 57 kg) is held in the rotation frame by the handling fixture. The rotation frame allows the generator to be rotated 360° and to be handled vertically with the inboard end up or down and horizontally. Figure 2 provides a more detailed view of the rotation frame and handling fixture, showing how the two interface. This interface allows the generator to be supported as well as installed and removed from the frame with the generator in either of its vertical positions or in the horizontal position. It also allows the generator to be rotated around its longitudinal axis when it is in the inboard end up position. Only one generator transport cart base and one rotation frame are required for each generator in the RPSF facility.

As a minimum, all handling fixtures consist of a block of aluminum machined to provide a groove straddled by a shoulder and a chamfered flange. The shelf plate and pivot arm on the rotation frame are positioned in the groove between the shoulder and the flange. As the slack is taken up when a clamp tightens the pivot arm, the handling fixture becomes firmly supported within the rotation frame. The flange supports the generator when the rotation frame is in the up position and the shoulder supports it when the frame is in the down position. Two other quick release clamps ensure the pivot arm stays secure. This is the interface that provides the major departure from previous handling techniques. The outboard handling fixture and inboard handling sling remain identical to the original design below the flanged end.

The inboard handling fixture replaces all other original inboard fixturing. Except for some specific moves, the generator will be handled at the inboard end. Once the inboard handling fixture is installed while the generator is still in the inert atmosphere assembly chamber (IAAC) where the fuel is loaded into the generator, the fixture stays installed until the generator is readied for shipping. This particular design removes the need to change the hardware constantly. The massive end provides a solid and structurally stiff interface with the test equipment during vibrational testing. The support shoulder provides a support for the generator in the down and horizontal positions as well as providing stiffness to the fixture itself during vibration testing.

TRANSPORT AND HANDLING OF THE RADIOISOTOPE THERMOELECTRIC GENERATORS

Once the unfueled generator is received in its shipping container, it will be examined for damage that might have occurred during shipping and then be stored in the shipping container until the fuel is to be loaded. The generator mounts to the shipping cart base assembly at its inboard end. The inboard end is the strongest part of the generator because it is this end that is mated to the satellite and is therefore the end preferred for handling. To install the generator in the IAAC, the outboard handling fixture will be attached to the outboard end. Using a crane connected to the outboard handling fixture, the generator will be lifted out of the base assembly and installed on the waiting transport cart (with the rotation frame in the up position). This will be done by lowering the generator such that the handling sling and rotation frame mating surfaces are up. Then either by rolling the cart to the generator or using the crane to move the generator horizontally, the generator and handling fixture will be slid into position in the rotation frame slot.

Once secured by the pivot arm and disengaged from the crane, the generator then will be rotated inboard end up. The inboard handling sling and fixturing, which supports the generator in the IAAC, will be installed and the rotation frame pivot arm released. At this point, the generator will be supported by both the crane and the rotation frame. The generator will be lifted slightly, and the cart can be rolled away from the generator. The rotation frame then will be rotated to the up position and the cart rolled back under the generator. Following the same procedure described above, the inboard handling sling will be positioned in the inboard frame, the pivot arm secured, and the crane hook disengaged. After the outboard handling fixturing is removed and support fixturing is attached, the generator will be transported into the IAAC.

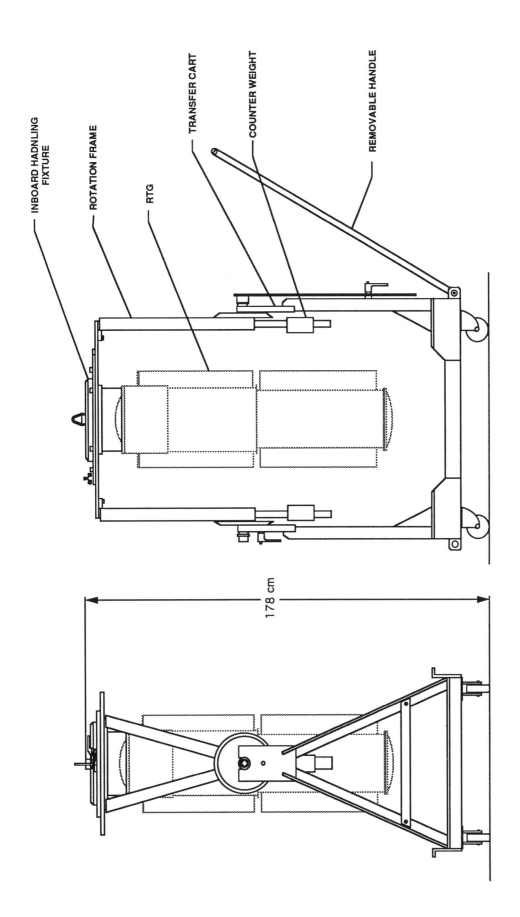

FIGURE 1. Radioisotope Thermoelectric Generator Transport Cart, Rotation Frame, and Inboard Handling Fixture.

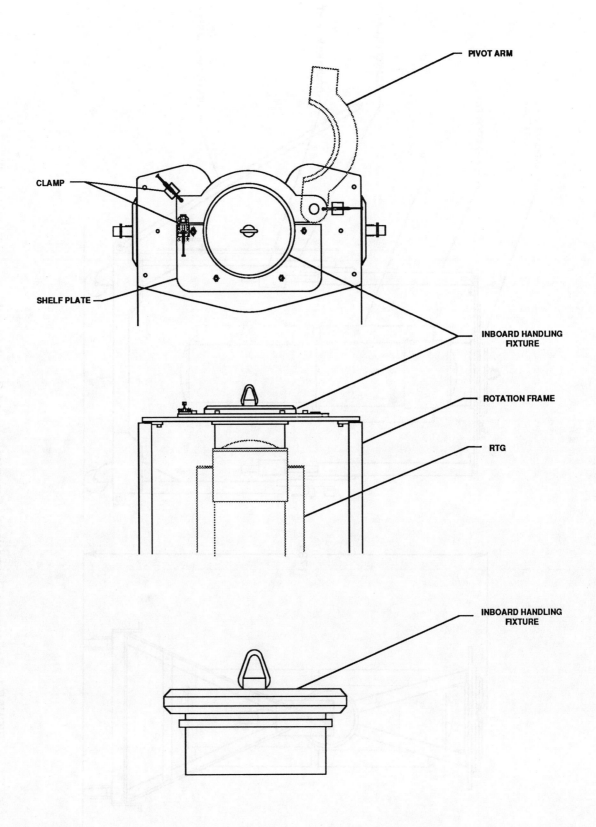

FIGURE 2. Handling Fixture and Rotation Frame Interface.

In the IAAC, the IAAC crane will slide the generator horizontally out of the rotation frame and cart. Once clear, the generator will be placed into the IAAC bell jar. Once the fuel has been loaded and the generator is resealed again, the generator then is referred to as a radioisotope thermoelectric generator, or RTG. To remove the RTG, the inboard handling fixture will be installed and the RTG transferred to the waiting transport cart.

Out of the IAAC, the RTG will be handled almost exclusively by the inboard handling fixture. After radiation dose measurements are taken with the RTG in both the vertical and horizontal positions, the outboard handling fixture will be reattached and the generator transported down the hallway to the vibration and mass properties room where it will be lifted from the cart by its outboard end and loaded onto the vibration table. The outboard fixture must be removed for these tests.

When the vibration testing is complete (both vertical and horizontal vibrational tests) the RTG will be transferred by crane (lifted by the outboard end) to the mass properties frame (identical to the rotation frame). The RTG will be connected to the frame at the inboard end. For mass properties, the outboard fixturing can remain on the RTG. Following the measurements, the RTG will be transferred by crane back to the transport cart and rotation frame and transported down the hall to magnetics testing. Here, the RTG and rotation frame will be lifted off the transport cart using a simple, manually controlled forklift. The RTG in the rotation frame will be placed on the magnetics table. The outboard fixturing remains on the RTG. After testing, the forklift again will be used to remove the RTG and rotation frame from the table. The forklift will either place the RTG back onto the transport cart or it can be used to transfer the RTG to the adjoining room for thermal vacuum testing.

For thermal vacuum testing, the RTG will be rotated to a horizontal position and the forklift will support the RTG until it is connected at the inboard and outboard fixtures to chain supports within the chamber. The pivot arm will be released, and the forklift will lower the rotation frame away from RTG. This process will be reversed when retrieving the RTG from the chamber. The generator will be rotated back to the vertical position and set onto the transport cart.

When transporting the RTG relatively long distances within the RPSF, the transport cart (with the RTG) will be placed inside a three-sided portable shielding unit. The portable shielding consists of 2-m-high stainless steel plate (1-cm-thick) encasing 15 cm of water. The cart has a ledge on the inside that interfaces with the angle on the lower part of the transport cart. A battery-powered lift truck will lift the shielding, and the shielding will pick up the transport cart. This way, the RTG can be transported along the corridors with minimal exposure to the operators walking along the side and behind the RTG. At the destination, the shielding will be lowered to the floor and the cart simply rolled out. A simple restraint ensures the RTG stays securely inside the shielding. This system allows the shielding to be brought up to the process room door to either pick up or drop off the generator. With the top, bottom, and one side open, enough air is allowed to circulate to keep the RTG properly cooled.

CONCLUSION

The system for handling and transporting RTGs within the RPSF will be an efficient and operator-oriented process. The generator interfaces have been standardized as much as practical to reduce the number of handling fixtures minimizing the number of varying and detailed procedures. The design allows a single rotation frame design to be used for making measurements as well as for transporting and, ultimately, storing the RTG while it is at the RPSF.

Acknowledgments

This work was sponsored by the U.S. Department of Energy, Office of Special Applications under contract number DE-AC06-87RL10930.

Many thanks to Ron Eschenbaum, Betty Carteret, and Dave Clark for their valuable input and reviews.

References

Alderman, C. J. (1990) *Assembly of Radioisotope Power Systems at Westinghouse Hanford Company*, Society of Women Engineers 1990 National Convention and Student Conference, New York, N.Y.

Amos, W. R. and C. J. Goebel (1990) "Assembly of Radioisotope Heat Sources and Thermoelectric Generators for Galileo and Ulysses Missions," in Proceedings, *8th Symposium on Space Nuclear Power Systems*, CONF-900104, M. S. El-Genk and M. D. Hoover, eds., University of New Mexico, Albuquerque, N.M.

Eschenbaum, R. C. and M. J. Wiemers (1990) "Design of Radioisotope Power System Facility," in Proceedings, *8th Symposium on Space Nuclear Power Systems*, CONF-900104, M. S. El-Genk and M. D. Hoover, eds., University of New Mexico, Albuquerque, N.M.

QUALITY ASSURANCE SYSTEMS EMPLOYED IN THE ASSEMBLY AND TESTING OF RADIOISOTOPE THERMOELECTRIC GENERATORS

William A. Bohne
EG&G Mound Applied Technologies, Inc.
Post Office Box 3000
Miamisburg, OH 45343-3000
(513) 865-4523

Abstract

Described are the quality assurance systems and techniques used at EG&G Mound Applied Technologies, Inc., for the assembly and testing of radioisotope thermoelectric generators (RTGs) for space and terrestrial power applications. The critical role that RTGs play in space and terrestrial missions, their inaccessibility for repair or replacement, and the use of radioactive materials for fuel require that every aspect of Mound's RTG programs, from the design of tools, fixtures, and facilities to the delivery of the unit to the end user, be under stringent and documented control systems. Moreover, these systems must be in compliance with the Department of Energy (DOE) orders governing the performance of non-weapons activities at contractor sites. This paper discusses some of the quality systems in place at EG&G Mound for RTG programs, describes the roles of the Operations and Quality Engineering personnel in the implementation of these systems, and shows how these systems interact with each other.

INTRODUCTION

For over 30 years, the Department of Energy's (DOE's) Mound Plant in Miamisburg, Ohio, has been involved in the assembly of radioisotope heat sources for a variety of programs. In the late 1970s this role was expanded to include the assembly and testing of radioisotope thermoelectric generators (RTGs). The first assignment in this area was to assemble and test the RTGs to be used on the National Aeronautics and Space Administration's (NASA) Galileo mission to Jupiter and the joint NASA/European Space Agency (ESA) Ulysses mission to the polar regions of the sun. This resulted in the development of a quality plan, along with the appropriate control systems, to satisfy DOE's non-weapon's quality orders, NASA's space quality philosophy, and the quality requirements of the systems contractor (General Electric Astrospace Division).

This paper describes the quality systems that were implemented at Mound for the Galileo and Ulysses RTGs. The success of these systems resulted in their subsequent use on Mound terrestrial RTG programs.

TOOLING AND FIXTURE DESIGN, PROCUREMENT, ACCEPTANCE, AND CONTROL

One of the major quality efforts was the establishment of a system for the design, procurement, acceptance, and control of tooling, fixtures, and hardware. This was necessitated by the concerns that the items used to assemble and test the RTGs were compatible with the RTG and its assembly/test environment, the items would be functional, all design attributes were correct, and traceability and control of the items would be maintained so as to prevent damage, substitution, inadvertent alteration, or unauthorized use.

One of the first steps in developing this system was the realization that not all items would require the same degree of control. Two categories of items, identified as critical and noncritical, were subsequently developed. It was further decided that Quality Engineering would establish and maintain files for all equipment determined to be critical, while the maintenance of documentation for the noncritical items would be the responsibility of Operations. Generally, any item that requires dimensional inspection, calibration, proofloading, material certifications, or special cleaning and inspection for acceptance is classified as a critical item, while those items that only require functional testing and standard cleaning (wipe with alcohol) are identified as noncritical items. The critical items are further subdivided into items that will be controlled by Mound's Material Control personnel and those items that will not be controlled (rack mounted control instrumentation) or cannot be controlled (permanently mounted room hoists, large handling fixtures) by Material Control.

The next step in the process was to establish the criteria for tool and fixture design reviews and approvals. Since Mound previously had not been involved in the fueling and testing of generators, it was agreed that Quality Engineering and the systems contractor would be required to review and approve the original issue of drawings, and that these drawings would be entered into Mound's formal drawing control system. It was also concluded that Quality Engineering and the systems contractor approve all changes to these drawings.

For the procurement and acceptance of items, a system was developed whereby all purchase requisitions had to be routed through Quality Engineering for review and approval. Under this system, the Quality Engineer makes a determination whether quality requirements have to be defined. If quality requirements are appropriate, the Quality Engineer defines the certifications or vendor documentation that is required, the in-house inspections and tests to be performed, and the type of cleaning required. Also, because all items to be used in the assembly or test operation are required to be delivered directly to Material Control, Quality Engineering has to initiate a routing sheet instructing Material Control how to handle the item to ensure that all identified inspections and tests are completed. For example, the routing sheet might instruct Material Control to submit vendor documentation received with the item to Quality Engineering while the item itself would be sent first to dimensional inspection, then released to Operations for functional and proof testing, then submitted to the Cleaning Lab for cleaning and inspection, and then placed in hold storage pending Quality review of inspection and test data. This system ensures that all movements and transfers of the item within the plant are documented, and allows Quality Engineering to review all documentation attesting to the functionality and configuration of the item prior to the acceptance of the item into "bonded stores", from which it can be issued for use in an assembly or test operation. Once issued for use, the item is controlled by the formal procedure governing the particular assembly or test operation. The tool files are continually updated as the items are routed through the inspection, calibration, and cleaning groups as the items are reprocessed in preparation for the next assembly or test activity.

CONFIGURATION CONTROL

Configuration control on RTG programs at Mound is accomplished through Mound's Engineering Change Notice (ECN) and Product Index (PI) systems. Under the ECN system, changes to controlled drawings, procedures, and specifications must be reviewed and approved by Operations and Quality Engineering, at the minimum. Approvals by the systems contractor or design agency, the DOE area office, DOE headquarters, Risk Management, Safety, and Health Physics could also be required depending upon the document being changed and the nature of the change.

The PI is a computerized listing of all the drawings, specifications, and procedures pertaining to an assembly or test operation. As changes are approved, the ECNs and new issue numbers are added to the PI by Operations. Operations and Quality Engineering verify that the current issue of these documents are being used on-line for the assembly or test operation. Quality Engineering also uses the PI to verify the acceptance of a tool or fixture has been performed to the current drawing issue. If a discrepancy between the PI, the drawing issue, and the acceptance documentation in the tool file is noted, either the PI must be corrected, the tool or fixture must be made to conform to the current drawing requirements, or the acceptance inspections and tests must be performed on the current configuration. Quality Engineering does have the right to waive repetition of the inspection and acceptance tests if an engineering determination is made that the change does not invalidate the previous tests. Such determinations must be documented in writing and are included as part of the documentation in the tool file.

MATERIAL REVIEW/FAILURE REVIEW BOARDS

Another facet of Mound's quality program are the review boards, the Material Review Board (MRB) and the Failure Review Board (FRB). MRB membership varies slightly with the various programs, but at a minimum includes Quality Engineering, Operations, the systems contractor or design agency, and the DOE area office. For flight RTG programs, the MRB membership also includes a Technical Liaison and a Risk Management representative.

The membership of the FRB also varies, and is dependent upon whether or not the board is a Mound FRB or a Design Agency FRB. The Mound FRB addresses unresolved Mound MRB actions and those failures relating to assembly and test operations conducted at Mound, and generally consists of management level personnel from Operations (who serves as chairman), Quality Engineering, the systems contractor or design agency, the DOE area office, DOE headquarters and, for flight programs, Risk Management. The Design Agency FRB addresses unresolved MRB actions and those failures that relate to the converter or design of the RTG. Membership is defined by the systems contractor, but generally includes a Mound representative. Both of these boards can only provide recommendations to the MRB, and the MRB has the option to accept, reject, or alter the recommendations provided.

NONCONFORMANCES AND CORRECTIVE ACTION

The primary system used for reporting nonconformances and documenting corrective action on RTG programs is the Problem/Failure Review (P/FR) system. This system is used to document and resolve anomalies, nonconformances, and failures related to tools, equipment, facilities, hardware, and product, and also provides a means for issuing instructions for the prevention of future occurrences of the problem.

On RTG programs, an anomaly is defined as a condition or result that deviates in excess of normal variation, as defined by the Operations and Quality engineers, but is still within specification limits. A nonconformance is defined as a condition or result that deviates from a requirement as stated in a drawing, specification, or procedure. Nonconformances are further subdivided into Class I and Class II nonconformances, with a Class I action being defined as a nonconformance that could adversely affect safety, performance, interchangeability, qualification status, or interface characteristics of the RTG. Class II actions are defined as being all other nonconformances. A failure is defined as an occurrence in which a component of the RTG breaks down or which causes the RTG performance to fall outside of acceptable limits.

On of the unique features of Mound's P/FR system is that it differentiates between nonconformances detected during preproduction activities and those nonconformances detected during production. Under the P/FR system, a nonconformance detected during production requires that operations proceed to the nearest safe-hold point, at which time operations cease and the problem is addressed by the MRB. Operations cannot resume until authorized by the MRB. A nonconformance detected during preproduction, on the other hand, does not involve a RTG, and therefore can be dispositioned by the Operations and Quality engineers unless the disposition is to use as is or to rework not to the drawing, or unless the nonconforming material will contact the RTG or become part of the RTG. In these instances the nonconformance is forwarded to the MRB. This feature of the system keeps the MRB from getting tied up with minor problems, such as decisions to rework tools and fixtures, but still ensures the integrity of the hardware and materials that are critical to the RTG assembly or test process.

The P/FR system also provides for the escalation of a problem to the FRB if the problem is determined to be a failure or cannot be resolved by the MRB. There are two failure review boards identified by the system, these being the Mound FRB and the Design Agency FRB.

The P/FR form has four parts (Figure 1). Part I of the form provides for the identification of the nonconforming item or process as well as the documentation of the problem encountered. Part II of the form, completed by Operations and concurred to by Quality Engineering, provides for an investigation into the cause of the problem along with a recommendation for the disposition of the item or process and a corrective action to prevent recurrence. Part II also includes a section for the verification of the completion of the disposition and corrective action by Quality Engineering for those anomalies and nonconformances which do not require review by the MRB. Part III of the form allows for MRB review of the problem, investigation, disposition, and corrective action, as well as the escalation of the problem to the FRB if the problem is determined to be a failure or can not be resolved by the MRB. Again, this section of the form includes provisions for the verification of the disposition and corrective action by the MRB chairman (the Quality Engineer) prior to the closeout of the form. The fourth part of the form is used for FRB actions. Since the FRB generally issues a report separate from the P/FR, this section of the form only identifies the FRB membership and the documents forwarded to the MRB for consideration.

PERSONNEL TRAINING

Another area that receives a considerable amount of attention is the area of personnel training. On RTG programs, a system was developed whereby the Operations and Quality Engineering representatives responsible for particular assembly and test operations develop training plans for their personnel. These training plans identify those segments of the assembly or test operation for which training is required, specify how the training is to be conducted, and identify the criteria (usually either demonstration of ability or a test) against which personnel are evaluated. Once personnel demonstrate their ability to perform specific operations, the supervisor of the group issues a letter qualifying that person for the operation. A copy of this qualification letter, or other similar documentation, is then posted in the work area. This qualification remains in effect until job performance warrants disqualification or if personnel have not performed a particular task within six months. This system has become the standard system for personnel training at Mound, meeting the latest DOE

EG&G MOUND APPLIED TECHNOLOGIES

Page _____ of _____

P/FR # _____

PROBLEM/FAILURE REVIEW

REPORT ORIGINATOR	REPORT DATE	P/F DATE OCCUR.	ITEM IDENTIFICATION/DWG. #/S/N	ISSUE # (Circle One) 0 1 2 3
ITEM DESCRIPTION			P/F NOTED DURING PROCEDURE/PARA	TIME

Part I

1. Specific Environment at the time of P/FR (mode and status of HS/RTG):

2. Description of Failure, Nonconformance or Anomaly:

3. Witness(es) _____

Part II

INVESTIGATION REPORT:
Interviewee(s) _____

4. Specification or Process Requirement (Include name and revision):

5. Summary of Investigation: ☐ Non-Conformance; ☐ Anomaly; ☐ Class I; ☐ Class II; ☐ Failure

6. Recommended Disposition:

7. Cause of Problem:

8. Recommended Corrective Action To Preclude Recurrence:

9. Safety Hazard:

10. Effectivity: ☐ This Item ☐ All Items   Unit _____   To MRB ☐ Yes ☐ No
                                                                      CAR   ☐ Yes ☐ No

OPERATIONS	DATE	QUALITY ENGINEER	DATE
CORRECTIVE ACTION/DISPOSITION VERIFIED/COMMENTS		QUALITY ENGINEER	DATE

ML-7353A (7-90)

FIGURE 1. PROBLEM/FAILURE REVIEW FORM.

EG&G MOUND APPLIED TECHNOLOGIES
PROBLEM/FAILURE REVIEW

Page _____ of _____
P/FR #
ISSUE # (Circle One)   0 1 2 3
DATE SUBMITTED TO MRB
DATE MRB MEETING HELD/TIME
CLASS

Part III

HS/RTG MRB ACTION
11. Interviewee(s) _____

12. Details of Investigation:

13. MRB Disposition: ☐ Interim   ☐ Final

14. MRB Corrective Action To Preclude Recurrence:

RISK MGMT.	DATE	TECH. LIAISON	DATE	LANL	DATE	FRB ☐ Yes ☐ No
						CAR ☐ Yes ☐ No
OPERATIONS	DATE	QUALITY ENGINEER	DATE	DOE/DAO	DATE	DESIGN AGENCY	DATE
DOE/OSA (Class I Only)	DATE	MGR., NUCLEAR TECH. (Class I Only)	DATE	SRS	DATE	_____	DATE

CORRECTIVE ACTION/DISPOSITION VERIFIED/COMMENTS:

MRB CHAIRMAN   DATE

Part IV

☐ Mound   ☐ Design Agency
FRB ACTION
15. MRB Members _____
    and Organizations _____

16. FRB Investigation Report and Additional Documents Submitted to MRB on _____ (Date).

DATE SUBMITTED TO RTG-FRB
DATE (S) FRB MEETING(S) HELD

ML-7353B (7-90)

FIGURE 1. PROBLEM/FAILURE REVIEW FORM (CONT'D).

orders regarding personnel training and certification. The system has been expanded to encompass not only "hands-on" training, but also training on administrative, safety, and health systems and requirements.

SYSTEM INTERACTIONS AND RELATIONSHIPS

An example of how some of these systems interact with each other is best illustrated by the flow diagram of Figure 2, which shows the path an RTG assembly tool might follow from its design inception through its use. Also identified is the related documentation that controls the procurement, acceptance, and use of the tool.

SUMMARY

The elements of Mound's RTG quality plan, which address procurement, personnel training, configuration control, and nonconformances, are systems which interact and support each other to form a DOE-accepted quality program. This program has been successful in the development and qualification of facilities, equipment, procedures, and processes to fuel and test one flight qualification RTG, four flight RTGs, and five terrestrial RTGs. The need for a space quality program plan at Mound, recognized in 1983, has resulted in a quality program whose basic features have remained unchanged despite a recent complex-wide realization of the need for non-weapons quality assurance and the issuance of recent DOE orders regarding the quality elements and systems required for a complete quality assurance program. This attests not only to the quality concepts and planning that went into the development of the Mound systems, but also points to the advanced nature of the space quality philosophy that now seems to be becoming an industrial standard.

Acknowledgments

Mound is operated by EG&G Applied Technologies for the U.S. Department of Energy under contract DE-AC04-88DP43495.

References

EG&G Mound (1991) <u>Quality Assurance Plan for Heat Source/Radioisotope Thermoelectric Generator Programs, Issue 6</u>, Technical Manual MD-10260, EG&G Mound Applied Technologies, Miamisburg, OH, 30 May 1991.

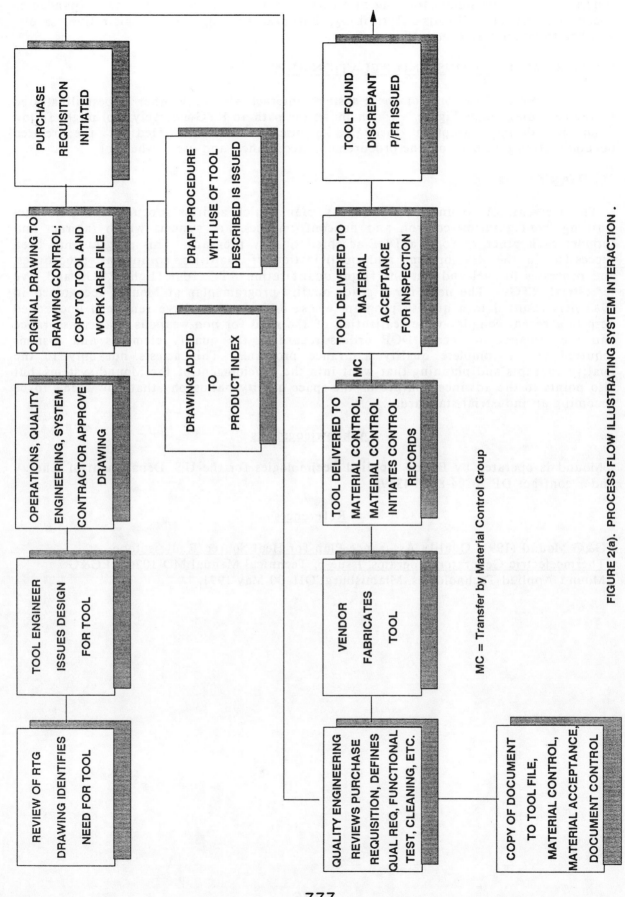

FIGURE 2(a). PROCESS FLOW ILLUSTRATING SYSTEM INTERACTION.

MC = Transfer by Material Control Group

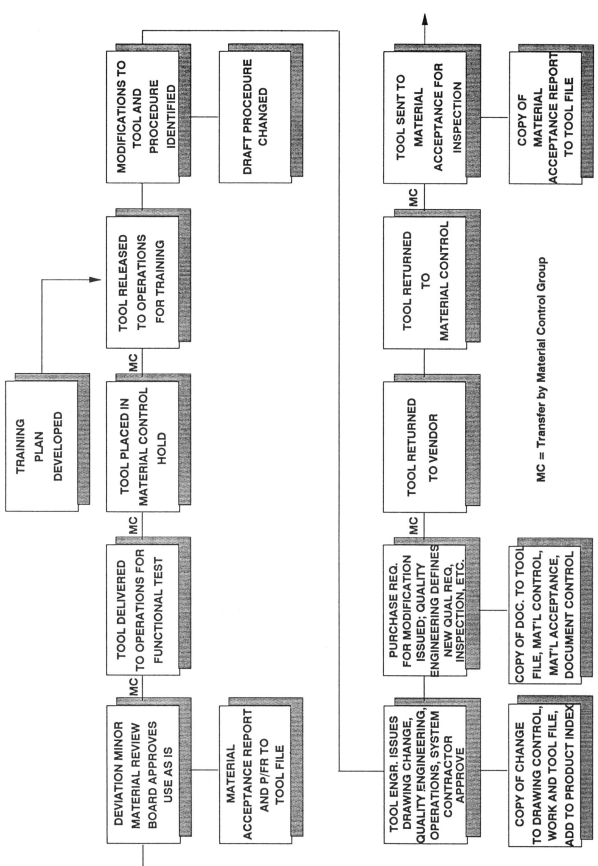

FIGURE 2(b). PROCESS FLOW ILLUSTRATING SYSTEM INTERACTION (cont'd).

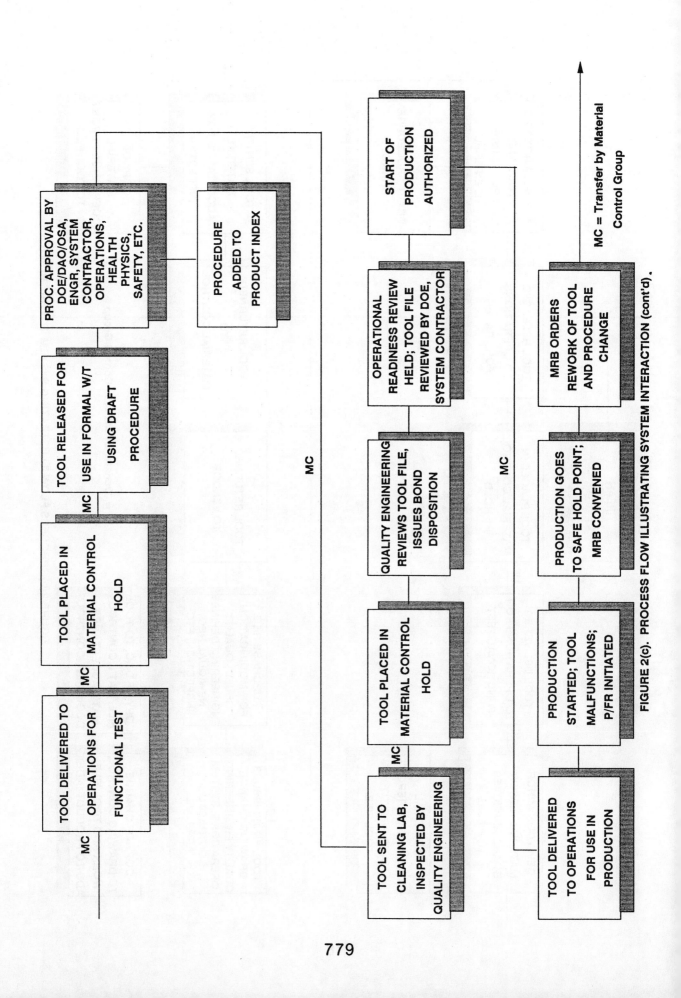

FIGURE 2(c). PROCESS FLOW ILLUSTRATING SYSTEM INTERACTION (cont'd).

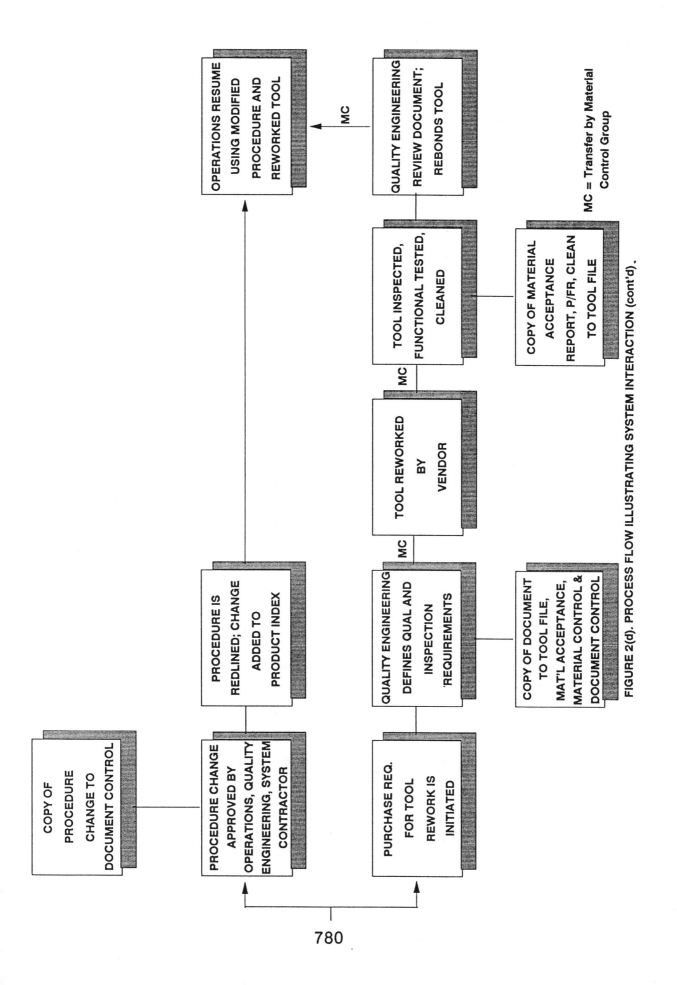

FIGURE 2(d). PROCESS FLOW ILLUSTRATING SYSTEM INTERACTION (cont'd).

LINER PROTECTED CARBON–CARBON HEAT PIPE CONCEPT

Richard D. Rovang and Maribeth E. Hunt
Rockwell International, Rocketdyne Division
6633 Canoga Avenue
Canoga Park, CA 91303
(818) 718-3382

Abstract

A lightweight, high performance radiator concept using carbon–carbon heat pipes is being developed to support space nuclear power applications, specifically the SP-100 system. Carbon–carbon has been selected as an outer structural tube member because of its high temperature and strength characteristics; however, this material must be protected from the potassium heat pipe working fluid. A metallic liner approach is being taken to provide this fluid barrier. Feasibility issues associated with this approach include materials compatibility, fabrication of the thin-walled liner, bonding the liner to the carbon–carbon tube, mismatch of coefficient of thermal expansion (CTE), carbon diffusion, and end cap closures. To resolve these issues, a series of test coupons have been fabricated and tested, assessing various liner materials, braze alloys, and substrate precursors. These tests will lead to a final heat pipe architecture, material selection, and component assembly.

INTRODUCTION

A program funded by the National Aeronautics and Space Administration-Lewis Research Center (NASA-LeRC) to develop an advanced SP-100 radiator has been undertaken by Rockwell International and its subcontractors — Science Application International Corporation (SAIC) and Delta-G. Overall objectives of this effort are to advance the state of the art and develop the technology base for large, high temperature radiator subsystems. More specifically, near-term objectives are to demonstrate advancements in radiator design, performance, packaging, weight, cost, life, reliability, and survivability. The basic radiator concept uses lightweight, high performance, carbon–carbon heat pipes, with integrally woven fins, potassium working fluid, and a metallic liner to protect the carbon–carbon tube from the working fluid.

In previous work (Rovang et al. 1991) under Phase III of the current contract, a chemical vapor deposition (CVD) process had been developed to provide a protective coating of niobium on the carbon–carbon, serving as the potassium isolation barrier. Although substantial progress was made, a defect-free coating over the entire inside tube surface was not achieved with the stationary CVD furnace approach. Because funding was not available at this time to develop the moving heat source CVD technique, an alternate barrier approach, using a free-standing inspectable liner, was selected. A discussion of the overall radiator concept, liner development issues, the technical approach, and results are presented.

OVERALL RADIATOR CONCEPT DESCRIPTION

Top level requirements for this high temperature, advanced radiator design include the following:

1. Absence of single-point failure modes;

2. Capability to reject 2.4 MW of thermal energy with an inlet temperature of 875 K to an effective sink temperature of 250 K; and

3. Ability to meet all safety and compatibility criteria and fit within two-thirds of the shuttle bay.

Radiator Design

To meet these requirements while minimizing weight, a carbon–carbon heat pipe radiator design was selected. The concept uses 12 independent lithium loops to transfer heat from the SP-100 thermoelectric generator to the heat pipe radiators. A tapered manifold is used to contain the transport lithium. The taper is proportioned to heat pipe condenser length; that is, the width of the manifold is equal to the respective evaporator lengths. The heat pipes use a potassium working fluid and operate well within the predicted entrainment, boiling, sonic, and capillary limits.

The radiator panels use carbon-carbon composite heat pipes with a thin metal liner on the inside diameter of the pipe to prevent interactions between the base materials and the heat pipe working fluid. The panel is designed to be easily integrated into the current SP-100 radiator configuration and is adaptable to a wide variety of radiator applications.

Heat Pipe Architecture

A cross section of the proposed heat pipe is shown in Figure 1. The two major elements are the carbon-carbon tube with integrally woven fins and the metallic liner. The carbon-carbon reinforcement fiber selected for the fabrication was Amoco T-300 carbon fiber. This polyacrylonitrile (PAN) fiber was selected because it exhibits an acceptable compromise among thermal conductivity, elastic modulus, and ease of handling and weaving. The finned tube was woven from 3000 filament yarn using an angle interlock fiber architecture. Good through-the-thickness thermal conductivity of the tube is achieved by the axial yarn weavers, which angle between the tube ID and OD. These axial yarns also increase the interlaminar shear and radial tensile strengths of the tube. The outer layer of hoop (fill) yarns in the tube forms the transverse yarns for the fins, thus ensuring good heat transfer from the tube to the fins.

Figure 1 shows a 1.27 cm fin; however, this fin width will vary with specific application and packaging optimizations. Under the current work, carbon-carbon tubes with 6.8 cm fins are typically produced, then trimmed to the desired width.

LINER DEVELOPMENT ISSUES

Although switching from a CVD barrier approach to an inspectable liner resolved one major issue (the ability to inspect the liner), most of the development issues associated with the CVD approach remain and several other issues have been added. Obviously, the liner material must be compatible with the potassium two-phase working fluid at the operating temperature of 875 K. Chemical corrosion, mass transport, mechanical erosion, and the combination of these effects must be considered. The material selected must be capable of being fabricated into defect-free tubes with a wall thickness of 50 to 75 µm (2 to 3 mils). Outside liner dimensions must be held to a prescribed tolerance to allow the liner to be inserted into the carbon-carbon tube without deforming (buckling) but with a fit close enough to minimize the amount of braze or filler required so that the heat pipe mass goals are not compromised. The fabrication process must also be sensitive to potential contamination of the metal. Weldability is an issue for most of the liner concepts because this is the preferred approach for attaching end closures.

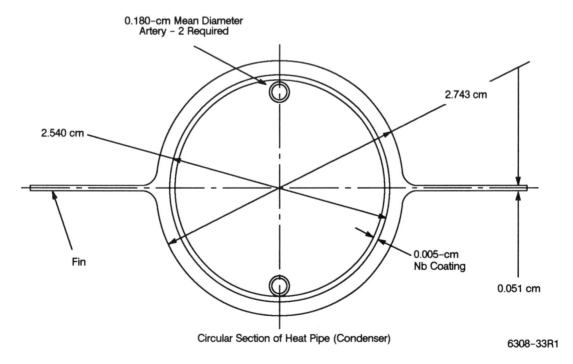

FIGURE 1. Potassium Heat Pipe Design.

Bonding of the liner to the carbon–carbon tube serves two purposes: it provides mechanical integrity to the system, and, to a large extent, it affects through-thickness thermal conductivity of the composite. The bonding method selected is a major factor in determining the heat pipe architecture. Three candidate bonding approaches are being investigated, including brazing, diffusion bonding, and a free-floating approach that relies on thermal expansion of the liner to provide sufficient contact pressure with the carbon–carbon tube wall. A common variable to be considered with these approaches includes the use of precursor coatings. Typically, a carbon–carbon surface is rather rough and slight deviations in the tube ID are expected. To provide a smooth surface and to correct any dimensional inconsistencies without excess breaking of fibers, a pyrolytic carbon coating can be used. This would be applied to the inside of the carbon–carbon tube, cured, then machined or honed to the appropriate finish and tolerance. This process improves fitup and contact area with almost no mass penalty, but, at the same time, it introduces more thermal resistance because of the insulating properties of the pyrocarbon.

A thin (several micrometers) metal CVD coating on the carbon–carbon substrate may improve the mechanical integrity of the braze joint and is necessary to affect a diffusion bond. In previous work, a precursor rhenium layer demonstrated excellent bonding to the carbon–carbon tube. This technique may be applicable with or without a pyrolytic carbon coating.

The interaction of carbon from the carbon–carbon substrate with the liner material is a major concern. In the previous CVD work, diffusion of carbon into the barrier coating was observed, producing a brittle carbide at the carbon–carbon coating interface. The question which then arises is whether the interface carbide layer provides an adequate diffusion barrier or if carbon diffusion continues at a significant rate over the 10-yr life of the radiator. Continued diffusion could have two deleterious effects: (1) conversion of the metallic liner into a carbide causing embrittlement and ultimate failure, or (2) diffusion of carbon to the surface of the liner where it would be picked up by the potassium working fluid and degrade heat pipe performance and service life.

APPROACH

A combination analytical and empirical approach is being taken to resolve the identified feasibility issues. Much of this work builds on results obtained from the referenced Phase III fabrication and testing experience. The goals of the Phase IV effort are to produce a prototypic 30.5 cm heat pipe section, demonstrating all required fabrication and assembly procedures and to verify the selected heat pipe architecture performance. To achieve these goals, a program consisting of four major activity areas has been undertaken, including architecture selection, component fabrication, heat pipe assembly, and testing.

Selection of a heat pipe architecture requires that various candidate liner materials be analyzed with respect to performance in the predicted heat pipe environment for the 10-yr radiator life. A downselect to two candidate liner materials is made based on this analysis. Trade studies and measurements are performed to determine the applicability of potential enhancement options, such as pyrolytic carbon coatings and a CVD Re coating. Assessment of the pyrolytic coatings involves investigation of methods to apply the coating, followed by experimental evaluation. Samples have been prepared and assessed for bond strength, machining capability, and compatibility with other processing steps.

A CVD rhenium coating has also been evaluated to determine its producibility and quality in combination with the pyrolytic carbon coatings. Sample coupons were fabricated, sectioned, examined, and measured for conductivity.

Sample specimens have been prepared to evaluate brazing materials and techniques. Parameters varied in this set of experiments include the use of the braze alloy, a Re precursor, a pyrolytic carbon coating, and alternate liner materials. These combinations have been assessed for bond strength, stability (which includes carbon diffusion estimates), thermal performance, mass, and fabrication capability. Diffusion bonded samples are also expected to be produced using the two selected liner materials and Re-coated carbon–carbon coupons.

In parallel with the bonding and architecture development tasks, carbon–carbon tubes with integrally woven fins are being fabricated. Enough tubing will be produced to provide samples for additional bonding and assembly experiments, plus material for the final heat pipe sections. A full description of the carbon–carbon fabrication process is given in (Rovang et al. 1991).

Liner fabrication techniques have been evaluated, downselections made, and test specimens produced, using the top rated candidate methods. Metallographic, chemical, and physical measurements are performed on these test specimens, leading to selection of the final liner fabrication process. Several full length liners will then be fabricated for additional testing and use in the final assembly.

A program to develop an optimum wick for this particular heat pipe will not be undertaken at this time. A standard type wick will be selected based on proven designs in similar application; that is, a 2.5 cm diameter tube, potassium working fluid, 825 K to 900 K operating temperature, and evaporator heat flux of 15 to 30 W/cm^2.

Integral with the design and fabrication of the liner, end cap closures must be developed along with appropriate welding methods. The end cap design should provide adequate access for the weld machine. A task to develop this procedure and qualify the weld has been planned.

With all the heat pipe components fabricated and inspected, assembly of the heat pipe will be initiated. Components are cleaned, the liner bonded, the wick installed, and end caps and fill tubes welded. Nondestructive examinations (NDE) will be performed at this point to ensure that all specifications are met and that the tube is leak tight. Filling of the heat pipes will be done in a vacuum, because this is a high temperature operation that precludes exposure of the carbon-carbon to oxygen. After filling, additional NDE will be performed to ensure proper fill and the integrity of the potassium containment barriers. The filled heat pipes then undergo initial checkout and validation again in a vacuum chamber at nominal operating temperatures.

RESULTS TO DATE

As the first step in selecting a liner material, an analytical assessment was made of the potential candidates, including nickel, titanium, niobium, molybdenum, tantalum, tungsten, and related alloys. Based on the physical, chemical, and metallurgical characteristics of these materials and the historical data base, two liner materials were selected for further development: niobium – 1%Zr and titanium. Thin foils of these materials were obtained for testing purposes.

With this liner material and representative carbon-carbon samples, 2.5 cm by 2.5 cm test coupons were assembled using various braze alloys. Alloys tested included copper ABA, Palcusil 15, silver ABA, Gapasil 9, Cusil ABA, Cusil, and Ticusil. The melting temperature of these alloys span a temperature range of 1080 to 1310 K. With 0.005 cm Ti liner material, the Cusil, Cusil ABA, and silver ABA showed good wetting of both the carbon-carbon and the Ti. Microstructure observations indicate that a small amount of Ti from the liner was alloying with the Cusil to facilitate wetting of the carbon-carbon. Figure 2 is a typical micrograph of the Cusil ABA sample, showing the carbon-carbon, braze, and liner. Figure 3 illustrates the good wetting capability of silver ABA around the edge of the carbon-carbon sample.

Palcusil 15 wet the carbon well, but some surface attack of the titanium was observed. Gapasil 9 showed no wetting of the carbon-carbon. Ticusil showed significant attack of the Ti and irregular features at the carbon-carbon surface. Copper ABA, the highest melting temperature braze alloy evaluated, aggressively attacked the Ti and carbon-carbon. This set of experiments led to the elimination of Palcusil 15, Gapasil 9, Ticusil, and copper ABA as candidate brazes.

Similar Nb-1Zr coupons were then fabricated using 125 μm (5 mil) liner material. Both brazes tested, Cusil ABA and silver ABA, showed good wetting and adhesion to carbon-carbon and to the Nb-1Zr. However, excessive residual stress caused by the thicker liner material caused debonding within the carbon-carbon. Successive tests with 25 μm (1 mil) Nb-1Zr foils also showed good adhesion and no debonding within the carbon-carbon.

To date, no carbon diffusion has been observed from the carbon-carbon surface into the braze alloy or liner material, as observed with the CVD process. Time at temperature is approximately the same for both processes, about 2 h; however, the CVD is performed at 1920 K, 720 K higher than silver ABA braze process. Thermal tests have been run with these brazed composites at temperatures of 922 K for 700 h.

CONCLUSIONS

Results of the braze experiments indicate that both Ti and Nb-1Zr metallic liners can be successfully bonded to a carbon-carbon substrate. Braze thicknesses of less than 20 μm provided sufficient adhesion, while maintaining a reasonable low mass impact on the radiator subsystem. From these experiments, successful fabrication and assembly of a carbon-carbon heat pipe is anticipated.

Acknowledgments

The work described in this paper was funded by NASA-LeRC under contract NAS3-25209, with A. Juhasz serving as project manager.

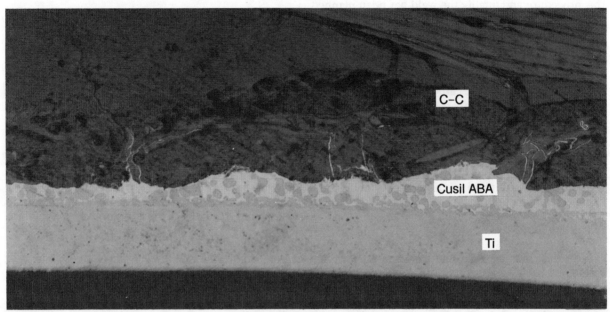

FIGURE 2. Photomicrograph (200x) Ti Liner, Cusil ABA Braze, Carbon–Carbon.

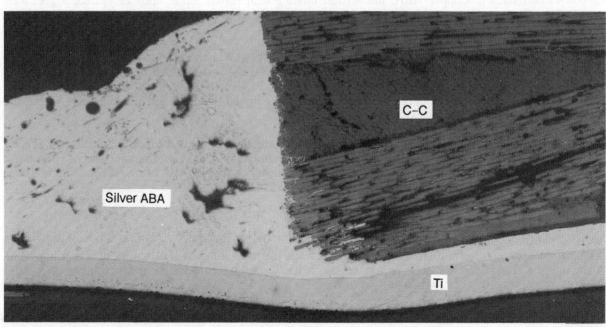

FIGURE 3. Photomicrograph (100x) Ti Liner, Silver ABA Braze, Carbon–Carbon.

References

Rovang, R. D. et al. (1991a), "SP-100 High Temperature Advanced Radiator Development," *Proceedings of Eighth Symposium of Space Nuclear Power Systems,* CONF-910116, M. S. El-Genk and M. D. Hoover, eds., American Institute of Physics, New York.

Rovang, R. D. et al. (1991b), *Advanced Radiator Concepts for SP-100 Space Power Systems, Phase III Final Report*, NASA Report CR-187170.

AN INVESTIGATION OF NATURAL CIRCULATION DECAY HEAT REMOVAL FROM AN SP-100 REACTOR SYSTEM FOR A LUNAR OUTPOST

Mohamed S. El-Genk and Huimin Xue
Institute for Space Nuclear Power Studies
Department of Chemical and Nuclear Engineering
The University of New Mexico
Albuquerque, NM 87131
(505) 277-5635/5442

Abstract

A transient thermal-hydraulic model of the decay heat removal from a 550 kWe SP-100 power system for a lunar outpost has been developed and used to assess the coolability of the system by natural circulation after reactor shutdown. Results show that natural circulation of lithium coolant is sufficient to ensure coolability of the reactor core after shutdown. Further improvement of the decay heat removal capability of the system could be achieved by increasing the dimensions of the decay heat exchanger duct. A radiator area of 10-15 m^2 would be sufficient to maintain the reactor core safely coolable by natural circulation after shutdown. Increasing the area of the decay heat rejection radiator or the diameter of the heat pipes in the guard vessel wall insignificantly affects the decay heat removal capability of the system.

INTRODUCTION

A permanent outpost on the moon is being considered by NASA for the early part of the next century. The lunar reference missions' objectives are prioritized as follows: (1) establish a lunar outpost with a long-term manned presence, (2) gain experience in working on planetary surfaces, (3) conduct science and manufacture resources, and (4) develop crew planning capabilities. The operation concept for the lunar outposts provides for the eventual decentralization of operation control, whereby planning, monitoring and control functions will shift from being earth-based to being shared by earth and the outpost (NASA 1989 and Bennett and Schnyer 1991).

The stationary power systems for the lunar outpost will be designed to meet the evolutionary growth in power demand ranging from the 10s to 100s of kilowatts electric. As the power demand increases for a constructible habitat, an SP-100 reactor, coupled to large dynamic engines, is emplaced to supply 550 kW_e (NASA 1989). A recent study suggested that a 550 kWe SP-100 power system with four Stirling or Brayton engines (with one standby) would be optimum (Harty et al. 1991) for reliability consideration.

OBJECTIVES

The objective of this paper is to investigate the decay heat removal capability of the 550 kWe SP-100 power system by natural circulation of the lithium coolant. A transient model that simulates the decay heat removal loop (DHRL) of the power system for a lunar outpost has been developed. The model is used to examine the effects of the following parameters on the decay heat removal capability of the system: (a) the surface area of the decay heat rejection radiator on the lunar surface, (b) the diameter of the guard vessel heat pipes, and (c) the elevation and dimensions of the decay heat exchanger (DHE) flow duct.

SYSTEM DESCRIPTION

The 550 kWe nuclear power system for a lunar outpost, which is currently being developed by Rockwell International Corporation for NASA (Harty et al. 1991), employs an SP-100 nuclear reactor and four dynamic energy conversion engines. The whole system, with the exception of the surface radiator, is emplaced in a cylindrical, excavated cavity in the lunar regolith. The primary coolant loop and the shadow shield for the reactor are located at the bottom of the cavity. The cavity is lined with a stainless-steel guard vessel, which is an integral part of the system. The guard vessel heat pipes are integrated to those of a dedicated radiator on the lunar surface for decay heat removal. In addition to the obvious structural and integration advantages, the guard vessel is used to transport: (a) the heat losses from reactor/primary cooling system and that deposited by neutrons and gammas during nominal power operation, and (b) the reactor decay heat after shutdown. Figure 1 presents a pictorial view of the power system and Figure 2 shows a line diagram of the decay heat removal model. As shown in Figure 1, the guard vessel fits closely around the reactor and the primary coolant loop to ensure that the reactor core will always be covered with lithium coolant in the event of a small leak in the primary loop or a partial or a total loss of

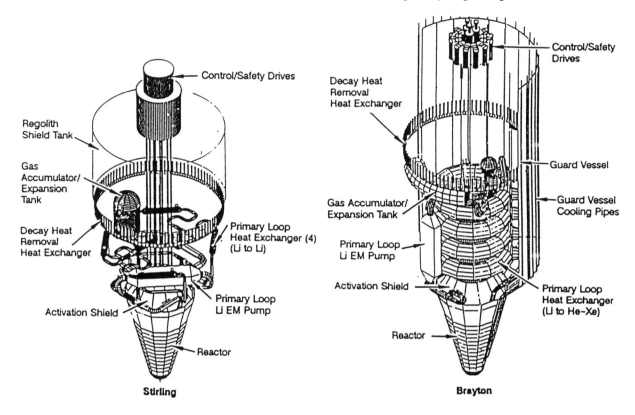

FIGURE 1. 550 kWe Lunar Surface Power System Reactor and Primary and Decay heat loops (Harty et al. 1991).

flow. During normal operation Electromagnetic (EM) pumps circulate the lithium coolant in both the primary and secondary loops. In the primary loop, the lithium coolant from the core is circulated through the primary side of an intermediate heat exchanger, then returned to the reactor core.

The decay heat exchanger (DHE) is installed on the primary side of the energy transport system. The primary loop and the DHE are hydrodynamically coupled both in the hot line and in the return line, via a venturi. During normal power operation, the venturi equalizes the pressure drop across the DHE for a zero net flow. Following a hypothetical Loss-of-Flow Accident (LOFA), forced flow through the venturi drops to zero, hence allowing natural circulation of the lithium coolant through the DHE under the effect of density difference in the hot and return pipes. The heat rejection from the DHE is accomplished by radiation to the inside wall of the guard vessel, then heat is transported by the heat pipes in the guard vessel to the decay heat radiator on the lunar surface. This radiator is independent of that used for waste heat rejection from the second side of the four Stirling or closed Brayton engines during normal power operation. As shown in Figure 1, the DHE is equipped with small sodium heat pipes to enhance the heat rejection capability of the decay heat removal loop (DHRL).

MODEL DESCRIPTION

A decay heat removal model, shown in Figure 2, consists of four coupled sub-models: (a) 2-D transient thermal model of the fuel pin, (b) transient natural circulation thermal-hydraulic model of the DHRL, (c) radiation heat rejection model of the DHE, and (d) a six-group point kinetics model of the reactor core. The latter includes reactivity feed-back due to changes in densities and temperatures of fuel material, cladding and coolant and the Doppler effect in fuel material (El-Genk et al 1992). The point kinetics model is used to investigate partial power operation of the power system with natural circulation cooling.

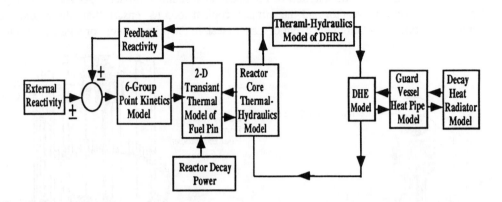

FIGURE 2. A Line Diagram of Decay Heat Removal Model of the 550kWe Power System for a Lunar Outpost.

The properties of the lithium coolant and those of the fuel, cladding and structure materials are taken to be temperature dependent. The radiation view factor between the DHE and the guard vessel is calculated as a function of the dimensions of both the DHE and the guard vessel (Rea, 1975). However, the effective surface emissivity of the DHE heat pipes and of the inside surface of the guard vessel are assumed constant in the analysis (see Table 1). The different sub-models are described in some details in the following sections.

2-D Transient Thermal Model of The Fuel Pin

The SP-100 reactor is a fast neutron spectrum reactor that is cooled by liquid lithium and fueled with UN fuel pins clad in Niobium-1% Zr. The spacing between the fuel pins in the reactor core is maintained using small diameter Niobium wires, which are wrapped along the length of the pins. The fuel pins are triangularly arranged in the reactor core at a small pitch-to-diameter ratio. The fuel pin and the flow channel are discretized into small axial segments where the radial transient heat conduction equations in the different regions of the pin are solved.

The axial temperature distributions are obtained by thermal-hydraulically coupling the different axial segments along the flow channel in the reactor core. The axial conduction in the flow channel and in the fuel pin are assumed negligible. This assumption is justified for small size axial segments; the size used in model was less than 4 cm. The gap conductance in the fuel pin is calculated using either a closed or an open gap model, which allows for fission gas release in the gap during reactor operation (El-Genk and Seo 1986). The volumetric heat generation in the fuel pin is allowed to vary in a cosine distribution in the axial direction. The heat transfer coefficient between the lithium coolant and the cladding is calculated using Jackson's correlation (Jackson 1955).

In each axial segment the physical properties of the coolant, cladding and fuel are evaluated at the average temperature in each region. The model calculates the radial and axial temperature distributions in the fuel and cladding regions as functions of time after reactor shutdown. The decay heat power in the fuel region is calculated using the decay heat curves recommended by Marr and Bunch (Marr et al. 1971) for fast spectrum reactors as a function of the operation power level and time after shutdown. The decay heat correlations are for a long reactor operation before shutdown (> six months). The mass flow rate and coolant temperature of the reactor core are determined from the coupling of the fuel pin thermal model with the thermal-hydraulic model of the DHRL.

The Thermal-Hydraulics Model of the DHRL

The natural circulation thermal-hydraulic model of the DHRL couples four components: the reactor core, the riser pipe (adiabatic section), decay heat exchanger (DHE), and the return pipe (adiabatic section). To calculate the coolant bulk temperature, the reactor core flow channels and DHRL components are discretized into small spatial segments. In the DHE duct, the heat transfer coefficient between the lithium coolant and the duct wall is determined using the Hartnett's correlations for liquid metal coolants (Hartnett et al. 1957). In calculating the radiative heat transfer coefficient between the outer surface of the DHE wall and the inside surface of the guard vessel, the temperature of the guard vessel wall is assumed equal to that of the evaporator of the vessel's heat pipes.

The overall heat balance equation of the DHRL is coupled to the DHRL's overall momentum balance equation to determine the transient coolant mass flow rate and temperatures in the DHRL. In addition to the pressure gain due to buoyancy, the momentum equation includes the pressure losses due to friction, pipe expansions and contractions and acceleration of the coolant in the reactor core. The friction losses in the core is calculated using the CRT model (Chiu, Rohsenow and Todreas 1978) and in the pipes and DHE duct the friction losses are calculated using Blasius relation for smooth walls.

Radiation Heat Rejection Model for the DHE

As indicated earlier, the DHE is radiatively coupled to the inside surface of the guard vessel. Initially, following reactor shutdown the temperature of vessel heat pipes is such that heat rejection by radiation to the ambient on the

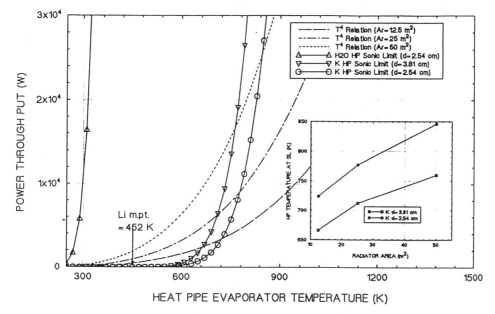

FIGURE 3. Effects of Guard Vessel Heat Pipe Diameter and Working Fluid on Decay Heat Rejection.

lunar surface is lower than the energy transported by the vessel heat pipes at the sonic limit. In this case, the decay heat removal from the reactor core is restricted by the surface area of the decay heat radiator. However, as the decay heat power decreases with time, both the guard vessel wall temperature and the sonic limit of the vessel heat pipes will decrease. Eventually, as the sonic limit of the heat pipe drops below that of the decay heat radiator, decay heat removal from the reactor core will be restricted by the sonic limit of the heat pipes.

Although the heat pipe sonic limit would slow the decay heat removal from the reactor core, it could prolong the time for the lithium coolant to cool down to its freezing temperature (El-Genk, Buksa and Seo 1987). Lithium freezing in the primary loop is not desirable because of the formation of voids during freezing could induce hot-spots in the reactor core during a subsequent startup. As delineated in Figure 3, for a decay hear radiator surface area of 25 m^2, the potassium heat pipes (d=2.54 cm) in the guard vessel wall reach their sonic limits at an evaporator temperature of 777 K. This transition temperature, from a radiator limited heat removal to a heat pipe limited heat removal, increases as the radiator area increases. Figure 3 demonstrates that water heat pipes are unsuitable for cooling the guard vessel wall because the wall temperature would initially exceed the critical temperature of water. Therefore, potassium heat pipes are a better choice for cooling the guard vessel wall.

RESULTS AND DISCUSSION

The overall momentum and energy balance equations of the DHRL, together with the characteristic equations of the decay heat radiator and of the heat pipe sonic limit are solved interactively using a finite difference method with a fully implicit time integrator. The coupled decay heat removal model calculates the coolant temperatures in the different components of the DHRL as well as the maximum temperatures of the fuel and cladding as a function of time after reactor shutdown.

The results for the base case parameters in Table 1 are presented in Figures 4 and 5 for up to 1000 seconds after reactor shutdown. As Figure 5 indicates, initially the mass flow rate increases rapidly due to buoyancy caused by the large temperature rise across the reactor core. For a radiator area of 25 m^2, the coolant mass flow rate peaks at about 0.13 kg/s, 20 seconds after reactor shutdown. Beyond this point, the coolant mass flow rate decreases rapidly to about 0.09 kg/s at 150 seconds after shutdown, then it decreases slowly with time. After reactor shutdown the combination of high decay power (see Figure 4) and low mass flow rate by natural circulation causes the maximum

TABLE 1. Base Case Parameters.

Parameter	Value	Parameter	Value
EXTERNAL PIPE AND DHE		**REACTOR CORE**	
Pipe height (m)	2.45	Effective core height (m)	0.3175
Pipe inner diameter (m)	0.107	Extrapolated core height (m)	0.4258
Aspect Ratio of DHE Duct	0.26	Effective core radius (m)	0.1764
Height of the DHE Duct (cm)	10	Inner radius of core vessel (m)	0.2546
Diameter of the DHE (m)	2.07	Total no. of fuel elements	1296
Pipe wall material	Nb-1% Zr	Fuel rod radius (m)	0.0037
Na heat pipe length (m)	0.3	Fuel rod lattice	triangular
DHE surface emissivity	0.8	pitch-to-diameter ratio(P/D)	1.07
Na heat pipe number	100	Fuel-cladding gap size (mm)	0.13
Na heat pipe diameter (cm)	2.75	Cladding material	Nb-1% Zr
GUARD VESSEL AND RADIATOR		Fuel material	UN
Guard vessel height (m)	3.53	Diameter of wire wrap (mm)	0.0548
Guard vessel inner diameter (m)	2.30	Coolant type	Lithium
Radiator view factor	1.0	**OPERATION PARAMETER**	
Emissivity of radiator	0.8	**BEFORE SHUTDOWN**	
Emissivity of guard vessel	0.8	Reactor thermal power (MW)	2.3
Radiator area (m^2)	25	Core inlet temperature (K)	1276
Number of heat pipe	45	Core exit temperature (K)	1355

temperature of the coolant, at the exit of the reactor core, and the maximum temperatures of the fuel and the cladding to increase initially very fast. As shown in Figure 4, all three temperatures peak almost at the same time, only 15 seconds after shutdown, then they decrease as the decay heat power in the reactor core decreases.

The results in Figure 4 show that while the maximum fuel temperature is only 1525 K, the maximum coolant temperature is as high as 1450 K, which is about 95 K higher than its value during full power operation of the SP-100 reactor at a thermal power of 2.3 MW. Figure 4 also shows that during the first 20 seconds after shutdown the decay heat rejection is restricted by the surface radiator (surface area of 25 m^2). At 20 seconds, the heat rejection capacity of the decay heat radiator equals the energy transport by the guard vessel's heat pipes at the sonic limit. Beyond this time, the former becomes higher than the latter and the heat rejection capability of the DHRL becomes restricted by the heat pipe sonic limit (see Figure 3) causing the temperature of the guard vessel wall to decrease very slowly with time.

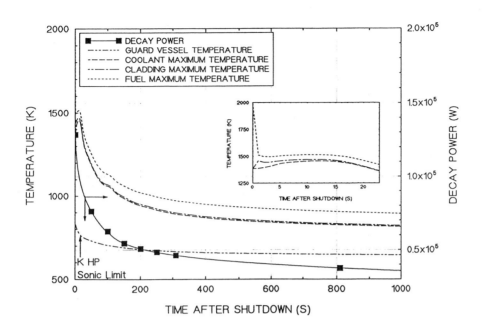

FIGURE 4. Calculated Reactor and Guard Vessel Temperatures as Functions of Time After a Reactor Shutdown.

Figures 4 and 5 show that at about 85 seconds after shutdown the coolant flow rate and coolant temperature undergo a small drop followed by a small increase at about 100 seconds after shutdown. This drop in the coolant flow rate and temperature marks the return of the cold coolant from the DHE after completing the first cycle by natural circulation in the DHRL.

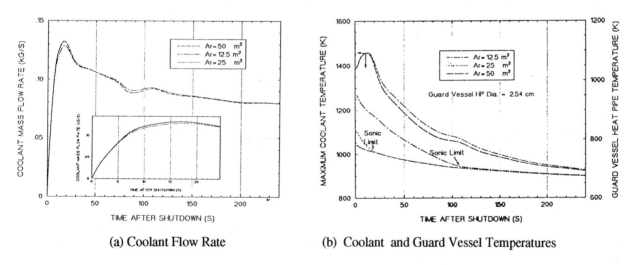

(a) Coolant Flow Rate (b) Coolant and Guard Vessel Temperatures

FIGURE 5. Calculated Coolant Flow rate and temperature for a Guard Vessel HP Dia. of 2.54 cm

Effects of Radiator Area and Guard Vessel Heat Pipe Diameter

Figures 5 and 6 show the effect of the surface area of the decay heat rejection radiator and the diameter of the guard vessel heat pipes on the vessel and coolant temperatures and the coolant flow rate after reactor shutdown. These Figures indicate that within the first 20 seconds after shut down, the coolant flow rate and the coolant temperature are independent of the radiator area and/or the diameter of the guard vessel heat pipes. However, at later times up to 200 seconds after shutdown, increasing the radiator area from 12.5 to 25 m², slightly lowers the coolant temperature; further increase in the decay heat rejection radiator to 50m² does not affect the coolant temperature. At times longer than 200 seconds after shutdown both the radiator area and the diameter of the guard vessel heat pipes insignificantly affect the coolant flow rate or the coolant temperature in the DHRL.

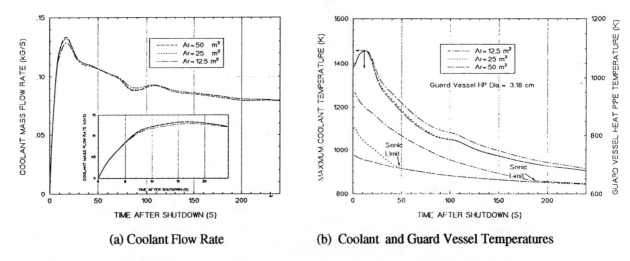

(a) Coolant Flow Rate (b) Coolant and Guard Vessel Temperatures

FIGURE 6. Calculated Coolant Flow rate and Temperature for a Guard Vessel HP Dia. of 3.81 cm.

Figures 5 and 6 also show that increasing the guard vessel heat pipe diameter does not affect the coolant temperature or flow rate in the DHRL, but reduces the guard vessel temperature and increases the time after shutdown, beyond which decay heat rejection is restricted by the sonic limit of the heat pipes. For a heat pipe diameter of 2.54 cm the sonic limit is reached at 20 seconds and 110 seconds after shutdown for a radiator area of 25 m² and 12.5 m², respectively (see Figure 5b). For the largest radiator area of 50 m², the radiator temperature immediately after shutdown was already below the sonic limit of the guard vessel heat pipes (see Figures 5b and 6b). When the heat pipe diameter in the guard vessel increased from 2.54 cm to 3.81 cm, the time to reach the sonic limit increased to 45 seconds and 190 seconds for a radiator area of 25 m² and 12.5 m², respectively (see Figures 5 b and 6b).

Effects of Elevation and Dimensions of the DHE Duct

Results delineated in Figures 7 and 8 demonstrate the effect of changing the elevation of the DHE and the dimensions of the DHE duct, respectively. While the former enhances natural circulation by increasing the pressure gain due to buoyancy, the latter increases flow circulation by reducing the pressure losses due to the friction in the duct. As shown in Figure 7, increasing the elevation of the DHE from 1.94 m to 2.94 m increases the maximum flow rate by about 17 percent (from 0.12 to 0.14 kg/s), but does not affect the time after reactor shutdown at which the coolant flow rate reaches its maximum value (~ 15 seconds). Although this increase in coolant flow rate insignificantly affects the maximum coolant temperature at the exit of the reactor core during the first 10 seconds, it slightly reduces it, by less than 10 K, at later times. Therefore, increasing the elevation of the DHE is not recommended since it does not improve the system coolability, but it would strongly impact the mass and size of the system, and hence the excavation and launch costs.

A better alternative to improve the decay heat removal capability of the power system, with a small impact on the excavation and launch costs, is to increase the dimensions of the DHE duct. For the same aspect ratio as the base case ($a/b = 0.26$), the affect of changing the height of the duct, b, on the coolability of the reactor after shutdown is investigated and the results presented in Figure 8. As this Figure indicates, increasing the duct height, up to 10 cm, strongly affects the coolability of the reactor core; beyond 10 cm the effect becomes negligible. Increasing the DHE duct dimensions not only increases the circulation rate of the coolant, hence the coolant heat transfer coefficient, but also increases the radiation surface area of the DHE to the inside surface of the guard vessel wall. The combination of high coolant heat transfer coefficient and large radiation surface area of the DHE reduces the maximum coolant temperature from 1611 K to 1456 K (about 155K) when the duct height is increased from 5 cm and 10 cm, respectively. A further increase in the duct height to 15 cm causes the maximum coolant temperature to decrease by less than 23 K (from 1456 to 1433 K). Therefore, it is recommended that for the base case aspect ratio, the DHE duct height be increased up to 10 cm as an effective option for enhancing the decay heat removal capability of the power system by natural circulation.

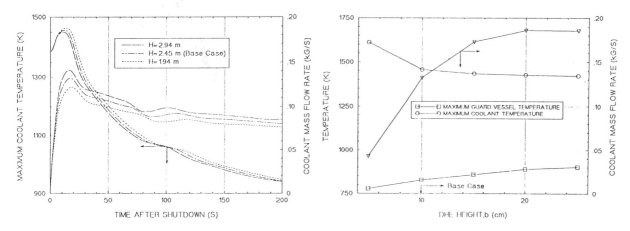

FIGURE 7. Effect of DHE Elevation on Decay Heat Removal Capability.

FIGURE 8. Effect of DHE Duct Dimensions on Decay Heat Removal Capability.

SUMMARY AND CONCLUSIONS

A transient thermal-hydraulic model is developed to investigate the coolability of the 550 kWe SP-100 power system for a lunar outpost by natural circulation of lithium coolant after a reactor shutdown. Results show that for the base case parameters listed in Table 1, natural circulation of Li coolant in the DHRL is capable of maintaining the reactor core coolable after shutdown. Results show that after reactor shutdown, the maximum fuel temperature is several hundred degrees below its value during full power operation while the coolant temperature is 95 K above its full power operation value. Increasing the dimensions of the DHE duct effectively enhances the coolability of the reactor core by increasing the coolant natural circulation, hence reducing both the coolant and fuel temperatures in the core.

Results also show that increasing the area of the decay heat radiator insignificantly affect the coolability of the DHRL; a radiator area in the order of 10-15 m^2, will be sufficient to maintain long term coolability of the power system by natural circulation. Analysis demonstrated that increasing the diameter of the heat pipes in the guard vessel does not influence the coolability of DHRL, but it lowers the guard vessel wall temperature.

Further improvement of the decay heat removal capability of the system could be achieved by increasing the dimensions of the DHE ducts rather than by increasing the area of the decay heat rejection radiator or the diameter of the heat pipes in the guard vessel wall. For the base case aspect ratio, it is recommended that the DHE duct height be increased from 5 cm to 10 cm. Such an increase in duct dimensions causes the maximum coolant temperature to decrease by as much as 155 K, which effectively enhances the coolability of the reactor core.

Increasing the elevation of the DHE is not recommended since it does not improve the system's decay heat removal capability, but it would strongly impact the mass and the size of the system, and hence the excavation and launch costs.

Acknowledgments

Research sponsored by NASA Lewis Research Center under Grant # NAG 3-992 to the University of New Mexico's Institute for Space Nuclear Power Studies. The authors wish to thank Mr. Richard Harty of Rockwell International Corp. and Mr. Robert Cataldo of NASA Lewis Research Center for their input and comments during the progress of the research.

References

NASA (1989) *Report of the 90-Day Study on Human Exploration of the Moon and Mars*, National Aeronautics and Space Administration, November 1989.

Bennett, G. L. and A. D. Schnyer (1991) "NASA Mission Planning for Space Nuclear Power," in *Proceedings of the 8th Symposium on Space Nuclear Power Systems*, CONF-910116, M. S. El-Genk and M. D. Hoover, eds., American Institute of Physics, New York, 1:#77-83.

Chiu, C., W. M. Rohsenow and N. E. Todreas (1978), Flow Split Model for LMFBR Wire Wrapped Assemblies, Coo-2245-56TR, MIT, Cambridge.

Harty, R. B., R. E. Durand, and L. S. Mason (1991) "Lunar Electrical Power System Utilizing the SP-100 Reactor Coupled to Dynamic Conversion Systems", *AIAA/NASA/OAI Conference on Advanced SEI Technologies, Paper No. AIAA 91-3520*, Clevelnad, OH (September 4-6, 1991).

Rea, S. N. "Rapid Method for Determining Concentric Cylinder Radiation View Factors", *AIAA Journal*, Vol.13, No. 8 (August, 1975)

Marr, D. R. and W. L. Bunch, "FTR Fission Product Decay Heat", HEDL-TIME 71-27, Hanford Engineering Development Laboratory (February 1971).

Hartnett, J. P. and T. F. Irvine, Jr., "Nussult Values for Estimating Turbulent Liquid Metal Heat Transfer in Noncircular Ducts", *A.I.Ch.E. Journal 3*, No.3, 313-317 (1957).

Jackson, C. B. et al "Liquid Metal Handbook", Atomic Energy Commission, Dept. of Navy, Washington, D.C. (July, 1955)

El-Genk, M. S. et al. (1987) "Load-Following and Reliability Studies of an Integrated SP-100 System", *Journal of Propulsion and Power*, Vol. 4, No.2, pp152-156.

El-Genk, M.S. and J. T. Seo (1986) " SNPSAM-Space Nuclear Power System Analysis Model," *Space Nuclear Power Systems 1986*, M.S. El-Genk and M. D. Hoover, eds., Orbit Book Company, Inc., Malabar, FL, Vol. 5, 111-123.

El-Genk, M.S. and J. T. Seo (1987) " Parametric Analysis of the SP-100 System Performance," *Space Nuclear Power Systems 1987*, M.S. El-Genk and M. D. Hoover, eds., Orbit Book Company, Inc., Malabar, FL, Vol. 7, 399-408.

El-Genk, M. S., H. Xue, C. Murray. and S. Chaudhuri (1992) " TITAM" Thermionic Integrated Transient Analysis Model: Load Following of a Single-Cell TFE", *Proceedings of the 9th Symposium on Space Nuclear Power Systems*, Albuquerque, NM, American Institute of Physics (10-16 January 1992).

ROTATING FLAT PLATE CONDENSATION AND HEAT TRANSFER

Homam Al-Baroudi
Department of Nuclear Engineering
Oregon State University
Radiation Center, C116
Corvallis, OR 97331-5902
(503) 737-7074

Andrew C. Klein
Department of Nuclear Engineering
Oregon State University
Radiation Center, C116
Corvallis, OR 97331-5902
(503) 737-7066

Abstract

A rotating flat plate condensation experiment has been built and tested to simulate a rotating heat rejection radiator for space nuclear reactor applications. The results obtained in this study give the relationships between the speed of a rotating cooled surface and other significant heat transfer parameters. Such design information and correlations are important to evaluate the feasibility of such designs for applications to space systems as well as to ground systems.

INTRODUCTION

The Bubble Membrane Radiator (BMR) is a heat rejection system in which a sphere or ellipsoid rotates about the minor axis of the sphere, vapor is introduced at the center of the sphere, and is condensed on the inside surface of the liner. The driving force for the vapor transport inside the bubble is the density gradient which is established as the vapor condenses on the liner surface. The heat of condensation is conducted through the thin metallic liner which is only required to contain the working fluid. A combination of radiation and conduction then transmits the condensate heat to the ultimate heat sink, space (Webb and Coombs 1988).

Since the BMR concept uses an artificial gravity imposed on the working fluid by means of the centripetal force to pump the fluid from the radiator, experimental and analytical studies have been initiated to understand the nature of fluid and heat transport under the conditions of rotation (Webb et al. 1989). A series of experiments have been devised to investigate a variety of phenomena which will be occurring. The first of these experiments involves the condensation of vapor on a rotating flat plate which is oriented normal to the Earth's gravity vector to reduce the effects of gravity as much as possible (Al-Baroudi et al. 1991). Future experiments will include construction of a full BMR simulator in preparation for possible reduced gravity flight opportunities. The purpose of this series of experiments is to attempt to understand the physics of condensation, fluid transport, and heat transfer under these conditions.

DESCRIPTION OF THE EXPERIMENTAL MODEL

A description of the experimental methods and apparatus used in the flat plate condensation experiment has been presented previously (Al-Baroudi et al. 1991). The primary components of this experiment consist of a large glass pipe section used to contain all of the components of the experiment, the rotating flat plate assembly, a variable speed motor, a vacuum pump, and a water vapor supply. Water is used as the working fluid in the experimental model. A revised schematic design of the experimental apparatus is shown in Figure 1.

EXPERIMENTAL RESULTS AND DISCUSSION

To date experiments have been conducted for 0, 200, and 300 rpm, with a plan to continue to expand the data base. The number of data points collected give preliminary indications on the behavior of some thermal and fluid parameters, but does not allow firm conclusions to be drawn. Significantly more data needs to be assembled.

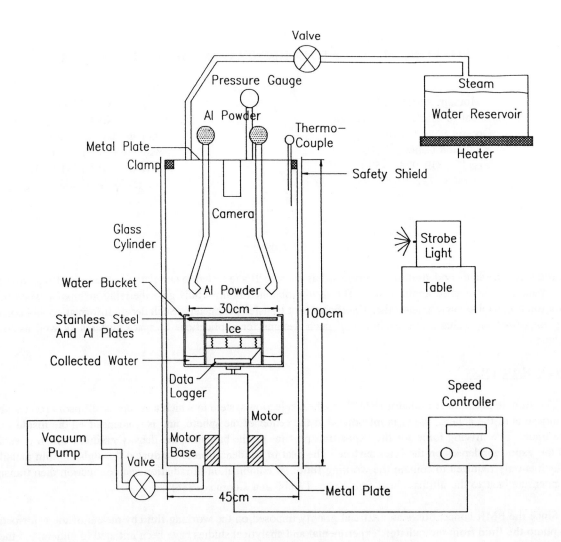

FIGURE 1. Flat Plate Condensation Experiment Schematic Diagram.

Heat Transfer Measurements

An Electronic Controls Design, Inc. Model 50 remote data logger is used in this experiment to collect temperature data from the inside of the rotating flat plate assembly. This data logger has five usable data channels. One is used to measure the temperature of the environment, two to measure the temperature of the top surface of the aluminum plate (note that they are located at different distances from the center of the plate), and two to measure the bottom temperature surface of the aluminum plate in contact with the ice block (note also that these two thermocouples are placed to correspond to the locations of the top thermocouples). The vacuum pump is turned on until the pressure in the system reaches 6 kPa. The rotating motor to be set to the desired speed. The steam is then introduced. A quasi-steady state condition is obtained after a short period of time after the introduction of steam to the chamber.

Each thermocouple is averaged over a chosen period of time where the quasi-steady state condition is observed and then the two thermocouples on the top surface of the aluminum plate are averaged to give one temperature value, while the same thing is done for the other two thermocouples on the bottom surface of the aluminum plate. The amount of heat transfer through the aluminum plate can then be calculated. If the heat transfer is assumed to be the same from the environment to the bucket, then the overall heat transfer can be found.

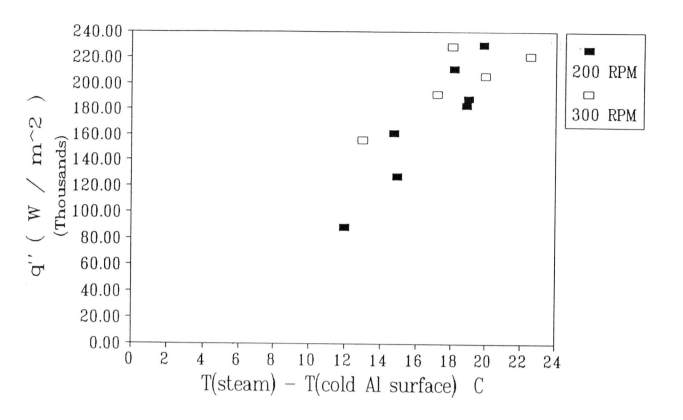

FIGURE 2. Removal Heat Flux as a function of Steam to Inside Test Assembly Temperature Difference for Plate Rotational Speeds of 200 and 300 rpm.

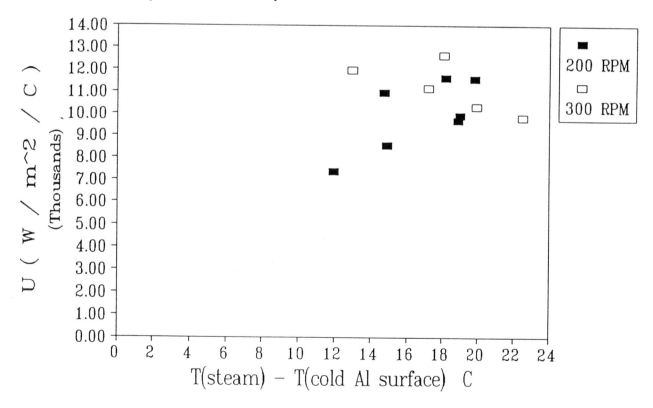

FIGURE 3. Overall Heat Transfer Coefficient as a function of Steam to Inside Test Assembly Temperature Difference for Plate Rotational Speeds of 200 and 300 rpm.

Preliminary results have been compiled for rotational speeds of 200 and 300 rpm. Figure 2 shows the measured heat flux for a number of tests at these rotational speeds as a function of the temperature difference between the environment and the cold aluminum surface. While there may be a slightly higher heat transfer rate at 300 rpm when compared to 200 rpm, there does not appear to be a significant difference in the measurements to date. In some of the 200 rpm experiments presented here it was observed that dropwise condensation occurred on the surface of the steel plate, rather than film condensation, thus possibly increasing the overall heat transfer rate. All of the data presented for 300 rpm, on the other hand, were the result of film condensation. The film condensation was achieved by treating the surface of the stainless steel plate. Figure 3 shows the calculated overall heat transfer coefficient obtained from these results.

Again, additional data needs to be collected to identify any trends, however the overall heat transfer coefficient appears to be slightly higher for 300 rpm than for 200 rpm. Further experimental procedure refinements are planned to attempt to reduce the scatter in the data.

Condensate Collection

The importance of the surface treatment of the stainless steel plate was observed in the amount of liquid condensate collected during a test run. Based on the observations during 200 and 300 rpm runs, droplet formation was noticed on condensation rather than the formation of a thin film of liquid. It was also noticed that condensate collection was considerably lower when droplets formed on the surface rather than a thin film of condensate. The surface tension of the droplet was strong enough to restrict the flow of liquid off of the plate. In order to get film formation and increased fluid transport, a surface treatment was used to break the surface tension of the droplets and allow the formation of a thin liquid sheet. There is a very wide scatter in the 200 and 300 data due to inconsistent methods used to treat the surface of the mirrored steel plate. Further studies are planned to reduce this scatter and improve the measurement techniques.

CONCLUSION

This study gives a preliminary view of the relationships between the speed of rotation and the heat transfer parameters of the BMR. This information will provide the base for designing new heat rejection systems utilizing the centerfugal force and the condensation phenomena.

Acknowledgments

This work was supported by the United States Department of Energy, grant number DE-FG07-89ER12901. Such support does not constitute an endorsement by DOE of the views expressed herein.

References

Al-Baroudi, H., A. C. Klein, and K. A. Pauley (1991) "Experimental Simulation of Bubble Membrane Radiator Using a Rotating Flat Plate," in *Proceedings of the 8th symposium on Space Nuclear Power Systems*, CONF-910116, M.S. El-Genk and M.D. Hoover, Eds., American Institute of Physics, New York, Vol. 2, pp. 723-727.

Antoniak, Z.I., W.J. Krotiuk, B.J. Webb, J.T. Prater, and J.M. Bates (1988) *Fabric Space Radiators*, PNL-6458, Battelle Pacific Northwest Laboratory, Richland, WA.

Webb, B.J. and E.P. Coombs (1988) *Rotating Bubble Membrane Radiator*, United States Patent, no. 4,789,517, 6 December 1988.

Webb, B.J., Z.I. Antoniak, A.C. Klein, K.A. Pauley (1989) "*Preliminary Design and Analysis of a Rotating Bbubble Membrane Radiator for Space Applications*", Draft report to NASA Johnson Space Center prepared by Battelle Pacific Northwest Laboratory under US DOE Contract DE-AC06-76RL01830, Richland, WA, March 1989.

HIGH TEMPERATURE THERMAL POWER LOOPS

Robert Richter
Jet Propulsion Laboratory
California Institute of Technology
4800 Oak Grove Drive, MS 157-102
Pasadena, CA
(818) 354-3253

Joseph M. Gottschlich
Wright Laboratory
Bldg 18A, Area B
WL/POOS-3
WPAFB, OH 45433-6563
(513) 255-6241

Abstract

The high temperature Thermal Power Loop (TPL) is presented as an attractive alternate to the electromagnetically pumped liquid metal loop as a thermal power transport device for space nuclear power systems. This two phase thermal power transport device has many performance advantages, specifically with respect to size, weight, and preservation of thermal potential. In this paper, difficulties associated with the thermal power transport in space nuclear power systems are considered. The TPL concept is described with its underlying thermodynamic principles. Specific TPL designs are presented with the procedure for their optimization. It is illustrated that development of specific TPLs can be highly cost effective as TPLs will provide substantial weight and cost savings for many future spacecraft. By generating its own pumping power the thermal power loop has the potential to significantly increase the specific power of future nuclear power systems.

INTRODUCTION

An electromagnetically pumped single phase Liquid Metal Loop (LML) as the thermal power transport device in a space nuclear power system has many inherent disadvantages. First, the electromagnetic pump presents a single point failure component and consumes a considerable amount of electric power which the nuclear conversion system generates with an efficiency of typically less than 10%. Secondly, because in the LML thermal energy is transported in the form of sensible heat, very high flow rates as well as large temperature differences are associated with such a thermal power transport device. The high flow rate leads to substantial pressure drops even with large diameter flow tubes. The pressure drops require a significant amount of pumping power which is supplied by a very inefficient electromagnetic pump. This pump receives its power from the power generation system of which it is a part. Since the power generating system itself consumes a large amount of its generated power, the specific system weight will be high. If an attempt is made to minimize the power consumption of the electromagnetic pump by increasing the internal diameter of the flow tubes, other aspects of the power system negate the gain, such as an increased temperature drop and a rise in weight of the tubes as well as of the circulating fluid.

Since in an out-of-core power conversion system the LML transfers thermal energy in the form of sensible heat from the reactor to the electric generating devices the available thermal potential, that is, the upper operating temperature, of the generating device is substantially lower than the thermal potential at which thermal power is generated. A system tradeoff has to be made between operating the electric generating devices at a lower efficiency or increasing the temperature of the reactor. The lower efficiency of the electric power generators requires more conversion devices adding weight to the system. A higher reactor temperature shortens the reactor lifetime and compromises its safety.

The replacement of a LML with a two phase flow device which converts thermal energy to kinetic energy for the circulation of the heat transfer fluid eliminates the consumption of electric power which is generated at a relatively low efficiency. Such a replacement would provide many additional advantages. First, the liquid metal inventory could be reduced substantially. This results in weight saving, decreases the power requirement for frozen startup and improves the reactor core reactivity. By its operating principle the passive two phase system is considerably less complex in its design and operation than the LML and is thereby more reliable.

Serious concerns arise when heat pipes are considered as alternatives to the LMLs. Heat pipes in nuclear power systems would be relatively large because of their large vapor core diameter. Thus, their use would result in an increased void fraction in the reactor. As known, core volume scales with a power greater than one as a function of void fraction and shield size is roughly directly proportional to core volume. Since the shield contributes a major fraction to the total weight of a reactor system, any benefit derived from the use of heat pipes will be negated by an increase in total system weight. Furthermore, safety will be compromised by the larger void fraction. In case of an accident during time of launch resulting in the reactor falling into the sea, the voids in the reactor will fill with water. This could cause the reactor to go critical. Clearly, any thermal transport system which is considered for replacing the LML must be optimized for its size, weight, and axial heat transport flux.

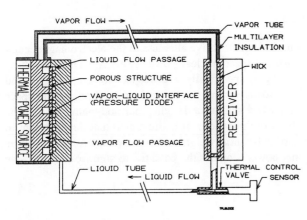

FIGURE 1. Thermal Power Loop.

The Thermal Power Loop (TPL) is designed to transport thermal energy in the form of latent heat of vaporization of the working fluid as in heat pipes and capillary pumped loops. It comprises a vapor tube, a liquid tube, an evaporator and a condenser as shown in Figure 1. By providing two separate tubes, the two phases of the working fluid, the vapor and the liquid phase, do not flow counter to each other as it occurs in heat pipes nor together as in capillary pumped loops. Furthermore, since the liquid phase does not flow through a wick structure, the total pressure drop suffered by the circulating fluid is considerably less in a TPL than in a heat pipe and the power transfer is much higher for the same vapor flow diameter.

THERMODYNAMIC ASPECTS OF A THERMAL POWER LOOP

The TPL transports thermal energy from a heat source to a heat receiver by circulating a working fluid in two phases, liquid and vapor. Any circulation of fluid requires pumping which consumes energy. This energy must be provided by a source external to the working fluid. The only source from which self pumping devices can extract energy is the heat source to which their evaporator is thermally coupled. The receiver is the heat sink to which the condenser is attached. The circulation of the working fluid in the TPL is possible only if thermal energy is converted to kinetic energy in a thermodynamic cycle.

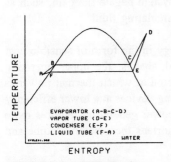

FIGURE 2. TPL Cycle.

The Temperature-Entropy diagram as shown in Figure 2 describes the thermodynamic cycle that supports the TPL operation. The working fluid enters the evaporator in its liquid state at a temperature T_1 (A). Its temperature is raised to temperature T_2 (B) at which it evaporates and expands from the liquid volume to the vapor volume. Some superheating can occur (C to D) depending on the design of the evaporator. The vapor flows under the influence of the pressure difference between the evaporator and condenser (D to E). This pressure difference is the result of a temperature difference between the vapor in the evaporator (C) and the vapor in the condenser (E), that is, $T_2 - T_3$. The liquid returns to the evaporator (F to A) at or below the temperature of the condenser. Thus, a pressure difference exists across the porous interface between liquid and vapor equal to the sum of the two pressure drops in the flow tubes.

The maximum pressure difference of the cycle can never exceed the pressure difference that can be supported by the fluid membranes of the porous structure, called the wick. This wick in the evaporator functions solely as a diode which separates the vapor from the liquid permitting liquid to flow towards the heat source from which it vaporizes.

The details of the underlying thermodynamic analysis which addresses the operation of two phase flow thermal energy transport devices were presented in References 2 and 3. The analysis identifies the absolute operating temperature and the difference between the vapor surface temperature and the liquid surface temperature of the porous structure in the evaporator as the two determining thermodynamic parameters. This result is restated here by the following relations (Richter and Gottschlich 1990):

$$\Delta p = p_{v2} - p_{v1} \tag{1}$$

where
- Δp is pressure difference, N/m²,
- p_{v2} is vapor pressure on the vapor side, N/m², and
- p_{v1} is vapor pressure on the liquid side, N/m².

The vapor pressure can be expressed to a good degree of accuracy by

$$p_v = C_v * \exp(-E_v/RT) \tag{2}$$

where
- E_v is latent heat of vaporization, J/mol,
- C_v is material constant, N/m², and
- R is universal gas constant, 8.3144 J/mol-K.

From relation 2, the required temperature difference for achieving the pressure difference of relation 1 is established by

$$\Delta T = (E_v/R) * \{1/\ln(C_v/p_{v2}) - 1/\ln(C_v/p_{v1})\} \tag{3}$$

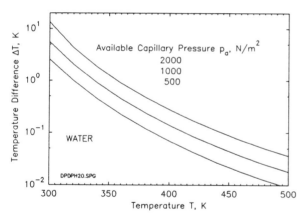
FIGURE 3. Temperature Difference for Water.

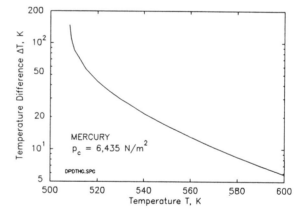
FIGURE 4. Temperature Difference for Mercury.

The interdependence between pressure difference and the two determining operating parameters, absolute operating temperature and temperature difference, is illustrated for water as the working fluid in Figure 3. From the illustration it is seen that the higher the operating temperature the smaller is the required temperature difference to achieve the desired pressure difference. Figure 4 illustrates that a minimum operating temperature is required for a TPL that uses mercury as its working fluid. Because of its high surface tension a pressure difference of $\Delta p = 6,435$ N/m² can be maintained across a porous structure. As shown in Figure 4 the operating temperature has to lie well above 507 K since below this temperature the temperature difference needs to be larger than the absolute temperature, which is naturally not possible.

The thermodynamic analysis (Richter and Gottschlich 1990 and 1991) clearly defines the operation of heat pipes and thermal power loops. It establishes that

1. The heat transport is a function of the temperature difference that can be maintained between the exit and entrance of the evaporator.

2. The operating temperature determines the temperature difference for a given heat transfer rate.

3. The "capillary pressure" establishes the maximum pressure difference that can be supported in a TPL or a heat pipe and at which the device can operate. A thermal power loop operates only to its maximum capacity of the "capillary pressure" if and unless a sufficient temperature difference exists that produces a pressure difference that is equal to the capillary pressure.

THERMAL POWER LOOP DESCRIPTION

The TPL concept was conceived and demonstrated by California Institute of Technology, Jet Propulsion Laboratory under a NASA program. The first TPL design was based on the thermodynamic principles as presented above. Every effort was made to optimize the design for the preservation of the available thermal potential and minimize the thermal difference across the evaporator.

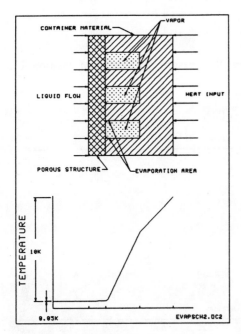

FIGURE 5. Heat Flux.

The heart of the TPL is the evaporator as shown in Figure 1. In this design four layers of wire cloth forms the porous structure that acts as a fluid diode. This diode functions only if a negative temperature gradient exists between the liquid on the downstream side and the vapor on the upstream side of the interface. Since heat is applied only to the vapor side of the evaporator a natural thermal gradient is established (Figure 5). Thus, heat is applied to the liquid solely at the vapor/liquid interface and liquid will never exceed the temperature of the interface.

The TPL has two distinct flow tubes, the vapor tube and the liquid tube. The volumes of these two tubes are separated by capillary barriers in the evaporator and the condenser. Since the vapor tube is maintained at a higher temperature than the liquid tube, the liquid phase of the working fluid is always driven by the vapor pressure in the vapor tube into the liquid tube.

Thermal power transfer control is easily accomplished with a TPL. Since in the TPL the vapor and liquid circulate in two distinct tubes, the thermal power transport can be regulated by a control valve as shown in Figure 1. Power control was demonstrated in the laboratory by a manual control valve in the liquid tube. If the reactor or the receiver have variable power requirements, a control valve in the loop can vary the flow rate for the desired thermal power transfer.

THERMAL POWER LOOP DESIGN

With the basic thermodynamic, fluid dynamic and heat transfer operating principles of a TPL established a thermal power loop can be designed and optimized for specific applications. The analysis of the thermal power loop is primarily concerned with the size and weight of the vapor and liquid tubes based on (1) the available pressure difference, (2) the total thermal power transfer, (3) the operating temperature, and (4) the distance between source and receiver.

The maximum supportable pressure difference across the liquid-vapor interface in the evaporator is determined by the surface tension of the working fluid and the pore size of the porous structure of which the fluid diode is constructed. The maximum pressure difference can be relatively large if a wick of very small pore size can be fabricated from a material that is compatible with the working fluid at the operating temperature. The supportable pressure difference must be equal or larger than the total pressure drop of the working fluid while flowing from the evaporator to the condenser in the vapor tube and from the condenser to the evaporator in the liquid tube. Because of the large flow areas and the short flow lengths which the porous structures in the evaporator as well as in the condenser present to the working fluid, the pressure drops in the evaporator and in the condenser must be considered only if the final design requires full optimization.

The total pressure drop in the TPL is determined by the three primary independent operating parameters and design parameters. The operating parameters are the operating temperature, the thermal power transfer and the distance between the source and the receiver. The design parameters are the selectable internal diameters of the tubes. The design analysis establishes a combination of internal tube diameters for the vapor and liquid tubes that limits the combined pressure drops to less than the available pressure difference.

A computer program has been developed which permits the design of specific TPLs based on stated operating requirements. The Mach number of the flow in the vapor tube was selected as the independent parameter. The higher the Mach number and thus the velocity of the vapor, the smaller is the internal diameter of the vapor tube but the larger will be the pressure drop for the vapor. Since only a limited total pressure drop can be tolerated for the flow in the entire thermal power loop, with increasing pressure drop in the vapor tube the pressure drop in the liquid tube has to be decreased by enlarging its internal diameter. This is shown in Figure 6. Within a range of Mach numbers any combination of internal diameters of the liquid and vapor tubes can satisfy the operation of the thermal power loop. The upper limit of Mach number is reached when the total available capillary pressure drop occurs in the vapor tube. At that limit the liquid tube would have to be infinitely large so that the liquid flow would occur without any pressure drop.

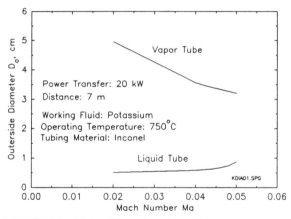

FIGURE 6. Tube Diameters of TPL.

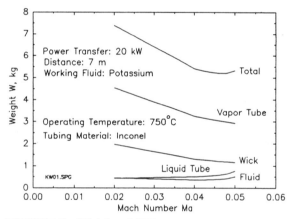

FIGURE 7. Weight of TPL.

Since weight is an important major system parameter, the lightest TPL design can be found as shown in Figure 7. The combined weight of the entire TPL includes the weight of the vapor and liquid tubes as determined by the internal diameters, the operating pressure, the operating temperature, the maximum allowable stress of the tube material and the specific weight of the tube material. The weight includes also the weight of the working fluid contained in the TPL.

Two TPLs were designed and optimized for two specific high temperature applications. Case #1 represents the requirements of a solar collector thermal power system for transferring 6 kW$_t$ over 4.57 meters at 1120 K (RICHTER 1974). Case #2 is the requirement of a nuclear power system for transporting 20 kW$_t$ over 7 meters at 1023 K. In both cases 100 mesh bolting cloth was used for constructing the wicks. For the nuclear system Figure 6 presents the combination of tube diameters while Figure 7 presents the weights as a function of Mach number in the vapor tube flow. The results are presented in Table 1. The weights shown do not include the weights of the evaporator and condenser as their designs depend on requirements which are not defined yet. The axial flux density of Case 1 is 2.6 kW/cm^2 over 4.57 meters and for Case 2 it is 2.3 kW/cm^2 over 7.0 meters.

TABLE 1. Operating Requirements and Designs of Two High Temperature Thermal Power Loops.

OPERATING REQUIREMENTS				
FLUID	MATERIAL	TEMPERATURE	LENGTH	POWER
SODIUM	INCONEL	1120 K	4.57 m	6 kW_t
POTASSIUM	INCONEL	1023 K	7.00 m	20 kW_t
THERMAL POWER LOOP DESIGNS						
FLUID	WEIGHT (kg)	VAPOR TUBE		LIQUID TUBE		AXIAL FLUX (kW_t/cm^2)
		I.D. (cm)	O.D. (cm)	I.D. (cm)	O.D. (cm)	
SODIUM	1.6	1.58	1.69	0.31	0.41	2.58
POTASSIUM	5.2	3.19	3.27	0.64	0.74	2.26

CONCLUSIONS

The thermal power loop has clear and obvious potential advantages over heat transfer devices which carry thermal energy in the form of sensible heat in space nuclear power thermal systems. A thermal power loop can save weight by requiring substantially less working fluid inventory, decrease overall volume, improve system reliability, simplify power control, increase overall system efficiency, substantially decrease start-up power, and minimize the power consumption of auxiliary equipment. The utilization of this new technology has the potential to significantly increase the specific power of future space nuclear power systems.

Acknowledgments

The research described in this paper was carried out by the Jet Propulsion Laboratory, California Institute of Technology, Pasadena, CA originally under contract with the National Aeronautic and Space Administration and later under contract with the US Air Force and by Wright Laboratory, Wright-Patterson Air Force Base under US Air Force funding.

References

Gottschlich, J. and Richter, R. (1991) "Thermal Power Loops," SAE Paper 911188, 1991 SAE Aerospace Atlantic, Dayton, Ohio, April 22-26, 1991.

Richter, R. (1974) Solar Collector Thermal Power System, Volume III Basic Study and Experimental Evaluation of Thermal Train Components, AFAPL-TR-74-89-3, Wright-Patterson Air Force Base, Ohio, November 1974.

Richter, R. and Gottschlich, J. (1990) "Thermodynamic Aspects of Heat Pipe Operation," AIAA-90-1772, AIAA/ASME 5th Joint Thermophysics Conference, June 1990.

NUCLEAR SAFETY POLICY WORKING GROUP RECOMMENDATIONS FOR SEI NUCLEAR PROPULSION SAFETY

Albert C. Marshall
Sandia National Laboratories
Division 6465
Albuquerque, NM 87185
(505) 272-7002

J. Charles Sawyer, Jr.
NASA Headquarters
600 Independence Ave. SW
Washington, DC 20546
(202) 453-1159

Abstract

Nuclear propulsion has been identified as an essential technology for the implementation of the Space Exploration Initiative (SEI). An interagency Nuclear Safety Policy Working Group (NSPWG) was chartered to recommend nuclear safety policy, requirements, and guidelines for the SEI nuclear propulsion program to facilitate the implementation of mission planning and conceptual design studies. The NSPWG developed a top level policy to provide the guiding principles for the development and implementation of the nuclear propulsion safety program and the development of Safety Functional Requirements. In addition, the NSPWG reviewed safety issues for nuclear propulsion and recommended top level safety requirements and guidelines to address these issues. Safety topics include reactor start-up, inadvertent criticality, radiological release and exposure, disposal, entry, safeguards, risk/reliability, operational safety, ground testing, and other considerations. All recommendations are in conformance with existing regulations and are formulated with the objectives of assuring safety rather than establishing prescriptive requirements.

BACKGROUND

The goals of the Space Exploration Initiative (SEI) include a return to the moon and a manned mission to Mars by 2019. Nuclear propulsion has been identified as a key technology to meet the SEI objectives. Both nuclear thermal propulsion (NTP) and nuclear electric propulsion (NEP) are being considered for the SEI. NTP systems provide thrust by heating a propellant (usually hydrogen) in a nuclear reactor and expanding the hot gases through a nozzle. NEP concepts convert the reactor thermal power to electrical power. The electrical power is then used to accelerate an ionized propellant through an electric thruster.

When nuclear propulsion was identified as a necessity for SEI, the National Aeronautics and Space Administration (NASA) held joint agency workshops, in June and July 1990, to explore options and requirements to initiate a new space nuclear propulsion program. The agencies represented included the Department of Energy (DOE) and the Department of Defense (DoD), as well as NASA. Based on these workshops, a joint agency Nuclear Safety Policy Working Group (NSPWG) was chartered to recommend a top level nuclear safety policy for the SEI nuclear propulsion program. The NSPWG was also asked to recommend safety requirements for a number of specific nuclear propulsion safety issues. In this paper, the emphasis will be placed upon the NSPWG recommendations. The safety issues were emphasized in another paper (Marshall and Sawyer 1991).

APPROACH

The development of the NSPWG recommendations took place in a series of ten meetings between November 1990 and September 1991. Existing national and international guidelines and regulations were reviewed (see, for example, U.S. Department of Energy 1984). In addition, pertinent previously developed safety policies were reviewed (see, for example, Wahlquist 1990). The NSPWG discussed and developed a top level safety policy. Each member was assigned principal responsibility for one or more of the safety issues to be addressed. The responsible member for each safety issue presented a discussion of the topic to the group. Discussions on each topic continued until a consensus was reached on the draft recommendations. Meeting minutes were mailed to a broad distribution, and several open sessions and presentations were given to obtain feedback from the technical community.

SCOPE

The NSPWG scope of activities can be understood by referring to the proposed safety program hierarchical structure illustrated in Figure 1. At the top of the hierarchy are the policies and mandatory requirements that are set by the existing agencies (DOE, NASA, and so on). Below this level is the SEI Nuclear Safety Policy that provides the guiding principles for the development and implementation of the safety program and safety requirements. The Safety Functional Requirements (for example, preclude inadvertent criticality) delineate the specific safety functions required of the system or program. The development of these policies is the responsibility of the government agencies. The contractors' responsibilities include the design specifications and system design and research.

The scope of the NSPWG activities is illustrated on the left side of Figure 1. It is expected that the recommended NSPWG Safety Policy will be adopted or adapted to formulate the Joint Agency Space Nuclear Safety Policy for the SEI space nuclear propulsion program. The NSPWG also recommended Safety Requirements and Guidelines for important safety issues and considerations. It is anticipated that the NSPWG-recommended Safety Requirements and Guidelines will be used to develop the Safety Functional Requirements for SEI nuclear propulsion.

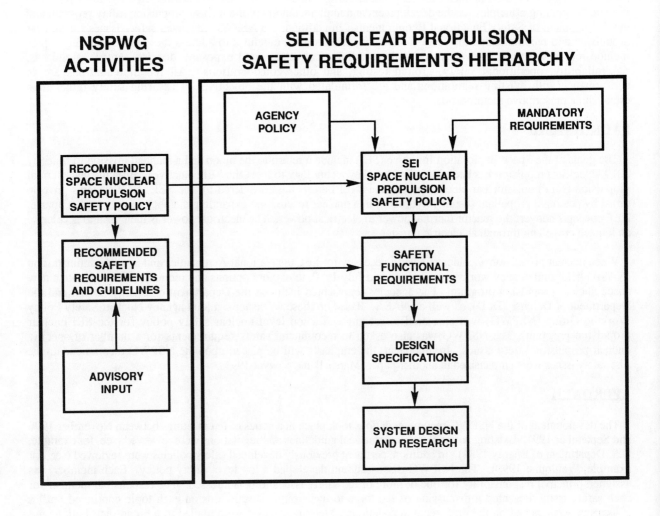

FIGURE 1. Nuclear Propulsion SEI Safety Policy and Requirements Hierarchy.

RECOMMENDED SAFETY POLICY

The NSPWG recommends the following safety policy to guide the development and implementation of the safety program and the development of functional safety requirements:

Recommended Space Exploration Initiative
Nuclear Propulsion Safety Policy

Ensuring safety is a paramount objective of the Space Exploration Initiative nuclear propulsion program; all program activities shall be conducted in a manner to achieve this objective. The fundamental program safety philosophy shall be to reduce risk to levels as low as reasonably achievable. In conjunction with this philosophy, stringent design and operational safety requirements shall be established and met for all program activities to ensure the protection of individuals and the environment. These requirements shall be based on applicable regulations, standards, and research.

A comprehensive safety program shall be established. It shall include continual monitoring and evaluation of safety performance and shall provide for independent safety oversight. Clear lines of authority, responsibility, and communication shall be established and maintained. Furthermore, program management shall foster a safety consciousness among all program participants and throughout all aspects of the nuclear propulsion program

In this context, safety is meant to include the health and safety of the public, program personnel, and mission crew, in addition to protection of the terrestrial and nonterrestrial environments. Safety also includes safeguarding nuclear systems and special nuclear materials from unauthorized use or diversion, and protection of property against accidental loss or damage. Although unstated in the policy, it is important to communicate openly with the public on both the benefits and the risks associated with the SEI program. The specific emphasis the program places on safety and the progress made toward achieving safety objectives must also be communicated.

RECOMMENDED REQUIREMENTS AND GUIDELINES

Recommended safety requirements were developed for reactor start-up, inadvertent criticality, radiological release and exposure, disposal, entry, and safeguards. Both quantitative and qualitative requirements have been recommended. Quantitative requirements for radiological exposures were recommended only when applicable established guidance could be cited. More specific and quantitative design requirements must be developed as the program matures and specific concepts and missions are defined. The development of all recommended requirements (and any subsequent modifications) must be compatible with the fundamental philosophy of reducing risk to as low as reasonably achievable.

Safety Guidelines are recommended for use by the program to establish program plans as well as design and operational requirements for the development of nuclear propulsion systems for SEI. Some of these guidelines may be useful for formulating additional Safety Functional Requirements. Safety guidelines are recommended for risk/reliability, operational safety, flight trajectory and mission abort, and space debris and meteoroid safety considerations. Safety guidelines are discussed in another report (Marshall et al. 1991)

The following safety requirements are recommended for use by SEI management and program safety to develop Safety Functional Requirements.

1. Reactor Start-Up

- The reactor shall not be operated prior to space deployment, except for low-power testing on the ground for which negligible radioactivity is produced.
- The reactor shall be designed to remain shutdown prior to the system achieving its planned orbit.

These requirements were based on the assumption that, because the need for suborbital reactor start-up had not been proposed for SEI, the topic of suborbital start-up was beyond the scope of our activities. These recommendations ensure that no hazard is presented from either direct radiation from a critical core or from the accidental release of fission products from power operation of a reactor prior to achieving orbit.

2. Inadvertent Criticality

- Inadvertent criticality shall be precluded for both normal conditions and credible accident conditions.

3. Radiological Release and Exposure

These requirements apply only to releases from reactors operating in space. Potential radiological releases on Earth are implicitly addressed under reactor start-up, inadvertent criticality, disposal, and entry. Guidance on radiological releases and exposure from reactor propulsion systems deployed in space are needed to protect the mission crew, space environment, and Earth environment. The protection of other space enterprises, such as other space missions involving astronauts, is encompassed by protection of space environments. In addition, the mission spacecraft must be protected from any adverse effects that could result from radiological releases. This guidance pertains to routine operation and potential accidents in space. Space in this context includes all regions beyond Earth's biosphere, including other celestial bodies. For routine operation and expected occurrences, maximum allowable doses can be preestablished. For accidents, deterministic requirements are inappropriate and probabilistic guidance is recommended. In all cases the principle of reducing radiological risk to as low as reasonably achievable should be used to enhance safety. Current NASA guidance for total whole body dose is 50 rem/year, of which 5 rem/year (see 29 CFR 1910.96 (U.S. Occupational Safety and Health Administration, 1989)) may be received from man-made on-board radiation sources. Since the natural background radiation dose to the crew could be quite high for some proposed SEI missions, the 5 rem/year dose limit from reactor radiation is both reasonable and prudent.

Routine Operations and Expected Occurrences

- Use 29 CFR 1910.96 dose limits for on-board radiological sources.
- Radiological releases from the spacecraft shall not impair its use.
- Radiological release from the spacecraft shall not contribute significantly, over an extended period of time, to any local space environment.
- Radiological release from the spacecraft shall have an insignificant effect on Earth.

Accidents

- The probability of accidents involving radiological release affecting the immediate or long term health of the crew shall be extremely low.
- For those accidents involving radiological release for which the crew is expected to survive, the radiological release shall not render the spacecraft unusable.
- The probability of a significant radiological effect from an accident on any local space environment over an extended period of time shall be extremely low.
- The consequence on Earth of a radiological release from an accident in space shall be insignificant.

The term "insignificant" as used here for radiological release means "much less than the value specified for terrestrial guidelines." An extremely low probability event is one that is not expected to ever occur during the execution of the SEI program. "Significant" means "greater than the most appropriate guideline or norm." An "extended period of time" is understood to mean "encompassing the time period of potential future space enterprises in the region of space under consideration."

4. Disposal

Disposal plans for space reactor systems and associated spacecraft components should be established before the propulsion system is deployed. Furthermore, the method of disposal must be safe; that is, the radioactive materials of the disposed reactor system must not endanger the public or the environment. Strategies for disposal must preclude entry by orbital decay in order to comply with the requirement for no planned Earth entry (next topic). The reactor system integrity must be protected for all normal conditions and credible accident conditions, including collisions with meteoroids and orbital debris, that could compromise the user's ability to dispose of the reactor system safely.

The recommended disposal requirements are as follows:

- Safe disposal of spent nuclear systems shall be explicitly included in Space Exploration Initiative mission planning.
- Adequate and reliable cooling, control, and protection for the reactor system shall be provided for all normal and credible accident conditions to prevent reactor system disruption or degradation that could preclude safe disposal.

5. Entry

Entry refers to the event in which a reactor system enters the atmosphere and/or impacts the surface of Earth or another celestial body.

- Planned Earth entry shall be precluded from mission profiles.
- Both the probability and the consequences of an inadvertent entry shall be made as low as reasonably achievable.
- Inadvertent entry through an atmosphere shall be essentially intact, or, alternatively, result in essentially full dispersal of radioactivity at high altitude.
- For an impact, radioactivity shall be confined to a local area to limit radiological consequences.
- The reactor shall remain subcritical throughout an inadvertent entry and impact.

6. Safeguards

Safeguards encompass all measures used to control and protect special nuclear materials.

- Positive measures shall be provided to control and protect the nuclear system and its special nuclear materials (SNM) from theft, diversion, loss, or sabotage.
- To the extent practicable, the design of the nuclear system shall incorporate features that enhance safeguards and permit proven safeguards methods to be employed.
- Positive measures or features shall be provided to facilitate timely identification of the status as well as the location and, if necessary, recovery of the nuclear system or its SNM.

Ground Test Safety Recommendations

The NSPWG also recommended the general type of safety validation that may be required for nuclear propulsion systems and provided guidelines for ground facility and equipment safety.

Ground Test Needs for Safety Validation

Control of hazards associated with space nuclear propulsion systems will require safety test information to validate analysis and to support demonstration of compliance with safety requirements. Early guidance on safety validation testing requirements is needed to permit identification of the scope of test facility requirements.

The ground test element of the safety program should focus on data required to assure that safety objectives for flight systems will be achieved. These data are necessary to obtain flight approval. Some of the data identified will also be useful to other major tasks, such as ground testing of propulsion reactors.

The flight safety tasks have been logically separated into four groupings: launch and deployment safety, operational safety, disposal safety, and inadvertent entry safety. Potential safety testing that should be considered during facility planning have been delineated (Marshall et al. 1991)

Ground Facility and Equipment Safety

The testing of space nuclear propulsion reactors will be conducted on government sites in accordance with DOE orders. The DOE orders provide requirements and guidance consistent with the recommended policy statement. The DOE orders will require some interpretation in applying them to specific testing. Interpretation of specific details should be done as part of design development and the independent safety review process.

Requirements for the control of radioactive material during normal, off-normal, and credible accident conditions have been developed for many types of ground test reactors. These generally can be applied to the testing of nuclear thermal propulsion (NTP) and nuclear electric propulsion (NEP) reactors. However, the necessity to exhaust the reactor coolant in the NTP presents a unique area of requirements that have only been considered during the Nuclear Engine for Rocket Vehicle Application (NERVA) programs and the Aircraft Nuclear Propulsion (ANP) program.

The issue of beyond-design-basis accidents also requires special attention and may have a strong influence on the approach selected for containment or confinement. The inherent capabilities of the nuclear thermal propulsion fuel to retain fission products may also influence the selection of technology development activities and safety features. Safety design activities must evaluate the value of risk-reducing design approaches and the practicality of implementing the approaches.

CONCLUSIONS

The Safety Policy, Requirements, and Guidelines, and other recommendations have been provided to guide the design and development of nuclear propulsion systems for SEI. The Safety Requirements and Guidelines are intended to provide initial direction for early trade studies and should be expected to be expanded and refined as a result of development work and design evolution. Timely adoption or adaptation of these recommendations should greatly facilitate the implementation of the nuclear propulsion program and will assure the safety of the public, mission staff, and the environment during all program activities.

Acknowledgments

This work was a joint agency, multilaboratory effort supported by the Department of Energy (DOE), the National Aeronautics and Space Administration (NASA), and the Department of Defense (DoD). The principal author is employed by Sandia National Laboratories, which is operated for the U.S. Department of Energy under contract DE-AC04-76DP00789. Although this effort was led by the DOE, the funding and execution of this work was provided by the participating laboratories and agencies.

The authors gratefully acknowledge the contribution of the other NSPWG panel members and the industrial advisor. The panel members and the advisor were equal contributors in this effort and dedicated themselves to work cooperatively toward the successful completion of this task. The other panel members include: R. Bari, Brookhaven National Laboratory; H. Cullingford, NASA Lunar Mars Exploration Program; A. Hardy, NASA Johnson Space Center; K. Remp, NASA Lewis Research Center; J. Lee and W. McCulloch, Sandia National Laboratories; G. Niederauer, Los Alamos National Laboratory; J. Rice, Idaho National Engineering Laboratory; and J. Sholtis, U.S. Air Force. N. Brown, General Electric Co., served as the NSPWG industrial advisor.

References

Marshall, A. C. and J. C. Sawyer, Jr. (1991) "A Recommended Interagency Nuclear Propulsion Safety Policy," presentation at AIAA/NASA/OAI Conference on Advanced SEI Technology, Cleveland, OH, 4-6 September 1991.

Marshall, A.C., et al. (1991) *Nuclear Safety Policy Working Group Recommendations on Nuclear Propulsion Safety for the Space Exploration Initiative*, (To be published.) September 1991.

U.S. Department of Energy (1984) *DOE Order 5480.4: Environmental Protection, Safety, and Health Protection Standards*, 15 May 1984.

U.S. Occupational Safety and Health Administration, Department of Labor (1989) *U.S. Code of Federal Regulations, 29 CFR 1910.96, Chapter VII: Ionizing Radiation*, Office of the Federal Register, National Archives and Records, 1 July 1989.

Wahlquist, E.J. (1990) "U.S. Space Reactor Programs," *The Anniversary Specialist Conference on Nuclear Power Engineering in Space, Transactions Part 2, Papers of International Specialist*, Obninsk, U.S.S.R., 15-19 May 1990.

IMPLICATIONS OF THE 1990 ICRP RECOMMENDATIONS ON THE RISK ANALYSIS OF SPACE NUCLEAR SYSTEMS

Bart W. Bartram
HALLIBURTON NUS
Environmental Corporation
910 Clopper Road
Gaithersburg, MD 20877-0962

Abstract

The International Commission on Radiological Protection (ICRP) published its latest recommendations regarding radiation protection in ICRP Publication 60 (ICRP 1990). The document represents a significant revision to ICRP guidance on radiation protection contained in earlier ICRP publications. It clarifies the applicability of ICRP guidelines to planned actions involving potential accidents, and reflects consideration of the latest studies on the health effects of ionizing radiation. Implications of the ICRP recommendations on the nuclear safety considerations and risk analysis of space nuclear systems are discussed herein. Those areas affected include the applicability of ICRP guidelines to potential accidents, the calculation of internal dose, the assumed age distribution of the exposed population, and the health effects of ionizing radiation.

INTRODUCTION

The ICRP has periodically published its recommendations on radiation protection since 1959. Guidance of the National Council on Radiation Protection and Measurements (NCRP), and Federal regulations and guidance from the U.S. Nuclear Regulatory Commission (NRC), U.S. Environmental Protection Agency (EPA), and U.S. Department of Energy (DOE) related to radiation protection reflect consideration of ICRP guidance. The principles of radiation protection adopted by ICRP in ICRP-1 and the concepts of internal dosimetry established in ICRP-2 in 1959 have been propagated throughout the nuclear industry, as modified by subsequent ICRP Publications (ICRP 1959a and b). Of particular note is the work of the ICRP Task Group on Lung Dynamics that incorporated the concept of particle size into internal dose modeling (ICRP 1966), and the 1977 recommendations of the ICRP contained in ICRP-26 that incorporated results of studies on the health effects of ionizing radiation into a risk-based approach to dose level guidelines and internal dosimetry (ICRP 1977). The efforts of the Task Group on Lung Dynamics and the risk-based concepts developed in ICRP-26 culminated in a major revision to internal dose modeling, as contained in ICRP-30 published in 1979 (ICRP 1979). Consideration of more recent health effects studies resulted in updated guidance presented in ICRP-45 in 1985 (ICRP 1985).

The latest recommendations of the ICRP, presented in ICRP-60, represent a significant revision to ICRP guidance on radiation exposure contained in ICRP-26 and subsequent related pulications. ICRP-60 clarifies the applicability of ICRP guidelines to planned actions involving potential accidents, and reflects consideration of the latest studies on the health effects of ionizing radiations by the National Academy of Sciences, contained in the BEIR-V Report (NAS 1990), and the United Nations Scientific Committee on the Effects of Atomic Radiation (UNSCEAR 1988). Implications of ICRP-60 are provided herein, with the focus on those aspects that warrant consideration with respect to nuclear safety and risk analysis of space nuclear systems. As such, only those guidelines related to public exposure are addressed, rather than occupational exposure. When deemed appropriate, direct quotes from the draft are used to ensure that ICRP intent is properly conveyed.

DEFINITION OF CONCEPTS

ICRP-60 clarifies radiation protection guidelines applicable to potential accidents by defining three exposure situations as follows:

- Preexisting Exposure Situations

    Situations in which the sources, the pathways, and the exposed individuals are already in place when decisions about control measures are being considered.

- Planned Exposure Situations

    Situations in which the choice of practices giving rise to an exposure has not been made. Depending upon the choice made, planned exposure is subject to some degree of control.

- Potential Exposure Situations

    Situations in which not all exposures occur as forecast. There may be accidental departures from the planned operating procedures, or equipment may fail. Such events can be foreseen and their probability of occurrence estimated, but they cannot be predicted in detail.

Accidents and emergencies which may arise from planned exposure situations are considered as potential exposure situations, but if they actually happen and result in contamination, they become preexisting exposure situations.

OVERALL APPROACH TO RADIATION PROTECTION

The system of radiological protection recommended by ICRP for planned and potential exposure situations is based on the following general principles:

- Justification of a Practice

    "No practice involving exposures to radiation should be adopted unless its use produces sufficient benefit to the exposed individuals or to society to offset the radiation detriment it causes."

- Optimization of Protection

    "In relation to any particular source within a practice, the magnitude of individual doses, the number of people exposed, and the likelihood of incurring exposures where these are not certain to be received should all be kept as low as reasonably achievable, economic and social factors being taken into account. This procedure should be constrained by restrictions on the doses to individuals (dose constraints), or the risks to individuals in the case of potential situations (risk constraints), so as to limit the inequity likely to result from inherent economic and social judgments."

- Individual Dose or Risk Limits

    "The exposure of individuals resulting from the combination of all the relevant sources should be subject to dose limits, or to some control of risk in the case of potential exposure situations, aimed at ensuring that no individual is deliberately exposed to radiation risks that are judged to be unacceptable in any normal circumstances."

In dealing with preexisting exposure situations, these principles still apply, but in a somewhat modified form as follows:

- Guidelines for Preexisting Exposure Situations

    The first principle, the justification of a practice, now applies to the process of intervention. It has to be shown that the proposed intervention will do more good than harm. The second principle, the optimization of protection, is applied by choosing the form, scale, and duration of the intervention. The third principle, the use of limits, is not relevant.

APPLICABILITY OF DOSE AND RISK LIMITS

ICRP-60 makes it clear under what conditions ICRP recommended dose limits are to be applied:

- "It will never be appropriate to apply dose limits to all types of exposure in all circumstances... It is not surprising that management, regulatory authorities, and governments all improperly set out to apply dose limits whenever possible, even when the sources are partly, or even totally beyond their control, and when the optimization of protection is the more appropriate course of action."

- "The dose limits recommended by the Commission are intended for use in planned exposure situations and have been chosen to exclude the doses from preexisting exposure situations. The use of these dose limits, or any predetermined dose limits, in preexisting exposure situations might involve measures that would be out of all proportion to the benefit obtained."

- "Dose limits do not apply directly in potential exposure situations. At the planning stage, they have to be replaced by risk limits, which take account of both the probability of incurring a dose and the detriment associated with that does if it were to be received. If the event does occur, the potential exposure becomes a preexisting condition.

Thus, ICRP is clear that only risk limits should be considered for potential accidents (potential exposure situations) resulting from planned actions, rather than dose limits.

RATIONALE FOR DEVELOPING DOSE AND RISK LIMITS

The discussions of dose and risk limits by ICRP reflect consideration of the results of recent health effects studies and the concept of an acceptable level of risk. In the draft, ICRP has made some subtle adjustments in levels of acceptable risk and recommended dose limits, as originally developed in ICRP-26 and reflected in ICRP-45, due to the results of recent studies that indicate the health effects of ionizing radiation are higher than previously thought. The basic rationale used in the draft is as follows:

Recommended Dose Limits

The dose limits for public exposure as previously stated in ICRP-45 were as follows:

> "The commission's present view is that the principal limit is 1 mSv in a year. (Note: 1 Sievert = 1 Sv = 100 rem; 1 mSv = 100 mrem). However, it is permissible to use a subsidiary dose limit of 5 mSv in a year for some years, provided that the annual average dose equivalent over a lifetime does not exceed the principal limit of 1 mSv in a year."

ICRP-60 has modified the dose limit and introduced a new term "effective dose" to replace "effective dose equivalent". ICRP states:

> "The Commission's previous recommendations provided for a principal limit on the average annual effective dose over a lifetime, with a further limit in any single year. This recommendation is still sound in principle, but the Commission has concluded that the very long averaging period gives excessive flexibility. It now recommends that the limit for public exposure as an average effective dose of 1 mSv per year over a period of 5 consecutive years. It is implicit in this limit that the constraints for design purposes should not exceed 1 mSv in a year".

Thus, the current recommended dose limit for public exposure is an effective dose of 1 mSv per year averaged over any 5 consecutive years. (Note: The subsidiary dose limit of 5 mSv in a year for some years mentioned in ICRP-45 has been eliminated). Of particular note is that for the general population, that includes children, a 70-year committed dose is assumed. For adults, a 50-year committed dose is used.

Results of Health Effects Studies

The individual probability of cancer fatality due to radiation exposure (risk coefficient) used in ICRP-26 was 10^{-2} Sv^{-1} (equivalent to 100 fatalities per million person-rem). This value is consistent with estimates in the 1972 BEIR-I Report and the 1977 UNSCEAR Report (NAS 1980 and UNSCEAR 1977) for low-LET (linear energy transfer) radiation such as gammas and betas. Based on results of more recent health effects studies contained in the 1988 UNSCEAR Report and 1990 BEIR-V Report, ICRP recommends the following nominal risk coefficients for the general population, including children: 5×10^{-2} Sv^{-1} for low doses and 10×10^{-2} Sv^{-1} for high doses.

In this case "low doses" are absorbed doses below 0.2 Gy (1 Gray = 1 Gy = 100 rads) and from higher absorbed doses when the dose rate is less than 0.05 Gy per minute. For gammas, this corresponds to 20 rem and 5 rem/minute, respectively.

The higher risk coefficient, 10×10^{-2} Sv^{-1}, results from interpreting data at high doses to extrapolate to lower doses. ICRP reduced this by a factor of 2 for low doses to allow for the reduced effect of cell death at low doses.

Recommended Risk Limit

For potential exposure situations (including potential accidents), ICRP considers the concept of a risk limit (defined as the product of the probability of incurring a dose times and the lifetime conditional probability of attributable death from the dose given that the dose had been incurred). A risk limit could be developed that is equal to the nominal fatality probability corresponding to the long-term dose limit, that is approximately 5×10^{-5} for public exposure. Although ICRP had developed such a value in the draft leading up to ICRP-60, ICRP stopped short of recommending a numerical risk limit in ICRP-60 as published.

RELATED INFORMATION ON OCCUPATIONAL EXPOSURE

For completeness, the analogous ICRP guidelines for occupational workers are as follows:

- Dose limit: 100 mSv in 5 years
  50 mSv in any 1 year
  (based on the 50-year committed dose)

- Risk coefficient: 4×10^{-2} Sv^{-1} for low doses
  8×10^{-2} Sv^{-1} for high doses

COMPARISON WITH FEDERAL GUIDANCE

Historically, Federal guidance on radiation exposure has closely followed that established by ICRP. The NCRP in NCRP-91 adopted guidelines very similar to those contained in ICRP-26 and -45 (NCRP 1987). However, NCRP-91 takes the concept of acceptable risk one step further by establishing a "Negligible Level of Individual Risk" (NLIR) of 10^{-7} per year, corresponding to 0.01 mSv (1 mrem) per year based on the risk coefficient used (10^{-2} Sv^{-1}). EPA and DOE has also established 1 mSv (100 mrem) per year as a long-term dose level guideline in areas contaminated by transuranics above which clean-up is indicated to as low as reasonably achievable.

Concepts developed in ICRP-26, including risk coefficients and organ risk weighting factors, were used in developing internal dose conversion factors contained in ICRP-30 (ICRP 1979). These factors were later adopted by EPA, DOE, and NRC for use in dose calculations.

IMPLICATIONS FOR SPACE NUCLEAR SYSTEM SAFETY ANALYSES

The ICRP recommendations and the results of recent health effects studies presented in ICRP-60 have implications with respect to future space nuclear system safety analyses, as summarized below:

Health Effects Estimates

ICRP-60 adopts higher risk estimates for the health effects due to low-LET ionizing radiation that previously used. This is an important consideration when dealing with gamma and beta radiation associated with fission-reactor systems, such as the SP-100 space reactor, under normal and postulated accident conditions. NUS Corporation (NUS) used a value of 1.8×10^{-2} Sv^{-1} for the risk assessment of SP-100 and other fission-reactor systems in Bartram (1984 and 1988). ICRP is now recommending 5×10^{-2} Sv^{-1} for low doses and 10×10^{-2} for high doses, reflecting the results of the BEIR-V Report.

ICRP-60 addresses health effects due to high-LET radiation (neutrons and alphas) but does not present an explicit risk coefficient due to the organ-dependent nature of the supporting studies. This type of risk coefficient is appropriate when the plutonium-238 fueled GPHS-RTG and its mission applications are considered. NUS developed a high-LET risk coefficient applicable to plutonium-238 dioxide in the Ulysses Final Safety Analysis Report (FSAR), Vol. III based on the results of the BEIR-IV Report (NUS 1990 and BEIR 1988). The BEIR-V Report and ICRP-60 adopts the results of the BEIR-IV Report for high-LET radiation without modification. The NUS estimate for the risk coefficient, consistent with the later estimate of the Interagency Nuclear Safety Review Panel (INSRP) in the Ulysses Safety Evaluation Report (SER), was 3.5×10^{-2} Sv^{-2} (NUS 1990 and INSRP 1990). Previous NUS analyses, such as the Galileo FSAR, used 1.8×10^{-2} Sv^{-2} for high-LET radiation (NUS 1989).

Internal Dose Factors

Internal dose conversion factors used by NUS in risk assessments for DOE in safety analyses of space nuclear systems are consistent with the methods in ICRP-30 (ICRP 1979). The ICRP-30 method incorporates organ risk weighting factors for low-LET radiation based on the BEIR-I Report in arriving at effective equivalents. ICRP-60 proposes new organ risk weighting factors based on the BEIR-V Report. Thus, the internal dose factors for radionuclides with low-LET radiation are expected to change. Two other observations are:

- ICRP 60 has adopted a 70-year dose commitment period for the general population, rather than 50 years now applied to the adult population only. This would increase doses by about a factor of 1.2 for long-lived radionuclides with slow rates of biological clearance.

- ICRP-30 addressed dose conversion factors for adults only. ICRP appears to be moving towards age-dependent dose factors, as evidence by the age-dependent discussions in the draft, and a new ICRP publication related to age-dependent dose factors (ICRP 1989).

Regardless of the changes in organ risk weighting factors for low-LET radiation outlined above, NUS in effect modified the organ risk weighting factors for high-LET radiation in the Ulysses FSAR to reflect the BEIR-IV Report results. However, incorporation of age-dependent dose factors and extension of the dose commitment period to 70 years would still need to be done. Such changes would cause individual doses, especially for children, to increase.

CONCLUSIONS

Based on the information presented above, those involved in the nuclear safety aspects and risk analysis of space nuclear systems should closely follow ICRP and NCRP developments with respect to new guidelines and publications, new developments with respect to low- and high-LET risk coefficients, organ risk weighting factors, and internal dosimetry models, especially age-dependent extensions.

Acknowledgments

This work has spondored by the U.S. Department of Energy under Contract No. DE-AC01-87NE32134.

References

Bartram, B. W., R. W. Englehart, S. R. Tammara, and A. Weitzberg 1984 *Comparative Risk Analysis of Selected Missions Utilizing Space Nuclear Electric Power Systems*, NUS-4083, NUS Corporation, Gaithersburg, MD.

Bartram, B. W. and A. Weitzberg (1988) *Radiological Risk Analysis of Potential SP-100 Space Mission Scenarios*, Prepared by NUS Corporation for NUS-5125, NUS Corporation, Gaithersburg, MD.

ICRP (1959a), *Recommendations of the International Commission on Radiological Protection*, ICRP Publication 1, International Commission on Radiological Protection, New York, NY.

ICRP (1959b) *Report of Committee II on Permissible Dose for Internal Radiation*, ICRP Publication 2, International Commission on Radiological Protection, New York, NY.

ICRP (1977) *Recommendations of the ICRP*, ICRP Publication 26 International Commission on Radiological Protection, New York, NY.

ICRP (1979) *Limits for the Intake of Radionuclides by Workers*, ICRP-30 (1979 and additions thru International Commission on Radiological Protection, New York, NY.

ICRP (1985) *Quantitative Bases for Developing a Unified Index of Harm*, ICRP Publication 45 International Commission on Radiological Protection, New York, NY.

ICRP (1989) *Age-Dependent Doses to Members of the Public from Intake of Radionuclides: Part I*, ICRP Publication 56 International Commission on Radiological Protection, New York, NY.

ICRP (1990) *1990 Recommendations of the International Commission on Radiological Protection*, ICRP Publication 60 International Commission on Radiological Protection, New York, NY.

INSRP (1990) *Safety Evaluation Report for the Ulysses Mission*, INSRP-90-01 thru 03 Interagency Nuclear Safety Review Panel, New York, NY.

NAS (1980) National Academy of Sciences, Committee on the Biological Effects of Ionizing Radiation, *Health Effects of Exposure to Low Levels of Ionizing Radiation*, BEIR-III Report (1980).

NAS (1988) *Health Risks of Radon and Other Internally Deposited Alpha Emitters*, BEIR-IV Report National Academy of Sciences, Committee on the Biological Effects of Ionizing Radiation, Washington, D.C.

NAS (1990) *Health Effects of Exposure to Low Levels of Ionizing Radiation*, BEIR-V Report National Academy of Sciences, Committee on the Biological Effects of Ionizing Radiation, Washington, D.C.

NCRP (1987) *Recommendations on Limits for Exposure to Ionizing Radiation*, NCRP Report 91 National Council on Radiation Protection and Measurements, Washington, D.C.

NUS (1990) NUS Corporation, *Final Safety Analysis Report for the Ulysses Mission*, Volume III, Books 1 and 2, *Nuclear Risk Analyses Document*, Prepared for U.S. Department of Energy, ULS-FSAR-004 and 005 (1990).

NUS (1989) *Final Safety Analysis Report for the Galileo Mission* Volume III, Books 1 and 2, *Nuclear Risk Analysis Document*, Prepared for U.S. Department of Energy NUS Corporation, Gaithersburg, MD.

UNSCEAR (1988) *Sources, Effects and Risks of Ionizing Radiation*, E.88.IX.7 United Nations Scientific Committee on the Effects of Atomic Radiation, New York, NY.

UNSCEAR (1977) *Sources and Effects of Ionizing Radiation*, E.77.IX.I United Nations Scientific Committee on the Effects of Atomic Radiation, New York, NY.

THE PECULIARITIES OF PROVIDING NUCLEAR AND RADIATION SAFETY
OF SPACE NUCLEAR HIGH-POWER SYSTEMS

Albert S. Kaminsky, Victor S. Kuznetsov, Konstantin A. Pavlov,
Vladimir A. Pavshoock, and Lev Ja. Tikhonov
Kurchatov Institute of Atomic Energy
Moscow 123182, USSR
(095) 196-9443

Valery T. Khrushch
Institute of Biological Physics
Moscow 123182

INTRODUCTION

The problem of nuclear and radiation safety providing is one of the key ones while creating space nuclear power sources of any type. The practical application of such sources greatly depends on the solution of this problem. In the USSR, the development and operation of space nuclear power objects are performed taking into account the known recommendations of the United Nations Committee on the Peaceful Uses of Space (Ponomarev-Stepnoi 1989, Bennett 1990). The main safety principle is the reactor subcriticality providing at all stages of its development, fabrication, and delivery to the working orbit as well as the exception of its unapproved reentry after the reactor operation is ceased due to normal or emergency shutdown. The safety problem is conceptually solved for nuclear high-power (multimegawatt) systems in the same way as for electric low-power systems (Griaznov et al. 1989, Griaznov et al. 1990). Some principles formulated in the paper of Griaznov (1990) are specified taking into account the modern trends and construction-physical peculiarities of powerful systems. As an example, a nuclear high-power system (NHPS) with electric power 2 MW is considered for a transfer spacecraft with a total power generation 10 MW·yrs.

NUCLEAR SAFETY PROVIDING

The main requirement of nuclear safety before the delivery of a spacecraft to the working orbit is the reactor subcriticality providing. However, transport incidents and accidents of a transfer vehicle at the delivery stage may lead to the reactor configuration change and the correspondent reactivity insertion. The most hazardous core configurations appear when it becomes packed or when the reactor internal cavities are filled with a hydrogeneous medium. The analyses show that the core uniform packing by 10% inserts 5% of positive reactivity, and the cavities filling and the reactor surrounding with hydrogeneous medium may insert up to 25% of positive reactivity.

The conservation of the reactor subcritical state must be provided by the optimum combination of passive and active means. The optimization is proceeded from the condition of deterioration minimization of weght-dimension and operational reactor parameters.

The following means are considered as possible reactivity compensators:

- Operational organs of the control and safety system;
- Supplementary neutron absorbers removed from the reactor before it is brought to power;
- Resonance absorbers in the core the effectivity of which is increased with the neutron spectrum softening in the case when hydrogeneous matters appear in the reactor;
- Special means (supplanters) in the reactor cavities excluding the core packing during its deformation (impacts) and filling it with hydrogeneous medium which are removed after the system delivery to the working orbit;
- Planned changing of the reactor configuration (reflector travelling blocks, the core division into moved apart parts on the base of the principle of the construction nonequal strength).

RADIATION SAFETY PROVIDING

According to the concept adopted in the USSR, radiation safety was provided by the on-board systems of transfer to the long life orbit and by the reactor dispersion in the case of these systems failure (Griaznov et al. 1989, Griaznov et al. 1990). The specification of this concept is necessary, first of all, because the admissibility of reentry of any radiational objects including dispersed ones is quite doubtful. In the second place, the use of disposal orbits of 600-800 km hight which were earlier considered admissible is undesirable because of the concentration increase of artificial objects at 350-1250 km hight. The circle orbits in the range of 4500-15,000 km, 25,000-30,000 km, and more than 50,000 km hight are considered to be the most suitable for disposal.

The technical realization of the specified concept is connected with the use of ground- and space-based disposal rocket systems. The on-board disposal system is expedient to be used for the delivery of the NHPS disposed part to the intermediate orbit of 400-1000 km hight, and then, auxiliary systems are to be used for the delivery of disposed parts to the orbit of disposal. The auxiliary system must also provide the delivery of disposed objects to the orbit of disposal in the case of the on-board system failure. The preliminary technical-economic analyses show that the interorbit transportation with the help of space-based shuttle and versatile means (both on the base of LPRE and NRE or ERP) at the final phases of their active operation using the resourses left after the accomplishment of other missions will be the least expensive. In this case, the NHPS delivery to the hight more than 50,000 km will be connected with additional expenditures not exceeding 30% of the cost of NHPS delivery to the the working orbit of 300 km hight (the base cost).

As a backup option, the application of the ground-based system of single use delivered in space for the interception of the NHPS disposed parts is realized. The application of this option will lead to the additional expenditures exceeding the base cost by 90-120%.

Nevertheless, there exists the limited by standard requirements possibility of accidental NHPS reentry.

Radiation safety requirements are provided by the series of qualitative and quantitative criteria adopted by experts on the international level.

As the main radiation safety criterion, the limitation of the individual effectiive exposure dose is adopted to be 1 mSv per year. The probability mean approach is commonly used while analysing the radiation situation.

Monte-Carlo calculation analysis shows that in the range of values of calculated density of radioactive particles deposition $\nu \simeq (10^{-5}...10^{-3}) 1/m^2$, the ratio of dose intensity, the probability of nonexceeding which is 99.9%, to the mean probability value is in the range 12 $\div$ 30. Taking into consideration this factor makes the approach to safety criteria providing quite stricter.

For the specific case where $\nu = 2.44 \cdot 10^{-4} m^{-2}$ (calculated density of satellite Kosmos-954 fragments fall (INFO-0006 1980)), the expected effective equivalent exposure dose depending on exposure time with the probability 99.9% will not exceed the values given in Table 1 with a total power generation 10 MW yrs (about 10^{25} fissions) and fragment mass of U-235 about 0.7 g.

TABLE 1. The Limited Values of Expected Individual Equivalent Exposure Dose (with the Probability 99,9%) with the Exposure 30-1000 Years for a Total Power Generation 10 MW·yrs and $\nu = 2.44 \cdot 10^{-4} m^{-2}$.

Exposure Time, yrs	30	100	300	500	1000
Expected Individual Effective Equivalent Exposure Dose, mSv per year	480	94	1	0.03	0.016

Thus, the exposure time, providing the fulfillment of the radiation safety main criterion, does not exceed 300 years for the considered reactor according to the adopted calculated fall density.

To-day, there is not any commonly used approach to the practical application of the notion "collective dose" characterizing the scale of common affecting. To develop this approach, it is necessary to take into consideration the following circumstances:

1. The mean, according to the expectation probability, collective effective equivalent dose of gamma-ray, beta, and internal exposure does not practically depend on the extent of dispersion of radioactive matter fell on the Earth surface with other equal conditions (reactor campaign, fission products exposure time);
2. The mean, according to the probability, expected collective effective equivalent dose, caused by nuclear fuel (uranium isotops), may serve as a base measure for the evaluation of the maximal acheived numerical values of radiation safety reliability indices which the public adopts simultaneously with the decision to use nuclear reactors in space;
3. The increase of fission products exposure time more than 1000 years does not practically lead to the reducing of the expected collective dose.

It should be noted that fission products activity with the exposure time 1000 years after the realization of energy generation 10 MW·yrs is compared to the natural activity of nonexposed to radiation fuel for the typical composition of highly enriched nuclear fuel of space systems.

CONCLUSIONS

The selection of composition of active and passive means of reactivity suppression and the application of space-based and ground-based rocket systems for averting radionuclide appearance in the Earth biological sphere may serve as the main premise during the further studies of nuclear and radiation safety of space NHPS. Approach specification is necessary to increase the reliability of radiation safety prediction during the contingency reentry of a nuclear reactor.

Acknowledgments

This research was conducted in close collaboration of Kurchatov Institute of Atomic Energy and Institute of Biological Physics.

References

Ponomarev-Stepnoi, N. N. (1989) "Nuclear Power in Space," Presentation to the 6-th Symposium on Space Nuclear Power Systems. Atomnaja Energia, v.66, No.6, 1989.

Bennet, G. L. (1990) "Soviet Space Nuclear Reactor Incidents: Perception versus Reality," Presentstion to the 7-th Symposium on Space Nuclear Power Systems, held in Albuquerque, NM, January 1990.

Griaznov, G. M., V. S. Nikolaev, V. I. Serbin, and V. M. Tugin (1989) "The Concept of Space NPS Radiation Safety and Its Realization at "Kosmos-1900" satellite," Atomnaja Energia, v. 66, No.6, 1989.

Griaznov, G. M., E. E. Zhabitinsky, V. S. Nikolaev, V. I. Serbin (1990) "The Concept of Nuclear and Radiation Safety of Space Thermionic Systems," Presentation to the 7-th Symposium on Space Nuclear Power Systems, held in Albuquerque, NM, January 1990.

INFO-0006 (1980) "Kosmos-954. The Location and Characteristics of Discovered Fragments," Canada, May 1980. A. M. Bludov and V. S. Nikolaev, Private translation from English, 125 pages.

SP-100 APPROACH TO ASSURE DESIGN MARGINS FOR SAFETY AND LIFETIME

A. Richard Gilchrist, Richard Prusa, Ernest P. Cupo, Michael A. Smith
and Neil W. Brown
General Electric Company, Astro Space Division
P. O. Box 530954
San Jose, CA 95153-5354
(408) 365-6459

Abstract

An innovative, cost-effective approach has been developed to validate lifetime of the SP-100 Space Reactor Power System (SRPS). Life-limiting failure mechanisms are identified and evaluated, and associated lifetime margins are demonstrated through a combination of analysis and accelerated feature testing. The approach is flexible and allows for tailoring to address the full scope of functions and technologies (including those specifically related to safety features) inherent in the SP-100 SRPS design. At various stages throughout the process, feedback to the SRPS design allows for adjustments that mitigate or eliminate concern over the identified failure mechanisms.

INTRODUCTION

The traditional approach to establishing or validating product lifetime is to base it on a statistical evaluation of long-term, multiple-sample life tests. While this approach may be straightforward and appropriate for high-volume production items (for example, light bulbs or commercial nuclear fuel), it is not feasible for more complex and/or expensive, small-scale production items such as space reactor power systems. Thus, an alternate approach is required that does not depend on standard statistical methods, but that provides sufficient evidence of lifetime capabilities to achieve a proper balance between cost and risk.

The SP-100 SRPS (Josloff et al. 1991, Pluta et al. 1989) is designed for a ten-year on-orbit lifetime, with seven years of full power operation. A new approach to lifetime validation (Figure 1) has been developed, using a structured combination of analysis and accelerated feature testing to determine lifetime margins for critical functions. The major elements of this approach are discussed in the following sections.

FAILURE MODES, EFFECTS AND CRITICALITY ANALYSIS

Failure Modes, Effects and Criticality Analysis (FMECA) is an established technique for identifying (1) potential design weaknesses through systematic, documented consideration of the likely ways in which a component can fail, (2) causes for each failure mode, and (3) the effects of each failure. FMECA has been applied iteratively as part of the SP-100 SRPS design process to identify critical failure possibilities for corrective action. The FMECA results are also used to identify potential life-limiting failure modes and associated failure mechanisms for further evaluation supporting lifetime validation.

A System/Summary FMECA and eight Subsystem FMECAs were completed prior to the SP-100 System Design Review (SDR). Subsequent to the SDR, these analyses have been supplemented by selective application of Failure Modes and Effects Analysis (FMEA) to address the impact of design changes and updates.

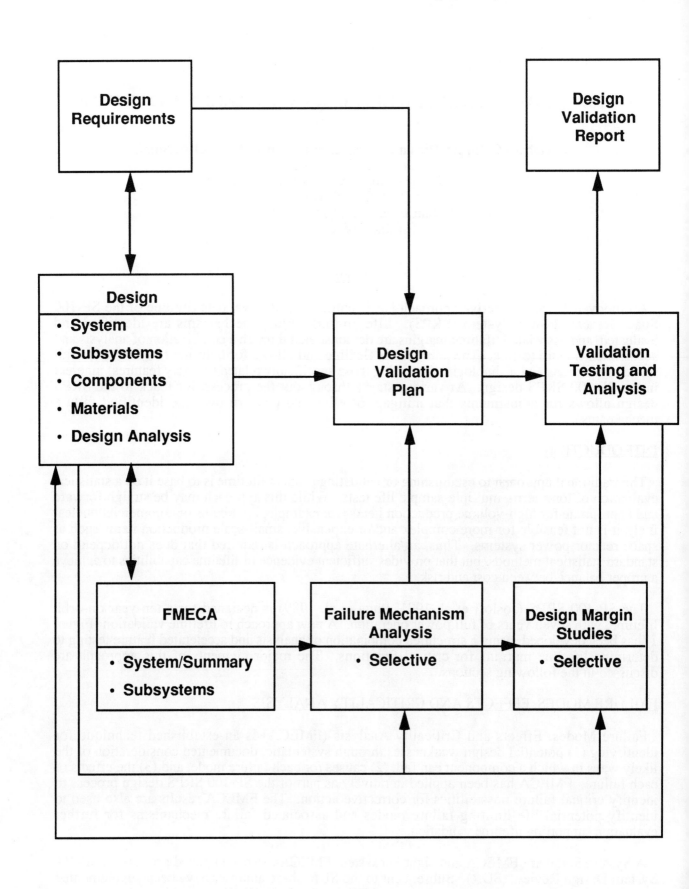

FIGURE 1. SP-100 Lifetime Validation Process.

FAILURE MECHANISM ANALYSIS

FMECA results are examined as the primary source for identification of failure mechanisms associated with life-limitng component failure modes. Additional mechanisms identified through activities such as critical review, testing, or analysis are also evaluated for impact on system lifetime. The failure mechanisms are then screened to focus attention on those related to SP-100 design concepts, material applications, or environmental conditions for which limited amounts of data are available. A failure mechanism is eliminated from further evaluation at this point if (1) it has no significant impact on system lifetime or (2) sufficient information already exists (or is being developed under other programs) to adequately quantify lifetime margins.

Failure Mechanism Analysis (FMA) is performed to provide a basis for understanding, characterizing, and/or mitigating life-limiting mechanisms. It is a structured process for: documenting the current state of knowledge for a given failure mechanism, including identification of affected component(s), failure processes, environmental factors, controlling parameters, design margins, potential means for mitigation, and plans for additional testing; assessing programmatic risk; and, where appropriate, defining information needs for further testing and analysis. FMA is largely qualitative but may be supported with quantitative evaluations.

FMA reports are prepared in a standard format (Figure 2), in accordance with specific guidelines for content, to provide a concise summary of pertinent information. Over sixty FMA reports have been completed to date, and additional ones are planned. They provide input to the Design Validation Plan as guidance for lifetime testing and analysis activities associated with the reactor, shield, heat transport, instrumentation and control, power conversion, and heat rejection subsystems. Most of the failure mechanisms under evaluation relate to degradation processes such as creep, self-welding, material compatibility/interaction, and irradiation effects.

DESIGN MARGIN STUDIES

Design Margin Studies build on FMA results to investigate selected failure mechanisms in greater detail and provide quantitative evaluation of lifetime margins. These studies identify the controlling parameters and functional relationships that determine lifetime/failure, evaluate available data, and determine the resulting margins to failure. Where results are incomplete or inconclusive, information needs are defined and requirements for test articles, test conditions, data acquisition, and data analysis are specified.

The Design Margin Studies are documented in formal reports, with details of format and content tailored to fit the specific failure mechanism. A small number of Design Margin Studies have been completed to date, indicating positive lifetime margins for key failure mechanisms. Additional studies are planned.

LIFETIME TESTING AND ANALYSIS

Failure Mechanism Analysis provides top level guidance for planning of subsequent lifetime testing and analysis, while Design Margin Studies provide specific, quantitative requirements. The focus of the testing and analysis activities is to provide a measure of component lifetime margins relative to the ten-year on-orbit lifetime requirement.

In general, lifetime testing and analysis are to be performed at the lowest feasible level of hardware assembly at which specified failure processes can be understood and measured. In many cases this will lead to testing of representative material samples. Degradation rates, conditions for failure, and/or design margins are predicted analytically based on established methodology and available data. Tests are performed where required to verify these analyses, and confirmatory endurance tests are performed selectively at higher levels of assembly. Wherever possible,

FAILURE MECHANISM ANALYSIS REPORT

FAILURE MECHANISM:

SUBSYSTEM VALIDATION ITEM:	SUBSYSTEM VALIDATION ITEM:
COMPONENT(S):	COMPONENT(S):
SUBSYSTEM VALIDATION ITEM:	SUBSYSTEM VALIDATION ITEM:
COMPONENT(S):	COMPONENT(S):

SECTIONS A-C	PREPARED BY:	DATE:
	REVIEWED BY:	DATE:
MANAGEMENT REVIEW	PROCEED WITH SECTION D? ☐ YES ☐ NO	
	APPROVED BY:	DATE:
SECTION D	PREPARED BY:	DATE:
	REVIEWED BY:	DATE:
SECTION E	PREPARED BY:	DATE:
	REVIEWED BY:	DATE:

FAILURE MECHANISM ANALYSIS REPORT (PAGE 2)

FAILURE MECHANISM:

A. FAILURE MECHANISM DESCRIPTION

B. CURRENT STATUS

C. SITUATION ASSESSMENT

FAILURE MECHANISM ANALYSIS REPORT (PAGE 3)

FAILURE MECHANISM

D. INFORMATION REQUIREMENTS

REFERENCES:

FIGURE 2. Failure Mechanism Analysis Report Format.

accelerated aging and prudent extrapolation of test data are used to minimize test duration.

Based on the Design Margin Studies and the results of lifetime testing and analysis, an overall assessment of SP-100 SRPS lifetime margins will be performed and documented in the Design Validation Report.

CONCLUSIONS

An innovative, cost-effective approach has been developed in which lifetime margins are demonstrated through a structured combination of testing and analysis to validate lifetime of the SP-100 SRPS without the need for large numbers of long-term tests.

Acknowledgments

This work is being performed as part of a contract (DE-AC03-86SF16006) between the U.S. Department of Energy, San Francisco Operations Office and the General Electric Company.

References

Josloff, A. T., D. N. Matteo and H. S. Bailey (1991) "SP-100 System Design and Technology Progress," in *Proc. 26th Intersociety Energy Conversion Engineering Conference*, held in Boston, MA, 4-9 August 1991.

Pluta, P. R., M. A. Smith and D. N. Matteo (1989) "SP-100, A Flexible Technology for Space Power from 10s to 100s of KWe," Paper No. 899287, in *Proc. of the 24th Intersociety Energy Conversion Engineering Conference*, held in Washington, DC, 6-11 August 1989.

A UNIQUE THERMAL ROCKET ENGINE USING A PARTICLE BED REACTOR

Donald W. Culver, Wayne B. Dahl, and Melvin C. McIlwain
Aerojet Propulsion Division
P.O. Box 13222
Sacramento, CA 95813-6000
(816) 355-2083

ASSESSMENT OF THE USE OF H_2, CH_4, NH_3, AND CO_2 AS NTR PROPELLANTS

Elezabeth C. Selcow, Richard E. Davis, Kenneth R. Perkins,
Hans Ludewig, and Ralph J. Cerbone
Brookhaven National Laboratory
Upton, NY 11973
(516) 282-2624

ENABLER-II: A HIGH PERFORMANCE PRISMATIC FUEL NUCLEAR ROCKET ENGINE

Lyman J. Petrosky
Westinghouse Advanced Energy Systems
P.O. Box 158
Madison, PA 15663
(412) 722-5110

THE TEXT OF THESE PAPERS ARE PRESENTED BEGINNING WITH PAGE 714 OF THESE PROCEEDINGS IN SESSION 18 "NUCLEAR THERMAL PROPULSION I"

NUCLEAR ELECTRIC PROPULSION
DEVELOPMENT AND QUALIFICATION FACILITIES

Dale Dutt
Westinghouse Hanford Company
P. O. Box 1970
Richland, WA 99352
(509) 376-9336

Jim Sovey
NASA Lewis Research Center
2100 Brookpark Road
Cleveland, OH 44135
(216) 977-7454

Keith Thomassen
Lawrence Livermore National Laboratory
7000 East Avenue
Livermore, CA 94550
(415) 422-9815

Mario Fontana
Oak Ridge National Laboratory
P. O. Box 2009
Oak Ridge, TN 37831
(615) 574-0273

Abstract

This paper summarizes the findings of a Tri-Agency panel; consisting of members from the National Aeronautics and Space Administration (NASA), U.S. Department of Energy (DOE), and U.S. Department of Defense (DOD); charged with reviewing the status and availability of facilities to test components and subsystems for megawatt-class nuclear electric propulsion (NEP) systems. The facilities required to support development of NEP are available in NASA centers, DOE laboratories, and industry. However, several key facilities require significant and near-term modification in order to perform the testing required to meet a 2014 launch date. For the higher powered Mars cargo and piloted missions, the priority established for facility preparation is: (1 thruster developmental testing facility, (2 thruster lifetime testing facility, (3 dynamic energy conversion development and demonstration facility, and (4 advanced reactor testing facility (if required to demonstrate an advanced multiwatt power system). Facilities to support development of the power conditioning and heat rejection subsystems are available in industry, federal laboratories, and universities. In addition to the development facilities, a new preflight qualification and acceptance testing facility will be required to support the deployment of NEP systems for precursor, cargo, or piloted Mars missions. Because the deployment strategy for NEP involves early demonstration missions, the demonstration of the SP-100 power system is needed by the early 2000s.

INTRODUCTION

The President's initiative to return humans to the surface of the moon and then to proceed with human exploration of Mars requires high performance propulsion systems for cargo and human transport. Nuclear electric propulsion (NEP) and nuclear thermal propulsion (NTP) have been identified as enabling technologies to support the Mars piloted mission. Six panels with members from the U.S. Department of Energy (DOE), National Aeronautics and Space Administration (NASA), and the U.S. Department of Defense (DOD) were formed to review technology status and formulate plans for nuclear propulsion development for a piloted mission to Mars in the 2014 to 2019 timeframe.

Re-establishment of the facilities required for development and to conduct performance testing will be one of the pacing activities for the NP Program. Thus, a panel was formed to review testing requirements, resulting facility requirements, available facilities, and, then, recommend a facilities strategy for the participating agencies. This panel formed two subpanels to review the facility needs and availability, and funding priority for NEP and NTP, respectively. The activities of the NEP Facilities Panel are summarized in the remainder of this paper.

PROCESS

The Nuclear Electric Propulsion Facilities Panel and the Nuclear Electric Propulsion Technology Panel have common members and often meet jointly to ensure the responsiveness of planning. The development/demonstration strategy which is advocated by the NEP Technology Panel is key to the facilities plan. NEP systems are comprised of five major subsystems, heat source (reactor), energy conversion, power conditioning, heat rejection, and thrusters (Figure 1). These subsystems are relatively independent and can be developed and lifetime or performance demonstrated at the subsystem level. Testing requirements for NEP technologies, components, and subsystems, were provided by the NEP Technologies and the Fuels and Materials Panels (Table 1).

TABLE 1. Technology Requirements

Subsystem	Interplanetary Precursor 100 kWe, 30-50 kg/kWe	Lunar/Mars Cargo 1-5 MWe, 10-20 kg/kWe	Mars Piloted 500-600 Day 10-20 MWe, 7-10 kg/kWe	Mars Piloted Quick Trip 10-40 MWe, 3-7 kg/kWe
Reactor	SP-100 to Technology Readiness Level (TRL)-5, 1996	SP-100 Growth or Advanced, 25 MWt, 2000	Multiple Units	Advanced Reactor System, TRL-05, 2006
Power Conversion	Thermal Electric, TRL-5, 1997	Dynamic, TRL-5, 2000	Light-Weight Dynamic, TRL-5, 2006	Upgrade, TRL-5, 2006
Radiator-Heat Rejection	Pumped Loop to TE/Heat Pipe Radiator, TRL-5, 1997	Heat Pipe Carbon-Carbon, TRL-5, 2000	Upgrade, TRL-5, 2006	Upgrade, TRL-5, 2006
Power Management and Distribution (PMAD)	Shielded from Radiation, Conventional Si Technology, TRL-5, 1998	Shielded Conventional, TRL-5, 2000	High Temperature, Advanced, TRL-5, 2006	Upgrade, TRL-5, 2006
Thruster	Magnetoplasmadynamic (MPD) or Ion Thruster, TRL-5, 1998, Because of Lifetime Test	MPD or Ion, TRL-5, 2000	MPD, Ion or Advanced, TRL-5, 2006	Upgrade, TRL-5, 2006

One of the major assumptions/agreements reached by the NEP Technology and Facilities Panels was that an integrated NEP system test cannot be properly performed on earth. The first test of the entire propulsion system will probably be conducted in space on a demonstration mission. Thus, NEP lends itself to an evolutionary developmental approach where initial precursor missions, such as interplanetary science probes, are conducted with relatively near-term technologies as demonstration tests. Later, higher power cargo and piloted missions will use NEP systems comprised of larger scale subsystems or advanced technologies.

The modularity/independence of NEP subsystems greatly reduced the demand on any individual facility. Most of the NEP subsystems have scalability issues that must be addressed by a rigorous testing program. Thus, all of the recommended facilities are sized to accommodate those components/subsystems needed for the higher power piloted Mars missions. Subsystem integration tests for the reactor/conversion and thruster/power conditioning will be accommodated in the proposed facilities.

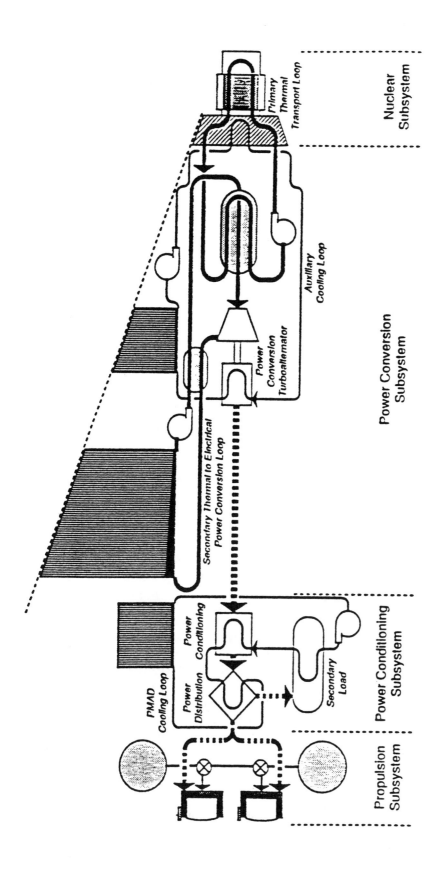

FIGURE 1. Nuclear Electric Propulsion System Schematic (from Doherty, 1991).

OVERVIEW OF AVAILABLE FACILITIES

The facility requirements to meet the testing requirements are provided in Table 2.

TABLE 2. Major Facility Requirements.

	Near Term	Cargo	Piloted A	Piloted B
Reactor	X	50 MWt ———————————————————>		
Power Conversion	X	2.5 MWe ——————>		5 MWe
Heat Rejection	X	*	*	*
PMAD	*	*	*	*
Thruster	< 0.5 MWe Component and Lifetime Testing (2 Facilities)	~ 2 MWe ———————————————————> Upgrade for Systems Tests		

* - No new facility required, available in industry or laboratories.
X - Test facilities supplied by SP-100 Program.

Facilities to Support Reactor Development

The SP-100 Flight System Qualification Program is focused on demonstrating technology and components for a 100 kWe power system with thermoelectric conversions. Resolution of the issues of SP-100 technology scalability and (assuming the technology scales), the applicability of the currently planned SP-100 ground test to the larger, higher powered systems required for 5-40 MWe piloted mission is key to the final facilities plan for reactor development.

In the event an advanced reactor technology is shown to provide significant mission performance benefits, an advanced fuels and materials technology program could be accommodated in existing DOE reactors. If the advanced concept is a liquid-metal-cooled fast reactor, the Fast Flux Test Facility and the Experimental Breeder Reactor-II can accommodate all planned testing. If the concept uses a gas-cooled Brayton cycle, a gas loop could be built for the Advanced Test Reactor.

The decision to conduct a large scale test (either with SP-100 or advanced technologies) would require an immediate commitment to develop a test facility to meet environmental, safety, and health requirements. The personnel at the SP-100 Test Site at Hanford provided cost and schedule estimates for modification of the facility to accommodate an order of magnitude increased thermal power (25 MWt). The 8- to 10-year construction schedule was accepted as typical for all potential reactor test sites. In addition to the Hanford Site, detailed reviews were conducted of the Idaho National Engineering Laboratory (INEL) Contained Test Facility, the Oak Ridge National Laboratory (ORNL) Experimental Gas Cooled Reactor, and the Sandia National Laboratories. Because of the perceived requirement for containment and remote locating, future and more detailed site reviews will focus on INEL, ORNL, and Hanford.

Facilities to Support Energy Conversion Development

The energy conversion technology used for the piloted Mars mission will most likely be a dynamic system (Brayton, Rankine, or Stirling). NASA and DOE laboratories have been active in energy conversion technology development since the 1960s. Historically, Brayton and Stirling conversion has been developed under the direction of NASA centers and Rankine under the direction of a DOE center.

A high temperature gas loop to support Brayton at ORNL was reviewed extensively, and NASA-Lewis has an ongoing Brayton development program. Brayton testing can be conducted by NASA and ORNL with minor upgrading of existing capabilities. If a full-scale Brayton development program were undertaken, it is likely that both the NASA and ORNL facilities would be used. The ORNL-led activities to demonstrate potassium Rankine in the 1960s and early 1970s provided a technology base. However, none of the facilities required for development of boilers, turbines, and subsystem testing exist today. A significant program would be required to re-establish those facilities, potentially by adapting liquid metal systems now on standby. System studies to determine if these facilities should be re-established are key to the NEP deployment schedule.

Facilities to Support Power Management and Distribution

Facilities for testing power conditioning and power management components and subsystems exist in university, industry, and government laboratories. No additional resource needs were identified by the panel.

Facilities to Support Thruster Development

Ion, magnetoplasmadynamic, and advanced thrusters issues are performance, lifetime, and scalability. There is a need for two test facilities; one for component development testing and a second for long term lifetime and performance testing. The key parameters for thruster testing facilities is the vacuum chamber size required for the megawatt-class thrusters.

Facilities at NASA-Lewis, Lawrence Livermore National Laboratory (LLNL), Oak Ridge National Laboratory, Arnold Engineering Development Center, Los Alamos National Laboratory, and Phillips Laboratories were reviewed for applicability. All the facilities reviewed will require some modification to accommodate the beam dump and achieve appropriate vacuum levels. Because of availability considerations, the facilities at NASA-Lewis, ORNL, and LLNL appear to be the most promising candidates (Sovey, 1991).

NASA-Lewis Research Center's Electric Power Laboratory houses two large space simulation chambers, tanks 5 and 6. Tank 5 is 4.6 m in diameter and 19 m long. Sufficient liquid helium or gaseous helium exists for 8-hour tests. Tank 6 is 7.6 m in diameter and 22 m in length. A NASA-funded rehabilitation of the tank is scheduled to be completed in January 1993. The tank will be capable of dissipating 0.35 MWe. The LLNL Magnetic Fusion Tandem Mirror Test Facility (MFTF) is 10.6 m in diameter and 55 m long. Large amounts of liquid helium storage capability, as well as 8 kW and 3 kW refrigeration/liquefier systems, are available onsite. Large magnets from the original program must be removed to allow use as a thruster test facility. The Large Coil Test Facility (LCTF) at ORNL is a 10.7-m-diameter, 9.1-m-high cylindrical chamber with a removable lid, vacuum capability, and liquid helium wall cooling; however, the helium cryosystems were last operated in 1986.

Facilities to Support Heat Rejection Development

A large number of vacuum chambers of appropriate size exist within industry, particularly aerospace companies. Although some additional heat rejection capability may be required, the anticipated costs are within the capability of a typical development program.

RECOMMENDATIONS

The subpanel recommendations are as follows:

1. Complete the SP-100 ground test and demonstration of thermoelectric conversion before the turn of the century to support flight demonstration and precursor missions;

2. Immediately start modifications to provide a thruster development test facility;

3. Immediately start modifications to provide a thruster lifetime test facility;

4. Complete system studies to determine the need for an advanced reactor system versus a scaled SP-100 reactor (the need for a new reactor test facility is dependent on results); and

5. Initiate the construction of a dynamic conversion test facility based on results of the systems in item 4.

The thruster, power conversion, and advanced reactor development facilities are highest priority (Table 3). The facility requirements to develop NEP appears to be manageable within current budget expectations.

TABLE 3. Priority and Availability of the Facilities to Support NEP.

New/Modification	NEP Facility Funding Priority	Existing Facilities
	HIGHEST	
Thruster Component Development		Near-Term Fuels Fabrication
Thruster Lifetime		Near-Term Reactor Ground Test
Dynamic Energy Conversion		Heat Rejection
Advanced Reactor Ground Test[a]		PMAD
		Fuels & Materials Examination
		Unirradiated Fuels & Materials Testing
	MEDIUM	
Advanced Fuel Fabrication		Materials Irradiation Testing
Fuel Element Testing Loops		Safety and Off-Normal Operations
Flight Test Support		
	LOWEST	
Advanced Thruster (Particles) Testing		Control Systems
		Shielding

[a] If reactor > 50 MWt is needed, depends upon mission analysis/choices.

Acknowledgments

The contribution of the following NEP Facilities Panel members Sam Bhattacharyya (ANL), John Dearien (INEL), Bob Holcomb (ORNL), and Mike Mahaffey (WHC), and the Air Force's Phillips Laboratories and Arnold Engineering Development Center provided valuable guidance and technical balance to the facilities panel.

References

Sovey, J. S. et al., (1991), Test Facilities for the High Power Electric Propulsion, AIAA 91-3499, American Institute of Aeronautics and Astronautics, September 1991.

Doherty, M. P., (1991), Blazing the Trailway: Nuclear Electric Propulsion and Its Technology Program Plans, AIAA 91-3441, American Institute of Aeronautics and Astronautics, September 1991.

THERMIONIC SYSTEM EVALUATION TEST (TSET)
FACILITY DESCRIPTION

Jerry F. Fairchild and James P. Koonmen
Phillips Laboratory
Space Power Branch
PL/STPP
Kirtland AFB, NM 87117-6008
(505) 272-7219

Frank V. Thome
Sandia National Laboratories
Division 6474
PO Box 5800
Albuquerque, NM 87185-5800
(505) 272-7218

Abstract

A consortium of US agencies are involved in the Thermionic System Evaluation Test (TSET) which is being supported by the Strategic Defense Initiative Organization (SDIO). The project is a ground test of an unfueled Soviet TOPAZ-II in-core thermionic space reactor powered by electrical heat. It is part of the United States' national thermionic space nuclear power program. It will be tested in Albuquerque, New Mexico at the New Mexico Engineering Research Institute complex by the Phillips Laboratory, Sandia National Laboratories, Los Alamos National Laboratory, and the University of New Mexico. One of TSET's many objectives is to demonstrate that the US can operate and test a complete space nuclear power system, in the electrical heater configuration, at a low cost. Great efforts have been made to help reduce facility costs during the first phase of this project. These costs include structural, mechanical, and electrical modifications to the existing facility as well as the installation of additional emergency systems to mitigate the effects of utility power losses and alkali metal fires.

INTRODUCTION

The Thermionic System Evaluation Test (TSET) is a unique opportunity for the US to test the complete balance of plant of an in-core thermionic space nuclear power system which will compress development processes and cut costs (Henderson 1991). The system is an unfueled Soviet TOPAZ-II thermionic space reactor which will be procured by SDIO. The approach is to include the USSR experience as a part of the total thermionics program (Figure 1) to advance thermionic system designs in the US faster and less expensive than could be accomplished by US efforts alone (Standley et al. 1991).

A wide variety of tests related to thermionic technology will be carried out at the TSET facility. Tests associated with TOPAZ-II will include, but not be limited to, its steady state operation, transient response, simulated solar flux variation, radiator degradation, and possibly coolant freeze/thaw. Included in the purchase of this system are several additional TOPAZ-II components and an electrically heated Thermionic Fuel Element (TFE) test stand. The additional components will be tested separately from the TOPAZ-II system tests. They will include the electromagnetic pump, cesium reservoir, control drum actuators, and moderator material. The TFE test stand will allow experiments to be conducted on emitter/collector materials, interelectrode gap pressure and thickness, electrode atmosphere, and insulators for single cell thermionic converters.

SDIO is funding the project and has given the responsibility for project execution to the Phillips Laboratory. The project is also being supported by Sandia National Laboratories, Los Alamos National Laboratory, and the University of New Mexico under the New Mexico Strategic Alliance agreement. Test facilities are located on the campus of the University of New Mexico at the New Mexico Engineering Research Institute (NMERI) complex. Many efforts have been made to reduce costs during setup and modifications to the test facilities. This paper will discuss the facility and associated modifications.

FACILITY DESCRIPTION

The TSET portion of the NMERI complex consists of two large high bay areas (Workspaces 1 and 2) and a large

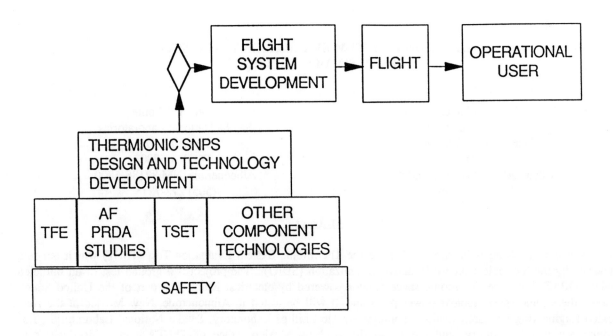

FIGURE 1. TSET Shown as a Part of the Total Thermionic Program.

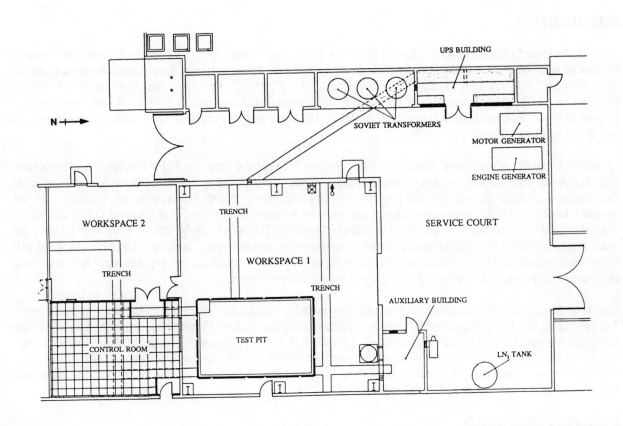

FIGURE 2. TSET Facilities at the NMERI Complex.

outdoor area (Figure 2). Workspace 1 will house TOPAZ-II, its vacuum chamber, a test scaffold to support TOPAZ and test equipment, an electrically heated single cell TFE test stand, a tungsten heater outgassing stand, and miscellaneous support equipment. Workspace 2 will house the control room (for control and data acquisition of TOPAZ-II) and a small machine shop. The outside area will contain an auxiliary building (for housing the cooling water system and vacuum pumps), an Uninterruptible Power Supply (UPS) building (for housing transformers, an UPS system with batteries, and breaker panels), a 300-kVA motor generator set (208V 60Hz input, 380V 50 Hz output), and a 300-kVA (380V 50 Hz) engine generator

Before testing can begin on the TOPAZ-II system, many facility modifications must be implemented. Some large scale modifications include a test pit, overhead crane, control room, and back-up power system. Changes to the NMERI complex began in July 1991, and are due to be completed by December 1991.

The TSET Facility includes a number of capabilities important to the conduct of TOPAZ-II tests, and the facility parallels an existing non-nuclear test facility at the Central Design Bureau for Machine Building in St. Petersburg, USSR (formerly Leningrad). Discussed below are some of the major facility modifications (Thome 1991). They include:

1. Building Modifications (pit, crane, UPS building, control room, and auxiliary building)
2. Test Scaffold/Test Stand
3. Electrical Power Generation and Distribution
4. Cooling Water System
5. Gas Systems (including LN_2)
6. Control System and Data Acquisition
7. Environmental/Safety Issues

BUILDING MODIFICATIONS

The building modifications are those structural changes that were altered at NMERI to accommodate the large TOPAZ test hardware.

Pit

One noticeable constraint of Workspace 1 on TOPAZ-II and its support equipment is the ceiling height of 6.71 m. TOPAZ's vacuum chamber has an overall height of 5.79 m and requires additional space (2.51 m) beneath it for the NaK coolant handling system. To accommodate this system a pit was designed that measures 5.49 m x 8.53 m x 4.57 m deep. The pit contains a large test scaffold, a trench/sump system to remove excess water, and safety/fire fighting equipment. Entering the pit on several sides are other trenches that contain power/instrumentation/control cabling, water piping, various gas piping, vacuum piping, and liquid nitrogen (LN_2) piping. The pit includes a major grounding system that will shunt all 300 kWe of deliverable power to the earth in the event of an electrical ground fault. The pit also accommodates the removal of up to 3.2 L/s of water that may leak into the pit from a cooling water or potable water leakage.

Crane

An overhead crane has been installed to facilitate assembly/disassembly of the test equipment. The crane is a bridge type with a tandem trolley design. The tandem trolleys allow simultaneous operation of both hoists. The crane is rated at 9.1 metric tonnes with a 5.74 m hook height. The overall crane system design allows for separate or simultaneous control of each trolley and their hooks. This capability requirement came from the need to disassemble the TOPAZ-II vacuum chamber sections where the sections must be lifted up and over the top of the TOPAZ-II satellite.

UPS Building

TOPAZ-II support equipment requires continuous power for testing to limit the number of rapid cooldown transients. The support equipment is driven by European power (50 Hz). To accommodate this power requirement an additional building has been constructed to house power conditioning equipment. The UPS building houses three 380V, single phase Soviet transformers, an UPS system with backup batteries, and breaker panels. This building is air conditioned and heated to provide a stable temperature year round.

Control Room

A control room has been constructed in Workspace 2. This room contains data acquisition (US and USSR) and control equipment. It also includes a distribution panel for test equipment electrical power. The room measures 6.60 m x 9.60 m x 2.74 m high with a 0.15 m raised floor. Two trenches run beneath the control room to the test pit carrying electrical cabling, one for power and the other for instrumentation. The control room is refrigeratively air conditioned and has a large window providing direct operator visibility of TOPAZ-II and Workspace 1.

Auxiliary Building

The auxiliary building is an existing structure located outside the northeast corner of Workspace 1. It contained the building air compressor and vacuum pump. This room has been altered and rearranged to also include two cooling water pumps and four Soviet roughing/fore vacuum pumps. This room isolates the more noisy equipment from Workspace 1 and is air conditioned and heated.

TEST SCAFFOLD/TEST STAND

The test scaffold is a large, three level stand that will support TOPAZ-II, its vacuum chamber, vacuum pumps, and electrical transformers (Figure 3). Its total load limit is approximately 25.4 metric tonnes with a live load of 11.5 kPa for the decking. The entire structure is an all bolted design which includes removable decking and railings. This design allows easy assembly/disassembly of the test equipment. The test scaffold was designed by the Soviets and nearly reproduces the structure in St. Petersburg. This arrangement reduces the need to design for new piping and pumping systems different from the Soviet Union and accommodate gravity draining of NaK, Cesium, and other liquids. This very large structure is bolted in the pit floor and is structurally sufficient to support all Soviet vacuum pumps including two large 10,000 L/s turbomolecular vacuum pumps. The stand has a stairway to the lower level and pit floor, an emergency exit ladder at the opposite end, and a ladder to the level above the building floor. The test scaffold is electrically grounded to the pit grounding system.

ELECTRICAL POWER GENERATION AND DISTRIBUTION

The electrically heated TOPAZ-II test system, along with its control system and all pumping equipment, are powered by standard European 3 phase, 380V 50 Hz power. The NMERI site power is 3 phase, 208V 60 Hz. The power needs for the TSET project requires the installation of a 300-kVA motor generator set (208V 60 Hz input, 380V 50 Hz output), an UPS system with a 5 minute battery backup and a 300-kVA, 380V 50Hz engine generator with a 24-hour fuel tank (Figure 4). The system power network allows for the complete loss of offsite power while continuing to operate TOPAZ indefinitely. The UPS with battery backup provides for a 5-minute period to bring the engine generator system on line. The complete power transfer system is automatic. This backup power supply system is required to minimize the number of thermal transients to TOPAZ-II which limit the lifetime of the satellite.

TSET power distribution is a reproduction of the St. Petersburg system. The major features include three variable transformers that supply power in four groups to stepdown heater transformers. The TOPAZ-II heaters (37 total) produce 85 kWt of heat nominally, but can be operated for short periods of time up to 135 kWt. Other electrical power is distributed to the Soviet vacuum pumps and to all control room instrumentation and control cabinets. The control room air conditioner is also powered at 380V 50 Hz to permit continued operation during

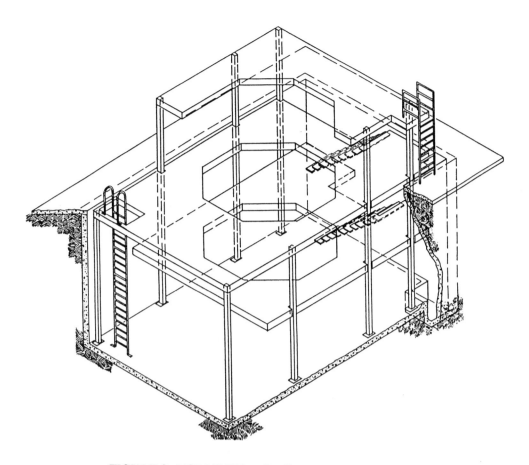

FIGURE 3. TOPAZ-II Test Scaffold (shown inside the pit).

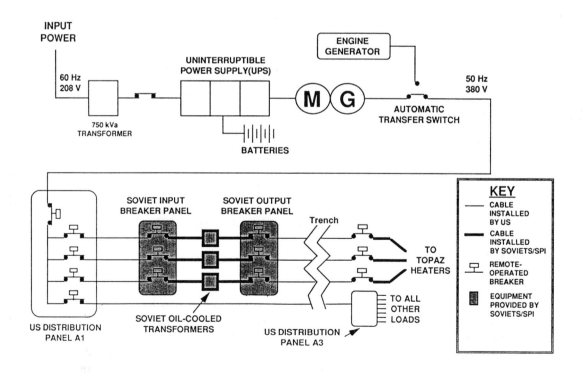

FIGURE 4. TSET Electrical Power and Distribution System.

loss of offsite power.

COOLING WATER SYSTEM

A cooling water system will be required to remove heat generated by TOPAZ-II (82 kWt nominally, 130 kWt maximum) plus additional heat produced by the TFE test stand, heater outgassing stand, and vacuum pumps. The normal cooling system requirement is 3.2 L/s for a maximum outlet temperature of 333 K on the vacuum chamber.

The cooling system consists of two multi-stage pumps in parallel, each capable of supplying a maximum flowrate of 3.2 L/s and a pump head of 600 kPa. Heat is rejected by an evaporative cooling tower rated at 250 kWt. It is oversized to cover high humid summer conditions in Albuquerque which may exist for several hours. The option also exists to increase the flowrate with both pumps working or the addition of potable city water. The pumps and cooling tower will be located in and over the auxiliary building with a sump tank in the northeast corner of Workspace 1.

The cooling system will include backup water from city utilities that can be initiated during losses of building power or pump failures. The loss of normal site 208V power will automatically result in the initiation of cooling water from the city potable water system at up to 3.2 L/s. This cooling path is a "once-through" system with the water being directed through the TOPAZ-II vacuum chamber water jacket and into the drain system. Various temperatures and pressure sensors monitor the system performance for control. A complete loss of cooling results in a modest heatup rate to the experiment if no actions are taken to reduce TOPAZ-II power. The operator has between 30 minutes and an hour to restore cooling or reduce power.

GAS SYSTEMS

The operation of TOPAZ-II requires multiple gas systems containing helium (pure and ultrapure), argon and carbon dioxide. These gases are used as cover gases for reactor components and the safe handling of alkali metals. In addition, all gas systems will include accumulators for relieving pressure fluctuations. The gases required are identical to those used in the Soviet Union and are reproduced at NMERI. These four gas systems are piped into separate headers through pressure regulators and terminate at the pit in the northeast trench where they will be connected to the TOPAZ-II systems.

A liquid nitrogen (LN_2) system is also included with gas systems to supply either liquid or gaseous nitrogen to absorption pumps, cold traps, helium leak detectors, and other loads to support TOPAZ-II testing.

CONTROL SYSTEM AND DATA ACQUISITION

The control room discussed above will contain the instrumentation and control cabinets supplied by the Soviets for the monitoring and control of TOPAZ-II tests. There are over ten cabinets supplied and are the racks used in St. Petersburg for testing. In addition to this system, a computer based (IBM 386) data acquisition and monitoring system will support the control room operators normal information system. Data collection will also be available to experimenters at their data collection stations in Workspace 1 and 2 and will parallel many of these same sensors. Data collection and distribution details are not available at this time.

ENVIRONMENTAL/SAFETY ISSUES

The TSET facility had to be assessed for its environmental impact before any construction modifications could begin. The Phillips Laboratory took the responsibility for writing the Environmental Assessment (EA) and acquiring local, state, federal, and USAF inputs before obtaining approval by the Air Force (DAF 1991). Once approved, some design requirements were implemented along with many personnel safety requirements. The process of evaluating the TSET facility determined the performance of the test to be a negligible impact on the environment. In general, the small quantities of TOPAZ-II working fluids, NaK and cesium, limit the magnitude of a potential incident. Other hazardous materials, including the lithium hydride shield and beryllium reflectors, are contained to prevent their release to the environment. Release of shield and moderator materials can only result from a catastrophic accident such as an aircraft crash or a major building fire, for which the liberation of these materials becomes a small part of the total accident. Other chemicals, such as acetone and diesel fuel, were also evaluated and were found to have negligible effects due to their small quantities. All wastes will be handled through appropriate university and USAF organizations.

To reduce the possible harm an alkali metal fire could present to the environment, an air washer will be installed to remove combustion products from Workspace 1 (Thome 1991). An alkali metal fire has the greatest potential of beginning in the test pit. Smoke detectors in the pit will activate an air washer system. An exhaust fan will be located in the pit, and is designed to pull smoke from the pit and discharge it to the air washer. From there, the smoke is "washed" with water and the contaminants are contained in the air washer.

To protect test personnel, safety equipment and systems have been designed into the facility. These include a video/intercom system, oxygen sensors and alarms within the pit, dry chemical extinguishers for alkali metal fires, and breathing apparatuses in the event of a fire or release of heavy gases (nitrogen, argon, and carbon dioxide).

CONCLUSIONS

Facility designs are completed and modifications are nearly complete. The TSET facility will provide all the resources necessary to conduct this unique test of an unfueled Soviet TOPAZ-II space nuclear power system. Testing of the Soviet hardware will help "jump start" lagging US thermionic reactor development (Henderson 1991). The facility, following the completion of TOPAZ-II testing, can be used for other types of testing of importance to the space and reactor communities.

Acknowledgments

Funding for this project is being provided by SDIO. Special thanks to NMERI for their assistance in reducing costs and construction time on this project, and to Kirtland AFB Civil Engineering Design Section for their help on the test scaffold. This project also received technical assistance from Dr. Glen Schmidt.

References

DAF (1991) *Environmental Assessment-Phillips Laboratory Thermionic System Evaluation Test at New Mexico Engineering Research Institute*, Department of the Air Force, Space Systems Division, Los Angeles AFB, CA.

Henderson, B. W. (1991) US Buying Soviet TOPAZ 2 to Boost Space Nuclear Power Program," *Aviation Week and Space Technology*, 14 January 1991, pp. 54-55.

Standley, V. H., D. B. Morris, and M. S. Schuller (1991) *Thermionic System Evaluation Test (TSET) Project Plan*, Phillips Laboratory/Space Power Branch, Kirtland AFB, NM[1].

Thome, F. V. (1991) "Specifications for the Creation of a Special Test Facility in the New Mexico Engineering Research Institute (NMERI) Building (University Blvd) in Preparation for TOPAZ (TSET) Test," Internal Document, Phillips Laboratory/Space Power Branch, Kirtland AFB, NM.

---

[1] Distribution authorized to U.S. Government agencies only; Proprietary Information, Aug 1991. Other requests for this document must be referred to PL/STPP, Kirtland AFB, NM 87117-6008. Research sponsored by SDIO with Phillips Laboratory acting as agent.

RESOLVING THE PROBLEM OF COMPLIANCE
WITH THE EVER INCREASING AND CHANGING REGULATIONS

Harley Leigh
Westinghouse Hanford Company
P.O. Box 1970, Mailstop N1-41
Richland, WA 99352
(509) 376-2972

Abstract

The most common problem identified at several U.S. Department of Energy (DOE) sites is regulatory compliance. Simply, the project viability depends on identifying regulatory requirements at the beginning of a specific project to avoid possible delays and cost overruns. The Radioisotope Power Systems Facility (RPSF) is using the Regulatory Compliance System (RCS) to deal with the problem that well over 1000 regulatory documents had to be reviewed for possible compliance requirements applicable to the facility. This overwhelming number of possible documents is not atypical of all DOE facilities thus far reviewed using the RCS system. The RCS was developed to provide control and tracking of all the regulatory and institutional requirements on a given project. WASTREN, Inc., developed the RCS through various DOE contracts and continues to enhance and update the system for existing and new contracts. The RCS provides the information to allow the technical expert to assimilate and manage accurate resource information, compile the necessary checklists, and document that the project or facility fulfills all of the appropriate regulatory requirements. The RCS provides on-line information, including status throughout the project life, thereby allowing more intelligent and proactive decision making. Also, consistency and traceability are provided for regulatory compliance documentation.

INTRODUCTION

All nuclear facilities are required by DOE Order 5480.5, *Safety of Nuclear Facilities* (DOE 1986), to perform a readiness review on all new facilities and most facilities that require a restart. To be successful, these reviews require an assessment of the state of regulatory compliance. How is this task accomplished on complex projects like the Radioisotope Power Systems Facility (RPSF)? The list of potentially applicable regulatory documents exceeds 1200. The number of requirements invoked by approximately 400 applicable documents is tens of thousands.

Currently there has been a change regarding what represents full compliance with federal regulations, DOE Orders, and requirements of other federal and state agencies. In the past, detailed reviews by experts in their specific fields of expertise would review for compliance and provide a statement that the document, procedure, system, facility, or process equipment was in compliance with the requirements. For today and future projects, only the line-by-line documentation of the specific requirements and how those requirements are met is the acceptable and successful approach. This is particularly true where fines and/or jail sentences may be invoked by law for noncompliance, even if inadvertent or unintentional.

One of the major problems in assessing regulatory compliance is not knowing all of the requirements of a specific operation or project. This problem is further compounded by the rapid changing of and additions to regulatory and institutional requirements. The next major consideration is how to accomplish the task of regulatory compliance to maintain startup schedules in a cost effective and timely manner.

The RPSF is located in the Fuels and Materials Examination Facility (FMEF) at the DOE Hanford Site in southeastern Washington State. The FMEF, completed in 1984, was designed and constructed as a Seismic Class I facility to be used for the examination of nuclear fuels and materials. As the DOE missions have evolved, so has the purpose of the FMEF. It is a building well-suited for RPSF purposes because it meets the environmental and safety requirements of a nuclear facility and currently is being modified to provide for the assembly of Radioisotope Thermoelectric Generators (RTG). The assembly process is based on one developed at Mound Laboratory (currently operated by EG&G Mound Applied Technologies Inc. (Mound)) on the DOE Mound Site in Miamisburg, Ohio. Westinghouse Hanford Company (Westinghouse Hanford) is designing and constructing the facility to ensure the heat source assemblies are produced in a manner that is in complete compliance with today's regulatory and institutional requirements.

SYSTEM METHODOLOGY

Although the RCS is being used for many different reviews or assessments, establishing regulatory requirements, and conducting operational readiness reviews (ORR), the system methodology is fairly constant. An RCS ORR flow chart (Figure 1) illustrates the interface between the activities performed in demonstrating and documenting facility readiness.

The initial phase, "Analysis," includes the preparation of a requirements document, a critical systems document, a compliance decision tree, and a review plan.

The requirements document contains the results of applicability assessment from a list of potentially applicable regulatory and institutional documents. Specific criteria are established, reviewed, and approved to perform the assessment. Each document then is reviewed against the criteria and established as either applicable or nonapplicable with a respective justification. The applicability is addressed for both the DOE site and the specific facility or project.

The critical systems document contains the results of an assessment to determine which systems (hardware, personnel, and programs) are critical to safe and reliable operations of a specific facility or project. Specific criteria are established, reviewed, and approved to perform the assessment. Then each system (usually taken from a safety analysis report) is reviewed against the criteria and established as either critical or noncritical with a respective justification.

The compliance decision tree is a graphic portrayal of all the critical systems. It shows which major activities under each system must be reviewed during the compliance review. The decision tree also is a graphic representation of how the following customized database will appear.

The review plan and procedures using the previous documents describe the entire review process including purpose, scope, organization, responsibilities, schedules, and success criteria. The plan is reviewed formally and approved by the appropriate organizations including the DOE field office.

The second phase, "Assessment," includes the modification of the existing RCS database, the preparation of the review checklists, the conduct of the compliance assessment by the line organizations, the initiation of an action tracking system from the existing RCS, the conduct of a document validation, the performance of a deficiency analysis, and the preparation of a plan and a critique plan for the conduct of a preoperational checkout.

The modification of the RCS database consists of extracting the specific requirements from the applicable regulatory and institutional documents as indicated in the requirements document, adding them to the existing requirements, and sorting them into the set major activities by critical system within the system.

The critical system checklists are produced in two phases. Not all of the requirements will be applicable to a specific facility or project even though the document is deemed applicable. Therefore, the checklists are printed in modules, and a final applicability assessment is conducted with justifications for nonapplicability. This further concentrates the requirements for a specific facility or project to the exact requirements that must be assessed for compliance. System notebooks then are printed by critical system with provisions for compliance documentation to be indicated for the each requirement within that system.

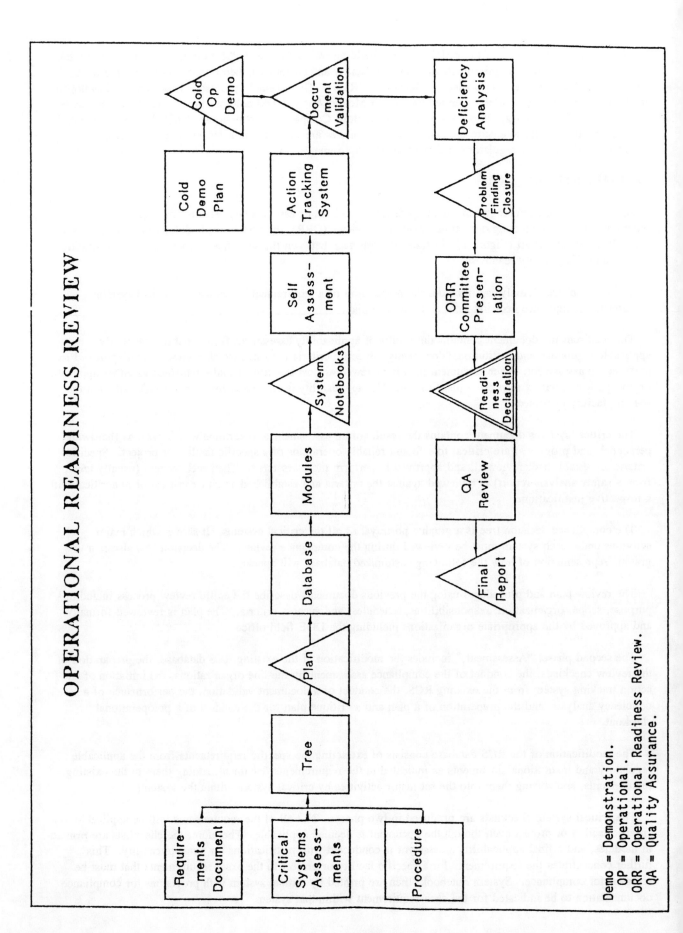

The actual compliance assessment is performed by transmitting the system notebooks to the line organizations, providing them with assistance as required, and requesting them to identify all available compliance documentation for the appropriate criteria by critical system. During this same period, an action tracking system is initiated from within the RCS. When the system notebooks have been completed, a document validation is conducted on a random sampling basis by an independent organization. Once all compliance information has been entered into the RCS, the actual state of compliance can be assessed and desired future compliance can be tracked. Also at this stage, external agency audit findings can be entered and tracked in the system. Further, resource and funding allocations can be identified and schedules can be prepared. If desired, the risk of noncompliance can be assessed.

A deficiency analysis is performed by collecting all noncritical deficiencies (for example, corrective actions from nonconformance reports, work orders, preventive maintenance tasks, procurement deviations, design modifications, calibrations, corrective actions from unusual occurrences or off-normal conditions, corrective actions identified from audits or appraisals, and nonessential open items) and assessing whether or not the total of all or a group of deficiencies against a particular critical system should be considered a finding. This step is accomplished because a large number of deficiencies or nonessential open items can be perceived as a lack of preparation for operational readiness.

A cold operational demonstration is conducted to ensure that the dynamics of a particular system work as designed in a total system. This step is a physical performance of a proposed operation including the use of all required facilities, personnel, and procedures through an approved plan but without the use of hazardous or radioactive materials. The demonstration also shall include performance of the proper response for simulated off-normal conditions. This step is not necessary if the RCS is being used for compliance assessment only. A plan and procedures is prepared for the demonstration. A critique plan also is prepared with a subsequent critique report following the performance of the demonstration.

The third phase is the closure of all findings from the various assessments, both internal and external. This is accomplished through both electronic and physical records within the RCS.

The fourth phase is the announcement of a desired state of compliance or readiness.

The last phase is the preparation of a final report summarizing all findings and corrective actions, identification of future plans, a detailed description of the assessment process used, and the final results. This phase also includes a final quality check and archival of all records.

DISCUSSION

The RCS is a tried and proven system. In addition to the FMEF/RPSF, it (the RCS) has been used previously or currently is supporting other DOE applications and projects with positive results. Currently, it is being used for: the ORR of the TRUPACT Loading Facility at the DOE Rocky Flats Plant, Idaho Falls, Idaho; the ORR of the Supercompactor and Repackaging Facility, also at the Rocky Flats Plant; the establishment of regulatory objectives for the DOE-HQ/EM Site Roadmaps; and the evaluation and acceptance of ORRs for DOE-HQ/EM (currently three facilities at the Savannah River Plant: the Defense Waste Production Facility, In-Tank Precipitation Facility, and 1-H Evaporator). This system also was used for Operational Readiness Review of the DOE Waste Isolation Pilot Plant (WIPP).

The following advantages and their inherent cost savings are the result of using RCS for a project.

- Provides a means to better protect personal and corporate liability.

- Provides objective, documented evidence of compliance by regulatory document, subject, or critical system on an individual requirement basis, thereby reducing future manpower requirements for the preparation of various audits and reviews by external agencies.

- Aids in obtaining required funding for specific compliance tasks since the exact state of desired compliance can be assessed based on resource and funding availabilities. More simply, aids in determining what state of compliance the customer or custodian is willing to accept.

- Provides preliminary, integrated, concise information for project documents.

- Aids in alleviating retrofits by establishing and obtaining customer approval of all project requirements at the beginning of the various project phases rather than during or after.

- Provides documented traceable compliance evidence throughout the project life.

- Provides accurate, traceable information for all implementing documents and procedures.

- Provides consistent systems and records configuration from conceptual through operations phases.

- Provides for an accelerated acceptance process and ORR.

- Enhances the ability to respond to regulatory audits and appraisals.

- Provides greater visibility and impacts of future regulatory requirements.

CONCLUSION

The RCS and the database developed for FMEF/RPSF critical systems were beneficial to the regulatory compliance of the RPSF. The RCS is a key tool for the technical experts in ensuring that an important requirement has not been overlooked in assessing the operational readiness of the facility. These requirements need to be known in the beginning of the project, not during construction or startup when identified requirements necessitate retrofits, subsequent delays, and cost overruns.

Acknowledgments

David Nearing, General Manager of the WASTREN, Inc., Hanford Division, and the WASTREN, Inc., staff are acknowledged for providing information on the Integrated Project Control Regulatory Compliance System and for their input and review. Frank Moore and John Williams of Westinghouse Hanford are acknowledged for their valuable input and reviews.

References

DOE (1986) *Safety of Nuclear Facilities*, DOE Order 5480.5, U.S. Department of Energy, Washington, D.C.

MODIFICATION OF HOT CELLS FOR GENERAL PURPOSE HEAT SOURCE ASSEMBLY AT THE RADIOISOTOPE POWER SYSTEMS FACILITY

Betty A. Carteret
Westinghouse Hanford Company
P.O. Box 1970, Mail Stop N1-42
Richland, WA 99352
(509) 376-8680

Abstract

Eight existing, unused hot cells currently are being modified for use in the Radioisotope Power Systems Facility (RPSF) to assemble ^{238}Pu-fueled heat sources for Radioisotope Thermoelectric Generators (RTGs). Four air atmosphere cells will be used for storage, decanning, and decontamination of the iridium-clad radioisotope fuel. The remaining four argon atmosphere cells will be used to assemble fuel and graphite components for production and packaging of general purpose heat source (GPHS) assembly modules, which provide heat to drive the thermoelectric conversion process in the generators. The hot cells will be equipped to perform remote and glovebox-type operations. They will provide shielding and contamination control measures to reduce worker radiation exposure to levels within current U.S. Department of Energy (DOE) guidelines. Designs emphasize the Westinghouse Hanford Company (Westinghouse Hanford) as low as reasonably achievable (ALARA) radiation protection policy.

INTRODUCTION

The Radioisotope Power Systems Facility (RPSF) is a new project currently being designed by Westinghouse Hanford and constructed by Kaiser Engineers Hanford for the U.S. Department of Energy (DOE). The facility is located in the Fuels and Materials Examination Facility (FMEF) on the Hanford Site in southeastern Washington State. Technology for assembly of radioisotope-fueled generators was developed by EG&G Mound Applied Technologies Inc. (Mound) in Miamisburg, Ohio. Westinghouse Hanford was selected by the DOE to design and construct a new facility to replace the aging Mound production line. The purpose of this project is to build a modern production facility that meets current environmental, safety, and health standards. The facility will provide the capability for assembly of ^{238}Pu-fueled Radioisotope Thermoelectric Generators (RTGs) (Figure 1) and other radioisotope-fueled power sources for space and terrestrial applications.

The RPSF is being constructed by modifying unused portions of the FMEF building to meet requirements for RTG production. The RPSF is comprised of six systems that provide production capability for all stages of the RTG assembly process, from receiving fuel and generator components to shipment of completed generators to DOE customers, primarily the National Aeronautics and Space Administration (NASA). A general discussion of RPSF and individual systems has been reported previously (Alderman 1990 and Eschenbaum and Wiemers 1990).

The six systems that make up the RPSF are organized as follows.

1. System 6100--Nonfuel Component Preparation,
2. System 6200--Fueled Component Storage and General Purpose Heat Source Assembly,
3. System 6300--Heat Source Assembly and Installation,
4. System 6400--Heat Source Assembly Testing,
5. System 6500--Transportation and Storage, and
6. System 6600--Production Integration and Certification System (Kiebel and Wiemers 1990).

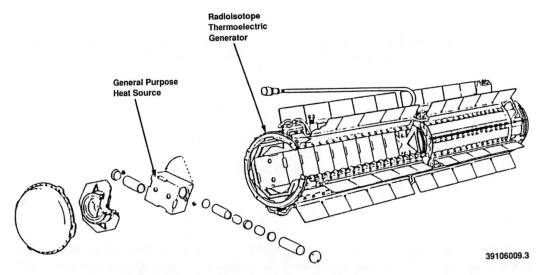

FIGURE 1. Radioisotope Thermoelectric Generator.

This paper describes the design of the System 6200 hot-cell assembly line for GPHS production. Eight interconnected hot cells will be used for storage of fuel clad canisters (FCCs), fueled clad (FC) decanning and decontamination, assembly of GPHS modules (Figure 2), and packaging of GPHS modules in welded canisters for further processing by System 6300.

SYSTEM DESIGN AND DEVELOPMENT

Construction activities for System 6200 are under way to outfit the hot cells for their new mission. Equipment currently is being installed in eight hot cells located on the lowest elevation of the FMEF building (Figure 3). The hot-cell assembly line will provide the capability to produce 240 GPHS modules annually. The high-density concrete walls, stainless steel liner, and some embedded piping and penetrations for these cells originally were constructed as part of the Breeder Reprocessing Engineering Test (BRET) program, which was canceled before completion. Although much of the hot-cell equipment was purchased, it never was installed in the cells (which have been maintained in excellent condition). Demolition of unneeded piping and support structures has been completed, and installation of equipment in the cells is in progress. The cells will be outfitted with a combination of original BRET equipment and new equipment purchased and fabricated specifically to meet the design requirements of System 6200. A hot-cell mockup is being used for equipment testing and design verification.

Detailed mechanical and electrical design and supporting analyses are being completed to support procurement, fabrication, and construction activities. Thermal analyses have been performed to ensure adequate cooling is provided for storage and processing of high-temperature fuel. Each FC produces 62.5 W of thermal energy and can become heated to incandescent temperatures during shipment or storage. Shielding calculations have been completed to support design of cell boundary components such as windows, manipulators, and airlock doors and to verify the adequacy of shielding provided by the existing cell liners and walls. Shielding calculations were based conservatively on a doubled neutron quality factor (Q=22). System 6200 shielding designs will allow the system to operate as a full-time occupancy zone (<.002 mSv/h (<0.2 mrem/h)) in all areas, except at glove-ported windows used during GPHS module assembly operations and at airlock doors during transfers (.02 mSv/h (<2 mrem/h)). Access to System 6200 operations areas will be restricted during those operations where dose rates exceed the Westinghouse Hanford criteria for full-time occupancy (.002 mSv/h to .005 mSv/h (0.2 mrem/h to 0.5 mrem/h)).

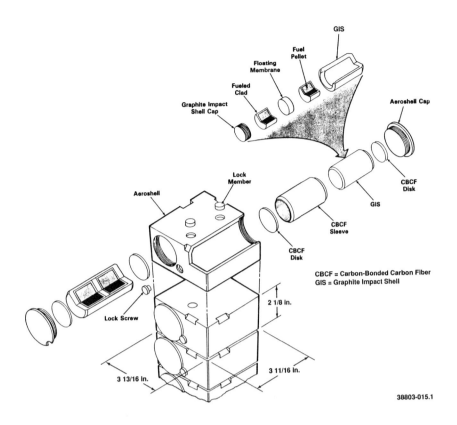

FIGURE 2. General Purpose Heat Source Module.

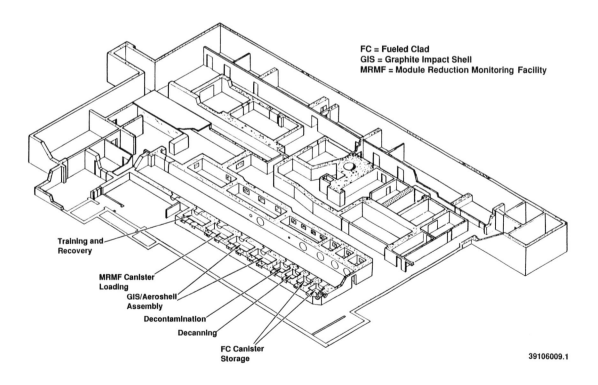

FIGURE 3. Radioisotope Power Systems Facility Level 1 (-35 ft).

SYSTEM DESCRIPTION

System 6200 is divided into six subsystems that perform various stages of the GPHS module assembly process. The subsystems are organized as follows:

1. Subsystem 6210--FCC Storage,
2. Subsystem 6220--FC Decanning,
3. Subsystem 6230--FC Decontamination,
4. Subsystem 6240--GPHS Assembly and Measurement,
5. Subsystem 6250--Module Reduction and Monitoring Facility (MRMF) Canister Loading, and
6. Subsystem 6260--Training and Recovery.

Subsystems 6210 (2 cells), 6220, and 6230 will be installed in 1.8 m x 1.8 m x 4.3 m air atmosphere cells with a single window and airlock. The remaining subsystems will be installed in 1.8 m x 4.3 m x 4.3 m argon atmosphere cells with two windows and a single airlock. Figure 3 shows the arrangement of the hot cell assembly line. Each subsystem will be outfitted with special tooling and fixtures required to perform its function. A discussion of the specific equipment requirements and the function of each subsystem is provided below.

General Cell Equipment

The listed general cell equipment is common to all cells:

- Master/slave manipulators and 1/2-ton bridge crane,
- Oil-filled shielding windows (Subsystem 6210 only),
- Glove-ported dry shielding windows and window workstation,
- Equipment transfer airlocks and between cell passthroughs,
- Lighting, closed-circuit television (CCTV) system, mirrors,
- In-cell alpha detection system (except Subsystem 6210),
- Fan coil units (chillers) and environmental monitoring,
- Air ventilation or argon recirculation system and high-efficiency particulate air filtration,
- Fire suppression system (air cells only),
- Elevated floor platform and in-cell equipment transport cart,
- Vacuum manifold for tooling and helium leak test system (except Subsystem 6210),
- Water-cooled, inert atmosphere storage chambers (Subsystems 6240, 6250, 6260), and
- Shadow shielding.

Subsystem 6210 The first two cells in the hot-cell line will be used for storage of fuel in graphite-packed FCCs. Each cell is designed with a capacity to store 36 FCCs (2 FCs per FCC). The FCCs will be transferred into the cells through equipment transfer airlocks and stored in numbered locations by manipulators. Two FCC storage racks and required tooling to handle and store FCCs will be provided in this cell. The design of the water-cooled storage rack will provide additional shielding for the ^{238}Pu fuel that emits primarily neutron radiation.

Subsystem 6220 The next cell will be used to remove the high temperature FCs from welded shipping canisters. Two FCCs will be received in a tray passed from subsystem 6210 through a between-cell passthrough with sealed doors on either side. To prevent the potential spread of airborne contamination, the passthrough will be evacuated and backfilled with argon before opening the door in the next cell. Manipulators and a remotely operated cutting fixture will be used to decan FCs. A dust collection unit will be installed to capture graphite fibers released from the canister packing material. This unit will be used to maintain cell cleanliness by preventing the spread of potentially contaminated particles within the cell. Special tooling and water-cooled fuel storage blocks will be provided for the decanning process. Following decanning, the fuel will be examined visually for surface damage, checked for cladding breaches using a helium leak test fixture, and then surveyed for smearable surface contamination. A remotely operated alpha detection station will be used to survey swipes taken from the surface of the fuel. Visual examination will be performed at the window workstation using a high-resolution, color CCTV camera.

Subsystem 6230 The fourth air atmosphere cell will be used to perform surface decontamination of FCs before assembly into graphite module components. Segregated FC storage areas will be provided for pre- and post-decontamination. Four FCs on a tray will be received through the passthrough and stored on a chilled block. Decontamination supplies will be received through the airlock. Manipulators, gloves, and special tooling and fixtures are used to perform FC decontamination. Acid decontamination solutions, water, wiping cloths, and ultrasonic baths will be used to decontaminate the FCs. An in-cell ventilation hood will be installed over the decontamination station to collect and neutralize water and neutralize acid vapors generated during high-temperature FC cleaning. This will provide a means of contamination control and prevent degradation of cell equipment by corrosive vapor. Following decontamination, a helium leak test, alpha survey, and visual inspection of each FC will be performed. Special manipulator, hand and vacuum operated tools for handling storage and transfer of FCs, and decontamination materials will be provided.

In an effort to minimize the generation of liquid waste, water and acid solutions will be brought into the cell in small-volume containers instead of piping liquids into the cell. Used decontamination solutions, which are classified as hazardous waste, will be collected into a container using an aspirator. Liquid waste will be sampled for plutonium content then passed out of the airlock for treatment and disposal in accordance with Washington State Department of Ecology requirements. After neutralization and stabilization, the final waste form will be classified as a radioactive but nonhazardous material. Preparation of decontamination materials and waste treatment operations will be performed in a laboratory hood in the operations corridor behind the cells.

Subsystem 6240 The next two identically outfitted argon atmosphere cells will be used to perform assembly of the GPHS modules. Four decontaminated FCs will be received through the passthrough and other graphite module components will be received through the airlock. Two window workstations in each cell will be arranged for the FC frit vent activation, graphite impact shell (GIS) assembly, and aeroshell fueling steps of the GPHS module assembly operation (see Figure 2). Window workstations are equipped with hand and vacuum operated tooling and fixtures. At the first workstation, a foil decontamination cover will be removed from the four FCs to activate a vent, and then two GISs will be assembled. After installing and torquing the GIS cap, the length will be measured with a digital gauge, then both GISs will be passed to the other workstation and placed on a chilled block. A manual, vacuum-operated GPHS assembly fixture will be used to insert the GIS assemblies into the aeroshell body and install the cap. After the cap is torqued, the module will be removed from the fixture and placed in a cooled storage location. Alpha surveys will be performed at each step of the assembly operation.

Subsystem 6250 This cell will provide the capability to load and weld GPHS modules into stainless steel canisters for shipment to and processing in the MRMF (System 6300). Two GPHS modules will be received through the passthrough, and MRMF canister components will be received through the airlock. The GPHS modules will be packaged into the MRMF canister with graphite-felt packing material. The canister will be placed into a remotely operated welding fixture, and a lid assembly with a flexible metal hose and valve will be welded onto the canister. A visual inspection will be performed on the weld, and then the canister will be backfilled with helium. The canister will be placed into a water cooled leak test chamber, and a helium leak test will be performed to ensure weld integrity. A screening alpha survey of the canister surface will be performed in the cell and verified by an ex-cell Health Physics survey before releasing the canister from the cell for transport to the MRMF System 6300.

Subsystem 6260 This subsystem will be used for training, temporary storage of in-process materials requiring disposition, and for repackaging damaged or defective materials and fuel. The cell will be outfitted with a canister welding fixture identical to that used in Subsystem 6250. An in-cell alpha survey station and helium leak test chamber will be provided for canister inspection. This cell also can be outfitted with duplicates of tooling and fixtures used in other subsystems for operator training.

CONCLUSION

The RPSF System 6200 will provide a modern production line for assembly of radioisotope-fueled components that meets current environmental, safety, and health standards for handling of radioactive materials. In addition to its current production mission, the system will provide the capability and flexibility for adaptation to additional missions requiring production or handling of radioisotope-fueled components. The hot-cell designs have emphasized environmental and worker safety by providing measures for control of contamination, exposure reduction, and waste minimization.

Acknowledgments

Andy Kee, Curt Dickey, and Richard Steele are acknowledged for their work performed as members of the System 6200 design team. Wayne Amos, Don Fleming, and Bob Falkner of EG&G Mound Applied Technologies Inc. are acknowledged for their assistance in providing technology transfer information on GPHS module assembly operations. Ron Eschenbaum, Carol Alderman, and Jim Fredrickson are acknowledged for their valuable input and reviews.

References

Alderman, C. J. (1990) *Assembly of Radioisotope Power Systems at Westinghouse Hanford Company*, WHC-SA-0881-FP, Westinghouse Hanford Company, Richland, WA.

Eschenbaum, R. C. and M. J. Wiemers (1990) *Design of Radioisotope Power Systems*, WHC-SA-0927-FP, Westinghouse Hanford Company, Richland, WA.

Kiebel, G. R. and M. J. Wiemers (1990) *An On-Line Information System for Radioisotope Thermal Generator Production*, WHC-SA-0925-S, Westinghouse Hanford Company, Richland, WA.

TORNADO WIND-LOADING REQUIREMENTS BASED ON RISK ASSESSMENT TECHNIQUES
(For Specific Reactor Safety Class 1 Coolant System Features)

Theodore L. Deobald, Garill A. Coles, and Gary L. Smith
Westinghouse Hanford Company
P.O. Box 1970
Richland, WA 99352
(509) 376-0176

Abstract

Regulations require that nuclear power plants be protected from tornado winds. If struck by a tornado, a plant must be capable of safely shutting down and removing decay heat. Probabilistic techniques are used to show that risk to the public from the U.S. Department of Energy SP-100 reactor is acceptable without tornado hardening parts of the secondary system. Relaxed requirements for design wind loadings will result in significant cost savings. To demonstrate an acceptable level of risk, this document examines tornado-initiated accidents. The two tornado-initiated accidents examined in detail are loss of cooling resulting in core damage and loss of secondary system boundary integrity leading to sodium release. Loss of core cooling is analyzed using fault/event tree models. Loss of secondary system boundary integrity is analyzed by comparing the consequences to acceptance criteria for the release of radioactive material or alkali metal aerosol.

INTRODUCTION

The SP-100 reactor is a Category B, liquid-sodium-cooled, fast flux nuclear reactor. The SP-100 Flight System Qualification (FSQ) Test Site is a U.S. Department of Energy (DOE) facility located on the Hanford Site near Richland, Washington (see Figure 1). The facility will provide an environment that closely simulates the vacuum and temperature conditions of space operation. In addition, the facility will provide reactor containment and the associated heat transport, vacuum, plant protection, support, and security systems necessary for safe and reliable operation of the reactor.

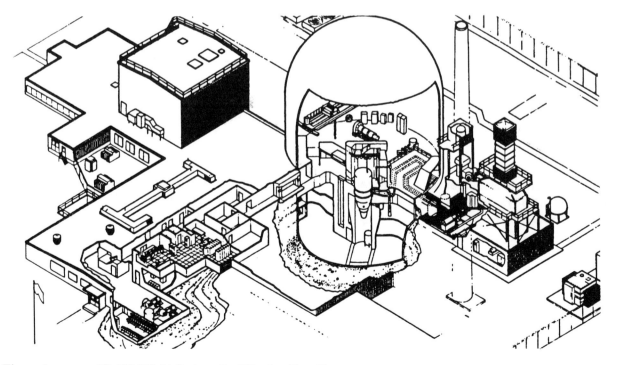

Figure 1. SP-100 Flight System Qualification Test Site.

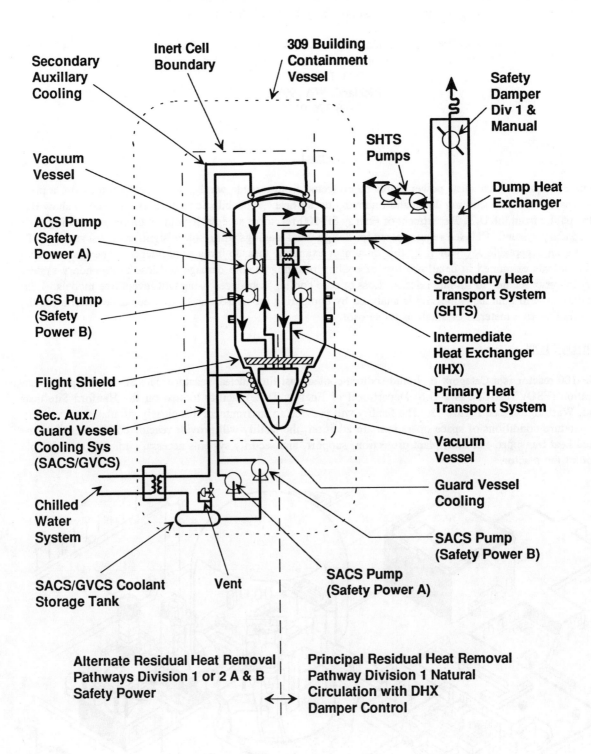

Figure 2. SP-100 Reactor Normal and Residual Heat Transport Systems.

TECHNICAL BASIS FOR RELAXATION OF TORNADO DESIGN

General Design Criterion 2 of Appendix A to 10CFR50, and Section C of Regulatory Guide 1.117 state that structures, systems, and components important to safety should be protected from tornado winds. This includes the reactor coolant boundary, those systems necessary to shut down the reactor and maintain its safe condition, and those systems whose failure could lead to excessive radioactive release.

Normally, safety class systems (such as primary and residual heat transport systems) would be protected against tornado winds. Analysis shows, however, that for the SP-100 FSQ, the regulatory requirements can be met without protecting all of the residual heat transport systems.

The evaluation demonstrates that an acceptable level of risk exists even though selected portions of the SHTS are not protected against tornado. This evaluation is an examination of accident scenarios initiated by a tornado. The accident scenarios examined are loss of cooling capacity leading to core damage and loss of SHTS integrity leading to release of sodium to the atmosphere.

LOSS OF CORE COOLING CAPACITY

Loss of capacity to cool the core was analyzed by employing Probabilistic Risk Assessment techniques. Accident sequences were constructed and analyzed using an event tree. This event tree includes a tornado as the initiating event followed by RHR system failures that would lead to core damage. Fault trees of the RHR systems were used to support the event tree. The analysis shows the probability of these accident sequences is low (2.3×10^{-9}/year).

The sequences in the event tree shown in Figure 3 begin with a tornado. The first branch point is the success or failure of the PHTS/SHTS. The SHTS is not fully tornado hardened; therefore, loss of PHTS/SHTS is assumed and this branch is given a failure probability of 1.0. The next branch point is success or failure of the ACS. If ACS fails, then core cooling fails. If the ACS is successful, then the SACS must be successful to prevent core damage. Accordingly, the last branch point shows the success or failure of SACS. The probabilities shown are determined from fault tree analysis.

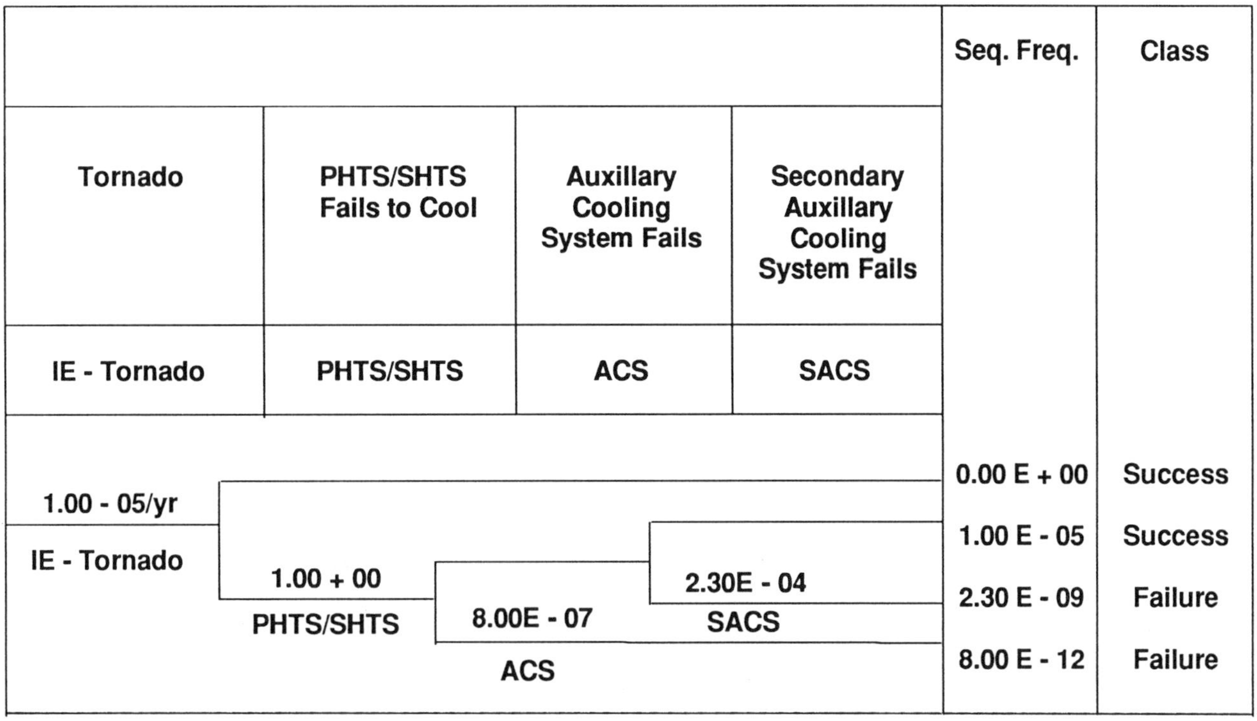

Tornado	PHTS/SHTS Fails to Cool	Auxillary Cooling System Fails	Secondary Auxillary Cooling System Fails	Seq. Freq.	Class
IE - Tornado	PHTS/SHTS	ACS	SACS		
1.00 - 05/yr IE - Tornado	1.00 + 00 PHTS/SHTS	8.00E - 07 ACS	2.30E - 04 SACS	0.00 E + 00	Success
				1.00 E - 05	Success
				2.30 E - 09	Failure
				8.00 E - 12	Failure

Figure 3. Sequence of Event Tree.

The frequency of the initiating event is derived from Figure 4. This figure is taken from a report by Coats and Murray (1985) and shows the wind hazard curve for the Hanford Site. This hazard curve is given as a function of wind speed and probability of exceedance in 1 year. The initiating event frequency chosen (1×10^{-5}/year) is slightly more than the frequency for a design basis tornado for non-reactor Class 1 facilities. The design basis tornado wind speed is dictated by the site design criteria (SDC 4.1, 1989). It is assumed that a tornado with wind speeds of this magnitude or greater would fail nonhardened portions of the SHTS.

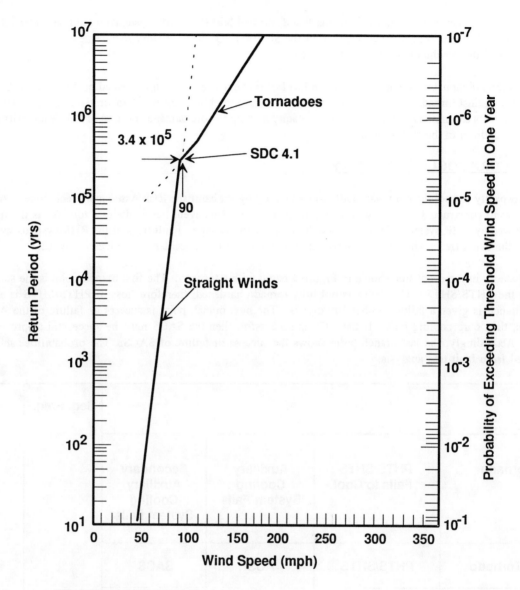

Figure 4.   Wind Speed Return Period.

Fault trees previously developed by Westinghouse Hanford Company (Westinghouse Hanford) and General Electric-Space Nuclear Engineering and Technology were used in this analysis to calculate system failure probability. To show the effects of a tornado on unhardened systems, these models are modified in the following manner: (1) no heat transfer is allowed via the SHTS, and (2) alternating current (AC) electric power to the ACS, SACS, and Chilled Water System (CWS) is failed.

When normal electric power is lost, the ACS and SACS pumps use backup battery power; however, the CWS would be lost because it has no backup power. Because this heat transfer pathway is not available, discharge of steam from the SACS storage tank will provide the initial heat transfer. Therefore, some decrease in the reliability of these systems is expected. When the residual heat has decayed sufficiently, passive cooling is adequate to cool the reactor.

The event tree in Figure 3 includes two sequences that end in failure. In the first sequence, the PHTS/SHTS fails after the tornado, the ACS is successful, but the SACS fails. The second sequence is similar except that the ACS fails and the state of SACS is, therefore, of no consequence. Accordingly, two probabilities are needed to complete the event tree: (1) the ACS is failed, and (2) the SACS is failed given that the ACS is successful.

These probabilities are determined by solving, or quantifying, the fault trees. To determine what effect loss of power would have on cooling, the fault trees were first quantified assuming that normal electrical power was not available and then quantified assuming that it was. The following is a summary of the results:

- SACS failure, normal power failed   2.3E-04
- ACS failure, normal power failed   8.0E-07
- SACS failure, normal power available   1.2E-04
- ACS failure, normal power available   8.0E-07
- SACS accident sequence frequency   2.3E-09/year
- ACS accident sequence frequency   3.2E-12/year

The first two probabilities listed are the system failure probabilities assuming normal electric power has failed. These are the probabilities that were incorporated into the event tree. The next two probabilities given are system failure probabilities when normal power is available. As shown, the difference in failure probability for ACS and SACS with and without normal electrical power is small. This indicates that these RHR systems are insensitive to loss of normal AC power, apparently as a result of the advantages of natural recirculation. The final two entries listed are the final sequence frequencies of the accidents of concern shown in the event tree in Figure 3. These accident frequencies are very low.

LOSS OF SECONDARY HEAT TRANSPORT SYSTEM INTEGRITY

Loss of SHTS integrity leading to release of sodium to the atmosphere was evaluated. Although SHTS sodium coolant is never in contact with the reactor fuel, low neutron fluxes result in some activation of the sodium and associated impurities. In addition, there is some permeation of tritium generated in the PCS lithium coolant to the SHTS.

A tornado-related failure of the SHTS could potentially release the sodium coolant and associated radioactive material. The combustion products of sodium are oxides and hydroxides of sodium. Should a release of the entire SHTS sodium coolant inventory occur, the potential radioactive material release would be small. The offsite dose, presented in Table 1, would be within regulatory acceptance criteria:

TABLE 1. Accident Offsite Dose Consequences.

Boundary	SHTS Loss Dose (rem)	10CFR100 Acceptance Criteria (rem)
Exclusion Area	0.003 whole body 0.004 lung 0.003 thyroid	20 whole body 150 thyroid
Low Population Zone	0.085 whole body 0.097 lung 0.085 thyroid	20 whole body 150 thyroid

Onsite personnel exposures would be within the Westinghouse Hanford occupational dose guidelines for radiation workers of 3 rem annually, 1.25 rem quarterly, and 1.25 rem weekly whole body exposure and the annual dose guideline of nonradiation workers of 0.1 rem. Additionally, it has been assessed that resulting site-boundary alkali metal aerosol concentrations of 7 mg/m^3 would be below the plant toxicological acceptance criteria of 10 mg/m^3.

CONCLUSIONS

After accounting for the proposed relaxation in design tornado/wind loading, it is concluded that the risk to the public from the SP-100 reactor during tornado-initiated events is acceptable. The risk from loss of core cooling accidents initiated by a tornado is acceptable based on the low frequency of occurrence (2.3×10^{-9}/yr). Risk from commercial reactors is accepted for much higher core damage frequencies (1×10^{-5}/yr). The risk from loss of secondary system boundary integrity is acceptable based on toxicological and radiological acceptance criteria.

Acknowledgments

The fault tree analyses that serve as a technical bases for this paper were performed by Suzanne E. Lindberg, Westinghouse Hanford; Robert H. Meichle, Westinghouse Hanford, is acknowledged for his independent safety review of this paper; and Jerry P. Wilson, Westinghouse Hanford, is acknowledged for his independent peer review of this paper.

References

Coats, C. W. and R. C. Murray (1985) "Natural Phenomena Hazards Modeling Project: Extreme Wind/Tornado Hazards Models for Department of Energy Sites," Rev. 1, UCRL-53526, Lawrence Livermore National Laboratory, Livermore, CA.

General Design Criterion 2 (1986) "Design Bases for Protection Against Natural Phenomena," of Appendix A, "General Design Criteria for Nuclear Power Plants," to 10CFR50, "Licensing of Production and Utilization Facilities," 11 April 1986.

SDC (1989) Department of Energy Hanford Project Standard (1989) Arch-Civil Design Criteria Design Loads for Facilities, SDC-4.1, Rev. 11, 6 September 1989.

Section C. Regulatory Position, Regulatory Guide 1.117 (1978) Tornado Design Classification, April 1978.

DESIGN AND EQUIPMENT INSTALLATION FOR RADIOISOTOPE POWER SYSTEMS FACILITY METALLIC CLEANING ROOM 162 SUBSYSTEM 6130

Harold E. Adkins and John A. Williams
Westinghouse Hanford Company
P. O. Box 1970, N1-42
Richland, WA 99352
(509) 376-5219

Abstract

A fully equipped cleaning and packaging line for cleaning nonradioactive metallic components and tools required in the assembly/disassembly of Radioisotope Thermalelectric Generators (RTG) is being designed and installed in room 162 of the Fuels and Materials Examination Facility (FMEF), which contains the Radioisotope Power Systems Facility (RPSF). The FMEF is a modern nuclear facility meeting current safety standards. Process equipment in cleaning room 162 includes a glovebox cleaning station with an airlock, an assembly and packaging workstation, and a downflow spray cleaning station. The glovebox cleaning station is used for controlled atmosphere cleaning of metallic components such as tools and canisters. The assembly and packaging workstation is used for tool and component cleanliness examination, inspection, assembly, and packaging. The downflow spray cleaning table is used for final particulate cleaning and inspection before packaging. The metallic cleaning line is planned for operation in FY 1993 as part of the RPSF.

INTRODUCTION

In order to meet new safety, health, and environmental requirements imposed on the manufacture and assembly of Radioisotope Thermalelectric Generators (RTGs), the Radioisotope Power Systems Facility (RPSF), one of the newest facilities available, has been designed and is being constructed to perform this function (Figure 1). In the RPSF, components and fuel for RTGs will be received and assembled into reliable power units suitable for launch into deep space. These RTGs require meticulous care and cleanliness in the loading of fuel, final assembly, and testing. Figures 2 and 3 show plan and elevation views of the metallic cleaning room, room 162, within the RPSF. The cleaning line is for cleaning canisters, fixtures, and tools that will come in contact with the RTGs. The cleaning process consists of transferring bagged components, tools, fixtures, and other items, to room 162 and placing them through the airlock into the glovebox. Inside the glovebox, parts and components are removed from bags and ultrasonically cleaned in one of two sizes of tanks. After cleaning, the parts are rinsed in the rinse tank and placed on an evaporative drying rack. Then they are removed from the glovebox through the airlock and moved to the assembly/inspection/packaging station. After cleanliness inspection and assembly (if required), they are sent to spray cleaning and final inspection. From there they are returned to the assembly/inspection/packaging station for bakeout preparation and/or packaging for storage. Tools, fixtures, canisters, and components will be cleaned until no particles of 0.005 in. or greater are observed during inspection.

Glovebox Cleaning Station

In the metallic cleaning line, the glovebox cleaning station will be used for general cleaning and degreasing of metallic tools and components. Ultrasonic cleaning will be the primary cleaning method used in this station. Two ultrasonic cleaning tanks, a deionized water and/or alcohol rinse tank, and an air drying rack constitute the major equipment used inside the glove-box. The ultrasonic cleaning tanks will contain methylene chloride or alcohol and will have capacities of 20.82 L and 9.5 L, respectively. The deionized water/alcohol rinse tank has the capacity of 3.8 L to 7.6 L. Capacities were selected to accommodate a range of parts; however, actual fluid volumes used will be controlled administratively for each required cleaning operation by established procedures. Methylene chloride and alcohol vapors will be controlled by exhausting the glovebox atmosphere to the FMEF

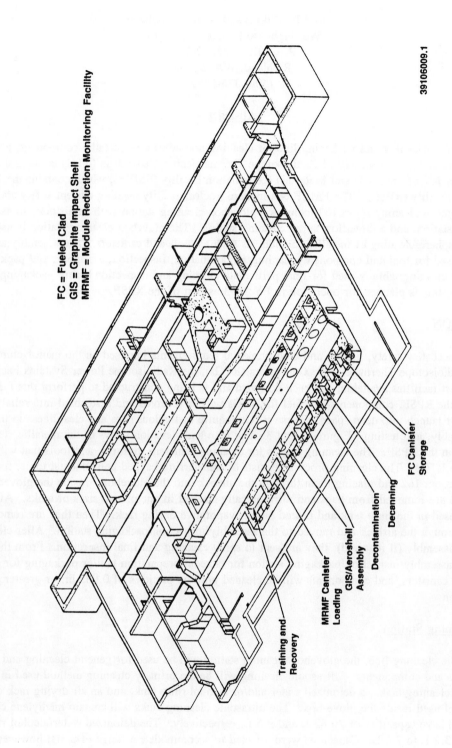

FIGURE 1. Radioisotope Power Systems Facility Level 1.

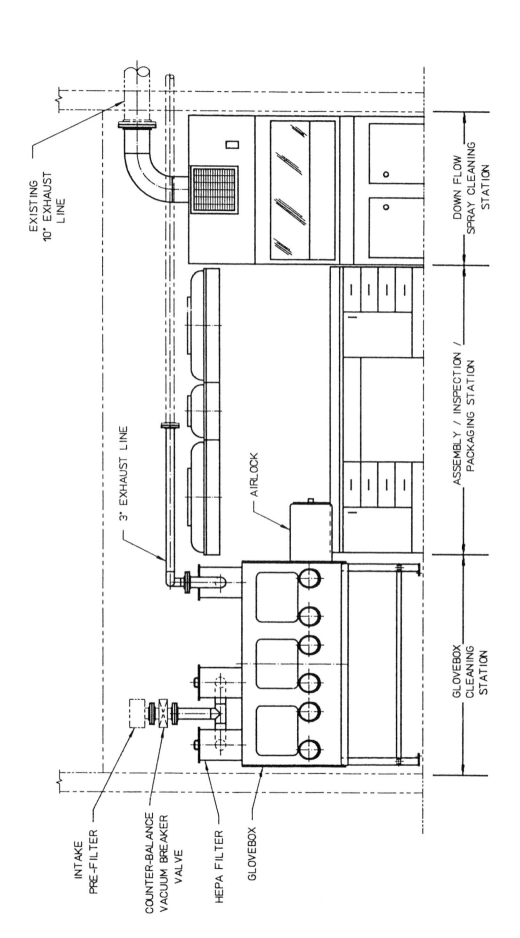

FIGURE 2. Metallic Cleaning Room 162 Equipment Layout.

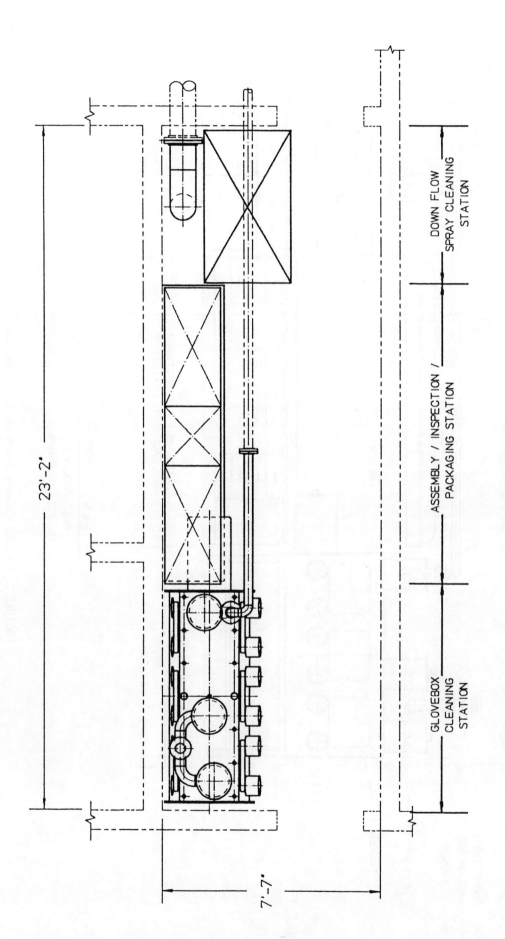

FIGURE 3. Top View of Room 162 Equipment Line.

exhaust system. When not in use during interim periods, the cleaning and rinse tanks will be provided with covers. Cleaning agents will not be stored permanently in cleaning tanks. Safety cans of limited volume will be used to store cleaning agents in the glovebox when not in use.

The glovebox air exchange rate will exceed one air change per minute and airflow is of sufficient quantity to meet the largest credible breach criteria. Drying of cleaned components and materials will be by the evaporative process on drying racks inside the glovebox. An airlock, which will be used to transfer components, supplies, materials, and cleaning agents into and out of the box, will be located at one end of the glovebox. The airlock will prevent perturbation of established glovebox airflow rate. Equipment too large to move through the airlock will be installed through a removable glovebox side window. Cleaning agents will be downloaded before operations necessitating window removal begin.

Assembly/Inspection/Packaging Station

The assembly/inspection/packaging station is a commercial laboratory cabinet module with a large table top. At this station, the cleaned parts that are to be inspected and packaged are removed from the glovebox through the airlock. Particulate contaminate control is accomplished by use of three high-efficiency particulate air (HEPA) filtered clean air modules located over the table at this station. Continuous downward airflow over the work area prevents other particles in the room from entering the work zone and being deposited on the cleaned parts or components. After the metallic parts have been removed from the airlock, they are inspected for cleanliness and sent to the downflow spray cleaning table for final inspection and final spray cleaning with Freon 22 (Freon is a trademark of E.I. duPont de Nemours and Company). The cleaned and inspected parts then are sent back to the assembly and inspection station for packaging.

Downflow Spray Cleaning Station

The downflow spray cleaning station equipment will be commercial grade and will include an overhead air module (fan and HEPA filter) for controlling particulate matter that otherwise may become attached to the components being cleaned and inspected. The air module has no safety function and serves only a process function. Total air makeup to the downflow table will be exhausted through the FMEF building exhaust system. Freon 22 used as a cleaning process is sprayed onto the component being cleaned by a trigger-operated spray gun fed from a portable disposable container source. The Freon 22 washdown effluent is filtered and the filters are visually inspected under a low-power microscope for particulates. The washdown/inspection is repeated until no particulates ≥ 0.005 in. are observed. The Freon 22 is exhausted with the makeup air to the building exhaust. The downflow table exhaust controls will be interlocked with the room air supply damper control, so that sufficient makeup air is supplied automatically to the room during table operation.

Cleaning Agents

The metallic components preparation process will use a solvent presently identified as methylene chloride and ethyl alcohol. Methylene chloride has been identified as a hazardous chemical, and ethyl alcohol presents a combustible liquid concern. The ventilation airflow is sufficient to keep the glovebox atmosphere below lower combustion level (LEL) concentrations.

Waste Streams

The waste streams that will come from the cleaning process are as follow.

- Alcohol. Ethyl alcohol will be used in volumes of ≤ 2 gal for cleaning components. Alcohol is covered when not in use.

- Methylene Chloride. Methylene chloride is used for removing hydrocarbons from the components and will be in quantities of ≤ 2 gal.

- Rinse Water. Water will be used for rinse and dilution after the cleaning cycle. It will be used in quantities of ≤2 gal.

- Freon Trichlorotriforomethane (TF). Freon TF is used in the downflow table as the final cleaning and inspection process. Each component is sprayed with Freon TF to remove small particulate from the surface. Freon TF is removed by the plant exhaust ventilation system.

The cleaning solutions are contained in tanks with covers. Liquids are drained and removed from the glovebox after determination of the cleaning life period. This will be a routine liquid removal and replacement with closed cans for each liquid. All liquids will be recycled if possible.

CONCLUSION

A new metallic cleaning room has been designed with state-of-the-art workstations for cleaning, examination, inspection, assembly, and packaging.

This will constitute a new, modern, safe metallic cleaning facility for RPSF components.

Acknowledgments

The authors would like to thank Dave Brown, Ron Eschenbaum, Jim Frederickson, Lou Goldmann, and Dan Webb for the valuable assistance and advice regarding this project. The RPSF is sponsored by the U.S. Department of Energy Office of Special Applications.

AUTHOR INDEX
(Bold Page Numbers Indicate Senior Authorship)

Abraham, Douglas S., 516
Adamov, E. O., **1060**
Adams, Steven F., **643**
Adkins, Harold E., 860
Afanasyeva, Irma V., **1274**
Akimov, Vladimir N., 1368
Al-Baroudi, Homam, **796**
Alderman, Carol J., **764**
Allen, Daniel T., 1359
Allen, George C., **692**
Amos, Wayne R., 749
Anderson, William G., **1162**
Andreev, Pavel V., 1368
Antar, Basil N., **1210**
Arinkin, F. M., 916
Armstrong, Robert C., 58
Ashe, Thomas L., **884**
Atwell, Jerry C., 1074
Bailey, Herbert S., 363
Bajgar, Clara, 326
Baker, Eric W., **1007**
Bamberger, Judith A., **228**, **544**, 675
Bankston, C. Perry., 516
Barnett, William, 177
Bartram, Bart W., **812**, **907**, **929**
Batyrbekov, G. A., 916
Baumann, Eric D., 401
Beam, Jerry E., 1162
Beaty, John S., **332**
Beaudry, Bernard J., 319
Begg, Lester L., **114**, 492
Bellis, Elizabeth A., 498, **1245**
Bennett, Gary L., 24, **383**, **662**
Bennett, Ralph G., 107
Bernard, John A., 562, **583**
Best, Frederick R., 1216
Bhattacharyya, Samit K., 604, **681**
Bitten, Ernest J., 1067
Boain, Ronald J., 78
Bohl, Richard J., 1089, 1097
Bohne, William A., **770**
Bond, James A., 343, **353**
Borshchevsky, Alex, 326
Braun, James F., 171, 758
Bremser, A. H., (Applied Tech. II)
Briese, John A., 572
Britt, Edward J., 120, 1368
Brittain, Wayne M., **743**

Brown, Neil W., 824
Bryhan, Anthony J., (Applied Tech. I)
Bryskin, Boris, **278**
Buckman, Jr., R. William, **150**
Buden, David, **91**, 648
Buksa, John J., **1089**, 1097
Busboom, Herb, (Applied Tech. I)
Butiewicz, David A., 1316
Butler, Robert E., 145
Bystrov, P. I., 916
Caldwell, C. S., (Applied Tech. II)
Carmack, William J., (Applied Tech. II)
Carteret, Betty A., **848**
Caveny, Leonard H., 643
Cerbone, Ralph J., 721
Choi, Chan K., 30
Chaudhuri, Shobhik, 1013
Choudhury, Ashok, (Applied Tech. I)
Clark, Dennis E., (Applied Tech. II)
Clark, John S., 692, 703
Clay, Harold, 298
Coleman, Anthony S., 433
Coles, Garril A., 854
Collett, John M., 1074
Collins, Frank G., 1210
Connell, Leonard W., **923**
Connolly, John F., **70**
Coomes, Edmund P., **675**, **878**
Cooper, R. H., 681
Cornwell, Bruce C., **1067**
Cowan, Charles, 532
Cross, Elden H., **1205**
Culver, Donald W., **714**
Cupo, Ernest P., 824
D'Annible, Dom, 58
Dagle, Jeffrey E., **234**, 655, 675, 878
Dahl, Wayne B., 714
Dalcher, Alfred W., **304**
Dandini, Vincent J., **240**
Davis, Monte V., 604
Davis, Paul R., 623
Davis, Richard E., 721
Dawson, Sandra M., 516
Deane, Nelson A., **97**
Deobald, Theodore L., **854**, 1067
Desplat, Jean-Luis, 312
Determan, William R., **504**, **510**, **1046**, 1237
Diachenko, Michael A., 1304

AUTHOR INDEX
(Bold Page Numbers Indicate Senior Authorship)

Dibben, Mark J., **1135**
DiStefano, J. R., (Applied Tech. I)
Dix, Terry E. 141
Dochat, George P., **894**
Doherty, Michael, **1183**
Dolan, Thomas J., **550**
Donovan, Brian, 617
Dorf-Gorsky, I. A., **35**
Dudenhoefer, James E., 894
Dudzinski, Leonard A., 389
Duffey, Jack, 58
Dugan, Edward T., **471**
Dunn, Charlton, **950**
Durand, Richard E., **1052**
Dutt, Dale S., **830**
Dzenitis, John M., **1226**
Eastman, G. Yale, 1170
El-Genk, Mohamed S., 410, 417, **787**, 955, **1013**, 1023
English, Robert E., **272**
Ezhov, Nikolai I., 58a
Fairchild, Jerry F., **836**
Fallas, T. Ted., **298**
Fisher, Mike V., 498
Fitzpatrick, Gary O., **1359**
Fleurial, Jean-Pierre, **326**, 332
Fontana, Mario, 830
Frederick, D. A., 164
Galbraith, David L., 1078
Gamble, Robert E., 208, 298
Gaustad, Krista L., 228
George, Jeffrey A., **130**, 389
Gernert, Nelson J., 1153
Gerwin, Richard, 1293
Gilchrist, A. Richard., 208, **824**
Gilland, James H., **1192**
Gilliland, Ken., 266
Giraldez, Emilio, **312**
Gizatulin, Sh. Kh., 916
Good, William A., 556
Goodwin, Gene M., 164
Gottschlich, Joseph M., 800
Grandy, Jon D., (Applied Tech. II)
Greenslade, David L., 594
Gryaznov, Georgy M., **58a**, **1268**, 1274, **1368**
Gunther, Norman G., 120, **604**
Hack, Kurt J., 389
Hagleston, Greg, 1046

Hammond, Ahmad N., 401
Hanan, Nelson A., 604
Hanlon, James C., 866
Harper, Jr., William B., 884
Hartless, Lewis C., 577
Hartman, Robert F., **177**
Harty, Richard B., **202**, **216**
Hatch, G. Laurie, 612
Hefferman, Timothy F., 638
Hemler, Richard J., **171**
Hendricks, J. W., (Applied Tech. I)
Hill, Wayne S., 1216
Hobbins, Richard R., (Applied Tech. II)
Hobson, Robert R., 120
Holdridge, Jeff, 58
Hooper, E. Bickford, 1287
Hoover, Darryl, 532
Horner, M. Harlan, 1052, **1237**
Horner-Richardson, Kevin, **629**
Horton, P., (Applied Tech. II)
Houts, Michael G., **462**, 1089, **1097**
Howell, Edwin J., **749**
Huber, William G., **64**
Hunt, Maribeth E., **222**, 781
Hunt, Thomas K., **1316**
Husser, Dewayne L., (Applied Tech. II)
Ivanenok, Joseph F., 937
Jacobson, Dean L., 53, 292
Jacox, Michael G., 1259
Jahshan, Salim N., **107**
Jeffries-Nakamura, Barbara, 1325, 1331
Jekel, Todd B., 1114
Jensen, Grant C., 298
Johnson, Gregory A., 216, 1199
Johnson, Richard W., **998**
Josloff, Allan T., **363**
Kahook, Samer D., 471
Kaibyshev, Vladimir, 417
Kaminsky, Albert S., **819**
Kammash, Terry, **1078**, 1083
Kangilaski, Mike, 145, 150 (Applied Tech. I)
Kawaji, Masahiro, 1210
Kelley, James H., **78**
Kelly, Charles E., 171
Kenny, Barbara H., **433**
Keshishan, Vahe, **141**
Khrushch, Valery T., 819
Kiebel, Gary R., **738**

AUTHOR INDEX
(Bold Page Numbers Indicate Senior Authorship)

Kim, Kwang Y., 629
Kim, Taewon, 1135
King, Donald B., 254
Kirillov, E. Y., 41
Kjaer-Olsen, Christian, 298, 304
Klein, Andrew C., 796, 1123, 1147
Klein, John W., 516
Knight, R. Craig, **594**
Kniskorn, Marc W., 923
Knocke, Phillip C., 516
Knoll, Dana A., **943**
Koonmen, James P., 836
Koroteev, Anatoly S., 1368
Kraus, Robert, 401
Krotiuk, William J., **47, 1231**
Kruger, Gordon B., 298
Kugler, Walter A., 343
Kuhl, Kermit D., **758**
Kull, Richard A., 343
Kummer, Jospeh T., 1316
Kuznetsov, Victor S., 819
Kwok, Kwan S., **562**
Lamp, Thomas R., **617**, 623
Lanning, David A., 462, 562
Lawrence, John W., **668**
Lawrence, Leo A., 312, **492**
Lazareth, Otto W., **967**
Lee, Celia, 1344
Lee, Deuk Yong, **53**
Lee, Keith, 298
Leigh, Harley, **843**
Lessing, Paul A. (Applied Tech II)
Lewis, Bryan R., 498
Lieb, David P., 612, **1344**
Lindemuth, James E., 1170
Lipovy, N. M., 916
Lititsky, V. A., 916
Lorenzo, Carl F., **446**
Ludewig, Hans, 721, 967
Lunsford, David W., 208
Luo, Anhua, **292**
Luppov, A. N., 35, 41
Lyon, William F., 655
Magera, Gerald G., **623**
Makenas, Bruce J., 492
Mantenieks, Maris A., 1279
Mansfield, Brian C., 866
Marcille, Thomas F., 97, 532

Marks, Timothy S., **1147**
Marshall, Albert C., **806**
Martin, Charles R., 556
Martinell, John, 692
Martinez, Carlos D., 304, (Applied Tech.I)
Martishin, Victor M., 1368
Mason, Lee S., **866**
Matteo, Donald N., **343**, 353, 363
Matthews, R. B., 681
Matus, Lawrence G., 246
McIllwain, Melvin C., **714**
McKissock, Barbara I., 866
McNeil, Dennis C., 749
McVey, John B., 1344, 1359
Mead Jr., Franklin B., 30
Meoller, H. H., (Applied Tech. II)
Merrigan, Michael A., **1038, 1338**
Metcalf, Kenneth J., **427**
Meyer, Ray A., 266
Miley, George H., **479**
Miller, Katherine M., 1226
Miller, Thomas J., **24**, 383
Mills, Joseph C., **504**, 1199
Miskolczy, Gabor, **612**, 1344
Morimoto, Carl N., **572**
Morley, Nicholas J., **955**
Morris, D. Brent, **1060a**, 1129
Moses, Jr., Ronald W., **1293**
Murray, Christopher S., 410, **417**, 1013
Musgrave, Jeffrey, L., 446
Myers, Ira T., **401**
Myers, Roger M., **1279**
Narkiewicz, Regina S., **1074**
Nealy, John E., **372**
Nechaev, Y. A., 41
Negron, Scott B., **577**
Neudeck, Philip G., **246**
Newkirk, Douglas W., 97
Nicitin, V. P., 35, **41**
Noffsinger, Kent E., 234, 675
North, D. Michael, 1205
O'Connor, Dennis, 1325, 1331
Ogawa, Stanley Y., 304
Ogloblin, B. G., 35, 41
Ohriner, Evan K., **164**
Olsen, C. S., 681
Otting, William D., 114, 510
Otwell, Robert, 532

AUTHOR INDEX
(Bold Page Numbers Indicate Senior Authorship)

Ovcharenko, M. K., 916
Pantolin, Jan E., 1316
Parks, James E., 1279
Parma, Edward J., 1103
Pauley, Keith A., **990**
Pavlov, Konstantin A., 819
Pavshoock, Vladimir A., 440, 819, 1304
Pawel, S., (Applied Tech.II)
Pawlowski, Ronald A., **1123**
Pelaccio, Dennis G., 937
Perekhozhev, V. I., 1060
Perkins, David, 692
Perkins, Kenneth R., 721
Peters, Ralph R., **1114**
Peterson, Jerry R., 177
Petrosky, Lyman J., **728**, 937
Pickard, Paul S., 1103
Pivovarov, O. S., 1060
Plumlee, Donald E., (Applied Tech II)
Ponomarev-Stepnoi, Nikolay, **440, 1304**, 1368
Poston, David I., **1083**
Potter, Donald L., 923
Powell, George E., 1351
Powell, James R., 967
Protsik, Robert, 208
Prusa, Richard, 824
Pupko, Victor Ya., 456, **916**
Ramey, R. R., (Applied Tech.)
Raskach, F. P., 916
Redd, Frank J., **1351**
Reinarts, Thomas R., **1216**
Reyes, Angel Samuel, 260
Rhee, Hyop S., **120**
Richter, Robert, **800**
Ring, Peter A., 150, (Applied Tech I)
Robbins, Daniel J., **1141**
Rodgers, Douglas N., **1310**
Rodriguez, Carlos D., 866
Rolfe, Jonathan L., 332
Rosenfeld, John H., **1170**
Rovang, Richard D., 222, **781, 1199**
Roy, Prodyot, 1310
Ruffo, Thomas J., 159
Rutger, Lyle L., 1
Ryan, Margaret A., 1325, 1331
Sager, Paul H., **1251**
Salamah, Samir, A., 1310
Santandrea, R. P., (Applied Tech. II)

Sawyer, Jr., J. Charles., 806
Sayre, Edwin D., **145, 159**
Scheil, Christine M., 937
Schmidt, Eldon, 967
Schmidt, Glen, 120
Schock, Alfred, **182**
Schoenberg, Kurt F., 1293
Schuller, Michael J., 1060a, 1129
Schuster, Gary, (Applied Tech. I)
Scoville, Nancy, 326
Segna, Donald R., 234, **655**
Selcow, Elizabeth C., **721**
Semenistiy, Victor L., 456
Serbin, Victor I., 58a, 1268, 1274
Shaubach, Robert M., **1153**
Sheftel, Leonid M., 58a, 1268
Shepard, Kyle, **58**
Shepard, Neal, 532
Shestyorkin, A. G., 916
Shin, Kwang S., 292
Shukla, Jaik N., 266, 572
Sievers, Robert K., 1316
Simonsen, Lisa C., 372
Sinha, Upendra N., 1074
Sinkevich, V. G., 41
Sinyutin, G. V., 35
Slivkin, Boris V., 58a, 1268
Smetannikov, V. P., 1060
Smith, Gary L., 854
Smith, Jr., Joe N., **638**
Smith, Michael A., 824
Sobolev, Yu. A., 916
Sovey, James, 830
Springer, Dwight R., 304
Standley, Vaughn H., 1060a, **1129**
Stewart, Samuel L., 97
Striepe, Scott A., 372
Stone, James R., **1177**
Suitor, Jerry W., **1325**
Sulmeisters, Tal K., 1259
Suo-Anttila, Ahti J., **1103**
Sutton, Paul D., 516
Switick, Dennis, **532**
Syed, Akbar, **266**, 572
Tammara, Seshagiri R., 907, 929
Talanov, S. U., 916
Taylor, Thomas C., **556**
Temmerson, Ian R., 208

AUTHOR INDEX
(Bold Page Numbers Indicate Senior Authorship)

Thomassen, Keith I., 830, **1287**
Thome, Frank V., **254**, 836
Thompson, Walter, 58
Thornborrow, John, 990
Tikhonov, Lev Ja., 819
Tilliette, Zephyr P., **901**
Titran, Robert H., 681
Tokarev, V. I., 1060
Tournier, Jean-Michel, **1023**
Trujillo, Vincent, 1038
Trukhanov, Yuri L., 58a, 1268
Truscello, Vincent C. **1**
Tschetter, Melvin J., **319**
Tukhvatulin, Sh. T., 1060
Tuttle, Ronald F., 1007, 1135, 1141
Underwood, Mark L., 1325, **1331**
Upton, Hugh A., **208**, 298
Usov, Veniamin A., 35, 41, 440, 1368
Vaidyanathan, Swaminathan, (Applied Tech. II)
Vandersande, Jan W., 326, 332
Van Hagan, Thomas H., **498**, 504, 1237
Vernon, Milton E., 1103
Vining, Cronin B., **338**
Volnistov, V. V., 916
Wadekamper, Donald C., (Applied Tech. II)
Walter, C. E., 681
Wang, Mei-Yu, 30

Ward, William C., **1338**
Warren, John, 692
Watts, Ken, 1259
Weitzberg, Abraham, 907
Wernsman, Benard R., **410**
Westerman, Kurt O., 577
Wetch, Joseph R., 120
Widman, Jr., Frederick, 1205
Wilcox, Reed E., **516**
Williams, John A., 860
Williams, Roger M., 1325, 1331
Wiltshire, Frank R., 298
Witt, Carl A., 1344
Witt, Tony, 312
Wong, C. Channy, 923
Wright, Steven A., 1103
Wutche, Thomas J., 114, 504
Xue, Huimin, 787, 1013
Yaspo, Robert, 304
Yelchaninov, A. A. 35
Yen, Chen Wan, 78
Zaritzky, Gennady A., 1274, 1368
Zerwekh, John R., (Applied Tech. I)
Zhabotinski, Evgeny E., 58a, 1268, 1274, 1368
Zheng, Cinian, 120
Zrodnikov, Anatoly V., **456**
Zubrin, Robert M., **979**, **1259**

(Note: Names of the presenters in the Applied Technology sessions are included in the author list, but written versions of their presentations are not available.)